Alexander Zimmermann

Characters of Groups and Lattices over Orders

Also of Interest

Algebra and Number Theory
Benjamin Fine, Anthony Gaglione, Anja Moldenhauer,
Gerhard Rosenberger, Dennis Spellman, 2017
ISBN 978-3-11-051584-8, e-ISBN (PDF) 978-3-11-051614-2,
e-ISBN (EPUB) 978-3-11-051626-5

Abstract Algebra. Applications to Galois Theory, Algebraic Geometry, Representation Theory and Cryptography
Celine Carstensen-Opitz, Benjamin Fine, Anja Moldenhauer,
Gerhard Rosenberger, 2019
ISBN 978-3-11-060393-4, e-ISBN (PDF) 978-3-11-060399-6,
e-ISBN (EPUB) 978-3-11-060525-9

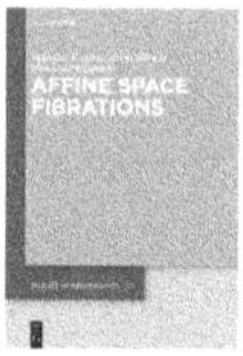

Affine Space Fibrations
Rajendra V. Gurjar, Kayo Masuda, Masayoshi Miyanishi, 2021
ISBN 978-3-11-057736-5, e-ISBN (PDF) 978-3-11-057756-3,
e-ISBN (EPUB) 978-3-11-057742-6

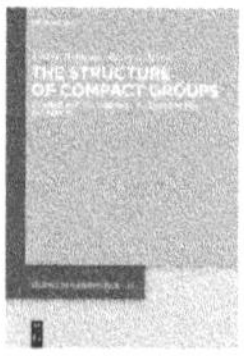

The Structure of Compact Groups, A Primer for the Student – A Handbook for the Expert
Karl H. Hofmann, Sidney A. Morris, 2020
ISBN 978-3-11-069595-3, e-ISBN (PDF) 978-3-11-069599-1,
e-ISBN (EPUB) 978-3-11-069601-1

Commutative Algebra
Aron Simis, 2020
ISBN 978-3-11-061697-2, e-ISBN (PDF) 978-3-11-061698-9,
e-ISBN (EPUB) 978-3-11-061707-8

Alexander Zimmermann

Characters of Groups and Lattices over Orders

From Ordinary to Integral Representation Theory

DE GRUYTER

Mathematics Subject Classification 2020
Primary: 20-01, 20C05, 16S34, 20C10, 16G30; Secondary: 11R29, 11R65

Author
Prof. Alexander Zimmermann
Université de Picardie
Dept. de Mathématiques
33 Rue St Leu
80039 Amiens
France
alexander.zimmermann@u-picardie.fr

ISBN 978-3-11-070243-9
e-ISBN (PDF) 978-3-11-070244-6
e-ISBN (EPUB) 978-3-11-070255-2

Library of Congress Control Number: 2021944862

Bibliographic information published by the Deutsche Nationalbibliothek
The Deutsche Nationalbibliothek lists this publication in the Deutsche Nationalbibliografie; detailed bibliographic data are available on the Internet at http://dnb.dnb.de.

Cover image: Isogood / Gettyimages
Typesetting: VTeX UAB, Lithuania
Printing and binding: CPI books GmbH, Leck

www.degruyter.com

Für Lotte und Simon

Preface

The mathematical model of symmetry is a group. In physics and chemistry, in particular in quantum mechanics, symmetry often can be realised by invariance of the model under a set of linear transformations of a vector space. A group representation is the link between these two worlds. After the initial ideas the theory became a well-established and far developed theory. Nowadays group representation theory is one of the most efficient and most popular tool to study groups. Many theories in mathematics use representations of groups in various situations. We may think of algebraic topology, combinatorics, number theory, Fourier analysis, random walks and many more. Besides this, representations of groups provides a very easily accessible and readily motivated tool to introduce many of the much needed and from time to time somewhat technical tools used in higher algebra. Also, many mathematical methods, such as algebraic geometry, topology, number theory, ring theory, commutative algebra and alike can be used in representations of groups. Since many tools used in more advanced parts of representation theory can be explained in representations of finite groups in a very explicit and somewhat elementary way, representations of groups is a very important step in the learning procedure for students aiming later to a deeper knowledge of representation theory. All these reasons explain why representations of groups is part of many curricula in universities around the world.

For the beginner the easiest part is definitively representation theory of finite groups over fields of characteristic 0, and more precisely over the complex numbers. From that point onwards there are many possibilities for a deeper development of the subject. A considerable number of good textbooks are available already dealing with character theory of finite groups. In practically all these textbooks the focus is given on a development towards modular representation theory, that is representations of groups over fields of finite characteristic. In contrast we try to emphasize links to number theory and develop the character theory of finite groups towards an introduction into group representations over the integers. We give a rather complete treatment of the representation theory over complex numbers for a beginners' course and include already there features which open the eyes towards representations of groups over the integers. Restricting a priori to fields of characteristic 0, or even more to complex numbers is quite restrictive, and in some sense misleading. Even the beginner's eyes acknowledge broader view and in many situations the case of representations over fields of finite characteristic, or over the integers explains well the more specific results which can be obtained over the complex numbers. A beginner will have difficulties to fully appreciate the semisimple case if he does not have an example at hand of a non semisimple algebra. This is the reason why the text leaves open as far as possible the base ring one works on. Of course, nevertheless, the main focus in the first chapters is on representations of finite groups over fields of characteristic 0, and mainly even the complex numbers, and character theory takes a large place. In particular, characters are a mixture between algebraic and analytic animals, and

https://doi.org/10.1515/9783110702446-201

we could not resist from giving an application of characters to a problem in analytic number theory, namely Dirichlet's theorem on primes in arithmetic progressions, in particular since this result is needed in the Dixon-Schneider algorithm to compute character tables algorithmically, and since we explain the mathematical content of the algorithm in a separate section.

Another feature is the fact that we do introduce group cohomology in degree 1 and 2 since I believe various cohomology theories a very essential method for all further studies in algebra. Nevertheless, we choose a quite non standard way, and use our study of invariant bilinear forms to introduce degree 1 group cohomology, and projective representations and Clifford theory to introduce degree 2 group cohomology. We hope to provide in this way a useful alternative approach to this exciting subject.

In the last chapter an introduction to integral representation theory of groups, of lattices over orders is given. We restrict to a short introduction which follows naturally the subjects treated in the first chapters. We include a self-contained and short introduction into some relevant notions of homological algebra, ring theory, and algebraic number theory used in the sequel. Then, orders and lattices over orders are introduced, and as main goals we prove the Jordan-Zassenhaus theorem on the finiteness of isomorphism classes of lattices in a given vector space, and Swan's example of a stably free but not free ideal in an integral group ring. Up to my best knowledge this is the first treatment of this example in a textbook.

As a natural continuation in a different direction we propose the monograph [Zim-14], which is more of research level and covers many important developments left out here, in particular the above-mentioned modular representation theory aspect, and can be used for subsequent studies. Besides this, Curtis-Reiner produced with [CuRe-82-86] a complete and extremely well-written 2-volumes work covering most of the classical material, and even beyond. Further reading on the integral representation theory aspect is also provided in research monographs, such as the books of Reiner, Roggenkamp, Huber-Dyson [ReiRo-79, Rei-75, RoHDVol1-70, RoVol2-70] or Kuku [Kuku-07].

A slightly more detailed description of the content of the chapters are as follows.

I start in the first chapter with the definition of an algebra and of a module. I do not restrict to the case of algebras over a field, since this small generalisation offers advantages in the sequel and is not really much more difficult. Then, the three basic results are developed: Krull-Schmidt theorem, Maschke's theorem and Wedderburn's theorem. These three provide already the theoretical and technical framework for representations of finite groups over fields of characteristic 0. Group rings are introduced on the way, and some elementary consequences for group rings are developed immediately.

The second chapter gives the definition of characters and proves some of the standard results, including Burnside's p^aq^b-theorem. It also contains a proof of the Dixon-Schneider algorithm to compute explicitly the character table. As an application of characters, which can be seen as basically analytic objects, we give a proof of

Dirichlet's theorem on the infinity of primes in arithmetic progressions, using Dirichlet's L-functions. Whenever possible I preferred the formulation and proof of statements in terms of modules rather than in terms of characters.

The next big chapter is to introduce tensor products, with induced modules in mind, and all annexed questions, such as Clifford's theory and Mackey's theorem. Also, the number theoretical ingredient in Burnside's $p^a q^b$-theorem is replaced by a purely algebraic consideration developed in this chapter. Moreover, Clifford's theorem can be interpreted easily in terms of projective representations, and the degree 2 group cohomology. We give a short introduction.

Then, in the fourth chapter the question of when invariant bilinear forms exist on simple modules is studied, and Frobenius Schur indicators are introduced. More interesting than the case of complex coefficients is the case of fields of characteristic 2. This is done using an approach due to Sin and Willems, introducing on the way group cohomology of degree 1.

The fifth chapter introduces character rings and Brauer induction formulae. The methods developed there provide a nice application of what is developed in chapter three. Further some of the results are needed for chapter eight.

These chapters four and in particular five may be considered as somewhat more technical, but they introduce important techniques, and what is more important, they provide non trivial applications and surprising statements. They also give the student an idea of what a possible sequel in the subject would look like. Though somewhat deeper, these two chapters do not use any further methods from other sources. There are approaches using deep methods from algebraic topology or algebraic geometry, but we chose an elementary approach.

Then, we proceed to more advanced subjects, while still staying in the continuity of the first chapters. Chapter six provides a short introduction to some technical features and methods from ring theory and homological algebra needed for the rest of the book, and providing also a short introduction to necessary theoretical knowledge to understand modern research problems in algebra.

Chapter seven then gives the necessary background in algebraic number theory as far as it is needed for the subsequent chapter eight.

This chapter eight then gives an introduction to representation theory of groups over the integers. The main direction we pursue is to introduce class groups of orders, which is at once also a first step into problems of algebraic K-theory. We show there the most interesting Jordan-Zassenhaus theorem, which implies that the class groups are finite in favorable situations, but also give an arithmetic formula, generalising Steinitz famous formula for ideal classes in number fields to orders, using idèles. A second highlight is the detailed presentation of Swan's famous example of an ideal in the integral group ring of a generalised quaternion group which is stably free but not free. At the end I included a short section explaining some results on Galois module structure of algebraic integers. This very deep part of algebraic number theory is highly developed, and proofs of the results are for too complicated for being detailed in this book. However the results I include use methods and results of virtually the entire book, so that this section answers pretty well the question on what the theory of this book is good for.

The last chapter gives solutions to selected exercises. In each chapter I singled out a specific exercise for which the reader will find a detailed solution in Chapter nine.

There are quite a number of textbooks on representation theory of groups, but I believe not so many taking the same place as the present one. Most texts are far more ambitious, such as Curtis-Reiner [CuRe-82-86] or Benson [Bens-91] for the module theory, or Isaacs [Isa-76] for the character theory. Some are more elementary, such as Alperin-Bell [AlBe-95] or Steinberg [Ste-12]. A very well-written and quite comprehensive book is Kowalski [Kow-14] and the book which served for many generations of mathematicians is Serre [Ser-78]. The book [Kow-14] also covers the representation theory of (infinite) compact groups, which is not easily available elsewhere. However, it is somewhat more elusive on quite a few aspects of the representation theory of finite groups. Moreover, except the encyclopedic Curtis-Reiner [CuRe-82-86] few other textbooks, if any, deal with integral representations and develops the theory of ordinary representations in direction of lattices over orders.

The reader I have in mind has a decent knowledge in linear algebra, knows Galois theory of finite degree field extensions in any characteristic and has an idea of basic properties on finite groups and on rings. At a few places some complex analysis is needed and some general topology. Starting from this, basically all results are fully proven, with exception of some results in Chapter eight, some features on Haar measures, and a few results of number theoretic nature.

I used the manuscript for many years in the Université de Picardie in a first year master course, and I benefited a lot from the comments of the students. Teaching the subject gives a rather precise idea at what places proofs have to be detailed, which parts of the concepts have to be motivated further, or if a subject is too difficult for a beginner's class. I wish to thank the students in the various generations during the years for asking questions, and not being satisfied with half complete answers. Also, this work would not have been possible without constant support of the algebra group of the Laboratoire Amiénois de Mathématique Fondamentale et Appliquée in its varying composition. I thank my own children for giving me a very clear idea of all the sorrows a student may have when the professor poses incomprehensible questions or asks to work through tons of material for the day after tomorrow.

Contents

1 Ring theoretical foundations

Zu meinem fünfundzwanzigsten geburtstag
schreib ich ein epos, groß, überwältigend...

Dachte ich

Nun schreib ich diesen bericht

Reiner Kunze, Gedichte @ 2001, S. Fischer Verlag Frankfurt am Main

On my twenty-fifth birthday
I would write an epos, great, overwhelming...

I thought

Now I am writing this report

Reiner Kunze, poetry;[1]

In mathematics a lot of material originates from the attempt of the humanity to understand the real world by abstraction. One of these processes is the concept of a group as a mathematical model for symmetry. Once defined by a system of axioms algebra in particular studies the properties of these axioms, getting thus properties of abstract groups. What is the consequence for the real world problem one started with? In particular in physics very often the status of the object of study is given by a vector in a vector space, usually over the real or complex numbers. The fact that the problem carries a symmetry is given by the fact that given a solution vector to the physical problem, such as a differential equation, then there is a transformation of the vector, often linear transformation T of the vector space such that for any solution ψ of the physical problem, also $T(\psi)$ satisfies the problem as well.

1.1 Algebras, actions and modules: definition of the basic objects

The main tool for passing from abstract groups to its realisation by matrices is the concept of an algebra and modules over an algebra. We start with algebras first.

1.1.1 Algebras

Throughout this section let K be a field and let R be a commutative ring.

Definition 1.1.1. An R-algebra A is a ring $(A, +, \cdot)$ with a ring homomorphism $R \longrightarrow Z(A)$ from R to the centre $Z(A) := \{a \in A \mid b \cdot a = a \cdot b \ \ \forall b \in A\}$ of A.

1 in der Übersetzung von Susanne Brennecke, translation by Susanne Brennecke.

https://doi.org/10.1515/9783110702446-001

In more explicit terms this just means that an R-algebra is a ring A, on which an operation $\bullet$ of a commutative ring R is defined, such that for all $r, s \in R$, $a, b \in A$ one has

$$\begin{aligned} r \bullet (a+b) &= (r \bullet a) + (r \bullet b) \\ r \bullet (ab) &= (r \bullet a)b \\ (r \bullet a)b &= a(r \bullet b) \\ 1_R \bullet a &= a \\ (r+s) \bullet a &= (r \bullet a) + (s \bullet a) \\ (rs) \bullet a &= r \bullet (s \bullet a) \end{aligned}$$

Taking $a = 1$ the last three equations are equivalent to the fact that one has a ring homomorphism $R \longrightarrow A$, and taking $b = 1$ the second equation is equivalent to the fact that the image of the homomorphism is in $Z(A)$. The first equation is a consequence of the distributivity in A and the second equation is a consequence of the associativity of the multiplication in A. The third, fourth, fifth and sixth equation for general a and b follow then by the associativity and distributivity in A.

Example 1.1.2.

1. As a consequence, for a field K, a K-algebra A is a K-vector space furnished with an additional associative bilinear form $A \times A \longrightarrow A$, the multiplication. Indeed, this point of view would give an equivalent definition. Moreover, a $\mathbb{Z}$-algebra is nothing but a ring, since there is a unique homomorphism $\mathbb{Z} \longrightarrow A$ for any ring A, and automatically the image of this homomorphism is in the centre of A.
2. Let K be a field and let V be a K-vector space. Let $A = End_K(V)$ be the set of K-linear endomorphisms of V. Then, this is a K-algebra, with additive ring structure being given by $(\varphi + \psi)(v) := \varphi(v) + \psi(v)$ for all $v \in V$, for all $\varphi, \psi \in A$, and with multiplicative structure given by composition of mappings. The homomorphism of K to the centre of A is given by sending $k \in K$ to the map $v \mapsto kv$ for all $v \in V$. In particular, using the standard identification of matrices with linear homomorphisms after having chosen a basis, if $V = K^n$ for some integer n, then the set of square $n \times n$ matrices is a K-algebra where the additive structure is given by componentwise addition of matrices and the multiplicative structure is given by matrix multiplication. If V is infinite dimensional, then $End_K(V)$ is infinite dimensional as well. In functional analysis one looks at subalgebras of those endomorphisms satisfying certain conditions of usually topological nature.
3. If L is a field extension of K, then L is a K-algebra. If D is a skew field with centre L, then D is an L-algebra. Moreover, if A is an L-algebra, then by restricting the mapping $L \longrightarrow Z(A)$ to a subfield K of L, one sees that A is also a K-algebra.

As usual, we often want to compare two algebra structures.

Definition 1.1.3. Let A be an R-algebra with structural mapping $\lambda_A : R \longrightarrow Z(A)$ and let B be an R-algebra with structural mapping $\lambda_B : R \longrightarrow Z(B)$. Then, a ring homomorphism $\varphi : A \longrightarrow B$ is an *algebra homomorphism* if $\varphi \circ \lambda_A = \lambda_B$. Two algebras A and B are isomorphic if there are algebra homomorphisms $\varphi : A \longrightarrow B$ and $\psi : B \longrightarrow A$ such that $\varphi \circ \psi = id_B$ and $\psi \circ \varphi = id_A$.

An *algebra epimorphism* is an algebra homomorphism which is a ring epimorphism. An *algebra monomorphism* is an algebra homomorphism which is a ring monomorphism. We use the term "algebra homomorphism" synonymous to the term "homomorphism of algebras", and likewise for epimorphism and monomorphism of algebras.

In more elementary terms let A and B be R-algebras, then $\varphi : A \longrightarrow B$ is an algebra homomorphism if and only if for all $a_1, a_2 \in A$ and $r \in R$ one has

$$\begin{aligned}\varphi(1_A) &= 1_B\\ \varphi(a_1 \cdot a_2) &= \varphi(a_1) \cdot \varphi(a_2)\\ \varphi(a_1 + a_2) &= \varphi(a_1) + \varphi(a_2)\\ \varphi(r \bullet a_1) &= r \bullet \varphi(a_1).\end{aligned}$$

Indeed, the first three equations are equivalent to the fact that φ is a ring homomorphism whereas the last equation for $a_1 = 1_A$ is equivalent to $\varphi \circ \lambda_A = \lambda_B$ in Definition 1.1.3. The case of general a_1 follows by associativity of the multiplication in A.

Lemma 1.1.4. *Let R be a commutative ring and let A_1 and A_2 be both R-algebras, then $A_1 \times A_2$ is an R-algebra as well such that the projection on each component is an algebra homomorphism.*

Proof. We first observe that for any two rings A_1 and A_2 we get $Z(A_1 \times A_2) = Z(A_1) \times Z(A_2)$. Now, if $\lambda_1 : K \longrightarrow Z(A_1)$ is the homomorphism defining the algebra structure of A_1, and if $\lambda_2 : K \longrightarrow Z(A_2)$ is the homomorphism defining the algebra structure of A_2, then

$$\lambda_1 \times \lambda_2 : K \ni k \mapsto (\lambda_1(k), \lambda_2(k)) \in A_1 \times A_2$$

defines an algebra structure on $A_1 \times A_2$. The fact that the projections on the components are algebra homomorphisms is clear. □

Remark 1.1.5.
- Since the composition of two ring homomorphisms is a ring homomorphism, the composition of two R-algebra homomorphisms is an R-algebra homomorphism.
- An algebra homomorphism which is invertible as a ring homomorphism is also invertible as an algebra homomorphism. This follows immediately from

$$(\varphi \circ \lambda_A = \lambda_B) \Rightarrow (\lambda_A = \varphi^{-1} \circ \varphi \circ \lambda_A = \varphi^{-1} \circ \lambda_B).$$

1.1.2 Modules

As we mentioned, a group can be seen as a mathematical model for symmetries, where the actual symmetry is given by an action of the group on some other structure. To be more precise it will be most convenient to use the language of algebras.

The basic example of an algebra is the endomorphism ring of a vector space of Example 1.1.2.(2). Now, most naturally, an endomorphisms α of V is a mapping $\alpha : V \longrightarrow V$ and so, it 'acts' by sending each $v \in V$ to $\alpha(v) \in V$.

This is our model.

Definition 1.1.6. Let A be a K-algebra. Then, an A-*module* (M, φ) is a K-vector space M together with a K-algebra homomorphism $\varphi : A \longrightarrow End_K(M)$. A *representation* of an algebra A is an A-module M. If A is a ring, an A-module M is an abelian group together with a ring homomorphism $A \longrightarrow End_{\mathbb{Z}}(M)$.

Here we denote by $Hom_{\mathbb{Z}}(M, N)$ the set of homomorphisms of abelian groups for M and N being two abelian groups. Usually, when no confusion can arise, we note $a \cdot m := \varphi_M(a)(m)$ for elements $a \in A$ and $m \in M$ in the situation of an A-module (M, φ_M).

Example 1.1.7.
- For a field K, a K-module is exactly the same as a K-vector space.
- For any ring A, the abelian group $\{0\}$ with only one element is an A-module, which will be denoted 0. The same holds of course if A is a K-algebra. This is of course the trivial case (do not confuse with the trivial module over a group, to be defined later).
- A $\mathbb{Z}$-module is just an abelian group.
- Let $\psi : A \longrightarrow B$ be an algebra homomorphism and let (M, φ_B) be a B-module. Then, $(M, \varphi_B \circ \psi)$ is an A-module. This follows directly from the definition since the composition of ring homomorphisms is again a ring homomorphism.
- Let K be a field. What is a $K[X]$-module M? We first observe that K is a subalgebra of $K[X]$. By the previous item M is first a K-module, and by the first item this is just a vector space. Hence, in order to define a $K[X]$-module structure on V, we need to pick a K-vector space V and an algebra homomorphism $K[X] \to End_K(V)$. The element $X \in K[X]$ is mapped to an endomorphism $\psi \in End_K(V)$, and X^n is mapped to ψ^n. Moreover, $p(X) = \sum_{i=0}^m a_i X^i \in K[X]$ is mapped to $p(\psi) := \sum_{i=0}^m a_i \psi^i \in End_K(V)$. Hence, a $K[X]$-module M is given by the action of a K-linear endomorphism ψ on a K-vector space M.
- An ideal of A is an A-module.

Definition 1.1.8. For any $a \in A$ the set $A \cdot a := \{b \cdot a |\ b \in A\}$ is a (left-) A-module by multiplication on the left. If $a = 1$, then we call this module the *regular* A-module.

Often it is convenient to work with algebras over a general commutative ring. We want to define A-modules for general R-algebras A. Later this will be even necessary. In order to do so we need to define first what we understand by a homomorphism of an R-module for R being a ring. For ordinary representation theory, this concept will not be used.

Definition 1.1.9. Let M be an R-module with structural mapping $\varphi_M : R \longrightarrow End_{\mathbb{Z}}(M)$ and let N be an R-module with structural mapping $\varphi_N : R \longrightarrow End_{\mathbb{Z}}(N)$. Then $\psi \in Hom_{\mathbb{Z}}(M,N)$ is a *homomorphism of R-modules* if $\varphi_N(r)(\psi(m)) = \psi(\varphi_M(r)(m))$ for all $m \in M$ and $r \in R$. Denote by $Hom_R(M,N)$ the set of R-module homomorphisms.

An *endomorphism* of an R-module M is an element in $Hom_R(M,M)$ and the set of endomorphisms of R-modules is denoted by $End_R(M)$.

A more explicit way to write the property of being an R-module homomorphism is to say that given two R-modules M and N a homomorphism of abelian groups $\psi : M \longrightarrow N$ is an R-module homomorphism if for all $r \in R$ and $m \in M$ one has $\psi(r \cdot m) = r \cdot \psi(m)$. Here we use the short hand notation of the remark after Definition 1.1.6. It is now clear that $End_R(M)$ is an R-algebra with structural homomorphism $\lambda : R \longrightarrow Z(End_R(M))$ given by $r \mapsto r \cdot id_M$.

Definition 1.1.10. Let R be a commutative ring and let A be an R-algebra. Then an A-module is an R-module M together with a homomorphism of R-algebras $A \longrightarrow End_R(M)$.

Of course, if R is a field or $R = \mathbb{Z}$, then Definition 1.1.10 and Definition 1.1.6 give the same concept.

Again we would like to compare modules.

Definition 1.1.11. Let A be an R-algebra and let (M, φ_M) and (N, φ_N) be A-modules. Then, an *A-module homomorphism* $M \longrightarrow N$ is a group homomorphism $\alpha : M \longrightarrow N$ such that for all $a \in A$ one has $\alpha \circ \varphi_M(a) = \varphi_N(a) \circ \alpha$. This is the same as saying $\alpha(a \cdot m) = a \cdot \alpha(m)$ for all $m \in M$ and $a \in A$.

For a K-algebra A and two A-modules M and N an A-module homomorphism $\alpha : M \longrightarrow N$ is K-linear.

Definition 1.1.12. For a K-algebra A and A-modules M and N, a homomorphism $\alpha : M \longrightarrow N$ of A-modules is

- an *epimorphism* if α is surjective,
- a *monomorphism* if α is injective,
- and an *isomorphism* if α is bijective.
- An *endomorphism* of M is a homomorphism $M \longrightarrow M$
- and an *automorphism* is a bijective endomorphism.
- A *submodule* of an A-module M is a subset N of M which is again an A-module, and such that the inclusion mapping $N \longrightarrow M$ is a monomorphism of A-modules.

- Two A-modules M and N are *isomorphic* if there is an isomorphism $\alpha : M \longrightarrow N$. One writes in this case $M \simeq N$.
- Denote by
  - $Hom_A(M,N)$ the set of homomorphisms $M \longrightarrow N$,
  - $End_A(M)$ the set of endomorphisms of M and
  - $Aut_A(M)$ the set of automorphisms of M.

Remark 1.1.13.
1. Observe that for a commutative ring R, an R-algebra A and an A-module M the set $End_A(M)$ is an R-algebra. The set $Aut_A(M)$ is its unit group.
2. Let A be a K-algebra and let M be an A-module. What precisely means that S is a submodule of M. By definition S is a subset, an A-module again, and the inclusion is a module homomorphism. The A-module M comes with a homomorphism $\varphi_M : A \to End_K(M)$. For each $a \in A$ the restriction of $\varphi_M(a)$ to S gives a map $A \to Hom_K(S,M)$. We claim that S is a submodule of M if and only if for each $a \in A$ the restriction $\varphi_M(a)|_S$ of $\varphi_M(a)$ to S gives a homomorphism $A \to End_K(S)$. Indeed, the fact that the inclusion of sets is an A-module homomorphism just means that the defining map φ_S is nothing than the restriction of $\varphi_M(a)$ to S for all $a \in A$.

We have the usual concepts of kernel, image and cokernel, as for vector spaces in linear algebra.

Definition 1.1.14. For an A-module homomorphism $\alpha : M \longrightarrow N$
- the kernel $\ker(\alpha)$ of an A-module homomorphism α is $\alpha^{-1}(0)$,
- the image $\mathrm{im}(\alpha)$ of an A-module homomorphism α is $\{\alpha(m) \mid m \in M\}$,
- the cokernel $\mathrm{coker}(\alpha)$ of an A-module homomorphism α is $N/\mathrm{im}(\alpha)$.

We get the usual properties.

Lemma 1.1.15. *Let A be an algebra and let M, N and L be A-modules.*
1. *The image* $\mathrm{im}(\alpha)$ *of a homomorphism $\alpha : M \longrightarrow N$ of A-modules is an A-submodule of N.*
2. *The kernel of α is an A-submodule of M.*
3. *If M has an A-submodule N, then M/N is an A-module as well.*
4. $\mathrm{im}(\alpha) \simeq M/\ker(\alpha)$.
5. *If α is an isomorphism of A-modules, then α^{-1} is an isomorphism of A-modules as well.*
6. *If $\beta : N \to L$ is a module homomorphism, then $\beta \circ \alpha$ is a module homomorphism.*

Proof.
1. For all $a \in A$ let $\varphi_{\mathrm{im}(\alpha)}(a) := \varphi_N(a)|_{\mathrm{im}(\alpha)}$. Since α is a module homomorphism,

$$\alpha \circ \varphi_M(a) = \varphi_N(a) \circ \alpha = \varphi_N(a)|_{\mathrm{im}(\alpha)} \circ \alpha.$$

The last equation holds since $\varphi_N(a)$ is evaluated on some image of α. Now, the left hand side of the equation shows that this is again an element in the image of α. Therefore $\mathrm{im}(\alpha)$ is a submodule of N by Remark 1.1.13.(2).

2. We put $\varphi_{\ker(\alpha)}(a) := \varphi_M(a)|_{\ker(\alpha)}$ for all $a \in A$. Now, $\alpha \circ \varphi_M(a) = \varphi_N(a) \circ \alpha$ for all $a \in A$ since α is an A-module homomorphism and hence if $m \in \ker(\alpha)$, also $(\alpha \circ \varphi_M(a))(m) = (\varphi_N(a) \circ \alpha)(m) = 0$ and hence $\varphi_M(a)(m) \in \ker \alpha$. This shows $\varphi_M(a)|_{\ker(\alpha)} \in End_K(\ker(\alpha))$ for all $a \in A$, which implies the statement by Remark 1.1.13.(2).
3. Since N is an abelian subgroup of M, and since for each $a \in A$ we have that $\varphi_M(a)$ is an endomorphism of M whose restriction to N is the endomorphism $\varphi_N(a)$, the endomorphism $\varphi_M(a)$ induces an endomorphism of M/N, which we denote by $\varphi_{M/N}(a)$. Since $\varphi_M(ab) = \varphi_M(a) \circ \varphi_M(b)$, for all $a, b \in A$, we also get $\varphi_{M/N}(ab) = \varphi_{M/N}(a) \circ \varphi_{M/N}(b)$, and likewise $\varphi_{M/N}(a + b) = \varphi_{M/N}(a) + \varphi_{M/N}(b)$. Again by the same reason $\varphi_{M/N}(1_A) = id_{M/N}$. This proves that $\varphi_{M/N} : A \to End_K(M/N)$ is an algebra homomorphism, and shows the statement.
4. Define $\overline{\alpha} : M/\ker(\alpha) \longrightarrow N$ by $\overline{\alpha}(m+\ker(\alpha)) := \alpha(m)$ for all $m \in M$. Since $\alpha(\ker(\alpha)) = 0$, this is well defined, and of course with image being equal to $\mathrm{im}(\alpha)$. The mapping $\overline{\alpha}$ is a monomorphism, since it is A-linear, α being A-linear, and since $\overline{\alpha}(m + \ker(\alpha)) = 0$ implies $\alpha(m) = 0$ which is equivalent to $m + \ker(\alpha) = 0$. Therefore $\overline{\alpha}$ is a monomorphism. Clearly $\mathrm{im}(\overline{\alpha}) = \mathrm{im}(\alpha)$ and hence $\overline{\alpha}$ is an epimorphism to $\mathrm{im}(\alpha)$.
5. $\alpha \circ \varphi_M(a) = \varphi_N(a) \circ \alpha$ implies immediately $\alpha^{-1} \circ \varphi_N(a) = \varphi_M(a) \circ \alpha^{-1}$ by conjugation with α.
6. For all $m \in M$ and $a \in A$ we get

$$\beta \circ \alpha(a \cdot m) = \beta(a \cdot \alpha(m)) = a \cdot \beta(\alpha(m)) = a \cdot (\beta \circ \alpha)(m)$$

and $\beta \circ \alpha$ is clearly a homomorphism of abelian groups.

This proves the lemma. □

1.1.3 Direct sums and products of modules

For a field K and two K-vector spaces V and W we can form $V \times W$, the vector space given by the direct product, which is usually denoted by $V \oplus W$. It is most important to note that this construction is equipped with two types of vector space homomorphisms. These are the embeddings

$$\begin{aligned} V &\xrightarrow{\iota_V} V \oplus W \\ v &\mapsto (v, 0) \\ W &\xrightarrow{\iota_W} V \oplus W \\ w &\mapsto (0, w) \end{aligned}$$

as well as

$$\begin{aligned} V \oplus W &\xrightarrow{\pi_V} V \\ (v, w) &\mapsto v \\ V \oplus W &\xrightarrow{\pi_W} W \\ (v, w) &\mapsto w \end{aligned}$$

The main property for the first construction is that if one has a third vector space U and linear maps $V \xrightarrow{\alpha} U$ and $W \xrightarrow{\beta} U$, then there is a unique linear map $V \oplus W \xrightarrow{\alpha+\beta} U$ such that $\iota_V \circ (\alpha + \beta) = \alpha$ and $\iota_W \circ (\alpha + \beta) = \beta$. For the second construction the corresponding property is that if there is a third vector space U and linear maps $U \xrightarrow{\sigma} V$ and $U \xrightarrow{\tau} W$, then there is a unique linear map $U \xrightarrow{(\sigma,\tau)} V \oplus W$ such that $(\sigma, \tau) \circ \pi_V = \sigma$ and $(\sigma, \tau) \circ \pi_W = \tau$.

The second concept generalises without significant change to infinite vector spaces. Let I be an infinite set and let V_i be a K-vector space for each $i \in I$. Then we may form the product $\prod_{i \in I} V_i$ and projections $\pi_j : \prod_{i \in I} V_i \to V_j$ of vector spaces, and again the property as in the finite case holds literally. However, the first concept of injections poses problems. Indeed, one would need to define an infinite sum "$\sum_{i \in I} \alpha_i$" of linear maps. This is not defined in general. The correct concept here is the coproduct

$$\coprod_{i \in I} V_i := \{(v_i)_{i \in I} : |\{i \in I : v_i \neq 0\}| < \infty\}.$$

In this case we may evaluate $\sum_{i \in I} \alpha_i$ on each element, since for each element $v = (v_i)_{i \in I}$ of $\coprod_{i \in I} V_i$ only finitely many of the components v_i are non zero, and we can sum the finitely many $\alpha_i(v_i)$ which are non zero.

We will define the same concepts for A-modules for an R-algebra A for a commutative ring R.

Let I be a non empty set and let for each $i \in I$ be M_i an A-module. Then the *direct product* $\prod_{i \in I} M_i$ is an A-module by setting

$$a \cdot (m_i)_{i \in I} := (a \cdot m_i)_{i \in I}$$

for every $(m_i)_{i \in I} \in \prod_{i \in I} M_i$ and $a \in A$. For finite sets I there is no difference to the direct sum construction. Indeed, we encounter the following problem. Suppose there is given an A-module N, an infinite index set I and for each $i \in I$ an A-module M_i. Of course, for every $i_0 \in I$ the projection

$$\begin{aligned} \prod_{i \in I} M_i &\xrightarrow{\pi_{i_0}} M_{i_0} \\ (m_i)_{i \in I} &\mapsto m_{i_0} \end{aligned}$$

is a homomorphism of A-modules since

$$\begin{aligned}\pi_{i_0}(a\cdot(m_i)_{i\in I}+(m_i')_{i\in I}) &= \pi_{i_0}((a\cdot m_i+m_i')_{i\in I})\\ &= a\cdot m_{i_0}+m_{i_0}'\\ &= a\cdot\pi_{i_0}((m_i)_{i\in I})+\pi_{i_0}((m_i')_{i\in I})\end{aligned}$$

for all $a \in A$.

Let I be a non empty set and suppose given for each $i \in I$ an A-module M_i. Then define the *coproduct* of the M_i; $i \in I$ to be those elements in the product which are 0 in all but a finite number of coordinates:

$$\coprod_{i\in I} M_i := \bigoplus_{i\in I} M_i := \left\{(m_i)_{i\in I} \in \prod_{i\in I} M_i : |\{i \in I : m_i \neq 0\}| < \infty\right\}.$$

Then for all $i_0 \in I$ there is a mapping

$$\begin{aligned} M_{i_0} &\xrightarrow{\iota_{i_0}} \coprod_{i\in I} M_i\\ m_{i_0} &\mapsto (m_i')_{i\in I}\end{aligned}$$

by putting $m_i' = 0$ if $i \neq i_0$ and $m_{i_0}' = m_{i_0}$.

1.1.4 Group algebras and modules over a group

We have seen that a module over a K-algebra A is given by an algebra homomorphism $A \longrightarrow End_K(M)$. In case of a group, we want to study representations as well. Here, staying in the philosophy expressed previously, we mean by a representation a concrete realisation of an abstract group by an action on a 'real world object', in our case a vector space. Since group elements have inverses, the same should be true for the representation of a group. So, a representation of a group is a homomorphism of the group to the group of invertible matrices.

Definition 1.1.16. Let G be a group and K be a field. A *representation of G over K of degree $n \in \mathbb{N}$* is a group homomorphism $G \longrightarrow GL_n(K)$. If the kernel of this homomorphism is all of G and $n = 1$, then we call the representation the *trivial representation.*

There are at least two obvious drawbacks of this concepts. First, it would be nice if we could phrase this concept in our setting of modules over an algebra. Second, we see that actually we would like to use at least some of the additional information the matrices provide. One of the most obvious ones is the fact that $GL_n(K)$ is the unit group of an algebra, the ring $Mat_{n\times n}(K)$ of degree n square matrices. Just as we can form the linear combinations of invertible matrices of the same size to get a (not necessarily invertible) square matrix, we can *formally* form K-linear combinations of group elements and get a K-algebra. This is the concept of a group ring.

Definition 1.1.17. Let G be a group and let R be a commutative ring. The *group ring* RG as R-module is the R-module $\oplus_{g\in G} R\cdot e_g$, where $R\cdot e_g = R$ for all $g\in G$. This becomes an R-algebra when one sets

$$\left(\sum_{h_1\in G} r_{h_1}e_{h_1}\right)\cdot\left(\sum_{h_2\in G} s_{h_2}e_{h_2}\right) := \sum_{g\in G}\left(\sum_{h_1\in G; h_2\in G; h_1h_2=g} r_{h_1}s_{h_2}\right)e_g.$$

Occasionally we shall call a group ring also a *group algebra* if we want to stress the algebra structure.

One should explicitly mention that in case G is infinite, the fact that RG is defined as a *direct sum* of $|G|$ copies of R as R-modules implies that an element $x = \sum_{g\in G} r_g e_g$ has only a finite number of non zero coefficients r_g. In other words for any $x = \sum_{g\in G} r_g e_g \in RG$ one gets

$$\left|\{g\in G \mid r_g \neq 0\}\right| < \infty.$$

Now, the group of invertible elements $(RG)^\times$ of RG contains G as a subgroup. Indeed, the mapping

$$\begin{aligned} G &\longrightarrow (RG)^\times \\ g &\mapsto 1\cdot e_g \end{aligned}$$

is a group monomorphism. Call G again the image of this homomorphism.

Using this we can formulate an important property of group rings, namely that RG is universal for having G as a subgroup of its unit group.

Lemma 1.1.18. *Let R be a commutative ring and let A be an R-algebra. Then for every group homomorphism $\alpha : G \longrightarrow A^\times$ there is a unique homomorphism of R-algebras $\beta : RG \longrightarrow A$ with $\alpha = \beta|_G$.*

Proof. We have to define $\beta(\sum_{g\in G} r_g e_g) := \sum_{g\in G} r_g\alpha(g)$ in order to get an R-linear mapping $RG \longrightarrow A$ with $\alpha = \beta|_G$. Moreover,

$$\begin{aligned} \beta\left(\left(\sum_{g\in G} r_g e_g\right)\cdot\left(\sum_{g\in G} s_g e_g\right)\right) &= \sum_{g\in G; h\in G}\left(\sum_{gh=k} r_g s_h\right)\alpha(k) \\ &= \sum_{g\in G; h\in G}\left(\sum_{gh=k} r_g s_h\right)\beta(e_k) \\ &= \beta\left(\sum_{g\in G} r_g e_g\right)\cdot\beta\left(\sum_{g\in G} s_g e_g\right) \end{aligned}$$

This shows the lemma. □

Lemma 1.1.19. *For any integer n and any representation $\varphi : G \longrightarrow GL_n(R)$ of G in R there is a unique RG-module structure $\hat\varphi$ on R^n such that $\hat\varphi|_G = \varphi$. Moreover, if R is a field, then* *the restriction of the structure map $\hat{\varphi} : RG \longrightarrow End_R(M)$ of an RG-module M induces a representation of G over R of dimension $dim_R(M)$.*

Proof. This follows immediately from Lemma 1.1.18. Indeed, $GL_n(R) = Mat_{n\times n}(R)^\times$ and so, by this lemma there is a unique ring homomorphism $RG \longrightarrow Mat_{n\times n}(R)$ restricting to φ. This shows the lemma. □

In elementary terms the previous lemma constructs the module structure on R^n in the following way. Suppose φ is a representation $\varphi : G \longrightarrow GL_n(R)$ of G in R. Then,

$$\hat{\varphi} : RG \ni \sum_{g\in G} r_g e_g \mapsto \sum_{g\in G} r_g \varphi(g) \in End_R(R^n)$$

is an R-algebra homomorphism.

Lemma 1.1.19 implies that there is no need to distinguish between modules over the group ring KG and K-representations of the group G.

Lemma 1.1.20. *Let ϕ and ψ be two representations of the same dimension over a field K of a group G and suppose that there is an invertible matrix T such that for all $g \in G$ one has $T \cdot \phi(g) = \psi(g) \cdot T$. Then, the modules M_ϕ and M_ψ which are induced by ϕ and ψ are isomorphic. If the modules M_ϕ and M_ψ are isomorphic as KG-modules, then there is an invertible matrix T such that $T \cdot \phi(g) = \psi(g) \cdot T$.*

Proof. Suppose there is a matrix T which conjugates $\phi(g)$ to $\psi(g)$ for all $g \in G$. Then, T yields a K-linear endomorphism $K^n \longrightarrow K^n$ which induces a KG-isomorphism between M_ϕ and M_ψ. Indeed, $T\phi(g)m = \psi(g)Tm$ for all $m \in K^n$ is exactly the equation needed for showing that T is KG-linear.

Suppose that there is an isomorphism $T : M_\phi \longrightarrow M_\psi$. This isomorphism is K-linear, hence given by an invertible matrix T, and since it is a homomorphism of KG-modules, $T\phi(g)m = \psi(g)Tm$ holds for all $m \in K^n$ and $g \in G$. Hence, $T\phi(g) = \psi(g)T$ for all $g \in G$. This shows the lemma. □

We would like to see some examples how this concept is going to work.

Example 1.1.21. Let $K = \mathbb{C}$ and $G := C_n$ the cyclic group of order n generated by c, say.

1. Let us study the 1-dimensional $\mathbb{C}C_n$-modules up to isomorphism. It is necessary and sufficient to define a group homomorphism $C_n \longrightarrow GL_1(\mathbb{C}) = \mathbb{C}^\times$. Since $c^n = 1$ in C_n, in order to be able to find images of c we need elements $x \in \mathbb{C}$ satisfying the same relation $x^n = 1$. Fix a primitive n^{th} root of unity ζ_n in $\mathbb{C}$. We get n possibilities, namely $\varphi_m(c^\ell) = (\zeta_n^m)^\ell = \zeta_n^{m\ell}$. We see that φ_{m_1} and φ_{m_2} lead to isomorphic representations if and only if $m_1 - m_2 \in n\mathbb{Z}$. Indeed, if $m_1 - m_2 \in n\mathbb{Z}$, then $\zeta_n^{m_1} = \zeta_n^{m_2}$ and so the representations are really equal. If one has a $\mathbb{C}$-linear isomorphism $\lambda : \mathbb{C} \longrightarrow \mathbb{C}$, then $\lambda(z) = xz$ for an element $x \in \mathbb{C} \setminus \{0\}$. If λ commutes with the action of C_n via φ_{m_1} on the left and via φ_{m_2} on the right, we need to have

$$x\zeta_n^{m_2} z = \lambda(c \cdot z) = c \cdot \lambda(z) = c \cdot xz = \zeta_n^{m_1} xz$$

for all z and so $\zeta_n^{m_1} = \zeta_n^{m_2}$, which implies $m_1 - m_2 \in n\mathbb{Z}$.
For example if $n = 3$, and $\zeta_3 =: j$ one has the following three possibilities.
(a) The first is $\varphi_1(c) = 1$ and therefore $\varphi_1(c^n) = 1^n = 1$ for all $n \in \mathbb{N}$.
(b) The second is $\varphi_j(c) = j$ and therefore $\varphi_j(c^2) = j^2$ as well as $\varphi_j(1) = 1$ of course.
(c) The third is $\varphi_{j^2}(c) = j^2$ and therefore $\varphi_{j^2}(c^2) = j^4 = j$ as well as $\varphi_{j^2}(1) = 1$ of course.

We observe that φ_j and φ_{j^2} are conjugate complex representations.

2. Let us study representations of dimension d. As before, for any representation φ one has that $\varphi(c) =: M_d$ is an invertible matrix of size d of order n; i. e. $M_d^n = 1$. Since M_d can be conjugate into Jordan normal form, and the power of a conjugate matrix is the conjugate of the power, one sees that M_d is diagonalisable with diagonal coefficients $\zeta_n^{u_1}, \zeta_n^{u_2}, \dots, \zeta_n^{u_d}$ for a vector $(u_1, u_2, \dots, u_d) \in \mathbb{Z}^d$. Since $\varphi(c^\ell) = M_d^\ell$, the representation is fixed by the choice of this single matrix M_d. By Lemma 1.1.20 the $\mathbb{C}C_n$-module M_φ which corresponds to the representation φ is isomorphic to the direct sum $\bigoplus_{j=1}^d M_{\hat{\varphi}_{u_j}}$.
   We classified all finite dimensional complex representations of cyclic groups.
3. In case of Example (2) the factors in the direct sum are all one-dimensional. In general this will not be the case. The reason can be that the field is smaller, or that the group is more complicated.
   - The field is too small: Consider the quadratic field extension $\mathbb{Q}(\zeta_3)$ of $\mathbb{Q}$ and an action of $C_3 = \langle c |\ c^3 = 1\rangle$ given by $c^i \cdot x := \zeta_3^i x$ for all $x \in \mathbb{Q}(\zeta_3)$ and all $i \in \mathbb{Z}$. Since $\zeta_3^3 = 1$ this defines a $\mathbb{Q}C_3$-module M. A $\mathbb{Q}C_3$-submodule N of M is of course at first place a $\mathbb{Q}$-sub vector-space of dimension zero, one or two. We suppose that N is not of dimension 0. Since $c \cdot x = \zeta_3 x$ for all $x \in N$, the submodule N is even a $\mathbb{Q}(\zeta_3)$-sub vector space of $\mathbb{Q}(\zeta_3)$. As a consequence, $M = N$, and so M cannot be written as a direct sum of two non zero submodules N_1 and N_2.
   - The group is more complicate: Consider the symmetric group $\mathfrak{S}_3$ of degree 3 and of order 6. As is well-known the permutation $(1\ 2\ 3)$ which sends 1 to 2, 2 to 3 and 3 to 1 generates a subgroup C of order 3 and hence of index 2 in $\mathfrak{S}_3$. As a consequence, C is a normal subgroup of $\mathfrak{S}_3$ with quotient C_2. The permutation $(1\ 2)$ which interchanges 1 and 2 and fixes 3 generates a subgroup D of order 2 of $\mathfrak{S}_3$. The groups C and D together generate $\mathfrak{S}_3$ since they intersect trivially and generate hence at least a group of order $2 \cdot 3 = 6$. Consider the $\mathbb{C}$-vector space M of dimension 2 on which the elements $(1\ 2)$ and $(1\ 2\ 3)$ act as follows.

$$(1\ 2) \mapsto \begin{pmatrix} -2 & -3 \\ 1 & 2 \end{pmatrix}$$

$$(1\ 2\ 3) \mapsto \begin{pmatrix} -2 & -3 \\ 1 & 1 \end{pmatrix}$$

We observe that using two abstract generators a, b of $\mathfrak{S}_3$ one may identify a with (1 2 3) and b with (1 2) and one observes that the relations of these two abstract generators are then $a^3 = b^2 = (ba)^2 = 1$. Observe that these are actually the relations of a dihedral group of a triangle; a reminder of the fact that $\mathfrak{S}_3$ is the symmetry group of the equilateral triangle. In order to verify that this actually gives an action of $\mathfrak{S}_3$ one just needs to verify

$$\text{that} \quad \begin{pmatrix} -2 & -3 \\ 1 & 2 \end{pmatrix} \quad \text{is of order 2,}$$

$$\text{that} \quad \begin{pmatrix} -2 & -3 \\ 1 & 1 \end{pmatrix} \quad \text{is of order 3}$$

$$\text{and that} \quad \begin{pmatrix} -2 & -3 \\ 1 & 2 \end{pmatrix} \cdot \begin{pmatrix} -2 & -3 \\ 1 & 1 \end{pmatrix} = \begin{pmatrix} 1 & 3 \\ 0 & -1 \end{pmatrix} \quad \text{is of order 2.}$$

But this is easily seen. If N is a $\mathbb{C}\mathfrak{S}_3$ submodule of M, then N must be 1-dimensional and hence a common eigenspace of the two matrices $\left(\begin{smallmatrix} -2 & -3 \\ 1 & 2 \end{smallmatrix}\right)$ and $\left(\begin{smallmatrix} -2 & -3 \\ 1 & 1 \end{smallmatrix}\right)$. As one readily verifies there is none. As a consequence, there is no direct sum decomposition of M as non zero $\mathbb{C}\mathfrak{S}_3$-submodules N_1 and N_2.

The module in this Example 1.1.21 (3) has a particular property, namely the fact that the image of the group G in the linear group defining the representation is isomorphic to G.

Definition 1.1.22. Let G be a group and let R be a commutative ring. An RG-module M is *faithful* if $\{g \in G \mid gm = m \forall\ m \in M\} = \{1\}$.

1.2 Maschke, Wedderburn and Krull-Schmidt

The phenomenon that a finite dimensional $\mathbb{C}G$-module M is a direct sum of smaller modules is quite typical.

Definition 1.2.1. Let K be a commutative ring and let A be a K-algebra. An A-module M is said to be

- *simple* if any non zero submodule N of M is equal to M.
- *semisimple* if there are an index set I and simple A-modules S_i, $i \in I$, such that $M \simeq \bigoplus_{i \in I} S_i$.
- If K is a field, then we say that a *K-algebra A is semisimple* if and only if every A-module is semisimple.

The module 0 is not considered as simple. This is just a technical exclusion which simplifies the statements later on.

Finite dimensional vector spaces are easier to handle than infinite dimensional ones. The corresponding concept for modules are finitely generated modules, as made precise in the next definition.

Definition 1.2.2. Let R be a ring and let M be an R-module. Then, M is *finitely generated* if there is an integer $n \in \mathbb{N}$ and an R-module epimorphism $\varphi : R^n \longrightarrow M$.

Of course, with this definition M is finitely generated if and only if there are elements $b_1, b_2, \ldots, b_n$ in M such that for every $m \in M$ there are $r_1(m), r_2(m), \ldots, r_n(m)$ in R with $m = \sum_{i=1}^n r_i(m)m$. One says that $b_1, b_2, \ldots, b_n$ generate M in this case.

Lemma 1.2.3. *A finitely generated A-module M is semisimple if and only if there are simple modules $S_1, \ldots, S_n$ such that $M \simeq S_1 \oplus \cdots \oplus S_n$ as A-modules.*

Proof. If M is finitely generated and semisimple then $M \simeq \bigoplus_{i\in I} S_i$ for an index set I and via an isomorphism $\varphi : M \to \bigoplus_{i\in I} S_i$. Since M is finitely generated, there is a finite set $\{m_1, \ldots, m_n\}$ such that $M = \sum_{i=1}^n Am_i$. Since each m_i belongs to the direct sum, there is a finite set $I_i \subset I$ such that $\varphi(m_i) \in \bigoplus_{j\in I_i} S_j$. Putting $J := \bigcup_{i=1}^n I_j$, we see that φ is an isomorphism $M \simeq \bigoplus_{i\in J} S_i$ with the finite set J. The converse is trivial. □

Example 1.2.4.
- Let $A = K = \mathbb{Z}$ the integers and as we have seen a $\mathbb{Z}$-module is just an abelian group. The simple $\mathbb{Z}$-modules are exactly the abelian groups $\mathbb{Z}/p\mathbb{Z}$ for a prime p in $\mathbb{Z}$.
- The simple $\mathbb{C}C_n$-modules are the one dimensional modules which we have determined in Example 1.1.21.(1).

We should note that if K is a field and if M is a finite-dimensional A-module over a K-algebra A, then of course M is semisimple if and only if $M = \bigoplus_{i=1}^n S_i$ for simple A-modules $S_1 \ldots, S_n$.

We shall first need a characterisation of semisimple modules. Semisimple modules M are precisely those such that all submodules N of M are direct factors. This is the subject of the following Proposition 1.2.6.

Remark 1.2.5. Since we are dealing with a priori infinitely generated modules and infinite index sets of arbitrary cardinal, the proof of this proposition is slightly more involved and uses Zorn's lemma. In fact, the standard definition for a module to be semisimple is precisely the statement of Proposition 1.2.6 below. Namely, a module is said to be semisimple if every submodule has a supplement. Our definition works particularly well in case of finite dimensional modules over some finite dimensional algebra (or slightly more generally in case of Noetherian modules, to be introduced in Section 6.1 below). In this simpler case an easier proof is possible. The reader may consult [Zim-14, Lemma 1.4.28] for this alternative approach.

Proposition 1.2.6. *Let K be a commutative ring, let A be a K-algebra and let M be an A-module. Then M is semisimple if and only if for all submodules N of M there is a submodule L of M such that $M = N \oplus L$.*

Proof. Suppose that for all submodules N of M there is a submodule L of M such that $M = N \oplus L$. Then let $x \in M \setminus \{0\}$ and $\mathfrak{T} := \{U \le M \mid x \notin U\}$. The set $\mathfrak{T}$ is partially ordered by inclusion and since $0 \in \mathfrak{T}$, we get $\mathfrak{T} \neq \emptyset$. We may therefore apply Zorn's lemma. Let I be a totally ordered set and let $(T_i)_{i\in I}$ be an increasing family in $\mathfrak{T}$. Then $T := \bigcup_{i\in I} T_i \le M$ and moreover, $T \in \mathfrak{T}$. Indeed, if $x \in T$, then $x \in T_{i_0}$ for some $i_0 \in I$, which contradicts $T_{i_0} \in \mathfrak{T}$. Zorn's lemma shows that $\mathfrak{T}$ contains maximal elements, and let P be such a maximal element in $\mathfrak{T}$. Then $P \le M$ and by hypothesis there is a submodule S of M with $P \oplus S = M$. We claim that S is simple. Since $x \notin P$, also $P \neq M$, and hence $S \neq 0$. Now $\overline{S} := M/P \simeq S$, and let $\overline{U}_1$ and $\overline{U}_2$ be two non zero submodules of $\overline{S}$. Then there are submodules U_1 and U_2 of M such that $P \le U_1$ and $P \le U_2$ with $U_1/P = \overline{U}_1$ and $U_2/P = \overline{U}_2$. Since $\overline{U}_1$ and $\overline{U}_2$ are both non zero, $U_1 \notin \mathfrak{T}$ and $U_2 \notin \mathfrak{T}$, by maximality of P. Hence $x \in U_1 \cap U_2$ and therefore $\overline{U}_1 \cap \overline{U}_2 \neq 0$. But by hypothesis there is a submodule V_1 of M such that $U_1 \oplus V_1 = M$. We may take $U_2 = P \oplus V_1$, and see that $x \notin U_2$. We obtain a contradiction to the maximality of P. Hence S is simple. Therefore M contains simple submodules. Let now $\mathcal{S}$ be the (non empty) set of simple submodules of M and put $M' := \sum_{S\in\mathcal{S}} S$. Then by hypothesis there is a submodule W of M such that $M = M' \oplus W$. Now, W again has the property that all submodules are direct factors. Indeed, $\overline{W} := M/M'$. If $\overline{V} = V/M'$ is a proper submodule of $\overline{W}$, then V contains M' and let C be a submodule of M with $V \oplus C = M$. Then $\overline{C} := (M' \oplus C)/M'$ is a submodule of $\overline{W}$, is a complement to $\overline{V}$ in $\overline{W}$, and is isomorphic to C. Hence W admits simple submodules S. But $S \in \mathcal{S}$ by definition, and therefore $W \cap M' \neq 0$, a contradiction. Hence $W = 0$. Let now $I_{\mathcal{S}}$ be the set of subsets of $\mathcal{S}$. This is again a non empty partially ordered set, and let $\mathcal{D} := \{J \subseteq I_{\mathcal{S}} \,;\mid \sum_{S\in J} S = \bigoplus_{S\in J} S\}$. The subsets of $\mathcal{S}$ of cardinal 1 are clearly in $\mathcal{D}$, and so $\mathcal{D} \neq \emptyset$. Zorn's lemma implies again that $\mathcal{D}$ contains maximal elements, and let J_m be such a maximal element. Let $M' := \sum_{S\in J_m} S = \bigoplus_{S\in J_m} S$. Again, M' is a submodule of M and by hypothesis there is a submodule W of M with $M' \oplus W = M$. If $W \neq 0$, by the argument above there is a simple submodule S_W of W, which is a contradiction to the maximality of J_m. Hence $\bigoplus_{S\in J_m} S = M' = M$ is therefore semisimple.

Let $M = \bigoplus_{i\in I} S_i$ be a semisimple A-module and let $0 \neq N \le M$, and suppose $N \neq M$. Let $\mathcal{J} := \{J \subset I \mid N \cap \sum_{j\in J} S_j = 0\}$. This is a partially ordered set by inclusion. Is $\mathcal{J} = \emptyset$, then $N = M$. Zorn's lemma shows that there is a maximal element in $\mathcal{J}$, and hence let J_m be such a maximal element. Let $M' := N + \bigoplus_{j\in J_m} S_j = N \oplus \bigoplus_{j\in J_m} S_j$. Suppose $M' \neq M$. Then there is $i_0 \in I$ with $S_{i_0} \not\le M'$. Since S_{i_0} is simple, $S_{i_0} \cap M'$ is a submodule of S_{i_0}, and since $S_{i_0} \not\le M'$ we get $S_{i_0} \cap M' = 0$. This contradicts the maximality of J_m, and hence $M' = M$. Therefore $\bigoplus_{j\in J_m} S_j =: W$ is a submodule of M with $W \oplus N = M$. □

1.2.1 Maschke's theorem

We show now one of the most important results in the representation theory of finite groups. The result actually states that if G is a finite group of order n, K is a field and $n \cdot K = K$, then any module of KG is semisimple.

Using Proposition 1.2.6 the concept of semisimplicity can be characterised by the fact that every submodule is a direct factor. We need to characterise this property in different terms.

Definition 1.2.7. An element φ in a ring with $\varphi \circ \varphi = \varphi$ is called *idempotent*.

The following Lemma gives the desired alternative definition of being a direct factor.

Lemma 1.2.8. *Let A be an R-algebra for a commutative ring R and let M be an A-module. Suppose there is an idempotent A-module endomorphism φ of M, then $M \simeq \operatorname{im}(\varphi) \oplus \ker(\varphi)$. Conversely, if $M \simeq L \oplus N$, then there is an idempotent endomorphism φ of M with $L = \ker(\varphi)$ and $N = \operatorname{im}(\varphi)$.*

Proof. By Lemma 1.1.15 $\operatorname{im}(\varphi)$ and $\ker(\varphi)$ are both A-submodules of M. Let $m \in \operatorname{im}(\varphi) \cap \ker(\varphi)$. Then, there is an $m' \in M$ such that $m = \varphi(m')$. Hence,

$$m = \varphi(m') = \varphi \circ \varphi(m') = \varphi(m) = 0.$$

Let $m \in M$. Then,

$$\varphi(m - \varphi(m)) = \varphi(m) - \varphi \circ \varphi(m) = \varphi(m) - \varphi(m) = 0$$

and therefore $m - \varphi(m) \in \ker(\varphi)$. As a consequence,

$$m = (m - \varphi(m)) + \varphi(m) \in \ker(\varphi) + \operatorname{im}(\varphi).$$

This proves the first part of the lemma.

If $M = L \oplus N$, then let π_N be the projection $M \to N$ and ι_N be the injection $N \to M$ both given by the universal property of the direct sum being a product, respectively a coproduct. Then $\varphi := \iota_N \circ \pi_N$ is the required nilpotent endomorphism since $\pi_N \circ \iota_N = id_N$. □

We now come to the statement of the main result of the section.

Theorem 1.2.9 (Maschke 1905). *Let G be a finite group and let K be a field. If the order of G is invertible in K, then KG is semisimple.*

Proof. Let M be a KG-module. Suppose there is a proper non zero submodule N of M. We will construct an idempotent endomorphism φ of M with $\ker(\varphi) = N$. Using Lemma 1.2.8 and Proposition 1.2.6 this shows that KG is semisimple.

By Lemma 1.1.15 the quotient $\tilde{L} := M/N$ is a KG-module as well. Denote by $\pi : M \longrightarrow \tilde{L}$ the mapping $\pi(m) = m + N$. We may choose a K-basis $\mathcal{N}$ of N which we may complete by vectors $\mathcal{L}$ to a K basis $\mathcal{M} = \mathcal{N} \cup \mathcal{L}$ of M. Then, $\{l + N | \ l \in \mathcal{L}\}$ is a K-basis of $\tilde{L}$. Denote by $\tilde{\rho}$ the K-linear mapping $\tilde{L} \longrightarrow M$ defined by $\tilde{\rho}(l + N) := l$ for any $l \in \mathcal{L}$. By definition, $\pi \circ \tilde{\rho} = id_{\tilde{L}}$. Moreover, π is a KG-module homomorphism, but there is however no reason why $\tilde{\rho}$ should be a KG-module homomorphism.

Define

$$\begin{aligned} \rho : \tilde{L} &\longrightarrow M \\ l + N &\mapsto \frac{1}{|G|} \sum_{g \in G} e_g^{-1} \tilde{\rho}(e_g \cdot (l + N)) \end{aligned}$$

Remark 1.2.10. Observe that we were supposing that $|G|$ is invertible in K, and so the term $\frac{1}{|G|} \in K$ makes sense. It is only here where we use this hypothesis.

Since $\tilde{\rho}$ is a well-defined homomorphism of K-vector spaces, this is true for ρ as well. Moreover, for any $l + N \in \tilde{L}$ one has

$$\begin{aligned} \pi \circ \rho(l + N) &= \pi\left(\frac{1}{|G|} \sum_{g \in G} e_g^{-1} \tilde{\rho}(e_g \cdot (l + N)) \right) \\ &= \frac{1}{|G|} \sum_{g \in G} e_g^{-1} \pi(\tilde{\rho}(e_g \cdot (l + N))) \\ &= \frac{1}{|G|} \sum_{g \in G} e_g^{-1} e_g \cdot (l + N) \\ &= l + N \end{aligned}$$

and so, $\pi \circ \rho = id_{\tilde{L}}$.

Furthermore, ρ is a KG-module homomorphism. We need to show that for any $l + N \in \tilde{L}$ and any $h \in G$ one has $\rho(e_h \cdot (l + N)) = e_h \cdot \rho(l + N)$. This is the case:

$$\begin{aligned} \rho(e_h \cdot (l + N)) &= \frac{1}{|G|} \sum_{g \in G} e_g^{-1} \tilde{\rho}(e_g \cdot (e_h \cdot (l + N))) \\ &= \frac{1}{|G|} \sum_{g \in G} e_g^{-1} \tilde{\rho}(e_{gh} \cdot (l + N)) \\ &= e_h \cdot \frac{1}{|G|} \sum_{g \in G} e_h^{-1} e_g^{-1} \tilde{\rho}(e_{gh} \cdot (l + N)) \\ &= e_h \cdot \frac{1}{|G|} \sum_{g \in G} e_{gh}^{-1} \tilde{\rho}(e_{gh} \cdot (l + N)) \\ &= e_h \cdot \frac{1}{|G|} \sum_{gh \in G} e_{gh}^{-1} \tilde{\rho}(e_{gh} \cdot (l + N)) \\ &= e_h \cdot \frac{1}{|G|} \sum_{g \in G} e_g^{-1} \tilde{\rho}(e_g \cdot (l + N)) \\ &= e_h \cdot \rho(l + N) \end{aligned}$$

Therefore, ρ is a homomorphism of KG-modules, and by Lemma 1.1.15 $L := \mathrm{im}(\rho)$ is a KG-submodule of M. Since $\pi \circ \rho = id_{\tilde{L}}$, one gets

$$(\rho \circ \pi) \circ (\rho \circ \pi) = \rho \circ (\pi \circ \rho) \circ \pi = \rho \circ id_{\tilde{L}} \circ \pi = \rho \circ \pi.$$

Moreover, since $\pi \circ \rho = id_{\tilde{L}}$, the mapping ρ is injective, $\ker(\rho \circ \pi) = \ker(\pi) = N$ and $\mathrm{im}(\rho \circ \pi) \simeq \mathrm{im}(\pi) = \tilde{L}$. Hence, $M \simeq N \oplus M/N$. This finishes the proof. □

We first illustrate this result on a small example.

Example 1.2.11. Let $G = \mathfrak{S}_n$ be the symmetric group on n letters $\{1, 2, \ldots, n\}$ and let K be any field of characteristic 0 or at least $n + 1$. Let N_n be an n-dimensional K-vector space, and fix a basis $\{b_1, b_2, \ldots, b_n\}$ in N_n. Then N_n becomes a $K\mathfrak{S}_n$-module if one defines $\sigma \cdot b_i := b_{\sigma(i)}$ for any $\sigma \in \mathfrak{S}_n$ and any $i \in \{1, 2, \ldots, n\}$. Of course, the K-sub vector space generated by $\sum_{i=1}^n b_i$ is invariant under the action of any $\sigma \in \mathfrak{S}_n$. Hence, this one-dimensional subspace T is a $K\mathfrak{S}_n$-subspace of N_n. Since any $\sigma \in \mathfrak{S}_n$ acts as identity, T is the trivial $K\mathfrak{S}_n$-module. By Maschke's theorem we know that $N_n \simeq T \oplus N_n/T$.

The hypothesis that n is invertible in K is actually equivalent to this decomposition. Indeed, T is simple, since T is 1-dimensional, generated by $b_1 + \cdots + b_n =: b$. Let α be an idempotent endomorphism of N_n as $K\mathfrak{S}_n$-module with $\mathrm{im}(\alpha) = T$. Hence, for all $x \in N_n$ there is $a_x \in K$ such that $\alpha(x) = a_x b$. In particular there are $a_i \in K$ with $\alpha(b_i) = a_i b$ for all $i \in \{1, \ldots, n\}$. Then

$$\sum_{i=1}^{n} a_b b_i = a_b b = \alpha(b) = \sum_{i=1}^{n} \alpha(b_i) = \sum_{i=1}^{n} a_i b_i.$$

Hence, $a_i = a_b$ for each $i \in \{1, \ldots, n\}$, using that $\{b_1, \ldots, b_n\}$ is a basis. Therefore, in the basis $\{b_1, \ldots, b_n\}$ the endomorphism α is given by a square $n \times n$ matrix having coefficients a_b in all positions. The square if this matrix is a square $n \times n$ matrix having coefficients $a_b^2 n$ in all positions. Since $\alpha^2 = \alpha$ we get $a_b = a_b^2 n$, and since $\mathrm{im}(\alpha) = T = Kb$, we have $a_b \neq 0$, and therefore $1 = a_b n$. This shows $nK = K$.

The main reason why Maschke's result is one of the most fundamental in the representation theory of finite groups is that the structure theory of finite dimensional (or more generally artinian) semisimple algebras is known in great detail.

However, a lot can be done also for some classes of infinite groups as well, namely locally compact groups. A group G is a topological group if G is a topological space such that multiplication and inversion are continuous maps. A topological group G is locally compact if G is a locally compact topological space. The main tool which we use for this class of groups is the so-called Haar measure. We only cite this quite deep result from harmonic analysis without giving the proof. Parts of the proof, which really is mostly a hard piece of fairly deep analysis, can be found in e. g. [Kow-14, Section 5.2]. The reader may skip this result in the first reading. It will be used in Section 7.6 for the proof of the Strong Approximation Theorem 7.6.9. When we use integrals we understand that we use the Lebesgue integral.

Theorem 1.2.12. *Let G be a locally compact topological group. Then*
- *There is a non zero Radon measure μ_r on G, which we call the* right Haar measure *on G, and a non zero Radon measure μ_ℓ on G, which we call the* left Haar measure *such that for any $g \in G$ and any measurable subset A of G $\mu_r(Ag) = \mu_r(A)$ and $\mu_\ell(gA) = \mu_\ell(A)$.*
- *For any two right (resp. left) Haar measures μ_1 and μ_2 on G there is a $c \in \mathbb{R}$ such that $\mu_1 = c \cdot \mu_2$.*
- *If G is compact, any right Haar measure is also a left Haar measure and there is a unique Haar measure μ on G with $\mu(G) = 1$, called the probability Haar measure. Then denoting $A^{-1} := \{x^{-1} \mid x \in A\}$ we have $\mu(A) = \mu(A^{-1})$.*

The left Haar measure μ_ℓ satisfies

$$\int_G f(gx)d\mu_\ell(x) = \int_G f(x)d\mu_\ell(x)$$

and the right Haar measure μ_r satisfies

$$\int_G f(xg)d\mu_r(x) = \int_G f(x)d\mu_r(x)$$

for any measurable non negative $f \in L^1(G, d\mu)$ and any $g \in G$. Note that if G is finite, then it is a compact topological group with the discrete topology, and the Haar measure is given by counting elements. With this measure it is possible to formulate Maschke's theorem for compact groups. It is not at all clear what precisely the group ring of an infinite group would be. However, properly defined it can be shown (cf. I. E. Segal [Seg-41]) that the group ring of a locally compact group which is either compact or abelian is semisimple.

1.2.2 The Krull-Schmidt theorem

For the moment we do not know anything about unicity of a decomposition of a semisimple module into its factors. This is the very important Krull-Schmidt theorem. It holds more generally replacing simple modules by indecomposable modules as factors and semisimple algebras by general finite dimensional algebras.

Definition 1.2.13. An A-module $M \neq 0$ is called *indecomposable* if whenever $M \simeq N \oplus L$, then either $N = 0$ or $L = 0$.

By definition semisimple indecomposable modules are always simple. In general however, indecomposable modules need not be simple.

Example 1.2.14. Let K be a field and let $A = K[X]/(X^2)$ the so-called ring of dual numbers. It is a K-algebra of dimension 2 over K, and the regular module is indecomposable but not semisimple. Indeed, the only non zero ideals of A are $X \cdot K[X]/(X^2)$ and A.

Theorem 1.2.15 (Krull-Schmidt). *Let K be a field and let A be a K-algebra. Let M be a finite dimensional A-module. Suppose there are indecomposable A-modules $M_1, M_2, \dots, M_m$ and $N_1, N_2, \dots, N_n$ such that*

$$\bigoplus_{j=1}^{m} M_j = M = \bigoplus_{i=1}^{n} N_i.$$

Then, $m = n$ and there is a permutation $\sigma \in \mathfrak{S}_n$ such that $M_j \simeq N_{\sigma(j)}$ for all $j \in \{1, 2, \dots, n\}$.

A remark should be made here. Actually, one needs at this place only that M is Noetherian (after Emmy Noether) and artinian (after Emil Artin). We do not really need this concept here in our context and refrain at this place from developing this richer theory since in general it presents quite some difficulties in the first reading. The proof of the Krull-Schmidt theorem is going to be divided into three lemmas. Each lemma is quite simple, but actually they are going to be interesting in their own right. These three lemmas are also valid under a weaker assumption. Actually the modules need to be Noetherian and artinian and the proof holds under this assumption as well.

Lemma 1.2.16 (Fitting). *Let K be a field, let A be a K-algebra and let M be a finite dimensional A-module. Then, for any endomorphism u of M there is a decomposition $M \simeq N \oplus S$ such that u induces a nilpotent endomorphism $u|_N$ of N and an automorphism $u|_S$ of S. One may take $N = \ker(u^m)$ and $S = \operatorname{im}(u^m)$ for a large enough integer m.*

Proof. Denote for any integer k the endomorphisms $u^k := u^{k-1} \circ u$ and $u^1 = u$.

- We get a sequence
$$\ker(u) \subseteq \ker(u^2) \subseteq \ker(u^3) \subseteq \cdots \subseteq M$$
Since the dimension of M is finite, there is an integer k_0 such that $\ker(u^{k_0}) = \ker(u^{k_0+1})$. Hence, for any $k \geq k_0$ one has $\ker(u^{k_0}) = \ker(u^k) =: N$.
- We get another sequence of submodules
$$M \supseteq \operatorname{im}(u) \supseteq \operatorname{im}(u^2) \supseteq \operatorname{im}(u^3) \supseteq \cdots \supseteq 0$$
which has to be actually finite, since M is finite dimensional. So, there is an integer k_1 such that $\operatorname{im}(u^{k_1}) = \operatorname{im}(u^{k_1+1})$ and then also for any integer $k \geq k_1$ one has $\operatorname{im}(u^{k_1}) = \operatorname{im}(u^k) =: S$.
- Put $k_2 := \max(k_1, k_0)$.
- We shall show that $u|_S$ is an automorphism of S. Indeed,
$$u(S) = u(\operatorname{im}(u^{k_2})) = \operatorname{im}(u^{k_2+1}) = \operatorname{im}(u^{k_2}) = S$$
and so, $u|_S$ is an epimorphism.

- We shall show that $u|_N$ is a nilpotent endomorphism of N. Indeed, if $n \in N$, then $n \in \ker(u^{k_2}) = \ker(u^{k_2+1})$. As a consequence,

$$0 = u^{k_2+1}(n) = u^{k_2}(u(n))$$

and so $u(n) \in \ker(u^{k_2}) = N$. By definition $(u|_N)^{k_0} = 0$. Hence, $u|_N$ is a nilpotent endomorphism of N.
- If $m \in \operatorname{im}(u^{k_2}) \cap \ker(u^{k_2}) = S \cap N$, then there is an $m' \in M$ such that $m = u^{k_2}(m')$. Since $m \in \ker(u^{k_2})$ one has

$$0 = u^{k_2}(m) = u^{2k_2}(m').$$

Hence, $m' \in \ker(u^{2k_2}) = \ker(u^{k_2})$. So, $m = u^{k_2}(m') = 0$.
- The previous statement gives that $N \cap S = 0$. Since $\ker(u) \subseteq N$, we obtain as well that $\ker(u|_S) = \ker(u) \cap S \subseteq N \cap S = 0$. Therefore $u|_S$ is injective, and hence an automorphism.
- Let $m \in M$. Then, since $\operatorname{im}(u^{k_2}) = \operatorname{im}(u^{2k_2})$, there is an $m' \in \operatorname{im}(u^{k_2})$ with $u^{k_2}(m) = u^{k_2}(m')$. Hence, $m = m' + (m - m')$ and $m - m' \in \ker(u^{k_2})$ whereas $m' \in \operatorname{im}(u^{k_2})$.

This finishes the proof of Fitting's lemma. □

In order to formulate the second lemma in a concise way we need another notion.

Definition 1.2.17. A ring A is *local* if the set of non invertible elements of A form a two-sided ideal.

Lemma 1.2.18. *Let A be a K-algebra and let M be an indecomposable A-module such that Fitting's lemma holds for all endomorphisms of M. Then, $End_A(M)$ is a local ring. Each endomorphism u of M is either nilpotent or bijective.*

Proof. Given an endomorphism u of A, then Fitting's Lemma 1.2.16 implies that there is a decomposition $M \simeq N \oplus S$ into submodules where $u|_N$ is a nilpotent endomorphism of N and $u|_S$ is an automorphism of S. Since M is indecomposable, either N or S is 0. If $N = 0$, then u is an automorphism. If $S = 0$, then u is nilpotent, and therefore not invertible.

Suppose now that u is not invertible. We need to show that for any endomorphism v of M, also $u \circ v$ and $v \circ u$ are not invertible. But since u is not invertible, and since M is indecomposable, $M = N$ and u is nilpotent. If $\ker(u) = 0$, then u is injective. Injective endomorphisms cannot be nilpotent. Hence $\ker(u) \neq 0$ and therefore $\ker(v \circ u) \supseteq \ker(u) \neq 0$. Therefore $v \circ u$ is not invertible. Since u is not invertible, u is nilpotent and there is an integer k such that $u^k = 0$. If u is surjective, $0 = u^k(M) = u^{k-1}(u(M)) = u^{k-1}(M) = \cdots = M$ and hence u cannot be surjective. Hence $u \circ v$ is not surjective, and therefore $u \circ v$ is not invertible.

We need to show that if u_1 and u_2 are not invertible endomorphisms, then u_1+u_2 is not invertible. Suppose to the contrary that u_1+u_2 is invertible. Define $v_1 := u_1\circ(u_1+u_2)^{-1}$ and $v_2 := u_2 \circ (u_1+u_2)^{-1}$. Then,

$$v_1 + v_2 = u_1 \circ (u_1+u_2)^{-1} + u_2 \circ (u_1+u_2)^{-1} = (u_1+u_2)\circ(u_1+u_2)^{-1} = id_M.$$

Since neither u_1 nor u_2 is surjective, neither v_1 nor v_2 is surjective. By the first paragraph of the proof, v_1 and v_2 are both nilpotent of order m, say. Hence,

$$\begin{aligned} id_M &= (id_M - v_2)\circ\left(id_M + v_2 + v_2^2 + v_2^3 + \cdots + v_2^{m-1}\right) \\ &= v_1 \circ \left(id_M + v_2 + v_2^2 + v_2^3 + \cdots + v_2^{m-1}\right) \end{aligned}$$

and therefore v_1 is surjective. Fittings lemma implies that $M = S$ with respect to v_1 and therefore v_1 is an automorphism of M. This is a contradiction to the fact that u_1 is not invertible. □

The next lemma is of a bit more technical nature.

Lemma 1.2.19. *Let M and N be two A-modules, and suppose that N is indecomposable. Let $u \in Hom_A(M,N)$ and $v \in Hom_A(N,M)$. Then,*

$$v \circ u \in Aut_A(M) \Rightarrow u \text{ and } v \text{ are isomorphisms}$$

Proof. $e := u\circ(v\circ u)^{-1}\circ v \in End_A(N)$.

$$\begin{aligned} e\circ e &= \left(u\circ(v\circ u)^{-1}\circ v\right)\circ\left(u\circ(v\circ u)^{-1}\circ v\right) \\ &= u\circ(v\circ u)^{-1}\circ(v\circ u)\circ(v\circ u)^{-1}\circ v \\ &= u\circ(v\circ u)^{-1}\circ v \\ &= e \end{aligned}$$

Since N is indecomposable, by Lemma 1.2.8 we get $e \in \{0, id_N\}$. As

$$\begin{aligned} 0 \neq id_M = id_M^2 &= (v\circ u)^{-1}\circ(v\circ u)\circ(v\circ u)^{-1}\circ(v\circ u) \\ &= (v\circ u)^{-1}\circ v\circ e\circ u \end{aligned}$$

we get $e = id_N$.

Since $id_N = e = u\circ(v\circ u)^{-1}\circ v$, the morphism u is surjective.

Since $v\circ u \in Aut_A(M)$, the morphism u is injective.

As a whole, u is an isomorphism, and since $v\circ u$ is an automorphism, v is an isomorphism as well. □

We are now ready to prove Theorem 1.2.15.

Not surprisingly we shall use induction on m. If $m = 1$, then M is indecomposable, and therefore $n = 1$ as well. Hence, $M_1 = M = N_1$.

Suppose the theorem is proven for $m-1$. Let M be an A-module and suppose there are submodules $M_1, M_2, \ldots M_m$ and $N_1, N_2, \ldots N_n$ of M such that

$$\bigoplus_{j=1}^{m} M_j = M = \bigoplus_{i=1}^{n} N_i.$$

Hence, for any $x \in M$ there are uniquely determined elements $e_i(x) \in M_i \subseteq M$ for all $i \in \{1, 2, \ldots, m\}$ such that $x = \sum_{i=1}^{m} e_i(x)$. Each of the mappings e_i is an idempotent endomorphism of M. Likewise, for any $x \in M$ there are uniquely determined elements $u_j(x) \in N_j \subseteq M$ for all $j \in \{1, 2, \ldots, n\}$ such that $x = \sum_{i=1}^{n} u_j(x)$. Each of the mappings u_j is an idempotent endomorphism of M.

Call $v_j := e_1 \circ u_j \in Hom_A(N_j, M_1)$ and $w_j := u_j \circ e_1 \in Hom_A(M_1, N_j)$ for all $j \in \{1, 2, \ldots, n\}$. As $\sum_{j=1}^{n} u_j = id_M$,

$$\left.\left(\sum_{j=1}^{n} v_j \circ w_j\right)\right|_{M_1} = \left.\left(\sum_{j=1}^{n}(e_1 \circ u_j \circ u_j \circ e_1)\right)\right|_{M_1}$$
$$= e_1 \circ \left.\left(\sum_{j=1}^{n} u_j\right) \circ e_1\right|_{M_1} = e_1|_{M_1} = id_{M_1}$$

and so by Lemma 1.2.18, there is a $j_0 \in \{1, 2, \ldots, n\}$ such that $v_{j_0} \circ w_{j_0} \in Aut_A(M_1)$. By Lemma 1.2.19, since N_{j_0} is indecomposable, v_{j_0} and w_{j_0} are both isomorphisms.

We still need to show that $M/M_1 \simeq M/N_{j_0}$.

The particular structure of w_{j_0} as a composition of projection-injections with respect to the direct sum decompositions, we see that we are in the following situation. Suppose given A-modules U, V, X, Y such that one has an isomorphism $U \oplus V \xrightarrow{\varphi} X \oplus Y$ which is given by

$$\varphi = \begin{pmatrix} \varphi_{U,X} & \varphi_{V,X} \\ \varphi_{U,Y} & \varphi_{V,Y} \end{pmatrix}.$$

Suppose further that $\varphi_{U,X}$ is an isomorphism. Then,

$$\tau = \begin{pmatrix} id_U & -\varphi_{U,X}^{-1}\varphi_{V,X} \\ 0 & id_V \end{pmatrix} \in Aut_A(U \oplus V).$$

Since τ is an automorphism and since φ is an isomorphism, the composition $\varphi \circ \tau$ is an isomorphism as well. But

$$\varphi \circ \tau = \begin{pmatrix} \varphi_{U,X} & \varphi_{V,X} \\ \varphi_{U,Y} & \varphi_{V,Y} \end{pmatrix} \cdot \begin{pmatrix} id_U & -\varphi_{U,X}^{-1}\varphi_{V,X} \\ 0 & id_V \end{pmatrix}$$
$$= \begin{pmatrix} \varphi_{U,X} & 0 \\ \varphi_{U,Y} & (\varphi_{V,Y} - \varphi_{U,Y}\varphi_{U,X}^{-1}\varphi_{V,X}) \end{pmatrix}$$

and hence,

$$(\varphi_{V,Y} - \varphi_{U,Y}\varphi_{U,X}^{-1}\varphi_{V,X}) : V \longrightarrow Y$$

is an isomorphism.

As a consequence we get an isomorphism

$$\bigoplus_{j=2}^{n} M_i \simeq M/M_1 \longrightarrow M/N_{j_0} \simeq \bigoplus_{j\neq j_0} N_j.$$

This finishes the proof of the Krull-Schmidt theorem. □

We mention some examples to show that this is really something special.

Example 1.2.20.

1. Let $A = K$ be a field and look at the module $M = \bigoplus_{i\in\mathbb{N}} K$ of a countably infinite direct sum of copies of K. Then

$$M \simeq M \oplus K \simeq M \oplus M.$$

   Of course, M is not finite dimensional.
2. The well-educated reader will know that for any square free positive integer d the ring of integers $\mathcal{O}_{\mathbb{Q}(\sqrt{-d})}$ of $\mathbb{Q}(\sqrt{-d})$ is not a principal ideal domain except some few small exceptions. On the other hand, the ideal class group of these rings are all finite. Hence, in almost all cases there is a non principal ideal $\mathfrak{a}$ such that there is an integer $n > 0$ with $\mathfrak{a}^n$ is principal. Since $\mathfrak{a}\oplus\mathfrak{b} \simeq \mathfrak{a}\cdot\mathfrak{b}\oplus\mathcal{O}_{\mathbb{Q}(\sqrt{-d})}$ by Steinitz' Theorem (Corollary 8.5.14 below), this gives examples of a Noetherian ring not satisfying the Krull-Schmidt theorem. An incidence of this phenomenon is the ring $\mathbb{Z}[\sqrt{-5}]$ in which one has two decompositions of 6 into prime elements

$$(1 - \sqrt{-5}) \cdot (1 + \sqrt{-5}) = 6 = 2 \cdot 3.$$

   Since both of the elements 2 and $(1+\sqrt{-5})$ are prime elements, the ideal generated by 2 and $(1 - \sqrt{-5})$ is not principal: If it was principal, then a generator a of the ideal would divide 2 and $(1+\sqrt{-5})$. Recall that the norm of an element $x + \sqrt{-5}y$ is $(x + \sqrt{-5}y)(x - \sqrt{-5}y) = x^2 + 5y^2$. The norm is multiplicative, and takes values in $\mathbb{Z}$ if $(x, y) \in \mathbb{Z}^2$. Since the norm of 2 is 4 and the norm of $(1 + \sqrt{-5})$ is 6, the norm of a would be 2. But $2 = x^2 + 5y^2$ does not have solutions $(x, y) \in \mathbb{Z}^2$. We will continue to study this example in Example 7.5.30 below.
   For more details on this subject one may consult Hasse's classical text [Has-50] or Curtis-Reiner's comprehensive book on representation theory [CuRe-82-86].
3. R. G. Swan shows in [Swa-62] that given the generalised quaternion group Q_{32} of order 32, then in $\mathbb{Z}Q_{32}$ there is an ideal $\mathfrak{a}$ which is not free as $\mathbb{Z}Q_{32}$-module, but

$$\mathfrak{a} \oplus \mathbb{Z}Q_{32} \simeq \mathbb{Z}Q_{32} \oplus \mathbb{Z}Q_{32}.$$

   Of course, $\mathbb{Z}$ is not a field. In a recent article [Nich-18] J. K. Nicholson showed that a smaller quaternion group will not work. Nicholson is generalising this way a classical result, known as Jacobinski's cancellation theorem. We refer to Section 8.6 for a detailed discussion.

1.2.3 Wedderburn's structure theorem on semisimple artinian rings

We have seen in Section 1.2.2 that given a finite group G and a field K in which the order of G is invertible, any module M is semisimple with an essentially unique decomposition into a direct sum of submodules. We do not know yet how to obtain a list of possibly occurring simple modules. This is the purpose of the present section.

Towards Wedderburn's theorem

The first step in this direction is the following statement.

Lemma 1.2.21 (Issai Schur). *Let K be a field and let A be a K-algebra. Let V be a simple A-module. Then, $End_A(V)$ is a skew-field containing K.*

Remark 1.2.22. Of course, it might happen that $End_A(V)$ is not only a skew-field but even a field. This is the case for example if V is the trivial KG-module of a group G.

Proof. Let $u \in End_A(V)$.
- By Lemma 1.1.15 the image $\mathrm{im}(u)$ is a submodule of V and $\ker(u)$ is another submodule of V.
- Since V is simple, the only submodules of V are 0 and V.
- If $\mathrm{im}(u) = 0$, then $u = 0$. Hence, u is either 0 or surjective.
- If $\ker(u) = V$, then $u = 0$. If $\ker(u) = 0$, then u is injective.

Hence, u is either 0 or an automorphism. By Remark 1.1.13 $End_A(V)$ is a K-algebra, and we just showed that any non zero element is invertible. This completes the proof. □

Example 1.2.23. Let Q_8 be the quaternion group of order 8. Recall that Q_8 is generated by three elements a, b, c subject to the relations $a^4 = 1$, $ab = c$, $a^2 = b^2 = c^2$. A representation of $\mathbb{Q}Q_8$ is the rational quaternion algebra $\mathbb{H}_{\mathbb{Q}}$, which is the $\mathbb{Q}$-algebra of dimension 4 with $\mathbb{Q}$-basis $\{1, i, j, k\}$. This becomes a $\mathbb{Q}$-algebra by the following multiplication rules:
- 1 is the neutral element of the multiplication
- $i \cdot j = k, j \cdot k = i, k \cdot i = j$
- $i^2 + 1 = j^2 + 1 = k^2 + 1 = 0$
- extend this $\mathbb{Q}$-bilinearly.

The verification that this indeed is an associative $\mathbb{Q}$-algebra is a straight forward, but slightly lengthy calculation. A multiplicative inverse of

$$x = x_1 \cdot 1 + x_i \cdot i + x_j \cdot j + x_k \cdot k$$

is

$$x^{-1} := \frac{1}{x_1^2 + x_i^2 + x_j^2 + x_k^2} \cdot (x_1 \cdot 1 - x_i \cdot i - x_j \cdot j - x_k \cdot k)$$

as one verifies by elementary multiplication. We show that $\mathbb{H}_\mathbb{Q}$ is a $\mathbb{Q}Q_8$-module by setting
a acts on $\mathbb{H}_\mathbb{Q}$ as multiplication *on the left* by i
b acts on $\mathbb{H}_\mathbb{Q}$ as multiplication *on the left* by j
c acts on $\mathbb{H}_\mathbb{Q}$ as multiplication *on the left* by k.

Since the relations of Q_8 are satisfied by definition in $\mathbb{H}_\mathbb{Q}$, this turns $\mathbb{H}_\mathbb{Q}$ into a $\mathbb{Q}Q_8$-module.

Moreover, since the multiplication in $\mathbb{H}_\mathbb{Q}$ is associative, multiplication *on the right* with elements in $\mathbb{H}_\mathbb{Q}$ gives a homomorphism $\mathbb{H}_\mathbb{Q} \longrightarrow End_{\mathbb{Q}Q_8}(\mathbb{H}_\mathbb{Q})$. Since $\mathbb{H}_\mathbb{Q}$ is a skew-field, this homomorphism is injective. Moreover, a $\mathbb{Q}$-linear endomorphism of $\mathbb{H}_\mathbb{Q}$ which has to commute with left multiplication with i, j and k, actually is an endomorphism of $\mathbb{H}_\mathbb{Q}$, which is $\mathbb{H}_\mathbb{Q}$-linear. Then, the image of the unit element determines every endomorphism by the fact that the endomorphism has to be linear with respect to the left-module structure of $\mathbb{H}_\mathbb{Q}$ over itself. Moreover, any image in $\mathbb{H}_\mathbb{Q}$ defines actually an endomorphism. So, $\mathbb{H}_\mathbb{Q} \longrightarrow End_{\mathbb{Q}Q_8}(\mathbb{H}_\mathbb{Q})$ is an isomorphism.

We see that actually skew-fields may occur.

Remark 1.2.24.
1. The so-called Brauer group $Br(K)$ classifies skew-fields D with centre K, up to a certain equivalence. The subgroup of $Br(K)$ consisting of those skew-fields D which occur as endomorphisms of simple KG-modules for some group G is called the Schur subgroup of $Br(K)$.
2. The centre Z of a skew-field is a field. Given an A-module M then M is an $End_A(M)$ right-module (i. e. instead of having an operation on the left of M, one has an operation on the right of M. The axioms are the same, except that for a left-module we ask $(ab)m = a(bm)$ for all $a, b \in A$, $m \in M$, whereas for a right-module we ask for $m(ab) = (ma)b$ for all $a, b \in A$, $m \in M$. If A is commutative, then there is no difference between these concepts. If A is non commutative, the concepts differ in general. For group rings RG they coincide, since $(gh)^{-1} = h^{-1}g^{-1}$ for all $g, h \in G$, and $g \cdot m =: m \cdot g^{-1}$ for all $g \in G$, $m \in M$ turns a left RG-module into a right RG-module.) Hence, a simple A-module S is an $End_A(S)$ right-module. Since Z is commutative, this defines a Z-module structure on S. So, simple modules are always vector spaces over a field.

The next lemma is almost trivial.

Lemma 1.2.25. *If K is a field and if A is a K-algebra, then if V and W are two non-isomorphic simple A-modules, then $Hom_A(V, W) = 0$.*

Proof. Let $u \in Hom_A(V, W)$ be a homomorphism. Then, $\ker(u)$ is a submodule of V, and so either $u = 0$ or u is injective. If u is injective, then $u(V)$ is a submodule of W isomorphic to V. Since W is simple $u(V) = W$. Since W is non isomorphic to V we get a contradiction. Hence, $u = 0$. □

Definition 1.2.26. Let R be a commutative ring and let A be an R-algebra. Then, the opposite algebra $(A^{op}, +, \cdot^{op})$ (or A^{op} for short) is A as R-module, and multiplication $\cdot^{op}$ defined by $a \cdot^{op} b := b \cdot a$ for all $a, b \in A$.

Remark 1.2.27. A left-module for an algebra A is a module in the sense of Definition 1.1.6. A right-module for an algebra A is a left-module for A^{op}. Observe that this coincides with the definition given in Remark 1.2.24.(2).

Lemma 1.2.28. *Let R be a commutative ring and let A be an R-algebra. Then, the endomorphism ring of the regular A-module ${}_AA$ is isomorphic to A^{op}.*

Proof. Let $\varphi \in End_A({}_AA)$. Then, $\varphi(1_A) =: a_\varphi$ and $\varphi(b) = \varphi(b \cdot 1_A) = b \cdot a_\varphi$ for all $b \in A$. Hence, every endomorphism φ of ${}_AA$ is given by multiplication on the right with an $a_\varphi \in A$.

Also, $a_\psi = a_\varphi$ for two endomorphisms ψ and φ of A, if and only if $\psi = \varphi$.

Moreover, ${}_AA \ni b \mapsto ba \in {}_AA$ is in $End_A({}_AA)$.

Finally, for all $b \in A$

$$\begin{aligned}(\varphi \circ \psi)(b) &= \varphi(\psi(b)) \\ &= (ba_\psi)a_\varphi \\ &= b(a_\psi a_\varphi) \\ &= ba_{\varphi\circ\psi}\end{aligned}$$

and so, $a_\psi a_\varphi = a_{\varphi\circ\psi}$. As a consequence, $End_A({}_AA) \simeq A^{op}$. □

Remark 1.2.29. The reason for this phenomenon is the fact that we write homomorphisms on the left. If we would write homomorphisms on the right, and compose them accordingly, we could avoid the use of the opposite algebra here. Nevertheless, the conventions are strong here and many readers might be shocked with the systematic use of mappings written on the right. Nevertheless, one should memorise that the appearance of the opposite algebra comes from the tradition and has no intrinsic reason.

Wedderburn's result and first consequences

We use the notation $Mat_{n\times n}(R)$ to denote the matrix ring of square matrices with coefficients in R. We are ready to formulate our main result of this subsection.

Theorem 1.2.30 (Wedderburn). *Let K be a field and let A be a finite dimensional semisimple K-algebra. Then, there is an integer $m \in \mathbb{N}$, skew-fields $D_1, D_2, \ldots, D_m$ with centre containing K, and integers $n_1, n_2, \ldots, n_m$ such that there is an isomorphism of algebras*

$$A \simeq \prod_{i=1}^{m} Mat_{n_i\times n_i}(D_i).$$

Moreover, the skew-fields D_i and the integers n_i for $i = 1, \dots, m$ are completely determined by A.

Proof. By Lemma 1.2.28 we know that the K-algebra A is isomorphic to the K-algebra $(End_A({}_AA))^{op}$.

Now, since A is semisimple, ${}_AA$ is isomorphic to a direct sum of simple modules

$$ {}_AA \simeq \bigoplus_{i=1}^{m} \bigoplus_{j=1}^{n_i} S_i $$

where we numbered the modules already in the way that $S_i \simeq S_j$ if and only if $i = j$, where there are exactly n_i different summands in the direct sum decomposition which are isomorphic to S_i, and so we expressed possible multiplicities by the second direct sum over j.

Hence,

$$ A \simeq \left(End_A\left(\bigoplus_{i=1}^{m} \bigoplus_{j=1}^{n_i} S_i \right) \right)^{op}. $$

But, by Lemma 1.2.25 $Hom_A(S_i, S_j) = 0$ whenever $i \neq j$ and so,

$$ A \simeq \prod_{i=1}^{m} \left(End_A\left(\bigoplus_{j=1}^{n_i} S_i \right) \right)^{op}. $$

But now, for A-modules X, Y, V, W one always gets

$$ Hom_A(X \oplus Y, V \oplus W) \simeq \begin{pmatrix} Hom_A(X, V) & Hom_A(Y, V) \\ Hom_A(X, W) & Hom_A(Y, W) \end{pmatrix} $$

Therefore,

$$ End_A\left(\bigoplus_{j=1}^{n_i} S_i \right) \simeq Mat_{n_i \times n_i}(Hom_A(S_i, S_i)) $$

with multiplication given by the usual matrix multiplication, and composition of mappings in $End_A(S_i)$. Schur's lemma 1.2.21 shows that $Hom_A(S_i, S_i) = D_i'$ is a skew-field with centre containing K. For the last step we observe that if D is a skew-field with centre containing K, then D^{op} is again a skew-field with centre containing K. Moreover, for two algebras A and B, since A and B commute in the direct product, we get $(A \times B)^{op} \simeq A^{op} \times B^{op}$. Finally, for any K-algebra B we get

$$ \begin{aligned} (Mat_{n \times n}(B))^{op} &\simeq Mat_{n \times n}(B^{op}) \\ M &\mapsto M^{tr} \end{aligned} $$

denoting by M^{tr} the transpose of a matrix. We just apply all of this to our situation, denoting $D_i := D_i'^{op}$ and get the requested isomorphism.

The unicity of the skew-fields D_i and the dimensions n_i for $i \in \{1, \ldots, m\}$ are a consequence of the fact that the skew-fields are the endomorphism algebras of the simple modules, that the Krull-Schmidt theorem implies the unicity of the decomposition of the regular module into a sum of simple submodules, and that the n_i are the multiplicities of the isomorphism classes of these simple modules as a direct factor of the regular representation. □

The Wedderburn structure theorem has a lot of consequences. We just mention a few here in order to illustrate the importance of this result.

Corollary 1.2.31. *A skew-field that is a finite dimensional vector space over an algebraically closed field is commutative. In particular if A is a finite dimensional semisimple K-algebra, and if K is algebraically closed, then there are integers $n_1, \ldots, n_m$ such that*

$$A \simeq \prod_{j=1}^{m} Mat_{n_j \times n_j}(K).$$

Proof. Indeed, let D be a finite dimensional skew-field over K and let $d \in D$. Since D is a finite dimensional K-vector space, by the Cayley-Hamilton theorem d has a minimal polynomial $P_d(X) \in K[X]$, that is a polynomial with leading coefficient 1 and smallest possible degree having d as root. Since K is algebraically closed, the degree of each irreducible polynomial is 1. Hence, $P = P_1 \cdot \cdots \cdot P_\ell$ for degree 1 polynomials P_i and $P(d) = P_1(d) \cdot \cdots \cdot P_\ell(d) = 0$. Hence, there is $i \in \{1, \ldots, \ell\}$ with $P_i(d) = 0$ and so $d \in K$. Therefore $D = K$. □

This applies in particular to group rings of finite groups G over an algebraically closed field K such that the order of G is invertible in K.

Corollary 1.2.32. *Let G be a finite group. Then there is a positive integer m and positive integers $n_1, n_2, \ldots, n_m$ so that $|G| = n_1^2 + n_2^2 + \cdots + n_m^2$ and these integers n_i for $i \in \{1, \ldots, m\}$ are the dimensions of the simple KG-modules for K being any algebraically closed field in which the order of G is invertible.*

Proof. Indeed, compare the dimensions of the algebras in Corollary 1.2.31. □

Corollary 1.2.33. *If A is semisimple, commutative, and finite dimensional over K then there are finite dimensional field extensions K_i over K such that $A \simeq \prod_{j=1}^{m} K_i$.*

Proof. Indeed, A being abelian, in the Wedderburn structure theorem no matrix ring can occur, nor a skew-field. □

Corollary 1.2.34. *Let A be a semisimple finite dimensional K-algebra and let S be a simple A-module. Then there is an $i \in \{1, \dots, m\}$ such that*

$$S \simeq S_i \simeq \begin{pmatrix} D_i \\ D_i \\ \vdots \\ D_i \end{pmatrix}_{n_i}$$

Proof. Indeed, if S is a simple A-module, take $s \in S \setminus \{0\}$. Then, $A \cdot s$ is a non zero A-submodule of S, such that then $S = As$ and therefore S is a quotient of ${}_AA$. Since ${}_AA$ is semisimple, there is an $i \in \{1, \dots, m\}$ such that $S \simeq S_i$. Since

$$S_i \simeq Hom_A({}_AA, S_i) \simeq Hom_A\left(\bigoplus_{i=1}^{m}\bigoplus_{j=1}^{n_i} S_i, S_i\right) \simeq \begin{pmatrix} D_i \\ D_i \\ \vdots \\ D_i \end{pmatrix}_{n_i}$$

we conclude. □

Since for every simple A-module S the endomorphism ring $End_A(S)$ is a skew-field and since S is naturally an $End_A(S)$-module, by putting $\varphi \cdot s := \varphi(s)$ for every $s \in S$ and $\varphi \in End_A(S)$, the simple A-module is an $End_A(S)$-vector space.

Corollary 1.2.35. *Let A be a finite dimensional semisimple K-algebra. Let S be a simple A-module. Then exactly $dim_{End_A(S)}(S)$ copies of S are direct factor of the regular A-module ${}_AA$.*

Proof. This is exactly the fact that in $Mat_{n \times n}(D)$ a simple direct factor corresponds to a column, and there are exactly as many columns as rows in a square matrix. □

Definition 1.2.36. A module M which is generated by a single element is called a *cyclic module*. A generator of the cyclic module is an element s such that As equals the entire cyclic module.

Remark 1.2.37. We see by the proof of Corollary 1.2.34 that simple modules are cyclic, with any non zero element being a generator. Also the regular module is cyclic, only units can be used as generators.

We now show that (direct products of) matrix rings over skew-fields are actually semisimple.

Definition 1.2.38. Let K be a field and let A be a K-algebra. We say that the *algebra A is simple* if for each two-sided ideal I of A we get $I = 0$ or $I = A$.

An example of an infinite dimensional simple algebra can be seen in [Zim-14, Example 1.4.32]. We shall now show that finite dimensional simple algebras are actually semisimple.

Proposition 1.2.39. *Let K be a field and let A be a finite dimensional simple K-algebra. Then A is semisimple.*

Proof. Let S be a simple A-left submodule of the regular module ${}_AA$, i. e. a minimal left ideal of A. Then define the left ideal $J := \sum_{a\in A} S \cdot a$ where the multiplication is taken inside A, which makes sense since $S \subseteq A$. Hence, by definition of J, we see that J is in fact a non zero two-sided ideal of A. Therefore $J = A$ since A is simple. However J is semisimple. Indeed, let $B_1 \subseteq A$ such that $L_1 := \sum_{b\in B_1} S \cdot b \neq A$, but such that for any $a \in A$ we get either $S \cdot a \subseteq L_1$ or $L_1 + S \cdot a = A$. Then, $S \cdot a$ is either 0 or isomorphic to the simple A-submodule S of A. Indeed $S \twoheadrightarrow S \cdot a$ is an epimorphism of A-modules and since S is simple, the kernel can only be all of S or 0.

Suppose $a \in A$ such that $L_1 + S \cdot a = A$. Then $S \cdot a \neq 0$ and therefore $L_1 \cap S \cdot a \leq S \cdot a$ and since $S \cdot a$ is simple, we get $L_1 \cap S \cdot a = 0$. Hence $A = L_1 \oplus S \cdot a$. Now, let $b \in B_1$ such that $S \cdot b \neq 0$, and let $B_2 \subseteq B_1$ such that $L_2 := \sum_{b\in B_2} S \cdot b$ has the property that for every $b_2 \in B_1 \setminus B_2$ we get $S \cdot b = 0$ or $L_2 + S \cdot b_2 = L_1$. By the very same argument as for L_1 we obtain that $L_2 \oplus S \cdot b_2 = L_1$. Proceeding by induction on the dimension of L we get that $J = \bigoplus_{j=1}^m S \cdot b_j$ for certain elements $b_j \in A$. Hence, the regular A-module is semisimple.

Let M be a finite dimensional A-module. Then there is a generating set $m_1, \ldots, m_t$ of M as A-module, and the map

$$\begin{aligned} ({}_AA)^t &\xrightarrow{\mu} M \\ (a_1, \ldots, a_t) &\mapsto \sum_{j=1}^t a_j m_j \end{aligned}$$

is an epimorphism of A-modules. Since the regular module ${}_AA$ is semisimple, also $({}_AA)^t$ is semisimple. Hence, in a notation explained later,

$$0 \longrightarrow \ker(\mu) \longrightarrow ({}_AA)^t \longrightarrow M \longrightarrow 0$$

is split and this shows that M is semisimple. For this it is worth mentioning a lemma. □

Lemma 1.2.40. *All submodules and quotients of semisimple modules are semisimple.*

Proof. Let S be a semisimple A-module and let T be an A-submodule of S. We need to show that T is semisimple as well. Let V be a submodule of T. Since T is a submodule of S, also V is a submodule of S. Hence, by Proposition 1.2.6 there is a submodule W of S such that $S = V \oplus W$ and a submodule R of S such that $S = T \oplus R$. We consider the quotient $T \simeq S/R$, and denote by π the natural map $S \to S/R$. Then

$$T \simeq S/R = (V \oplus W)/R \simeq V \oplus (W + R)/R$$

and hence V is a direct factor of T. By Proposition 1.2.6 we get that T is semisimple as well. The proof for quotients is dual. □

In the literature the following concept occurs in this concept frequently. For completeness we give the definition. Recall that the centre of an algebra is defined to be

$$Z(A) := \{a \in A \mid \forall b \in A : ba = ab\}.$$

It is clear that for a K-algebra A we always have $K \subseteq Z(A)$.

Definition 1.2.41. A *central simple K-algebra A* is a simple K-algebra A with $Z(A) = K$.

Remark 1.2.42. Let D be a skew field. Then $Mat_{n\times n}(D)$ is a simple algebra. Indeed, let M be a non zero element of an ideal I of $Mat_{n\times n}(D)$. Let $d_{i,j} \neq 0$ be a non zero coefficient of M. Then we may multiply from the right and from the left with a matrix, containing only one non zero entry, namely 1, in the i-th row and column, and likewise with a matrix with only one coefficient non zero, actually being 1, in the j-th row and column. Multiply these matrices from the left and from the right on M to get an element of I with only one non zero coefficient, namely $d_{i,j}$ in position (i, j). Then by multiplying permutation matrices from the left and from the right, we may produce matrices in I with only one non zero coefficient, namely $d_{k,\ell}$ in position (k, ℓ) for any choice of k and ℓ. Linear combinations of these matrices will give any matrix of $Mat_{n\times n}(D)$, and so $I = Mat_{n\times n}(D)$. Therefore $Mat_{n\times n}(D)$ is simple.

A consequence is that algebras described by Wedderburn's theorem are all semisimple.

Example 1.2.43.

1. Let $K = \mathbb{Q}$ and $G = C_3$ the cyclic group of order 3. Then, let ζ_3 be a root in $\mathbb{C}$ of the polynomial $X^2 + X + 1$ and consider the field $\mathbb{Q}(\zeta_3)$. A generator c of C_3 acts on $\mathbb{Q}(\zeta_3)$ by $c \cdot x = \zeta_3 x$ and then this defines a $\mathbb{Q}C_3$-module structure on $\mathbb{Q}(\zeta_3)$. Again, just the same proof as in Example 1.2.23 shows that $End_{\mathbb{Q}C_3}(\mathbb{Q}(\zeta_3)) \simeq \mathbb{Q}(\zeta_3)$. Hence, field extensions really occur as endomorphism algebras of simple modules. If $\mathbb{Q}(\zeta_3)$ would not be simple, since $\mathbb{Q}C_3$ is semisimple, the endomorphism algebra $End_{\mathbb{Q}C_3}(\mathbb{Q}(\zeta_3))$ would contain a non trivial idempotent, which does not exist in a field $\mathbb{Q}(\zeta_3)$.
   Moreover, we always get the trivial $\mathbb{Q}C_3$-module as a simple module. In total we found two simple modules. One, the trivial module of dimension 1 and second, the module $\mathbb{Q}(\zeta_3)$ of dimension 2. Now, $\mathbb{Q}C_3$ is a $\mathbb{Q}$-vector space of dimension 3. As a consequence, as $\mathbb{Q}C_3$-modules
   $$\mathbb{Q}C_3 \simeq \mathbb{Q} \oplus \mathbb{Q}(\zeta_3)$$
   Taking the endomorphism ring, one gets that
   $$\mathbb{Q}C_3 \simeq \mathbb{Q} \times \mathbb{Q}(\zeta_3)$$
   as algebras as well.

2. Let $K = \mathbb{C}$ or more generally a field containing $\mathbb{Q}(\zeta_3)$ and let $G = C_3$ again. Then, there are three ways for a possible operation of c on K. As we already have seen in Example 1.1.21.(1) the operation of c as 1, as ζ_3 or as ζ_3^2 yield all of them simple modules K_1, K_{ζ_3} and $K_{\zeta_3^2}$. Let $\varphi : K_{\zeta_3^i} \longrightarrow K_{\zeta_3^j}$ for $i, j \in \{1, 2, 3\}$ be a non zero homomorphism of KC_3-modules. Since φ is K-linear, $\varphi(\zeta_3^i) = \zeta_3^i \cdot \varphi(1)$. Since φ is C_3-linear, $\varphi(\zeta_3^i) = \varphi(c \cdot 1) = c \cdot \varphi(1) = \zeta_3^j \cdot \varphi(1)$. Hence, $i = j$.
We see that the simple $\mathbb{Q}C_3$-module $\mathbb{Q}(\zeta_3)$ splits into the semisimple KC_3-module $K_{\zeta_3} \oplus K_{\zeta_3^2}$ when one adjoins a third root of unity to the field $\mathbb{Q}$.
3. Actually, this generalises to all cyclic groups of order $n < \infty$. We turn back to Example 1.1.21.(1). There we have determined n simple $\mathbb{C}C_n$-modules $\mathbb{C}_{\zeta_n^i}$ for $i \in \{1, \ldots, n\}$ and the argument just above shows that they are mutually non isomorphic. Indeed, let $\varphi : \mathbb{C}_{\zeta_n^i} \longrightarrow \mathbb{C}_{\zeta_n^j}$ be an isomorphism, then

$$\zeta_n^j \varphi(1) = \varphi(c \cdot 1) = \varphi(\zeta_n^i) = \zeta_n^i \varphi(1)$$

and so $i = j$. As a consequence, $\bigoplus_{j=1}^n \mathbb{C}_{\zeta_n^j}$ is a direct factor of the regular $\mathbb{C}C_n$-module. Since $\bigoplus_{j=1}^n \mathbb{C}_{\zeta_n^j}$ and $\mathbb{C}C_n$ both have dimension n over $\mathbb{C}$ they are isomorphic as $\mathbb{C}C_n$-modules. Taking endomorphism algebras, one gets

$$\mathbb{C}C_n \simeq \underbrace{\mathbb{C} \times \cdots \times \mathbb{C}}_{n \text{ copies}}$$

as algebra and the simple modules correspond to the factors in the product.
4. From Example 1.1.21.(3) we know a 2-dimensional representation of the symmetric group $\mathfrak{S}_3$ of degree 3 over the integers $\mathbb{Z}$. We consider the representation as a representation over $\mathbb{Q}$. We have seen that the corresponding module $M_\mathbb{Q}$ does not have a non zero submodule, since any submodule would have to be of dimension one, and therefore a common eigenspace of all of the 6 matrices of images of group elements $\sigma \in \mathfrak{S}_3$. However, no such space exist.
Now, $\mathfrak{A}_3 = C_3$ the alternating group of degree 3 is a normal subgroup of $\mathfrak{S}_3$ of index 2. So, there is a group epimorphism $\mathfrak{S}_3 \longrightarrow C_2$, and any simple $\mathbb{Q}C_2$-module is also a simple $\mathbb{Q}\mathfrak{S}_3$-module. We know that $\mathbb{Q}C_2 \simeq \mathbb{Q} \times \mathbb{Q}$ corresponding to two simple one-dimensional modules. But, since the above 2-dimensional simple module yields a direct factor $Mat_{2\times2}(\mathbb{Q})$ of $\mathbb{Q}\mathfrak{S}_3$, we found all of the group ring $\mathbb{Q}\mathfrak{S}_3$:

$$\mathbb{Q}\mathfrak{S}_3 \simeq \mathbb{Q} \times Mat_{2\times2}(\mathbb{Q}) \times \mathbb{Q}.$$

Since we know the precise representations of each group element $g \in \mathfrak{S}_3$, we are able to view the image of $\mathbb{Z}\mathfrak{S}_3$ in $\mathbb{Q} \times Mat_{2\times2}(\mathbb{Q}) \times \mathbb{Q}$. Multiplication of the matrices, and considering of the second one-dimensional factor as the sign representation, and the first one as the trivial representation gives

$$id \mapsto \left(1, \left(\begin{array}{cc} 1 & 0 \\ 0 & 1 \end{array}\right), 1\right)$$

$$(1\,2) \mapsto \left(1, \begin{pmatrix} -2 & -3 \\ 1 & 2 \end{pmatrix}, -1\right)$$

$$(1\,2\,3) \mapsto \left(1, \begin{pmatrix} -2 & -3 \\ 1 & 1 \end{pmatrix}, 1\right)$$

$$(1\,3\,2) \mapsto \left(1, \begin{pmatrix} 1 & 3 \\ -1 & -2 \end{pmatrix}, 1\right)$$

$$(1\,2\,3) \cdot (1\,2) = (2\,3) \mapsto \left(1, \begin{pmatrix} 1 & 0 \\ -1 & -1 \end{pmatrix}, -1\right)$$

$$(1\,2) \cdot (1\,2\,3) = (1\,3) \mapsto \left(1, \begin{pmatrix} 1 & 3 \\ 0 & -1 \end{pmatrix}, -1\right)$$

Now, one observes that all of these couples of matrices are included in the set

$$\left\{\left(d_0, \begin{pmatrix} a_1 & b_1 \\ c_1 & d_1 \end{pmatrix}, a_2\right) \in \mathbb{Z} \times Mat_{2\times 2}(\mathbb{Z}) \times \mathbb{Z} \mid \begin{array}{l} a_2 - d_0 \in 2\mathbb{Z}; \\ a_1 - d_0 \in 3\mathbb{Z}; \\ a_2 - d_1 \in 3\mathbb{Z}; \\ b_1 \in 3\mathbb{Z} \end{array}\right\}$$

Moreover,

$$(1\,2\,3) + (1\,3) - (1\,2) - (1\,3\,2) \mapsto \left(0, \begin{pmatrix} 0 & 0 \\ 1 & 0 \end{pmatrix}, 0\right)$$

$$(1\,2\,3) + (2\,3) - (1\,2) - (1\,3\,2) \mapsto \left(0, \begin{pmatrix} 0 & -3 \\ 0 & 0 \end{pmatrix}, 0\right)$$

$$(1\,3) + (1\,2\,3) + (2\,3) - (1\,2) - (1\,3\,2) \mapsto \left(1, \begin{pmatrix} 1 & 0 \\ 0 & -1 \end{pmatrix}, -1\right)$$

$$id \mapsto \left(1, \begin{pmatrix} 1 & 0 \\ 0 & 1 \end{pmatrix}, 1\right)$$

$$(1\,3) + (2\,3) + (1\,2\,3) \mapsto \left(3, \begin{pmatrix} 0 & 0 \\ 0 & -1 \end{pmatrix}, -1\right)$$

$$(1\,3\,2) + (1\,2\,3) \mapsto \left(2, \begin{pmatrix} -1 & 0 \\ 0 & -1 \end{pmatrix}, 2\right)$$

These elements generate the above set. Hence, the image of $\mathbb{Z}\mathfrak{S}_3$ in the Wedderburn decomposition of $\mathbb{Q}\mathfrak{S}_3$ is described by

$$\mathbb{Z}\mathfrak{S}_3 \simeq \left\{\left(d_0, \begin{pmatrix} a_1 & b_1 \\ c_1 & d_1 \end{pmatrix}, a_2\right) \in \mathbb{Z} \times Mat_{2\times 2}(\mathbb{Z}) \times \mathbb{Z} \mid \right.$$
$$\left. a_2 - d_0 \in 2\mathbb{Z}; a_1 - d_0 \in 3\mathbb{Z}; a_2 - d_1 \in 3\mathbb{Z}; b_1 \in 3\mathbb{Z}\right\}$$

Definition 1.2.44. A field K is called a *splitting field* for a finite dimensional K-algebra A if the endomorphism ring of every simple A-module is isomorphic to K. The finite dimensional K-algebra A is called split semisimple if A is semisimple and K is a splitting field for A. An A-module M is absolutely simple if the module M_L is a simple A_L-module for all extension fields L over K.

Remark 1.2.45. The proof of Wedderburn's theorem shows that Corollary 1.2.31 and Corollary 1.2.32 are true under the weaker assumption that K is a splitting field for the algebra.

1.3 Group rings as semisimple algebras

The goal is now to determine these data of Wedderburn's theorem if possible entirely in terms of data of the group G and the field K.

We start with the number of simple KG-modules. Recall that the centre of an algebra A is given by

$$Z(A) := \{a \in A \mid \forall b \in A : ba = ab\}.$$

The centre of an R-algebra A is a commutative R-algebra. The definition implies immediately that two isomorphic algebras A and B have isomorphic centres:

$$A \simeq B \Rightarrow Z(A) \simeq Z(B).$$

The isomorphism of the centres is the restriction of the isomorphism $A \simeq B$ to the subset $Z(A)$. Moreover, if A_1 and A_2 are two K-algebras, then

$$Z(A_1 \times A_2) \simeq Z(A_1) \times Z(A_2).$$

Lemma 1.3.1. *For any algebra A one gets*

$$Z(Mat_{n\times n}(A)) = \left\{ z \cdot \begin{pmatrix} 1 & 0 & & \cdots & 0 \\ 0 & & \ddots & \ddots & \vdots \\ \vdots & & \ddots & \ddots & 0 \\ 0 & \cdots & & 0 & 1 \end{pmatrix} \;\middle|\; z \in Z(A) \right\}$$

Proof. Indeed, the inclusion "$\supseteq$" is clear. The other inclusion comes from the fact that first, any element in $Z(Mat_{n\times n}(A))$ must commute with any element of the form $b \cdot \begin{pmatrix} 1 & 0 & & \cdots & 0 \\ 0 & & \ddots & \ddots & \vdots \\ \vdots & & \ddots & \ddots & 0 \\ 0 & \cdots & & 0 & 1 \end{pmatrix}$ and hence

$$Z(Mat_{n\times n}(A)) \subseteq Mat_{n\times n}(Z(A)).$$

Moreover, the fact that elements in the centre of $Mat_{n\times n}(A)$ commute with elementary matrices implies the statement. □

Corollary 1.3.2. *Let K be a splitting field for a finite dimensional semisimple K-algebra A. Then, $dim_K(Z(A))$ is the number of isomorphism classes of simple A-modules.*

Proof. This is an immediate consequence of Lemma 1.3.1, Corollary 1.2.31 and Remark 1.2.45. □

Is it possible to express the number of isomorphism classes of simple KG-modules by pure group theoretic terms? This is actually the case. For an element $g \in G$ let $C_g := \{hgh^{-1} | \; h \in G\}$ be the conjugacy class of G.

Lemma 1.3.3. *Let R be a commutative ring and let G be a finite group. Then,*

$$\left\{ \sum_{h\in C_g} e_h \middle| \; g \in G \right\}$$

is an R-basis of Z(RG).

Proof. Since RG is a free R-module with basis $\{e_g | \; g \in G\}$, the set $\{\sum_{h\in C_g} e_h | \; g \in G\}$ is clearly linearly independent. Moreover,

$$x = \sum_{g\in G} x_g e_g \in Z(RG) \Leftrightarrow \forall h \in G : e_h \cdot x \cdot e_h^{-1} = x$$

But,

$$\begin{aligned} e_h \cdot x \cdot e_{h^{-1}} &= \sum_{g\in G} x_g e_h e_g e_{h^{-1}} \\ &= \sum_{g\in G} x_g e_{hgh^{-1}} \\ &= \sum_{g\in G} x_{h^{-1}gh} e_g \end{aligned}$$

and therefore, $x \in Z(RG)$ if and only if $e_h \cdot x \cdot e_h^{-1} = x \Leftrightarrow x_g = x_{h^{-1}gh}$ for all $h \in G$. This proves the lemma. □

We come to our first result which can be seen as a central statement and which has considerable practical importance. The proof does not pose any problems anymore.

Theorem 1.3.4. *Let G be a finite group and let K be a field in which the order of G is invertible. Suppose that K is a splitting field for KG. Then the number of simple KG-modules up to isomorphism is equal to the number of conjugacy classes of G.*

Proof. Lemma 1.3.3 and Corollary 1.3.2 prove the statement. □

The above result used that K is a splitting field for KG.

Definition 1.3.5. Let G be a finite group. A field K is a splitting field for G if K is a splitting field for KG.

We come to the question which are the splitting fields for G.

Lemma 1.3.6. *Let G be a group whose elements are all of finite order, let K be a field and let M be a KG-module. Denote by n_g the order of the element g in the group G. Suppose that n_g is invertible in K and suppose that the polynomial $X^{n_g}-1 \in K[X]$ splits into linear factors. Let $\varphi_M : G \longrightarrow Aut_K(M)$ be the group homomorphism which is induced by the KG-module structure on M. Then, the matrix $\varphi(g)$ is diagonalisable.*

Proof. We give three independent proofs here. The first one uses Maschke's theorem, the second one uses a standard argument on polynomials the last one is an elementary computation with matrices.

First proof. We consider $C := \langle g\rangle$ the cyclic group generated by $g \in G$. By Maschke's theorem KC is semisimple, and hence M restricted to C is semisimple. Since K is a splitting field for C by hypothesis, there is a K-basis of M such that g acts by diagonal matrices on M.

Second proof. Since $g^{n_g} = 1$, we get that $\varphi_M(g)$ divides $X^{n_g}-1$. Since n_g is invertible in K, the polynomial $X^{n_g} - 1$ has only simple roots. Indeed, its formal derivative only has the root 0, and 0 is not a root of $X^{n_g}-1$. Hence, all eigenvalues of $\varphi_M(g)$ are simple, and therefore $\varphi_M(g)$ is diagonalisable.

Third proof. By assumption, $(\varphi_M(g))^{n_g}$ is the identity matrix. Let L be an algebraically closed field containing K. Let $\epsilon \in L$ be an eigenvalue of $\varphi_M(g)$ and let $v \in M_L$ be an eigenvector of $\varphi_M(g)$ with value ϵ. Then,

$$v = (\varphi_M(g))^{n_g}(v) = \epsilon^{n_g} v$$

and therefore, ϵ is a root of $X^{n_g} - 1$. Our assumption implies that all roots of this polynomial lie already in K and therefore one may choose $v \in M$. We need to show that the eigenspace to each of the eigenvalues is the same as the multiplicity of ϵ in the characteristic polynomial. Over L the matrix $\varphi_M(g)$ is conjugate to a matrix in Jordan normal form. Let

$$\begin{pmatrix} \epsilon & 1 & 0 & \dots & 0 \\ 0 & \ddots & \ddots & \ddots & \vdots \\ \vdots & \ddots & \ddots & 1 & 0 \\ \vdots & & 0 & \epsilon & 1 \\ 0 & \dots & \dots & 0 & \epsilon \end{pmatrix}$$

be a Jordan block of size m. Now, the n_g-th power of this block is

$$\begin{pmatrix} \epsilon^{n_g} & \binom{n_g}{1}\epsilon^{n_g-1} & \binom{n_g}{2}\epsilon^{n_g-2} & \dots & \binom{n_g}{m-1}\epsilon^{n_g-m+1} \\ 0 & \ddots & \ddots & \ddots & \vdots \\ \vdots & \ddots & \epsilon^{n_g} & \binom{n_g}{1}\epsilon^{n_g-1} & \binom{n_g}{2}\epsilon^{n_g-2} \\ \vdots & & \ddots & \epsilon^{n_g} & \binom{n_g}{1}\epsilon^{n_g-1} \\ 0 & \dots & \dots & 0 & \epsilon^{n_g} \end{pmatrix}$$

where we define binomial coefficients $\binom{u}{v}$ to be 0 as soon as $v > u$. Since $n_g \neq 0$ in K, in order to get that this is the identity matrix, one needs to have $m = 1$. So, all Jordan blocks are of size 1, all eigenvalues are in K and so the matrix $\varphi_M(g)$ is diagonalisable over K. □

1.4 Exercises

Exercise 1.1. Let K be a field and let $K[X]$ be the ring of polynomials in one indeterminate with coefficients in K. Let I_4 be the principal ideal of $K[X]$ generated by $X^4 - 1$.

a) Suppose that the characteristic of K is different from 2. Show that $K[X]/I_4$ is semisimple. Find all simple $K[X]/I_4$-modules. Distinguish the cases when $X^2 + 1$ is irreducible in $K[X]$ or not.
b) Suppose that the characteristic of K is 2. Find the simple $K[X]/I_4$-modules. Is the algebra $K[X]/I_4$ semisimple? Show that $K[X]/I_4$ is isomorphic to the group algebra KC_4 where C_4 is the cyclic group of order 4. Find the invertible elements of KC_4 still supposing that K is a field of characteristic 2.

Exercise 1.2. Let K be a field and let A be a K-algebra. Let M be an A-module and let $B := End_A(M)$ the algebra of A-linear endomorphisms of M.

a) Show that the K-vector space M is a B-module if one puts $f \cdot m := f(m)$.
b) For each A-module M one defines $ann_A(M) := \{a \in A \mid a \cdot m = 0 \forall m \in M\}$. Show that $ann_A(M)$ is a 2-sided ideal of A and show that M is an $A/ann_A(M)$-module if one puts $\overline{a} \cdot m := a \cdot m$ for all $\overline{a} = a + ann_A(M) \in A/ann_A(M)$.
c) Let $\overline{A} := A/ann_A(M)$. Show that $End_A(M) = End_{\overline{A}}(M)$.
d) Show that $ann_{\overline{A}}(M) = \{0\}$.
e) Let $\{x_1, \dots, x_n\}$ be a system of generators of M as B-module; i. e. $M = \sum_{j=1}^n B \cdot x_j$. Show that

$$\begin{aligned} \overline{A} &\longrightarrow M^n \\ \overline{a} &\mapsto (\overline{a}x_1, \dots, \overline{a}x_n) \end{aligned}$$

is a monomorphism of left A-modules.

Exercise 1.3. Let A be a K-algebra. The *Jacobson radical* $\mathrm{rad}(A)$ is the intersection of all maximal left ideals of A. A two-sided ideal I of A is said to be nilpotent if there is $n \in \mathbb{N}$ with

$$\underbrace{I \cdot \dots \cdot I}_{n \text{ facteurs}} = 0.$$

The nil radical $nil(A)$ of A is defined to be the sum of all nilpotent two-sided ideals of A.

a) If A is finite dimensional, show that $nil(A)$ is nilpotent as well.
b) Show that $\mathrm{rad}(A)$ is a two-sided ideal of A.
c) If $x \in \mathrm{rad}(A)$, show that $1-x$ admits a left inverse t. (NB: Else, consider the principal left ideal generated by $(1-x)$, which is contained in a maximal left ideal in case $(1-x)$ is not invertible.)
d) Show that $1-(1-t)$ has a left inverse u. Deduce that t is a two-sided inverse of $1-x$.
e) Suppose that $1-ax$ has a left inverse for all $a \in A$. Show that $xS = 0$ for each simple A-module S.
f) Show

$$\{x \in A \mid 1 - ax \text{ has a left inverse for all } a \in A\} = rad(A).$$

g) Let A be a finite dimensional K-algebra over a field K.
   g1) Show that there is $m \in \mathbb{N}$ with $\mathrm{rad}(A)^k = \mathrm{rad}(A)^{k+1}$ for each $k \geq m$.
   g2) If $rad(A)^m \neq 0$, show that there is a minimal left ideal I_0 with $\mathrm{rad}(A)^m \cdot I_0 \neq 0$.
   g3) Show there is $a \in I_0$ with $\mathrm{rad}(A)^m a \neq 0$, and deduce $I_0 = \mathrm{rad}(A)^m \cdot a$.
   g4) Show there is $x \in \mathrm{rad}(A)$ with $(1-x)a = 0$. Deduce $a = 0$ and $\mathrm{rad}(A)^m = 0$. Show $\mathrm{rad}(A) \subseteq nil(A)$.

Exercise 1.4. Let K be a field and let

$$A := \left\{ \begin{pmatrix} \lambda & \mu \\ 0 & \nu \end{pmatrix} \in Mat_{2\times 2}(K) \;\middle|\; \lambda, \mu, \nu \in K \right\}.$$

a) Show that A is a K-algebra if one defines the multiplicative law the matrix multiplication, and the additive law the sum component by component sum of matrices.
b) Let

$$P_1 := \left\{ \begin{pmatrix} \lambda \\ 0 \end{pmatrix} \;\middle|\; \lambda \in K \right\} \subseteq K^2$$

and

$$P_2 := \left\{ \begin{pmatrix} \lambda \\ \mu \end{pmatrix} \;\middle|\; \lambda, \mu \in K \right\} \subseteq K^2.$$

Show that matrix multiplication on the column vectors defines the structure of an A-module on P_1 and on P_2.

c) Show that P_1 is a submodule of P_2. Denote in the sequel $S_2 := P_2/P_1$.
d) Show that S_2 is not isomorphic to P_1 and that S_2 is not isomorphic to P_2.
e) Show that S_2 and P_1 are two simple A-modules.
f) Let M be a simple A-module. Show that there is a left ideal I of A such that $M \simeq A/I$.
g) Consider a possible left ideal I of A, the K-dimension of I, and elements of I of a specific form, to show that a simple A-module is either isomorphic to S_2, or else isomorphic to P_1.

Exercise 1.5.

a) Let R be a (not necessarily commutative) algebra, let I be a left ideal of R and let M be a simple R-module. If I is simple as an R-module and if $I \cdot M \neq 0$, show that $I \simeq M$ as R-modules. (Hint: Show that $I \cdot M$ is a submodule and finally take a well-chosen element $m \in M$ which produces an isomorphism.)

In the sequel we say that a simple ideal is a left ideal I of R which is a simple R-module.

b) Let R be a (not necessarily commutative) algebra, let I be a left ideal of R and let L_I be the sum of left ideals J of R which satisfy $J \simeq I$ as R-modules. If I_1 and I_2 are simple ideals of R and if $I_1 \not\simeq I_2$ as R-modules, then $L_{I_1} \cdot L_{I_2} = \{0\}$, where one computes the product in R.

Suppose in the sequel that R is the sum of simple ideals of R.

c) Show that for each simple left ideal I of R one gets that L_I is a two-sided ideal of R.
d) Let E be a set of representatives of the isomorphism classes of simple ideals of R. If R is semisimple in the sense of Proposition 1.2.6 that any submodule admits a supplement, use the arguments in the proof of Proposition 1.2.6 to show

$$R = \sum_{I \in E} L_I.$$

e) Show that for each $I \in E$ there is $e_I \in L_I$ such that $1 = \sum_{s \in E} e_s$.
f) Show that E is finite, that $e_I^2 = e_I$ acts as a neutral element by multiplication on L_I for each $I \in E$, and that

$$R \simeq \prod L_{I_s}$$

as algebras.

Exercise 1.6. Let K be an algebraically closed field and let n, m be two integers $n > 1$, $m > 1$. Let $A_{n,m} = K[X, Y]/I$ be the algebra of polynomials in two variables modulo the ideal I generated by $\{X^n, Y^m\}$.

a) Considering a K-vector space V of dimension $k < \infty$. Further, consider the set $P_k^{(n,m)}$ of pairs $(\alpha, \beta) \in End_K(V) \times End_K(V)$ of K-linear endomorphisms of V such that $\alpha \circ \beta = \beta \circ \alpha$ and $\alpha^n = 0 = \beta^m$.
   a1) Show that an element p in $P_k^{n,m}$ defines an A-module M_p of dimension k if we put $X \cdot v := \alpha(v)$ and $Y \cdot v := \beta(v)$ for all $v \in V$.
   a2) Let M be an A-module of dimension k. Define p_M in $P_k^{(n,m)}$ such that $M_{p_M} = M$.
   a3) Show that the A-modules of dimension k are in bijection with the points of $P_k^{(n,m)}$.

b) If $p = (\alpha, \beta) \in P_k^{(n,m)}$, show that α has an eigenvector v_α of eigenvalue λ_α and β has an eigenvector v_β of eigenvalue λ_β. Determine λ_α and λ_β.

c) If $k = 1$ show that $P_k^{(n,m)} = \{(0,0)\}$.

d) Let M_1 and M_2 be two A-modules of dimension k and let $p_1 = (\alpha_1, \beta_1)$ and $p_2 = (\alpha_2, \beta_2)$ be two elements of $P_k^{(n,m)}$ with $M_{p_1} = M_1$ and $M_{p_2} = M_2$. Show that the A-modules M_1 and M_2 are isomorphic if and only if there is an automorphism γ of the K-vector space V, where we denote by V the K-vector space on which we define the module structure M_1 and M_2, and such that $\alpha_1 \circ \gamma = \gamma \circ \alpha_2$ and $\beta_1 \circ \gamma = \gamma \circ \beta_2$.

e) Let $k = 2$. Use Jordan canonical form to show that there is $c \in K$ such that $\alpha = c \cdot \beta$ or $\beta = c \cdot \alpha$.

Exercise 1.7. Let G be a finite group.

a) If G is of prime order p, show that

$$\mathbb{C}G \simeq \underbrace{\mathbb{C} \times \mathbb{C} \times \cdots \times \mathbb{C}}_{p \text{ copies}}.$$

b) Suppose that G is not abelian and $|G| = 12$.
   b1) Show that $G/[G,G]$ cannot be of order 8.
   b2) Show that one of the following possibilities can occur:

$$\mathbb{C}G \simeq Mat_{2\times2}(\mathbb{C}) \times Mat_{2\times2}(\mathbb{C}) \times \mathbb{C} \times \mathbb{C} \times \mathbb{C} \times \mathbb{C}$$

$$\mathbb{C}G \simeq Mat_{3\times3}(\mathbb{C}) \times \mathbb{C} \times \mathbb{C} \times \mathbb{C}.$$

   b3) Deduce from the first case that the Sylow 3-subgroup of G is normal. Deduce from the second case that the Sylow 2-subgroup of G is normal.

Exercise 1.8. Let G be a finite group and let H be a subgroup of G.

a) Show that $e_H = \frac{1}{|H|} \sum_{h \in H} h$ is an idempotent of $\mathbb{Q}G$.

b) Show that e_H is in the centre of $\mathbb{Q}G$ if and only if H is normal in G.

c) Show that $\mathbb{Q}G \cdot e_H$ is a $\mathbb{Q}G$-left module and that $\mathbb{Q}G \cdot e_H$ is isomorphic to the permutation module $\mathbb{Q}(G/H)$.

d) If H is normal in G, show that $\mathbb{Q}G \cdot (1 - e_H)$ is a two-sided ideal of $\mathbb{Q}G$. Show in this case $\mathbb{Q}G/(\mathbb{Q}G \cdot (1 - e_H)) \simeq \mathbb{Q}G \cdot e_H$, and $\mathbb{Q}G \cdot e_H \simeq \mathbb{Q}(G/H)$ as $\mathbb{Q}$-algebra.

Exercise 1.9. Let G be a finite group and let V be a $\mathbb{C}$-vector space of dimension $n < \infty$. Suppose that V has a $\mathbb{C}$-basis B such that for all $b \in B$ and all $g \in G$ we have $g \cdot b \in B$.

a) Show that the trivial $\mathbb{C}G$-module $\mathbb{C}$ is a submodule of V.
b) Let for each $b \in B$ be $H_b := \{g \in G \mid g \cdot b = b\}$. Show that H_b is a subgroup of G. Find a necessary and sufficient criterion for the existence of a $b \in B$ such that B can be identified with G/H_b.
c) Let H and N be two subgroups of G. Let V_H be the $\mathbb{C}$-vector space with basis G/H and let V_N be the $\mathbb{C}$-vector space with basis G/N. The group G acts on V_H by left multiplication on the basis elements, and likewise for V_N. Find a necessary and sufficient criterion on H and N for the existence of a $\mathbb{C}G$-module homomorphism $\psi : V_H \longrightarrow V_N$ with $\psi(g \cdot H) = g \cdot N$.
d) Let H and N be two subgroups of G. Find a necessary and sufficient criterion on H and on N such that there is a $\mathbb{C}G$-module homomorphism $\varphi : V_H \longrightarrow V_N$ with $\varphi(G/H) \subseteq G/N$. Express in terms of H and N when φ is an isomorphism.
e) Let H and N be two subgroups of G. Show

$$dim_{\mathbb{C}}(Hom_{\mathbb{C}G}(V_H, V_N)) \geq 1.$$

Exercise 1.10 (with solution in Chapter 9). Let $\mathfrak{A}_4$ be the alternating group of degree 4 and let K be a field of characteristic 0. The symmetric group $\mathfrak{S}_4$ of degree 4 contains $\mathfrak{A}_4$ and we consider $\mathfrak{A}_4$ as subgroup of $\mathfrak{S}_4$. Let V be a K-vector space of dimension 4. Fix a basis $\{b_1, b_2, b_3, b_4\}$ of V. For all $\sigma \in \mathfrak{A}_4$ and each $i \in \{1, 2, 3, 4\}$ define $\sigma \cdot b_i := b_{\sigma^{-1}(i)}$.

a) Show that V is a $K\mathfrak{A}_4$-module by this action.
b) Show that the trivial $K\mathfrak{A}_4$-module is a submodule T of V.
c) Compute the eigenspaces of the action $(1\ 2)(3\ 4)$ on V and compute the eigenspaces of the action of $(1\ 2\ 3)$ on V.
d) Is there a $K\mathfrak{A}_4$-submodule W of V with $dim_K(W) = 1$ and $W \neq T$?
e) Decompose V in a direct sum of indecomposable $K\mathfrak{A}_4$-submodules.

Exercise 1.11. Let K be a field and let G be a finite group. Denote

$$I(KG) := \left\{ \sum_{g \in G} x_g g \;\middle|\; x_g \in K \text{ et } \sum_{g \in G} x_g = 0 \right\}$$

and for each normal subgroup N of G

$$I_N(KG) := \left\{ \sum_{g \in G} x_g g \;\middle|\; x_g \in K \text{ and for all } gN \in G/N \text{ we have } \sum_{gn \in gN} x_{gn} = 0 \right\}$$

a) Show $I_G(KG) = I(KG)$.
b) Let N be a normal subgroup of G. Show that the group homomorphism $\alpha : G \longrightarrow G/N$ induces a homomorphism of K-algebras $\hat{\alpha} : KG \longrightarrow K(G/N)$ with $\hat{\alpha}|_G = \alpha$. Prove $\ker \hat{\alpha} = I_N(KG)$ and deduce that $I_N(KG)$ is a two-sided ideal of KG.

c) Compute $\dim_K(I_N(KG))$ as function of the order of G and the order of N.
d) Describe the KG-module $KG/I(KG)$, namely its dimension and the operation of G on $KG/I(KG)$.
e) Let $N := [G,G]$ be the derived subgroup of G. Let K be an algebraically closed field of characteristic 0. Show that the ideal $I_N(KG)$ is equal to the product of those Wedderburn components of KG which are of dimension strictly larger than 1.
f) Let N be a normal subgroup of G. Show that $I_N(KG)$ is generated as a left ideal of KG by $\{n - 1_G \mid n \in N\}$; where we denote by 1_G the neutral element of G.
g) Let N be a normal subgroup of G. Find a K-basis of $I_N(KG)$.

Exercise 1.12. Let K be a (commutative) field of characteristic different from 2 and let $a, b \in K$. On the four dimensional K vector space K^4 we fix a K-basis $\{1, i, j, k\}$ and define a multiplicative law $\cdot : K^4 \times K^4 \longrightarrow K^4$ by:

$$\begin{array}{ll} & i \cdot i = a1 \\ 1 \cdot 1 = 1 & j \cdot j = b1 \\ 1 \cdot i = i \cdot 1 = i & k \cdot k = -ab1 \\ 1 \cdot j = j \cdot 1 = j & i \cdot j = -j \cdot i = k \\ 1 \cdot k = k \cdot 1 = k & i \cdot k = = -k \cdot i = aj \\ & j \cdot k = = -k \cdot j = bi \end{array}$$

on the basis elements, and then extended K-bilinearly on all of K^4.
a) Show that we obtain a K-algebra structure on K^4 this way. We shall denote this K-algebra by $(\frac{a,b}{K})$.
b) Show that the K sub-vector space generated by 1 is in the centre of $(\frac{a,b}{K})$.
c) What is a more common name for $(\frac{-1,-1}{\mathbb{R}})$?
d) Let $q := x1 + yi + zj + wk \in (\frac{a,b}{K})$ for $x, y, z, w \in K$ and denote $\overline{q} := x1 - yi - zj - wk$. Show that $q \cdot \overline{q} = (x^2 - ay^2 - bz^2 + abw^2) \cdot 1 \in K \cdot 1$.
e) Show that the map

$$\begin{aligned} \left(\frac{a,b}{K}\right) &\longrightarrow K \\ q &\mapsto q \cdot \overline{q} \end{aligned}$$

defines a quadratic form N on K^4.
f) If a or b or $-ab$ is a square in K, show that there are $q \in (\frac{a,b}{K}) \setminus \{0\}$ with $N(q) = 0$.
g) Let $K = \mathbb{R}$ and suppose that $a < 0$ as well as $b < 0$. Show that $(\frac{a,b}{K})$ is a skew field. For a general K of characteristic different from 2, do there exist $(a, b) \in K$ such that $(\frac{a,b}{K})$ is a commutative field?
h) For a field K such that there is $x \in K$ with $x^2 = a$, and $y \in K$ with $y^2 = b$ consider the matrices

$$\sigma_1 := \begin{pmatrix} 1 & 0 \\ 0 & -1 \end{pmatrix}, \quad \sigma_2 := \begin{pmatrix} 0 & 1 \\ 1 & 0 \end{pmatrix}, \quad \sigma_3 := \begin{pmatrix} 0 & -x \\ x & 0 \end{pmatrix}.$$

Show that we get an isomorphism of K-algebras by putting

$$\begin{aligned}\left(\frac{a,b}{K}\right) &\longrightarrow Mat_{2\times 2}(K)\\ i &\mapsto -x\sigma_1\\ j &\mapsto -y\sigma_2\\ ij &\mapsto xy\sigma_1\sigma_2\end{aligned}$$

Remark 1.4.1. For $a = b = -1$ we call the above matrices the Pauli matrices. They actually arise in quantum physics when one studies the spin of elementary particles. Using Pauli matrices it is also possible to introduce the Minkowski four dimensional space used in Einstein's special relativity theory.

The *Hilbert symbol* (a, b) associates to two non zero elements a, b in K the value -1 if $(\frac{a,b}{K})$ is a skew field, and the value 1 if not. If the algebra $(\frac{a,b}{K})$ is not a skew field, then the algebra is isomorphic to a 2×2 matrix algebra over K.

2 Characters

Es wächst hienieden Brot genug
Für alle Menschenkinder,
[...]
Den Himmel überlassen wir
Den Engeln und den Spatzen

Heinrich Heine: Deutschland ein Wintermärchen.

Enough bread grows here on earth,
For all mankind's nutrition,
[...]
To the angels and the sparrows,
We leave Heaven and its Gods.[1]

Heinrich Heine: Germany. A Winter's Tale.

We come to a concept which is one of the oldest topics in representation theory, and which is one of the main objects of interest in the subject. The main objective is to phrase all indecomposable $\mathbb{C}G$-modules in terms of an ordered set of complex numbers. This ordered set of complex numbers actually is called the character table and has truly miraculous properties. Moreover, it opens the path to analytic methods. An instance of this phenomenon is the theory of Dirichlet L-series, which we shall explain, and which shows a famous result in number theory. The result is still elementary in its formulation, but the proof uses this passage from algebra to complex analysis.

2.1 Definitions and basic properties

As usual the trace $trace(N)$ of a square matrix N is the sum of the diagonal coefficients of this matrix.

Definition 2.1.1. Let G be a group, let K be a field and let $\varphi : G \longrightarrow GL_n(K)$ be a representation of G which affords the KG-module M. Then $\chi_M = trace \circ \varphi : G \longrightarrow K$ is the *character* of G afforded by φ. The number n is the *degree* of the character.

Remark 2.1.2. A character is hence a mapping $G \longrightarrow K$. Let N_1 and N_2 be two square matrices of the same size. Then $trace(N_1N_2) = trace(N_2N_1)$ and so, the trace of a conjugate to a matrix N is equal to the trace of N. As a consequence, let $g, h \in G$, and let χ be a character of G, afforded by some representation φ. Then,

$$\chi(h \cdot g \cdot h^{-1}) = trace(\varphi(h)\varphi(g)\varphi(h)^{-1}) = trace(\varphi(h)\varphi(h)^{-1}\varphi(g)) = \chi(g).$$

So, the characters are constant on conjugacy classes of G.

1 translation by Joseph Massaad.

https://doi.org/10.1515/9783110702446-002

Lemma 2.1.3. *Let G be a finite group of exponent m and let K be a field such that the order of G is invertible in K. Let $g \in G$ an element of order m. If K contains a primitive m-th root of unity then for every character χ of G the value $\chi(g)$ is the sum of $\chi(1)$ m-th roots of unity.*

Proof. Let χ be afforded by a representation φ. By Lemma 1.3.6 if K contains a primitive m-th root of unity then φ is diagonalizable and the main diagonal entries are m-th roots of unity. Moreover, $\varphi(1)$ is the identity matrix and so, $\chi(1)$ is the dimension of the module which is induced by φ. This proves the statement. □

Lemma 2.1.4. *Let $K = \mathbb{C}$, let G be a finite group and let χ be a character.*
1. *For all $g \in G$ we have $\chi(g^{-1}) = \overline{\chi(g)}$.*
2. *For all $g \in G$ we have $|\chi(g)| \leq \chi(1)$.*
3. $\ker(\chi) := \{g \in G | \chi(g) = \chi(1)\}$ *is a normal subgroup of G. If φ is a representation which affords χ, then* $\ker(\chi) = \ker(\varphi)$.

Proof. Let φ be a representation which affords χ. Again by Lemma 1.3.6 we see that $\varphi(g)$ is diagonalizable with diagonal entries being m-th roots of the unity. In the field of complex numbers the inverse of a root of unity equals the complex conjugate. This shows the first statement.

The second statement is the triangle inequality of the absolute value of a sum of complex numbers in connection with the fact that the absolute value of a root of unity is 1.

For the third statement observe that in case one has equality in the equation of the second statement, then $\varphi(g)$ has to be a scalar multiple of the identity matrix. Let x be this scalar, which is clearly a root of unity. Then, $\chi(g) = x \cdot \chi(1)$ and $g \in \ker(\chi) \Leftrightarrow g \in \ker(\varphi)$. This shows the third statement as well. □

Lemma 2.1.5. *Let K be a field, let G be a group and let M, M_1 and M_2 be KG-modules. Let χ_M be the character afforded by M, χ_{M_1} be the character afforded by M_1 and χ_{M_2} be the character afforded by M_2. Then,*

$$M \simeq M_1 \oplus M_2 \Rightarrow \chi_M = \chi_{M_1} + \chi_{M_2}.$$

Proof. Let φ be the representation which is induced by M, φ_1 be the representation which is induced by M_1 and φ_2 be the representation which is induced by M_2. Suppose that $M \simeq M_1 \oplus M_2$. Choose a basis $\mathcal{B} = \mathcal{B}_1 \cup \mathcal{B}_2$ of M such that $\mathcal{B}_1$ is mapped via this isomorphism to a basis of M_1 and $\mathcal{B}_2$ is mapped via this isomorphism to a basis of M_2. Then,

$$\varphi = \begin{pmatrix} \varphi_1 & 0 \\ 0 & \varphi_2 \end{pmatrix}$$

and the statement follows. □

Definition 2.1.6. The character of a simple module is called an *irreducible character*. The character of the trivial module is the *trivial character*.

In the literature one finds the expression of an irreducible module of a group G synonymous to a simple module. We only use the terminology of a simple module.

Remark 2.1.7. Let G be a finite group and put $K = \mathbb{C}$. Suppose that the order of G is invertible in K. Then, KG is semisimple, and so every KG-module is a direct sum of its simple submodules. Let $\chi_1, \chi_2, \ldots, \chi_m$ be the characters of the isomorphism classes of the simple KG-modules. By Theorem 1.3.4 m is also the number of conjugacy classes of G. Let $C_1, C_2, \ldots, C_m$ be the conjugacy classes of G. We are going to build a square matrix of complex numbers called the *character table*. The rows are parameterised by the irreducible characters of G and the columns are parameterised by the conjugacy classes of G. The coefficient in the i-th line and j-th column is the value $\chi_i(C_j)$. By convention we assume that χ_1 is the trivial character and C_1 the conjugacy classes of the identity element in G.

In the sequel we are going to elaborate various arithmetic properties of the character table.

2.2 Class functions, linking characters to the semisimple group algebra

Proposition 2.2.1. *Let K be a splitting field for the finite group G and suppose that the order of G is invertible in K. Let $\chi_1, \chi_2, \ldots, \chi_m$ be the irreducible characters of G with values in K. Then, $\{\chi_1, \chi_2, \ldots, \chi_m\}$ is K-linear independent in the space of functions $G \longrightarrow K$.*

Proof. Wedderburn's theorem gives a decomposition

$$KG \simeq \prod_{j=1}^{m} Mat_{n_j \times n_j}(K)$$

and to every direct factor $Mat_{n_j \times n_j}(K)$ corresponds a simple module S_j and an irreducible character χ_j with values in K. Since $KG \simeq End_{KG}(KG)^{op}$ by Lemma 1.2.28 the endomorphism given by projection on the i-th component and re-injection of the i-th component into the direct product is a central idempotent endomorphism and corresponds hence to a central idempotent element $e_i \in KG$. Now,

$$\chi_j(e_i) = \delta_{i,j} \cdot \chi_i(1),$$

where $\delta_{i,j}$ is the Kronecker symbol, being 0 whenever $i \neq j$ and 1 if $i = j$. In case K is of characteristic 0 we are already finished, but in case K is of finite characteristic we need to exclude that $\chi_i(1) = 0 \in K$. As we see in $\prod_{j=1}^{m} Mat_{n_j \times n_j}(K)$, for all $i \in \{1, \ldots, m\}$ the central idempotent e_i corresponds to the identity matrix in the i-th component. This is a sum of $\chi_i(1)$ non central idempotents $\epsilon_{i,j}$ for $j \in \{1,\dots,\chi_i(1)\}$, namely the matrix with 0 coefficients everywhere, except on the j-th diagonal entry. Usually, these idempotents are not unique. Actually, since they are not central, conjugation with any invertible matrix u of the Wedderburn decomposition of KG fixes e_i since e_i is central, but will usually not fix the summands $\epsilon_{i,j}$ in the sum decomposition $e_i = \sum_{j=1}^{\chi_i(1)} \epsilon_{i,j}$. We obtain this way another decomposition of e_i into a sum of idempotents. In any case, since the elements $\epsilon_{i,j}$ are idempotents, they correspond to idempotents in KG and since $KG\epsilon_{i,j}$ is isomorphic to the simple KG-module S_i, $\chi_i(\epsilon_{k,j}) = \delta_{k,i}$ for all $j \in \{1,\dots,\chi_i(1)\}$. This proves the proposition. □

A first corollary is important.

Corollary 2.2.2. *Let V be a simple KG-module and let $e^2 = e \in Z(KG)$ and e is not the sum of two other non-zero central idempotents in KG. Then $\chi_V(e) = 1$ if $eV \neq 0$ and $\chi_V(e) = 0$ if $eV = 0$.*

Proof. This follows actually from the argument in the proof of Proposition 2.2.1 since any of these idempotents e correspond to the idempotents e_i in the proof of the Proposition. □

We shall prove a little more in Proposition 2.2.1. We have seen that a complex valued character of G is a function $G \longrightarrow \mathbb{C}$ that is constant on conjugacy classes:

$$\forall g, h \in G : \quad g = xhx^{-1} \Rightarrow \chi(g) = \chi(h)$$

The irreducible characters of a group G form a basis of the $\mathbb{C}$-vector space of functions $G \longrightarrow \mathbb{C}$ which are constant on conjugacy classes. The space of these functions is called the space of class functions of G.

Definition 2.2.3. Let K be a field and let G be a group. Then, the K-vector space

$$\begin{aligned} CF_K(G) := \{f : G \longrightarrow K \mid \forall g, h \in G :& \\ &\text{if } \exists x \in G : g = xhx^{-1} \text{ then } f(g) = f(h)\} \end{aligned}$$

is the space of K-valued *class functions* of G.

A K-valued character is trivially a class function. What is more astonishing is that every class functions is basically a K-linear combination of characters.

Proposition 2.2.4. *Let G be a finite group and let K be a splitting field for G. If the order of G is invertible in K, then the set of irreducible characters of G is a K-basis for the space of class functions.*

Proof. Proposition 2.2.1 shows that the set of K-valued characters is a linearly independent set. We know by Theorem 1.3.4 that the number of isomorphism classes of simple KG-modules is equal to the number of conjugacy classes of G. This is trivially equal to the dimension of $CF_K(G)$. This proves the proposition. □

Characters can be used to express the central primitive idempotents of KG, that is the units of the matrix rings which are direct factors of the group ring in the Wedderburn components:

$$KG \simeq \prod_{i=1}^{m} Mat_{n_i \times n_i}(D_i)$$

and then

$$e_i = \left(0, 0, \dots, 0, \begin{pmatrix} 1 & 0 & & \dots & 0 \\ 0 & \ddots & & \ddots & \vdots \\ \vdots & \ddots & & \ddots & 0 \\ 0 & \dots & & 0 & 1 \end{pmatrix}_{n_i \times n_i}, 0, \dots, 0, 0 \right)$$

where the unit matrix is the i-th component.

Definition 2.2.5. More abstractly, a *central idempotent* in an algebra A is an element $e^2 = e \in Z(A)$, where we denote as usual $Z(A)$ the centre of A. Moreover, an idempotent $e \neq 0$ is *primitive* if whenever $e = e_1 + e_2$ and $e_1^2 = e_1$ and $e_2^2 = e_2$, then either $e = e_1$ or $e = e_2$. A primitive central idempotent is therefore an idempotent in the centre of the algebra and which is primitive as idempotent in the centre of the algebra.

Lemma 2.2.6. *Let A be a commutative finite dimensional K-algebra. Put*

$$E := \{e \in Z(A) |\ e \text{ is a primitive idempotent} \}.$$

Then, $\sum_{e \in E} e = 1$.

Proof. Clearly $1 \in A$ is an idempotent. If 1 is not indecomposable, write $1 = e_1 + e_2$. Again if one of the idempotents e_1 or e_2 is not indecomposable, write them again as sum of two. This recursion stops at some point since if not there would be a sequence

$$e_1, e_{1,1}, e_{1,1,1}, \dots, e_{1^{(n)}}, e_{1^{(n+1)}}, \dots$$

of successive refinements of idempotents such that $e_{1^{(n)}} \cdot e_{1^{(n+1)}} = e_{1^{(n+1)}}$ for all n. This produces a sequence $A \cdot e_{(1)^n}$ of ideals of A. Since A is finite dimensional, there is an n_0 such that $e_{1^{(n_0+k)}} = e_{1^{(n_0)}}$ for all $k \in \mathbb{N}$. Hence, one can write

$$1 = e_1 + e_2 + \dots + e_m$$

where $e_1, e_2, \dots, e_m$ are all primitive idempotents. Now, if f is a primitive idempotent not being one of the $e_1, e_2, \dots, e_m$. Then,

$$f = f \cdot 1 = fe_1 + fe_2 + \dots + fe_m.$$

Since A is commutative, $(fe_i)^2 = fe_i$ for all i and since $e_i = fe_i + (1 - f)e_i$, one gets that $fe_i = e_i$ or $fe_i = 0$. Since f is primitive as well, one gets $f = fe_i$ or $fe_i = 0$. Hence, $f = e_i$ for the unique case when $fe_i = e_i$. □

Corollary 2.2.7. *Let A be a finite dimensional K-algebra. Then in A there is a unique decomposition $1 = e_1 + e_2 + \cdots + e_s$ into central primitive idempotents and if $E = \{e_1, e_2, \ldots, e_n\}$ and f is a central primitive idempotent, then $f \in E$.*

Proof. This follows immediately from Lemma 2.2.6 after passing from A to the centre of A. □

Call χ_i the irreducible character of the simple KG-module S_i for which $e_iS_i = S_i$. This defines χ_i in a unique way. Indeed, since e_i is central, e_iS_i is a KG-submodule of S_i and since S_i is simple, $e_iS_i = S_i$ or $e_iS_i = 0$. Since $1 = e_1 + \cdots + e_s$ and $e_ie_j = \delta_{i,j}$, if $e_iS_i = S_i = e_jS_i$, we get $S_i = e_iS_i = e_ie_jS_i = 0$ and hence there is a unique $i \in \{1, \ldots, s\}$ for which $e_iS_i = S_i$.

Proposition 2.2.8. *Let K be a splitting field for G and let $|G|$ be invertible in K. Then*

$$e_i = \frac{\chi_i(1)}{|G|} \sum_{g \in G} \chi_i(g^{-1})g.$$

Proof. Denoting χ_{KG} the character of the regular KG-module KG, we know that

$$\chi_{KG} = \sum_{\chi \in Irr_K(G)} \chi(1)\chi$$

by Wedderburn's theorem. Indeed, by Corollary 1.2.35 to Wedderburn's theorem and the fact that K is a splitting field, for every $\chi \in Irr_K(G)$ the decomposition of the regular module KG has exactly $\chi(1)$ direct factors isomorphic to the simple module S which induces χ. Moreover, $\chi(1)$ is exactly this number in case we are dealing with splitting fields.

Write $e_i = \sum_{g \in G} k_g g$. For $i \neq j$ the element e_i acts on S_j as 0, and so also e_ih acts as 0 on S_j. Therefore, the character value $\chi_j(e_ih) = 0$ whenever $i \neq j$. Moreover, e_i acts as 1 on S_i and therefore $\chi_i(e_ih) = \chi_i(h)$ for each $h \in G$. Hence

$$\chi_{KG}(e_ih) = \sum_{\chi \in Irr_K(G)} \chi(1)\chi(e_ih) = \chi_i(1)\chi_i(e_ih) = \chi_i(1) \cdot \chi_i(h).$$

On the other hand, one can compute $\chi_{KG}(g)$ explicitly for every $g \in G$. In fact, G is a basis for KG. Now, $(\exists h \in G : gh = h) \Rightarrow g = 1$ and so multiplication permutes this basis without fix-point for every $g \neq 1_G$. Since $dim_K(KG) = |G|$, one has $\chi_{KG}(1_G) = |G|$ and $\chi_{KG}(g) = 0$ for every $g \in G \setminus \{1_G\}$. As a consequence

$$\chi_{KG}(e_ih) = \sum_{g \in G} k_g \chi_{KG}(gh) = k_{h^{-1}}|G|.$$

Therefore,

$$k_h = \frac{1}{|G|}\chi_{KG}(e_ih^{-1}) = \frac{1}{|G|}\chi_i(1)\chi_i(h^{-1})$$

which implies

$$e_i = \frac{\chi_i(1)}{|G|} \sum_{g \in G} \chi_i(g^{-1})g.$$

This proves the proposition. □

2.3 Multiplicities of simple submodules; orthogonality relations

We shall introduce a non degenerate symmetric bilinear form on $CF_K(G)$ for which we are going to see that $Irr_K(G)$ is an orthonormal basis. Of course this can be done just by defining the scalar product by this property. However, we want to be able to compute the values of the bilinear form in terms of the group only and this is why we take another path.

Lemma 2.3.1. *Let G be a finite group and let K be a field.*

- *Then, for every KG-module V the space of linear forms $Hom_K(V, K)$ is a KG-module when one defines for every $g \in G$ and every $f \in Hom_K(V, K)$*

$$(g \cdot f)(v) := f(g^{-1}v) \ \ \forall v \in V.$$

  The module $Hom_K(V, K) =: V^*$ is called the dual module to V.
- *For two KG-modules V and W the space $Hom_K(V, W)$ is a KG-module if one puts for every $g \in G$ and every $f \in Hom_K(V, W)$*

$$(g \cdot f)(v) := g \cdot f(g^{-1}v) \ \ \forall v \in V.$$

Proof. We prove only the second statement. The first statement follows from the second by considering K as the trivial KG-module.

Now, $1_G \cdot f = f$ is clear for all $f \in Hom_K(V, W)$, and for all $f \in Hom_K(V, W)$, $g, h \in G$ and $v \in V$ one has

$$\begin{aligned}((g \cdot h) \cdot f)(v) &= (g \cdot h) \cdot f((g \cdot h)^{-1} \cdot v) \\ &= (g \cdot h) \cdot f((h^{-1} \cdot g^{-1}) \cdot v) \\ &= g \cdot (h \cdot f(h^{-1} \cdot (g^{-1} \cdot v))) \\ &= g \cdot ((h \cdot f)(g^{-1} \cdot v)) \\ &= (g \cdot (h \cdot f))(v)\end{aligned}$$

We extend as usual the action K-linearly and obtain the statement. □

Lemma 2.3.2. *Let G be a group and let K be a field. Then for every KG-module M we get $Hom_{KG}(K, M) \simeq M^G$, the vector space of G-fixed points of M. In particular for two KG-modules V and W we get*

$$Hom_{KG}(V, W) \simeq Hom_{KG}(K, Hom_K(V, W)) \simeq (Hom_K(V, W))^G.$$

Proof. Actually, we shall see that there is an isomorphism

$$Hom_{KG}(V,W) \xrightarrow{\Phi} Hom_{KG}(K, Hom_K(V,W)).$$

This is a special case of a more general relation, the adjointness between two different functors (cf. e. g. [Zim-14, Chapter 3]). We are going to meet another instance of this soon when we discuss induced representations. Nevertheless, we do not need to know this deeper theory here. The proof can be given in very elementary terms.

Actually, $Hom_{KG}(V,W)$ is the subspace of those elements in $Hom_K(V,W)$ which are G-fixed under the G-action defined in Lemma 2.3.1. Indeed, for $f \in Hom_K(V,W)$ one has

$$\forall_{v\in V, g\in G} : (g\cdot f)(v) = gf(g^{-1}v) = f(v) \Leftrightarrow f \in Hom_{KG}(V,W).$$

Moreover for any KG-module M one has $Hom_{KG}(K,M)$ is the fixed point set $M^G := \{m \in M \mid g\cdot m = m \forall g \in G\}$ of M. Indeed, for any $m \in M^G$ the mapping $f(k) := km$ for all $k \in K$ is KG-linear. On the other hand side, $f(1) = m$ implies

$$gm = gf(1) = f(g\cdot 1) = f(1) = m$$

for all $g \in G$ since K is the trivial KG-module. Hence,

$$Hom_{KG}(K, Hom_K(V,W)) \simeq (Hom_K(V,W))^G = Hom_{KG}(V,W).$$

This shows the lemma. □

Now, let V and W be two simple KG-modules and let K be a splitting field for G. Then, $V \simeq W$ if and only if the trivial KG-module is a submodule of $Hom_K(V,W)$. We hence need to look for a criterion when the trivial module is a submodule of another module.

Remark 2.3.3. Let M be a KG-module and suppose the order of G is invertible in K. Put $\frac{1}{|G|}\sum_{g\in G} g =: e_1$ and observe that $e_1^2 = e_1 \in Z(KG)$ and that $e_1 \cdot M$ is a trivial submodule of M. Indeed,

$$h\cdot e_1 = h\cdot \frac{1}{|G|}\sum_{g\in G} g = \frac{1}{|G|}\sum_{g\in G} hg = \frac{1}{|G|}\sum_{hg\in G} hg = e_1.$$

Moreover, if $m \in M^G$, then $e_1 m = m$.

We want to stress that the arguments do not use any character theory and the arguments are valid even for commutative rings K such that the order of G is invertible in K.

We continue by asking what is the character value of $g \in G$ on the space $Hom_K(V,W)$. We choose a basis $B_{V,g}$ of V consisting of eigenvectors for the action of g and a basis $B_{W,g}$ of W consisting of eigenvectors for the action of g. This is possible because of Lemma 1.3.6. We get a basis of $Hom_K(V, W)$ by the mappings

$$\{f_{b_V,b_W} \mid b_V \in B_{V,g}; b_W \in B_{W,g}\}$$

where $f_{b_V,b_W}(b_V) = b_W$ and $f_{b_V,b_W}(b'_V) = 0$ whenever $b'_V \neq b_V$. Choosing this basis we see that

$$\chi_{Hom_K(V,W)}(g) = \chi_V(g^{-1})\chi_W(g)$$

for this $g \in G$. Now, in order to compute the character value of each $g \in G$ we are free to choose a basis, which is appropriate for the computation. The result does not depend on this choice. So, for two simple KG-modules V and W we get

$$\begin{aligned} V \simeq W \Leftrightarrow \chi_{Hom_K(V,W)}(e_1) &= \frac{1}{|G|}\sum_{g\in G}\chi_{Hom_K(V,W)}(g) \\ &= \frac{1}{|G|}\sum_{g\in G}\chi_V(g^{-1})\chi_W(g) \\ &\neq 0. \end{aligned}$$

Since V and W are simple, $Hom_{KG}(V, W)$ is of dimension at most 1, and so $\chi_{Hom_K(V,W)}(e_1) \in \{0, 1\}$ (see Corollary 2.2.2).

Let now V and W be any two KG-modules. Suppose that the order of G is invertible in K and suppose that K is a splitting field for G. Then, V and W are both semisimple by Maschke's theorem. By Lemma 2.1.5 we see that $\frac{1}{|G|}\sum_{g\in G}\chi_V(g^{-1})\cdot\chi_W(g)$ is the square of the number of common simple direct factors of V and W.

We proved the following

Theorem 2.3.4. (Orthogonality relations for characters) *Let G be a finite group and let K be a field. Suppose that the order of G is invertible in K and suppose that K is a splitting field for G. Define for two class functions χ_1 and χ_2*

$$(\chi_1, \chi_2) := \frac{1}{|G|}\sum_{g\in G}\chi_1(g)\cdot\chi_2(g^{-1}).$$

Then the set of irreducible characters $Irr_K(G)$ is an orthonormal basis on the vector space $CF_K(G)$ of K-valued class functions.

Proof. The fact that $Irr_K(G)$ is a basis for $CF_K(G)$ is Proposition 2.2.4. The fact that this basis is orthonormal is shown in the remarks preceding the theorem. □

The orthogonality relations of characters in Theorem 2.3.4 become orthogonality relations with respect to the columns of the character table when interpreted in an appropriate way.

For the moment we shall only give the second orthogonality relations for the field of complex numbers. We know that $\chi(g^{-1}) = \overline{\chi(g)}$ for every $g \in G$ and every character χ of a representation over $\mathbb{C}$.

Let T be the character table seen as square matrix of size $s \times s$ where $Irr_{\mathbb{C}}(G)$ is of cardinality s, where $\{C_1, C_2, \ldots, C_s\}$ are the conjugacy classes of G and where the i-th row of T is $(\chi_i(C_1), \chi_i(C_2), \ldots, \chi_i(C_s))$. Put

$$D(G) := \begin{pmatrix} \frac{|C_1|}{|G|} & 0 & \cdots & \cdots & 0 \\ 0 & \frac{|C_2|}{|G|} & 0 & & \vdots \\ \vdots & 0 & \ddots & \ddots & \vdots \\ \vdots & & \ddots & \ddots & 0 \\ 0 & \cdots & \cdots & 0 & \frac{|C_s|}{|G|} \end{pmatrix}.$$

Theorem 2.3.4 says that

$$T \cdot D(G) \cdot \overline{T}^{tr} = \mathbf{1}$$

where $\mathbf{1}$ is the unit matrix. Then, the matrix $D(G)$ is invertible as square matrix over $\mathbb{C}$. Hence, since $Gl_s(\mathbb{C})$ is a group,

$$\overline{T}^{tr} \cdot T = D(G)^{-1}.$$

We proved the following statement.

Theorem 2.3.5 (Orthogonality relations for conjugacy classes). *Let G be a finite group, let $\{C_1, C_2, \ldots, C_s\}$ be the conjugacy classes of G and let $Irr_{\mathbb{C}}(G) = \{\chi_1, \chi_2, \ldots, \chi_s\}$ be the irreducible complex characters of G. Then,*

$$\sum_{\chi \in Irr_{\mathbb{C}}(G)} \chi(C_i)\overline{\chi(C_j)} = \delta_{i,j} \frac{|G|}{|C_i|}$$

Remark 2.3.6.

- Actually, we did not really need to use $\mathbb{C}$ as ground field. It was only needed that there is a field automorphism $K \ni x \mapsto \overline{x} \in K$ which has the property that for all $|G|$-th roots of unity ζ one gets $\zeta \cdot \overline{\zeta} = 1 \in K$.
- There is a much more general version for the orthogonality relations. This is a theorem due to Fossum [Fos-71].

We are going to see how these relations may be used to compute character tables whenever one has only partial information on the character table by other means.

Example 2.3.7. Let $\mathfrak{A}_4$ be the alternating group of order 12. We want to compute the complex character table of $\mathfrak{A}_4$. We know that there is a normal 2-Sylow subgroup S

formed by the double transpositions and the identity element. The quotient is cyclic of order 3 and therefore one gets 2 conjugacy classes, one for the element (1 2 3) and another one for its inverse (1 3 2). The double transpositions form another conjugacy class of order 3, and the identity element is a fourth conjugacy class. Since $12 = 3^2 + 1 + 1 + 1$ and since C_3 is a quotient of $\mathfrak{A}_4$ the degrees of the irreducible complex characters of $\mathfrak{A}_4$ are 1, 1, 1 and 3. The three degree 1 characters come from the three 1-dimensional simple $\mathbb{C}C_3$-modules. We denote by C_g the conjugacy class of $g \in \mathbb{A}_4$ and by ζ a primitive 3-rd root of 1 in $\mathbb{C}$, i. e. $1 + \zeta + \zeta^2 = 0$. Therefore we may complete already the character table as follows.

$\mathfrak{A}_4$	$\{1\}$	$C_{(1\,2)(3\,4)}$	$C_{(1\,2\,3)}$	$C_{(1\,3\,2)}$
size of class	1	3	4	4
χ_1	1	1	1	1
χ_ζ	1	1	ζ	ζ^2
χ_{ζ^2}	1	1	ζ^2	ζ
χ_3	3	x_1	x_2	x_3

In order to fill the third line one just needs to apply the orthogonality relations with respect to the conjugacy classes of Theorem 2.3.5. Indeed, the first column is orthogonal to each of the others, and so we get the equations

$$1 \cdot 1 + 1 \cdot 1 + 1 \cdot 1 + 3 \cdot x_1 = 0$$
$$1 \cdot 1 + 1 \cdot \zeta + 1 \cdot \zeta^2 + 3 \cdot x_2 = 0$$
$$1 \cdot 1 + 1 \cdot \zeta^2 + 1 \cdot \zeta + 3 \cdot x_2 = 0$$

which implies $x_1 = -1$ and $x_2 = x_3 = 0$. The character table is therefore

$\mathfrak{A}_4$	$\{1\}$	$C_{(1\,2)(3\,4)}$	$C_{(1\,2\,3)}$	$C_{(1\,3\,2)}$
size of class	1	3	4	4
χ_1	1	1	1	1
χ_ζ	1	1	ζ	ζ^2
χ_{ζ^2}	1	1	ζ^2	ζ
χ_3	3	−1	0	0

The fact that $x_2 = x_3 = 0$ can be deduced as well from the known value of $x_1 = -1$ and the fact that

$$1 = \langle \chi_3, \chi_3 \rangle = \frac{1}{12}(3^2 + 3 \cdot (-1) \cdot (-1) + 4 \cdot x_2 \cdot \overline{x_2} + 4 \cdot x_3 \cdot \overline{x_3})$$

or from the fact that the second and the third column are of length 1:

$$1 + \zeta \cdot \overline{\zeta} + \zeta^2 \cdot \overline{\zeta^2} + x_2 \cdot \overline{x_2} = \frac{12}{4}$$

and

$$1+\zeta^2\cdot\overline{\zeta^2}+\zeta\cdot\overline{\zeta}+\chi_3\cdot\overline{\chi_3}=\frac{12}{4}.$$

It is possible as well to decompose characters into irreducible characters. We consider $\mathfrak{A}_4$ as subgroup of $\mathfrak{S}_4$ and $\mathfrak{S}_4$ acts on a 4-dimensional complex vector space by permuting the basis elements of a fixed chosen basis B according to the permutation of the indices: $B=\{b_1,b_2,b_3,b_4\}$ and $\sigma\cdot b_i:=b_{\sigma(i)}$ for all $i\in\{1,2,3,4\}$. This gives a 4-dimensional $\mathbb{C}\mathfrak{A}_4$-module P. We compute its character:

$$\chi_P=(4,\ 0,\ 1,\ 1)$$

since the value of such a character on σ is clearly the number of non zero main diagonal entries of the representing matrix, and this is the number of fixed points of the action of $\mathfrak{A}_4$ on $\{1,2,3,4\}$. Now,

$$\begin{aligned}(\chi_P,\chi_1)&=\frac{1}{12}(4+0+4\cdot1+4\cdot1)=1\\(\chi_P,\chi_\zeta)&=\frac{1}{12}(4+0+4\cdot\zeta+4\cdot\zeta^2)=0\\(\chi_P,\chi_{\zeta^2})&=\frac{1}{12}(4+0+4\cdot\zeta^2+4\cdot\zeta)=0\\(\chi_P,\chi_3)&=\frac{1}{12}(4\cdot3+0+0+0)=1\end{aligned}$$

Using now Theorem 2.3.4 and Lemma 2.1.5 one gets that $P\simeq K\oplus M_3$ where K is the 1-dimensional trivial $\mathfrak{A}_4$-module and M_3 the unique 3-dimensional simple $\mathfrak{A}_4$-module.

The fact that P is a sum of two non isomorphic simple modules can also be seen from the fact that if $\chi_P=\sum_{j=1}^m n_j\chi_j$ where $\chi_j\in Irr_{\mathbb{C}}(\mathfrak{A}_4)$ for all j,

$$\sum_{j=1}^m n_j^2=\sum_{j=1}^m n_j^2(\chi_j,\chi_j)=(\chi_P,\chi_P)=\frac{1}{12}(4^2+4\cdot1+4\cdot1)=2.$$

This example makes clear that it is possible to obtain for small groups the irreducible characters by these elementary considerations and the orthogonality relations. When the groups become bigger, it becomes quite complicated, even impossible to get the character table in this way. There is a very well working algorithm, the Dixon-Schneider algorithm, which is well suited to compute character tables on a computer for already quite complicated groups just by solving linear equations and eigenvalue problems. We shall present the mathematical principle of this algorithm in Section 2.4.4.

An easy remark is very useful. Let G and H be two groups and let $\varphi:G\longrightarrow H$ be a group homomorphism. If M is a KH-module, then M becomes a KG-module as well by putting for all $m\in M$ and $g\in G$ the action of g on m as $g\cdot m:=\varphi(g)m$. The axioms for

M being a KG-module are directly implied by φ being a group homomorphism and M being a KH-module. The same holds when A and B are two K-algebras, M is a B-module and $\varphi : A \longrightarrow B$ is a homomorphism of algebras. Then, M becomes an A-module by $a \cdot m := \varphi(a)m$ for every $m \in M$ and $a \in A$. Moreover, if φ is surjective, then M is a simple A-module whenever M is a simple B-module. Indeed, M is a simple B-module if and only if it is cyclic with respect to every non zero $m \in M$ as generator. Hence, this is the case if $M = B \cdot m$ for every $m \in M \setminus \{0\}$ and this implies $M = B \cdot m = \varphi(A) \cdot m = A \cdot m$ for every $m \in M \setminus \{0\}$, since φ is surjective. Call this process 'transport of structure from B to A'.

This implies a first attempt for small groups is however the following easy lemma.

Lemma 2.3.8. *Let G be a finite group and let K be a splitting field for G in which $|G|$ is invertible. Then, the number of characters of degree 1 for G equals the cardinality $|G/[G,G]|$ of the largest abelian quotient of G.*

Proof. Since the natural quotient homomorphism $\pi : G \longrightarrow G/[G,G]$ defines a G-module structure on each $K(G/[G,G])$-module by the remark preceding the lemma. Let M be a simple $KG/[G,G]$-module, then M is a simple KG-module as well, again using the remark preceding the lemma. Since $G/[G,G]$ is commutative, $KG/[G,G]$ is a commutative algebra. Wedderburn's theorem then implies that $KG/[G,G]$ is a direct product of copies of fields, and since K is a splitting field, the remark preceding the lemma implies as well that K is a splitting field for $G/[G,G]$. Hence, we get at least $|G/[G,G]|$ characters of degree 1. If G has a character χ of degree 1, then χ actually is a homomorphism of groups $G \longrightarrow K^\times$ from G to the multiplicative group of K since the trace function of a matrix ring of 1×1-matrices is the identity. Hence, χ factors through π, that is there is a mapping $\overline{\chi} : G/[G,G] \longrightarrow K$ such that $\chi = \overline{\chi} \circ \pi$. Hence, this one-dimensional character is a character coming from transport of structure of $KG/[G,G]$. This proves the lemma. □

A nice and elementary argument concerning the number of groups having n irreducibles characters uses an argument of Landau [Land-1903]. I learned the use of Landau's argument from an article of D. Passman [Pas-10].

Proposition 2.3.9. *Let K be an algebraically closed field of characteristic 0 and let $s \in \mathbb{N}$. Then there is a function $f(s)$ which depends only on s such that if KG has exactly s isomorphism classes of simple modules then $|G| \leq f(s)$.*

Proof. We know that the number of isomorphism classes of simple KG-modules is equal to the dimension of $Z(KG)$ and this is equal to the number of conjugacy classes $C_1, \dots, C_s$ of G. Now, $|C_i| \cdot |C_G(c_i)| = |G|$ for all $i \in \{1, \dots, s\}$ and for any choice of $c_i \in C_i$. Since

$$|G| = \sum_{i=1}^{s} |C_i|,$$

this gives

$$1=\sum_{i=1}^{s}\frac{1}{|C_G(c_i)|}$$

for $|C_G(c_i)| \leq |G|$ for all i. The set of positive integral solutions

$$(z_1,\ldots,z_s) \in \mathbb{N}^s \cap [1,|G|]^s$$

of the equation

$$1=\sum_{i=1}^{s}\frac{1}{z_i}$$

is a finite number $f(s)$. This shows the statement. □

2.4 Character degrees, Burnside's theorem and the Dixon-Schneider algorithm

A major application is the most famous theorem of Burnside's saying that groups of order divisible by only two primes are solvable. Though the statement does not mention at all any representations or characters the by far easiest proof of this statement uses character theory in a very essential way.

2.4.1 Integral elements in rings, rings of integers

We need some elementary notions in algebraic number theory. The reason for this is the following property that we have seen recently in Lemma 1.3.6, namely that the image of any group element $g \in G$ is conjugate to a diagonal matrix having only roots of unity in the main diagonals. Algebraic number theory has a general concept for this kind of elements.

Definition 2.4.1. Let R be a ring and let S be a ring containing R as a subring. An element $s \in S$ is *integral over R* if s commutes with all elements in R and if there is a polynomial

$$P(X)=X^n+a_{n-1}X^{n-1}+a_{n-2}X^{n-2}+\cdots+a_1X+a_0 \in R[X]$$

such that

$$P(s)=s^n+a_{n-1}s^{n-1}+a_{n-2}s^{n-2}+\cdots+a_1s+a_0=0.$$

The ring S is called integral over R if all of its elements are integral over R.

Example 2.4.2. Of course, R is integral over R since every $r \in R$ is root of $X - r \in R[X]$. Another example is $\mathbb{Z}[\zeta]$ where ζ is an n-th root of unity over $\mathbb{Q}$. Of course, ζ is a root of $X^n - 1 \in \mathbb{Z}[X]$. So, ζ is integral over $\mathbb{Z}$. We shall see that all elements in $\mathbb{Z}[\zeta]$ are actually integral over $\mathbb{Z}$, or in other words we shall prove that $\mathbb{Z}[\zeta]$ is integral over $\mathbb{Z}$. We shall see further properties on this ring in Lemma 7.1.5 below, when we will need a more precise statement.

Lemma 2.4.3. *Let R be a ring and let S be a ring containing R as a subring. Then, an element $s \in S$ which commutes with all elements of R is integral over R if and only if the smallest ring $R[s]$ containing R and s is a finitely generated R-module.*

Proof. If s is integral over R, there is a polynomial $P(X) \in R[X]$ with leading coefficient 1 and s as a root. Hence, there is an integer $n \in \mathbb{N}$ and elements $r_0, r_1, \ldots, r_{n-1}$ in R such that $s^n + r_{n-1}s^{n-1} + \cdots + r_1 s + r_0 = 0$. As a consequence,

$$R[s] = \{r_{n-1}s^{n-1} + \cdots + r_1 s + r_0 \mid r_0, r_1, \ldots, r_{n-1} \in R\}.$$

Indeed, this is a ring since for every $m \geq n$ the element s^m is an R-linear combination of the elements $s^{n-1}, s^{n-2}, \ldots, s, 1$. Hence, two elements of the form $r_{n-1}s^{n-1} + \cdots + r_1 s + r_0$ multiply and sum up again to an element of this form. Now, $1, s, s^2, \ldots, s^{n-1}$ generates $R[s]$.

If $R[s]$ is a finitely generated R-module, then there are elements $x_0, x_1, x_2, \ldots, x_m$ of $R[s]$ such that every other element $y \in R[s]$ is an R-linear combination of these elements $x_0, x_1, x_2, \ldots, x_m$. Now, $R[s]$ is an image of $R[X]$ sending X to s and extending this multiplicatively and hence elements in $R[s]$ are polynomials in s and coefficients in R. Therefore, there are $P_0, P_1, \ldots, P_m \in R[X]$ such that $x_i = P_i(s)$ for every $i = 0, 1, \ldots, m$. Let

$$n := max(deg(P_0), deg(P_1), \ldots, deg(P_m)) + 1$$

and then

$$s^n = r_0 P_0(s) + r_1 P_1(s) + \cdots + r_m P_m(s)$$

for some elements $r_0, r_1, \ldots, r_m$ in R. This implies that the degree of

$$r_0 P_0(X) + r_1 P_1(X) + \cdots + r_m P_m(X)$$

is at most $n - 1$ and

$$X^n - (r_0 P_0(X) + r_1 P_1(X) + \cdots + r_m P_m(X))$$

is a polynomial with leading coefficient 1 and s as a root. This proves the lemma. □

It can happen that all R-submodules of a finitely generated R-module M are finitely generated. This is the case for example when M and R are both finite dimensional K-vector spaces and R acts K-linearly on M. Another example for this is when R is a free abelian group of finite rank. Then, whenever M is a finitely generated R-module as well, the structure theorem of abelian groups of finite rank shows that every submodule of M is finitely generated.

Another instance are so-called Noetherian modules which are of fundamental importance in the deeper understanding of higher algebra. We refer to Section 6.1 for a short introduction. An elementary but more detailed treatment of this concept can be obtained from e. g. [Lang-84] or [Zim-14, Chapter 1].

Since for the moment we work over finite dimensional vector spaces we do not need this concept yet. In order to maintain the presentation at a level that is as accessible as possible, we shall introduce this concept when it is appropriate.

Lemma 2.4.4. *Let R be a subring of a ring S, suppose that every R-submodule of any finitely generated R-submodule of S is finitely generated and let $s_1, s_2 \in S$ be two elements of S which commute with all elements of R and which satisfy $s_1s_2 = s_2s_1$. If s_1 and s_2 are both integral over R, then $s_1 - s_2$ and s_1s_2 are integral over R as well.*

Proof. Since s_1 and s_2 are integral over R, by Lemma 2.4.3 $R[s_1]$ is finitely generated over R and $(R[s_1])[s_2]$ is finitely generated over $R[s_1]$. Hence, $(R[s_1])[s_2]$ is finitely generated over R. Since $R[s_1, s_2] = (R[s_1])[s_2]$ by definition, $s_1 - s_2 \in R[s_1, s_2]$ and also $s_1s_2 \in R[s_1, s_2]$. Since $R[s_1 - s_2]$ is an R-submodule of $R[s_1, s_2]$, and since $R[s_1, s_2]$ is finitely generated over R, also $R[s_1 - s_2]$ is finitely generated over R. Likewise $R[s_1s_2]$ is finitely generated over R. Lemma 2.4.3 proves the statement. □

Definition 2.4.5. Let S be a commutative ring and let R be a subring of S. Suppose that every R-submodule of S is finitely generated. Then $\mathrm{algint}_R(S) := \{s \in S \mid s \text{ is integral over R}\}$ is a ring, the *ring of algebraic integers* over R in S. If $R = \mathbb{Z}$ then $\mathrm{algint}_{\mathbb{Z}}(S)$ is the ring of algebraic integers.

2.4.2 Group rings and integrality; character degrees

We first see that forming the group ring of a finite group is an integral operation.

Proposition 2.4.6. *Let R be a ring formed by algebraic integers over $\mathbb{Z}$ and let G be a finite group. Then, all elements of RG are algebraic integers over $\mathbb{Z}$.*

Proof. Let $s = \sum_{g\in G} r_g g \in RG$. Then, put $S := \mathbb{Z}[r_g | g \in G]$ the smallest subring of R that contains all the coefficients that appear in s. Since all the coefficients are elements of R and all elements of R are algebraic integers, S is an abelian group of finite rank. As G is finite, $s \in SG$ is an element in the ring SG whose additive structure is an abelian group of finite rank. Hence $\mathbb{Z}[s] \subseteq SG$ and $\mathbb{Z}[s]$ is an abelian group of finite type. By Lemma 2.4.3 the element s is integral. □

Lemma 2.4.7. *Let G be a finite group. Then for every character χ of a complex representation of G and every $g \in G$ the value $\chi(g)$ is an algebraic integer.*

Proof. Let φ be the representation of G with character χ. Let $g \in G$. Since the field of complex numbers is algebraically closed, by Corollary 1.2.31 we may apply Lemma 1.3.6 to see that $\varphi(g)$ is conjugate to a diagonal matrix where all the coefficients are roots of 1. Roots of 1 in $\mathbb{C}$ are algebraic integers by Example 2.4.2. Hence, by Lemma 2.4.4 the character value $\chi(g)$ is integral over $\mathbb{Z}$. □

A final auxiliary lemma is needed.

Lemma 2.4.8. *Let K be a finite extension of $\mathbb{Q}$. Then* $\mathrm{algint}_{\mathbb{Z}}(K) \cap \mathbb{Q} = \mathbb{Z}$.

Proof. It is clear that $\mathbb{Z} \subseteq \mathrm{algint}_{\mathbb{Z}}(K) \cap \mathbb{Q}$ since every integer n is a root of $X - n \in \mathbb{Z}[X]$.

Conversely, let $u = \frac{n}{m}$ be integral over $\mathbb{Z}$ and suppose that $gcd(n, m) = 1$. Then u is a root of the polynomial $X^d + a_{d-1}X^{d-1} + \cdots + a_0$ for $a_i \in \mathbb{Z}$ for all $i \in \{0, \ldots, d-1\}$. Hence

$$\frac{n^d}{m} = -a_{d-1}n^{d-1} - a_{d-2}n^{d-2}m - \cdots - a_0 m^{d-1} \in \mathbb{Z}.$$

But this implies that m divides n^d, which contradicts the fact that $gcd(n, m) = 1$. □

We may use these properties to show that character degrees of irreducible characters divide the group order.

Theorem 2.4.9. *Let G be a finite group, let K be a splitting field in $\mathbb{C}$ of G and let S be a simple KG-module. Then $dim_K(S)$ divides $|G|$.*

Proof. By Proposition 2.2.8 the primitive central idempotent e_S in KG with $e_S S = S$ satisfies

$$\frac{|G|}{dim_K(S)} e_S = \sum_{g \in G} \chi_S(g^{-1})g$$

for χ_S being the character of S. By Lemma 2.4.7 this is in RG where R is the ring of algebraic integers in K.

Now, the mapping $Ke_S \longrightarrow K$ defined by $q \cdot e_S \longrightarrow q$ is an isomorphism of rings. So, since $(|G|/dim_K(S))e_S$ is integral over $\mathbb{Z}$, also $|G|/dim_K(S)$ is a rational number which is integral over $\mathbb{Z}$. Using Lemma 2.4.8 we get $|G|/dim_K(S) \in \mathbb{Z}$, which proves the statement. □

It is possible to give more precise statements on the character degree. Actually one can prove with basically the same method that the dimension of a simple KG-module for K being a splitting field of G of characteristic 0 divides the index of the centre of G in G. We are going to prove an even more precise statement, Ito's theorem, in Section 3 as a consequence of the very important Clifford theory.

2.4.3 Burnside's $p^a q^b$-theorem

Starting approximatively from 1960 towards 1985 a gigantic project was initiated to classify the finite simple groups up to isomorphism. The classification is an extremely complex proof and can be seen as one of the major achievements in algebra in the 20th century.

Definition 2.4.10. A group $G \neq 1$ is *simple* if there is no normal subgroup of G except the group G and the one element group $\{1\}$.

A group is *solvable* if there is a chain of subgroups

$$1 = N_0 \leq N_1 \leq N_2 \leq \dots N_n \leq N_{n+1} = G$$

of G such that N_i is normal in N_{i+1} and such that N_{i+1}/N_i is abelian for all $i \in \{0, 1, \dots, n\}$.

Let p be a prime number. A finite *p-group* is a group of order p^n for some n.

In some sense the class of solvable groups and the class of simple groups are the two extremes in the class of finite groups. In general it is quite difficult to show for a group for which only little information is known that it is solvable. Nevertheless, for a finite group there are conditions on the order which imply that a group is solvable.

Lemma 2.4.11. *Let G be a (not necessarily finite) group with a solvable normal subgroup N and with a solvable quotient G/N, then G is solvable.*

If G is solvable, then also all the quotients of G and all the subgroups of G are solvable.

Proof. Let N be a normal solvable subgroup and let G/N be solvable. We need to show that G is solvable. Let $\pi : G \longrightarrow G/N$ be the canonical projection. Then, since G/N is solvable, there is a chain of subgroups $1 = N/N \leq N_1/N \leq \dots \leq N_n/N \leq N_{n+1}/N = G/N$ of G/N which are successively normal in one another and such that $(N_{i+1}/N)/(N_i/N) \simeq N_{i+1}/N_i$ is abelian. Hence, we get a chain of subgroups $1 \leq N \leq N_1 \leq N_2 \leq \dots N_n \leq N_{n+1} = G$ such that N_{i+1}/N_i is abelian for all $i \geq 1$. Since N is solvable, there is a chain of subgroups $1 = M_0 \leq M_1 \leq \dots M_m \leq M_{m+1} = N$ such that M_i is normal in M_{i+1} and such that M_{i+1}/M_i is abelian. As a whole, G is solvable.

Let G be solvable and let S be a subgroup of G. Then, there is a chain of subgroups

$$1 = N_0 \leq N_1 \leq N_2 \leq \dots N_n \leq N_{n+1} = G$$

of G such that N_i is normal in N_{i+1} and such that N_{i+1}/N_i is abelian for all $i \in \{0, 1, \dots, n\}$. Then,

$$1 = N_0 \leq N_1 \cap S \leq N_2 \cap S \leq \dots N_n \cap S \leq N_{n+1} \cap S = S$$

is a chain of subgroups of S, one normal in the next, and $(N_{i+1} \cap S)/(N_i \cap S)$ is a subgroup of N_{i+1}/N_i. Since N_{i+1}/N_i is abelian, $(N_{i+1} \cap S)/(N_i \cap S)$ is abelian as well.

Let G be solvable and let M be a normal subgroup. Then, there is a chain of subgroups

$$1 = N_0 \leq N_1 \leq N_2 \leq \ldots N_n \leq N_{n+1} = G$$

of G such that N_i is normal in N_{i+1} and such that N_{i+1}/N_i is abelian for all $i \in \{0, 1, \ldots, n\}$. Let $\pi : G \longrightarrow G/M$ be the natural projection. Then,

$$1 = \pi N_0 \leq \pi N_1 \leq \pi N_2 \leq \ldots \pi N_n \leq \pi N_{n+1} = G/M$$

is a chain of subgroups of G/M such that $\pi(N_{i+1})/\pi(N_i)$ is a quotient induced by π of N_{i+1}/N_i. Since N_{i+1}/N_i is abelian, so is $\pi(N_{i+1})/\pi(N_i)$. This shows the lemma. □

Proposition 2.4.12. *Finite p-groups are solvable.*

Proof. As a first step one shows

Lemma 2.4.13. *The centre of a finite p-group is non trivial.*

Proof of Lemma 2.4.13. Let G be a p-group. Then, G acts on G by conjugation and the equivalence classes are precisely the conjugacy classes. An orbit is a conjugacy class, and the well-known orbit lemma implies that the size of the orbit of $x \in G$ equals $|G|/|C_G(x)|$. Here, $C_G(x)$ is the centraliser of x, i. e. the subgroup of formed by those elements of G which commute with x. Since $C_G(x)$ is a subgroup of G, the orbits under the conjugacy action are of length $p^{s(x)}$ for some integer $s(x)$ and $s(x) = 0 \Leftrightarrow x \in Z(G)$. Since orbits of two elements coincide or are disjoint, G is the disjoint union of the orbits, and so $|G| = \sum p^{s(x)}$. Since $|G| \equiv 0 \bmod p$, and since $s(1) = 0$, there must be at least $p - 1$ elements $x_1, \ldots, x_{p-1} \in G \setminus \{1\}$ with $s(x_i) = 0$ for all i. □

We continue the proof of Proposition 2.4.12 by induction on $|G|$. If $|G| = p$, then G is cyclic of order p and hence abelian. Trivially, abelian groups are solvable.

We assume that p-groups of order strictly less than $|G|$ are solvable. Since $|Z(G)| > 1$ by Lemma 2.4.13, and since $Z(G)$ is abelian, and since $G/Z(G)$ is solvable by induction, one sees that G is solvable by Lemma 2.4.11. This proves the Proposition. □

Can one do better? What about groups whose order is divisible by only 2 or only 3 primes?

Example 2.4.14. The symmetric group $\mathfrak{S}_5$ of order $120 = 2^3 \cdot 3 \cdot 5$ is not solvable. Our proof follows Artin [Art-72].

First we shall prove that when G is a subgroup of $\mathfrak{S}_m$, for $m \geq 5$ containing all 3-cycles and when N is a normal subgroup of G with abelian quotient G/N, then N contains all 3-cycles.

Let $(a\ b\ c)$ be any 3-cycle. We shall use the right multiplication of cycles on symbols, i. e. $a \cdot (a\ \ b\ \ c) = b$ and so $(1\ \ 2\ \ 3) \cdot (3\ \ 4\ \ 5) = (1\ \ 2\ \ 4\ \ 5\ \ 3)$ for example. Then $\{1, 2, 3, \ldots, m\} \setminus \{a, b, c\} \supseteq \{d, e\}$. Put $x := (d\ \ c\ \ a)$ and $y := (a\ \ e\ \ b)$. Then,

$x \cdot y \cdot x^{-1} \cdot y^{-1} = (a\ b\ c)$ and $x \cdot y \cdot x^{-1} \cdot y^{-1}$ is mapped to $1_{G/N}$ since G/N is abelian. So, $(a\ b\ c) \in N$.

Suppose now $\mathfrak{S}_m$ is solvable, then there is a chain of normal subgroups

$$1 = N_0 \leq N_1 \leq N_2 \leq \ldots N_n \leq N_{n+1} = \mathfrak{S}_m$$

of $\mathfrak{S}_m$ such that N_i is normal in N_{i+1} and such that N_{i+1}/N_i is abelian for all $i \in \{0, 1, \ldots, n\}$. Since $\mathfrak{S}_m$ contains all 3-cycles, so does N_n by what was shown before. Hence, this is true for N_{n-1} as well and by induction all of the groups $N_1, N_2, \ldots, N_n$ contain all the 3-cycles. But, $(1\ 2\ 3)$ and $(1\ 2\ 4)$ do not commute and so N_1 is not commutative. Contradiction.

Burnside's theorem closes the gap between p-groups and groups whose order is divisible by 3 primes. It is most remarkable that no simple proof of this result is known that does not use character theory in an essential way. A complicated proof without representation theory was given by Goldschmidt [Gol-70] and Matsuyama [Mat-73], and independently by Bender [Bend-72]. I am grateful to Martin Hertweck for drawing my attention to [Gol-70, Mat-73].

Theorem 2.4.15 (Burnside). *Let p and q be two prime numbers and let G be a group of order $p^a q^b$ for two integers a and b. Then G is solvable.*

We first need a preliminary statement.

Proposition 2.4.16. *Let G be a finite simple non abelian group and let $\chi : G \longrightarrow \mathbb{C}$ be a complex irreducible character of G. Let C be a conjugacy class of G different from $\{1\}$. Then,*

$$gcd(|C|, \chi(1)) = 1 \Rightarrow \chi(C) = 0 \text{ or } \chi \text{ is the trivial character.}$$

As usual we denote by gcd the greatest common divisor of two integers. Using this proposition the proof of Theorem 2.4.15 is not difficult. We shall give the proof of Proposition 2.4.16 after the proof of Theorem 2.4.15 which follows below.

Proof of Theorem 2.4.15. Let G be a group of minimal order such that the theorem is false.

Then, G is not solvable. Actually G is simple. Indeed, since if N is a non trivial normal subgroup of G, then G/N is again a group of order $p^{a'} q^{b'}$ for $a' \leq a$ and $b' \leq b$, where $a' + b' < a + b$ and by the minimality of the order of G one gets G/N is solvable. By the same reason, N is solvable and by Lemma 2.4.11 one gets that G is solvable. This is a contradiction to our hypothesis that the theorem is not true for G.

Let P be a Sylow p-subgroup of G. By Lemma 2.4.13 the centre $Z(P)$ is not trivial. Let $c \in Z(P) \setminus \{1\}$ and let C be the conjugacy class of c in G. Since c is central in P, the centraliser $C_P(c)$ contains P. Moreover, since G is simple, $Z(G) = \{1\}$ and so in particular $c \notin Z(G)$. Therefore, $C_G(c) \neq G$. Hence, p does not divide $|C| = |G|/|C_G(c)|$ and there is an integer $0 < b' \leq b$ such that $|C| = q^{b'}$.

Let S be a simple $\mathbb{C}G$-module. Suppose S is not the trivial module. Then, Lemma 2.3.8 implies that S is of dimension at least 2 over $\mathbb{C}$ since else there is a non trivial linear representation, which has a non trivial kernel. This kernel is a non trivial normal subgroup, which contradicts the fact that G is simple.

Suppose moreover $\chi_S(c) \neq 0$. Then Proposition 2.4.16 implies that q divides $\chi_S(1)$ and therefore, there is an integer $n_S \in \mathbb{N}$ such that $\chi_S(1) = q \cdot n_S$.

Let $S_1, S_2, \ldots, S_m$ be a list of representatives of the isomorphism classes of simple $\mathbb{C}G$-modules. Let S_1 be the trivial $\mathbb{C}G$-module. We compute

$$\begin{aligned} 0 = \chi_{\mathbb{C}G}(c) &= \sum_{i=1}^{m} \chi_{S_i}(c) dim_{\mathbb{C}}(S_i) \\ &= 1 + \sum_{i=2}^{m} \chi_{S_i}(c) dim_{\mathbb{C}}(S_i) \\ &= 1 + \sum_{i=2}^{m} \chi_{S_i}(c)\chi_{S_i}(1) \\ &= 1 + qz \end{aligned}$$

for some $z = \sum_{i=2}^{m} dim_{\mathbb{C}}(S_i)\chi_{S_i}(c)$. By Lemma 2.4.7 z is an algebraic integer.

But, the ring of algebraic integers in $\mathbb{C}$ is a ring R and qR is a non trivial ideal using Lemma 2.4.8. Therefore, $0 = 1 + qz$ does not have a solution $z \in R$. This contradiction proves the theorem. □

For the proof of Proposition 2.4.16 we need a technical lemma.

Lemma 2.4.17. *Let G be a finite group and let K be a splitting field of G. Suppose that the order of G is invertible in K and that S is a simple KG-module. Let C be a conjugacy class of G. Then, $|C|\frac{\chi_S(C)}{\chi_S(1)}$ is integral over $\mathbb{Z}$.*

Proof of Lemma 2.4.17. Let e_S be the primitive central idempotent of KG that acts as identity on S. Then, e_SKG is a full matrix ring over K and so $Z(e_SKG) = K$. Therefore, there is a $k \in K$ such that

$$e_S \cdot \sum_{g \in C} g = ke_S.$$

By Proposition 2.2.8 one has

$$e_S = \frac{\chi_S(1)}{|G|} \sum_{g \in G} \chi_S(g^{-1})g$$

and so

$$\frac{\chi_S(1)}{|G|}\Big(\sum_{g \in C} g\Big)\Big(\sum_{h \in G} \chi_S(h^{-1})h\Big) = k \cdot \frac{\chi_S(1)}{|G|} \sum_{g \in G} \chi_S(g^{-1})g.$$

We expand the term on the left and compare the coefficients of 1_G on the left and on the right to get

$$|C|\chi_S(C) = \chi_S(1)k.$$

Proposition 2.4.6 shows that $e_S \cdot \sum_{g\in C} g = k \cdot e_S$ is integral. Therefore, $k = |C|\frac{\chi_S(C)}{\chi(1)}$ is integral since multiplication by e_S gives a projection of $Z(KG)$ to a subring of $\mathbb{C}$. This proves the lemma. □

Proof of Proposition 2.4.16. Let S be a simple $\mathbb{C}G$-module such that $\chi_S = \chi$. Put $a := \frac{\chi(C)}{\chi(1)}$.

1. We are going to show that a is integral over $\mathbb{Z}$ and that $|a| \leq 1$ in $\mathbb{C}$.
   Since $gcd(\chi(1), |C|) = 1$ there are integers x and y such that $x \cdot |C| + y \cdot \chi(1) = 1$. We multiply this equation by a to get

$$\frac{\chi(C)}{\chi(1)} \cdot |C| \cdot x + \chi(C) \cdot y = a.$$

   Lemma 2.4.17 shows that $\frac{\chi(C)}{\chi(1)} \cdot |C| \cdot x$ is integral. Lemma 2.4.7 shows that $\chi(C)y$ is integral over $\mathbb{Z}$ and Lemma 2.4.4 shows that a is integral. Part 2) of Lemma 2.1.4 shows that $|a| \leq 1$.
2. We shall show that $|a| = 1$.
   Since a is algebraic, there is a minimal polynomial $P(X)$ over $\mathbb{Q}$. Let $P(X) \in \mathbb{Q}[X]$ be the minimal polynomial of a over $\mathbb{Q}$. Then there are elements $a = a_1, a_2, \ldots, a_n$ in $\mathbb{C}$ such that

$$P(X) = (X - a)(X - a_2) \cdots (X - a_n).$$

   Let ζ_m be a primitive m-th root of unity over $\mathbb{Q}$ and let Γ be the Galois group of $\mathbb{Q}(\zeta_m)$ over $\mathbb{Q}$. Then Γ is abelian; actually Γ is isomorphic to the group of invertible elements in $\mathbb{Z}/m\mathbb{Z}$.
   Put $K := \mathbb{Q}(a) \subseteq \mathbb{Q}(\zeta_m)$. Let Σ be the automorphism group of $\mathbb{Q}(a)$ over $\mathbb{Q}$. Then restriction of a field automorphism induces a homomorphism $\rho : \Gamma \longrightarrow \Sigma$. Since Γ is abelian, every subgroup of Γ is normal, and by the principal theorem of Galois theory ρ is surjective and K is a Galois extension of $\mathbb{Q}$. As P is the minimal polynomial over $\mathbb{Q}$ and as K is a Galois extension, $K \simeq \mathbb{Q}[X]/P(X)$. Moreover, again by Galois theory the group Σ acts transitively on the roots of P.
   Therefore, for each $i \in \{2, 3, \ldots, n\}$ there is a $\sigma_i \in \Gamma$ such that

$$\sigma_i(a) = a_i = \frac{\sigma_i(\chi_S(C))}{\chi_S(1)}.$$

   Since σ_i permutes the primitive m-th roots of unity and since $\chi_S(C)$ is a sum of $\chi_S(1)$ m-th roots of unity,

$$|a_i| = \frac{|\sigma_i(\chi_S(C))|}{\chi_S(1)} \leq 1$$

   for all i.

Now, $P(X) \in \mathbb{Q}[X]$ and $P(0) = (-1)^n \cdot a \cdot a_1 \cdot \dots \cdot a_n$. From the first step we use that a is integral over $\mathbb{Z}$ and so we deduce that all the elements $a_1, a_2, \dots, a_n$ are integral as well. Hence, $P(X) \in \mathbb{Z}[X]$. In particular, $P(0) \in \mathbb{Z}$. But,

$$1 = |P(0)| = |a| \cdot |a_2| \cdot \dots \cdot |a_n|$$

and so all the Galois conjugates $a, a_2, \dots, a_n$ are of absolute value 1.

3. We show that S is the trivial module.
Let H be the cyclic group generated by $c \in C$. Suppose S is not the trivial module. By Lemma 1.3.6 we know that there is a basis B of S consisting of eigenvectors of c. Moreover, B is of cardinality $dim(S)$, which is at least 2 since a simple group does not have a non trivial one-dimensional representation. Hence, all elements of H act via diagonal matrices on S. Therefore for each $b \in B$ the module $S_b := \mathbb{C}b$ is a $\mathbb{C}H$-module and when we consider S as $\mathbb{C}H$-module only,

$$S \simeq \bigoplus_{b \in B} S_b$$

as $\mathbb{C}H$-modules. Now $|a| = 1$, as it was shown in the previous step,

$$\chi_S(1) = |\chi(c)| = \left|\sum_{b \in B} \chi_{S_b}(c)\right|.$$

We know that each S_b is one-dimensional, and so $|\chi_{S_b}(c)| = 1$ for all $b \in B$. We get that either $\chi_{S_b}(c) = 1$ for all $b \in B$ or there is a root of unity ζ such that $\chi_{S_b}(c) = \zeta$ for all $b \in B$. In any case c acts on S by multiplication with either 1 or ζ, since it acts this way on the basis B. If c acts on S by multiplication by 1, then c is in the kernel of the homomorphism $\varphi : G \longrightarrow Gl_{|B|}(\mathbb{C})$ whose trace function is the character χ. Since G is simple, c acts as ζ on S. Again using that G is simple, we get that φ is injective. Hence G is a subgroup of $Gl_{|B|}(\mathbb{C})$ and c is identified with the diagonal matrix having ζ in the main diagonal. This matrix is central in $Gl_{|B|}(\mathbb{C})$ and so c is central in G. Contradiction.

This shows the proposition. □

2.4.4 The mathematical principle of the Dixon-Schneider algorithm

In this section we shall present another way to compute the character table of a finite group in a deterministic and algorithmic way. This method actually uses basically work of Burnside [Bur-11] of the beginning of the 20-th century.

Wedderburn's theorem shows that for a finite group G one has

$$\mathbb{C}G \simeq \prod_{i=1}^{s} Mat_{n_i \times n_i}(\mathbb{C})$$

as algebras. The centre of a matrix algebra are the multiples of the identity matrix:

$$Z(Mat_{n\times n}(K)) = \left\{ \lambda \cdot \begin{pmatrix} 1 & 0 & & \dots & \dots & 0 \\ 0 & 1 & 0 & & & \vdots \\ \vdots & 0 & \ddots & & \ddots & \vdots \\ \vdots & & \ddots & & \ddots & 0 \\ 0 & \dots & \dots & & 0 & 1 \end{pmatrix} \middle| \lambda \in K \right\}$$

Hence $Z(\mathbb{C}G) \simeq \mathbb{C}^s$ and by Lemma 1.3.3 a basis is given by $\{\sum_{g\in C_h} g \mid h \in G\}$ where $C_h := \{xhx^{-1} \mid x \in G\}$ is the conjugacy class as usual.

We abbreviate $K_h := \sum_{g\in C_h} g$ for every $h \in G$. Let G be the disjoint union of the conjugacy classes $C_{g_1}, C_{g_2}, \dots, C_{g_s}$:

$$G = C_{g_1} \coprod C_{g_2} \coprod \cdots \coprod C_{g_s}$$

Now, $Z(\mathbb{C}G)$ is an algebra and so there are $c_{i,j}^k \in \mathbb{C}$ such that

$$K_{g_i} \cdot K_{g_j} = \sum_{k=1}^{s} c_{i,j}^k K_{g_k}$$

The special structure of the elements K_g shows that in this case

$$c_{i,j}^k = \left|\{(x,y) \in C_{g_i} \times C_{g_j} \mid x \cdot y = g_k\}\right|.$$

Now, each element K_{g_i} is central in $\mathbb{C}G$. Let $e_1, e_2, \dots, e_s$ be the central primitive idempotents of KG. Recall that by Corollary 2.2.7 these are uniquely defined by this property. We claim that for all $m, i \in \{1, 2, \dots, s\}$ there are $\omega_m(C_{g_i}) \in \mathbb{C}$ such that $K_i e_m = \omega_m(C_{g_i})e_m$. Indeed, this follows from the fact that

$$K_i e_m \in Z(\mathbb{C}Ge_m) = Z(Mat_{n_m\times n_m}(\mathbb{C})) = \mathbb{C} \cdot \begin{pmatrix} 1 & 0 & & \dots & 0 \\ 0 & \ddots & & \ddots & \vdots \\ \vdots & \ddots & & \ddots & 0 \\ 0 & \dots & & 0 & 1 \end{pmatrix}.$$

But then, applying traces to the equation $K_i e_m = \omega_m(C_{g_i})e_m$, using that the trace of each element ge_m; $g \in C_{g_i}$ equals $\chi_m(C_{g_i})$, one gets

$$\chi_m(1) \cdot \omega_m(C_{g_i}) = n_m \cdot \omega_m(C_{g_i}) = |C_{g_i}| \cdot \chi_m(C_{g_i}).$$

Definition 2.4.18. The class functions $\omega_m : G \longrightarrow \mathbb{C}$ are the irreducible *central characters* of G.

Remark 2.4.19. As we have seen, irreducible central characters are just scalar multiples of the irreducible characters. Immediately one sees by this that the irreducible central characters form a basis for $CF_{\mathbb{C}}(G)$ since by Proposition 2.2.4 this is true for $Irr_{\mathbb{C}}(G)$ as well. In particular $\omega_i \neq \omega_j$ whenever $i \neq j$.

Lemma 2.4.20. *Let G be a finite group, let $C_1, \dots, C_s$ be the conjugacy classes of G and let $\omega_1, \omega_2, \dots, \omega_s$ be the irreducible central characters of G. Denote by $c_{i,j}^k$ the structure constants of the centre of $\mathbb{C}G$ with respect to the basis $\{\sum_{g \in C_i} g \mid i = 1, \dots, s\}$. Then for every $m \in \{1, 2, \dots, s\}$ the vector*

$$\hat{\omega}_m := \begin{pmatrix} \omega_m(C_1) \\ \omega_m(C_2) \\ \vdots \\ \omega_m(C_s) \end{pmatrix}$$

is an eigenvector for the matrix $(c_{i,j}^k)_{1 \le j,k \le s}$ to the eigenvalue $\omega_m(C_i)$.

Proof. Multiplying the equation $K_{g_i} \cdot K_{g_j} = \sum_{k=1}^s c_{i,j}^k K_{g_k}$ from above by $e_m = e_m^2$ one gets

$$\omega_m(C_i) \cdot \omega_m(C_j) = \sum_{k=1}^{s} c_{i,j}^k \omega_m(C_k)$$

This proves the lemma. □

Corollary 2.4.21. *The values of the irreducible central characters are algebraic integers.*

Proof. Indeed, $\omega_m(C_i)$ is an eigenvalue of the matrix $(c_{i,j}^k)_{j,k}$. Now, the remark of the beginning of this section shows that

$$c_{i,j}^k = \left|\{(x, y) \in C_{g_i} \times C_{g_j} \mid x \cdot y = g_k\}\right| \in \mathbb{N}.$$

But, then

$$\det\left(\lambda \cdot \begin{pmatrix} 1 & 0 & \dots & 0 \\ 0 & \ddots & \ddots & \vdots \\ \vdots & \ddots & \ddots & 0 \\ 0 & \dots & 0 & 1 \end{pmatrix} - \begin{pmatrix} c_{i,1}^1 & c_{i,1}^2 & \dots & c_{i,1}^s \\ c_{i,2}^1 & & & c_{i,2}^s \\ \vdots & & & \vdots \\ c_{i,s}^1 & \dots & \dots & c_{i,s}^s \end{pmatrix}\right) \in \mathbb{Z}[\lambda]$$

is a polynomial in λ with coefficients in $\mathbb{Z}$ and with leading coefficient 1. Moreover, by Lemma 2.4.20 the value $\omega_m(C_i)$ is a root of this polynomial. This shows the corollary. □

Usually eigenvectors are not unique, even up to scalars, but here considering all the s matrices $(c_{i,j}^k)_{1 \le j,k \le s}$ together, there is only one decomposition into eigenspaces.

Lemma 2.4.22. *The set of eigenvectors* $\{\hat{\omega}_1, \hat{\omega}_2, \ldots, \hat{\omega}_s\}$ *is the only set of vectors whose linear spans* $\{\mathbb{C}\hat{\omega}_1, \mathbb{C}\hat{\omega}_2, \ldots, \mathbb{C}\hat{\omega}_s\}$ *are simultaneous eigenspaces for all the matrices* $(c_{i,j}^k)_{1\le j,k\le s}$ *for* $i \in \{1, 2, \ldots, s\}$.

Proof. By Remark 2.4.19 the family $\{\hat{\omega}_1, \hat{\omega}_2, \ldots, \hat{\omega}_s\}$ is a linear independent family of s vectors.

Moreover, the mapping

$$\begin{aligned} Z(\mathbb{C}G) &\xrightarrow{\gamma} Mat_{s\times s}(\mathbb{C}) \\ K_{g_i} &\mapsto (c_{i,j}^k)_{1\le j,k\le s} \end{aligned}$$

is a ring homomorphism. Indeed, this is actually the definition of structure constants: The basis elements multiply exactly as the structure constant matrices. Therefore, by the very definition of a $Z(\mathbb{C}G)$-module, γ induces a $Z(\mathbb{C}G)$-module structure on $\mathbb{C}^s$. Call Γ this $Z(\mathbb{C}G)$-module of dimension s given by γ.

But,

$$Z(\mathbb{C}G) \simeq \underbrace{\mathbb{C} \times \mathbb{C} \times \cdots \times \mathbb{C}}_{s \text{ factors}}$$

is semisimple, and therefore this representation induced by γ is semisimple as well. We already have s different eigenvectors $\{\hat{\omega}_1, \hat{\omega}_2, \ldots, \hat{\omega}_s\}$ which give a decomposition of Γ into a direct sum of one-dimensional submodules

$$\Gamma = \mathbb{C}\hat{\omega}_1 \oplus \mathbb{C}\hat{\omega}_2 \oplus \cdots \oplus \mathbb{C}\hat{\omega}_s.$$

The matrices $(c_{i,j}^k)_{1\le j,k\le s}$ act on the space $\mathbb{C}\hat{\omega}_m$ by multiplication with their eigenvalues, that is $\omega_m(C_i)$. We have however that

$$\omega_m(C_i) = \omega_n(C_i) \forall 1 \le i \le s \Rightarrow m = n$$

by Remark 2.4.19.

This shows the lemma. □

We are ready to give a description of the way the Dixon-Schneider algorithm computes the character table of a group.

Let G be a finite group. First one computes the conjugacy classes $C_1, C_2, \ldots, C_s$ of G and from this the basis elements

$$K_1 = \sum_{g\in C_1} g, K_2 = \sum_{g\in C_2} g, \ldots, K_s = \sum_{g\in C_s} g$$

of the centre of $\mathbb{C}G$. Then the structure constants $c_{i,j}^k$ of the centre of $\mathbb{C}G$ with respect to the basis $K_1, K_2, \ldots, K_s$ are computed. This can be done by multiplying explicitly two of the elements K_i and K_j and solve the linear equation to express the result in the basis $\{K_1, K_2, \ldots, K_s\}$ by Gauss' matrix reduction algorithm.

How to find the class of the neutral group element 1 from the knowledge of the structure constants $c_{i,j}^k$? We know that $K_iK_1 = K_i$ for all i, and K_1 is characterised by this equation. Hence, there is a unique $k_0 \in \{1, 2, \dots, s\}$ such that $c_{i,k_0}^j = \delta_{i,j}$. This k_0 corresponds to the $1 \in G$. We shall rearrange the conjugacy classes such that $k_0 = 1$.

In a second step one looks for inverses. For every $i \in \{1, 2, \dots, s\}$ there is a unique $i^* \in \{1, 2, \dots, s\}$ such that $x \in C_i \Rightarrow x^{-1} \in C_{i^*}$. This can be detected by the knowledge of the structure constants $c_{i,j}^k$ only. Actually, $c_{i,j}^1 \neq 0 \Rightarrow j = i^*$. To see this choose $g_i \in C_i$. Then,

$$K_iK_j = \Big(\sum_{g \in G/C_G(g_i)} gg_ig^{-1}\Big) \cdot \Big(\sum_{h \in G/C_G(g_j)} hg_jh^{-1}\Big)$$

has a coefficient of K_1 different from 0 exactly when there are two elements $g, h \in G$ such that $gg_ig^{-1} \cdot hg_jh^{-1} = 1$. This implies $g_ig^{-1}h = g^{-1}hg_j^{-1}$ and therefore g_i is conjugate to g_j^{-1}.

The knowledge of the structure constants $c_{i,j}^k$ and the order of the group G even implies the knowledge of the size of the conjugacy classes. In fact,

$$\begin{aligned} c_{i,i^*}^1 &= \big|\{(x, y) \in C_{g_i} \times C_{g_{i^*}} \mid x \cdot y = 1\}\big| \\ &= \big|\{(\overline{g}, \overline{h}) \in G/C_G(g_i) \times G/C_G(g_{i^*}) \mid gg_ig^{-1} \cdot hg_{i^*}h^{-1} = 1\}\big| \\ &= \big|\{(\overline{g}, \overline{h}) \in G/C_G(g_i) \times G/C_G(g_{i^*}) \mid gg_ig^{-1} = hg_ih^{-1}\}\big| \\ &= \big|C_G(g_i)\big| \\ &= \frac{|G|}{|C_{g_i}|} \end{aligned}$$

We still need an efficient way to compute the eigenspaces of the matrices $(c_{i,j}^k)_{1 \le j,k \le s}$. A method to turn this into an efficient computation on a computer was actually given by Dixon [Dix-67] for the first time, reducing the computation into a computation in a finite prime field. Schneider [Sch-90] then gave a more efficient way to compute the eigenspaces. The reduction to the eigenspace problem is due to Burnside.

How can one compute the character degrees from the structure constants $c_{i,j}^k$? For this we use the orthogonality relations of the columns of the character table Theorem 2.3.5. Actually,

$$\begin{aligned} \sum_{i=1}^s \frac{\omega_m(C_i)}{|C_i|} \cdot \overline{\omega_m(C_i)} &= \sum_{i=1}^s \frac{\chi_m(C_i)}{\chi_m(C_1)} \cdot \frac{|C_i| \cdot \overline{\chi_m(C_i)}}{\chi_m(C_1)} \\ &= \frac{1}{(\chi_m(C_1))^2} \sum_{g \in G} \chi_m(g)\chi_m(g^{-1}) \\ &= \frac{|G|}{(\chi_m(C_1))^2} \end{aligned}$$

and so $\chi_m(C_1)$ can be computed once $|G|$ is known, and the eigenvectors $\hat{\omega}_1, \hat{\omega}_2, \dots, \hat{\omega}_s$ of the matrices $(c_{i,j}^k)_{1 \le j,k \le s}$. Observe that since eigenvectors are unique only up to a scalar one needs to normalise one coefficient. Since for our central characters ω_m one gets $\omega_m(C_1) = 1$, one normalises the eigenspaces in the way that the first coefficient is always 1.

This completes the preliminaries we need to perform the Dixon-Schneider algorithm:

1. The input is the group order and the structure constants $c_{i,j}^k$; $i, j, k \in \{1, 2, \ldots, s\}$.
2. Determine the class of the element 1_G. Reorganise the data such that this class is the first index C_1.
3. Determine the bijection $\{1, \ldots, s\} \ni i \mapsto i^* \in \{1, \ldots, s\}$.
4. Determine $|C_i|$ for all $i \in \{1, \ldots, s\}$.
5. Determine the eigenvectors $\hat{\omega}_1, \hat{\omega}_2, \ldots, \hat{\omega}_s$ of the matrices $(c_{i,j}^k)_{j,k\in\{1,2,\ldots,s\}}$ and normalise the eigenvectors in such a way that $\omega_m(C_1) = 1$, or in other words such that the first coefficient of $\hat{\omega}_m$ is 1 for all $m \in \{1, 2, \ldots, s\}$.
6. Compute the character degrees $\chi_m(C_1)$.
7. The irreducible characters are given by $\chi_m(C_i) := \omega_m(C_i)\chi_m(C_1)/|C_i|$.

We come to Dixon's reduction, how to perform efficiently step (5), (6) and (7).

Put $M_i := (c_{i,j}^k)_{1\le j,k\le s}$ for all $i \in \{1, 2, \ldots, s\}$. By Lemma 2.1.3 we know that the character values are sums of $|G|$-th roots of unity. On the other hand, the multiplicative group of a finite field is cyclic.

Hence, denoting by $\mathbb{F}_p$ the prime field with p elements, for every prime p with $|G|$ divides $p - 1$ one has a surjective ring homomorphism $\mathbb{Z}[\zeta_{|G|}] \longrightarrow \mathbb{F}_p$. Recall that we denote by $\zeta_{|G|}$ a primitive $|G|$-th root of unity in $\mathbb{C}$.

Look for a generating element $z \in \mathbb{F}_p^*$, that is a generating element z of the multiplicative group of $\mathbb{F}_p$. Then an explicit surjective ring homomorphism is given by

$$\begin{aligned} \theta : \mathbb{Z}[\zeta_{|G|}] &\longrightarrow \mathbb{F}_p \\ 1 &\mapsto 1 \\ \zeta_{|G|} &\mapsto z^u \end{aligned}$$

where $u := \frac{p-1}{|G|}$. For all $\lambda \in \{1, 2, \ldots, p-1, p\}$ compute a basis of $\ker(M_i - \lambda 1)$ modulo p, where 1 denotes the unit matrix.

One knows that each M_i is diagonalisable, and so there is a basis consisting of eigenvectors.

Suppose we have written

$$\mathbb{F}_p^s = \bigoplus_{i=1}^k V_i$$

where V_i are simultaneous eigenspaces of $M_1, M_2, \ldots, M_r$ and $r < s$.

For each i_0 with $dim_{\mathbb{F}_p} V_{i_0} > 1$ choose a decomposition into eigenspaces of M_{r+1}. At latest when $r = s$ each of the spaces V_i is one-dimensional.

Normalise the basis of the eigenspaces (using step 6 above) in a way that

$$\chi_m(C_1)^2 = \frac{|G|}{\sum_{i=1}^{s} \frac{1}{|C_i|}\omega_m(C_i)\omega_m(C_{i^*})}.$$

Since $p > |G| > (\chi_m(C_1))^2$ the value $\chi_m(C_1)$ is fixed by its congruence modulo p.

Hence

$$\chi_m(C_i) \equiv \frac{1}{|C_i|}\omega_m(C_i)\chi_m(C_1) \bmod p.$$

How to get $\chi_m(C_i)$ from $\chi_m(C_i) \bmod p$?

Put $n_m := \chi_m(C_1)$. We know that $\chi_m(x) = \sum_{i=1}^{n_m} \zeta^{s_i}$ for ζ being a fixed $|G|$-th root of unity and $s_i \in \mathbb{N}$. Moreover, $\chi_m(x^n) = \sum_{i=1}^{n_m} \zeta^{n \cdot s_i}$. Express $\chi(x) = \sum_{j=0}^{|G|-1} \mu_j \zeta^j$ with non negative integers μ_j. We need to determine these coefficients μ_j. Observe that

$$\sum_{j=0}^{|G|-1} \zeta^{j \cdot t} = \begin{cases} \sum_{j=0}^{|G|-1} (\zeta^t)^j = \sum_{i=0}^{|G|-1} 1 = |G| & \text{if } |G| \text{ divides } t \\ 0 & \text{else} \end{cases}$$

This computation yields

$$\mu_j = \frac{1}{|G|} \sum_{j=0}^{|G|-1} \chi(x^k)\zeta^{-jk}$$

which gives modulo p

$$\mu_j \equiv \frac{1}{|G|} \sum_{j=0}^{|G|-1} \theta(\chi(x^k))z^{-jk} \bmod p$$

Since $\mu_j \le \chi_m(C_1) \le \sqrt{|G|} < p$, then μ_j is the preimage of the value $\frac{1}{|G|} \sum_{j=0}^{|G|-1} \theta(\chi(x^k))z^{-jk}$ in $\mathbb{N}$ in the range between 0 and p.

There is one problem still to solve. Can one choose a big enough prime p such that $|G|$ divides $p - 1$? The answer is Dirichlet's theorem on the distribution of primes in arithmetic progressions. This will be the subject of Section 2.5.

2.5 Application to number theory; Dirichlet's theorem on arithmetic progressions

In this section we shall give an application of character theory to elementary number theory. We shall prove Dirichlet's theorem dating from 1837 which shows that in any arithmetic progression $P_{r,m} := \{mn + r \mid n \in \mathbb{N}\}$ for $r, m \in \mathbb{Z} \setminus \{0\}$, $gcd(r, m) = 1$ there is an infinite number of primes. The proof we are going to present is due to de la Vallée-Poussin from 1896. Since characters are complex valued functions, it seems to be natural to use them for arguments using real and complex analysis. Dirichlet's theorem is an example. Except Proposition 2.5.3, which was used in Section 2.4.4, the results of this section are not needed in the sequel anymore. The proof of Proposition 2.5.3 uses only elementary algebra and makes no use of complex analysis. The general case however uses many nice arguments from complex analysis. Nevertheless, I feel that this section gives a nice and quite surprising way of seeing representation theoretic arguments applied in mathematical subjects not very close to representation theory a priori. Moreover, the proof we present here is a very beautiful interplay between complex analysis in one variable and group theory. Since those parts which do use the complex analysis are not needed in the sequel there should not occur any difficulty for readers who do not share my enthusiasm with this method. Actually this application was historically one of the main driving engines to develop representation theory.

Theorem 2.5.1 (Dirichlet). *Let r and m be non zero integers without a non trivial common divisor. Then there is are infinitely many prime numbers of the form $mn+r$ for $n \in \mathbb{N}$.*

The proof uses in an essential way some character theory of abelian groups. This seems to be quite astonishing since the statement has not really any link to representation theory.

First we shall deal with the case $r = 1$. This can be done by a very elementary argument, and actually this is also the case we need for the Dixon-Schneider algorithm in Section 2.4.4. We shall use a variant of the proof given in Washington's book [Was-97].

Remark 2.5.2. The proof we shall give below for the case $r = 1$ is a modern adaption of Euclid's proof for the infinity of the set of primes. The proof we shall give for general r is due to de La Vallée Poussin. His proof is a huge sophistication of Euler's proof for the divergence of the series $\sum_{p \text{ prime}} \frac{1}{p}$, and hence the infinity of the set of primes.

Proposition 2.5.3. *Dirichlet's theorem 2.5.1 holds for $r = 1$.*

Proof of the case $r = 1$. Recall that for every $m \in \mathbb{N}$ we have

$$X^m - 1 = \prod_{d|m} \Phi_d(X)$$

for irreducible polynomials $\Phi_d(X) \in \mathbb{Z}[X]$, the cyclotomic polynomials (see for example Lang's book on general algebra [Lang-84]). Moreover, $\Phi_d(0) = 1$ for all $d > 1$. We suppose that we already know prime numbers $p_0, p_1, \dots, p_{r-1}$ for $r \geq 0$ such that $p_i \equiv 1 \bmod m$ for all i. (The index set was chosen such that the case of no integer with the required property known is included.) We shall construct p_r with this property. Let

$$x_r := b \cdot m \cdot p_0 \cdot p_1 \cdot \dots \cdot p_{r-1}$$

where b is a sufficiently large integer such that whenever for a complex number z one has $\Phi_m(z) \in \{1, -1\}$ then $|z| \leq bm$. The existence of such a b follows from the fact that the two polynomials $\Phi_m(X) + 1$ and $\Phi_m(X) - 1$ only have finitely many roots. This condition implies $\Phi_m(x_r) \notin \{1, -1\}$ since $|x_r| > bm$. If $\Phi_m(x_r) = 0$, then $x_r^m - 1 = 0$, using that $\Phi_m(X)$ divides $X^m - 1$ and therefore $|x_r|^m > (bm)^m > bm > 1$. A contradiction. Hence $\Phi_m(x_r) \neq 0$. Since $\Phi_m(X) \in \mathbb{Z}[X]$ there is a prime number p_r dividing $\Phi_m(x_r)$. As x_r is a multiple of m, one gets $\Phi_m(x_r) \equiv \Phi_m(0) = 1 \bmod m$ and so $p_r \nmid m$.

We claim that if for an integer x a prime p divides $\Phi_m(x)$ but not m, then $p \equiv 1 \bmod m$. Indeed, $p|\Phi_m(x)$ implies $x^m \equiv 1 \bmod p$. For every divisor $d > 1$ of m one has

$$(x^m - 1)/(x^{\frac{m}{d}} - 1) = 1 + x^{\frac{m}{d}} + \cdots + x^{(d-1)\frac{m}{d}}.$$

If p divides $x^{m/d} - 1$, then $(x^m - 1)/(x^{\frac{m}{d}} - 1) \equiv d \bmod p$. Now, $\Phi_m(X)$ divides $(x^m - 1)/(x^{\frac{m}{d}} - 1)$ whenever $d > 1$ and so there is a polynomial $G_{m,d} \in \mathbb{Z}[X]$ with

$$(x^m - 1)/(x^{\frac{m}{d}} - 1) = \Phi_m(x) \cdot G_{m,d}(x)$$

which implies that p divides d. Since p divides d and d divides m, the prime p divides m, which was excluded. Therefore, p does not divide $x^{m/d} - 1$ for every proper divisor $d > 1$ of m. Hence, x is of order m in $\mathbb{Z}/p\mathbb{Z}$. Since the unit group of $\mathbb{Z}/p\mathbb{Z}$ is of order $p - 1$ this implies that m divides $p - 1$. As a consequence $p \equiv 1 \bmod m$.

This applies in particular to our prime p_r. Since $x_r \equiv 0 \bmod p_0 p_1 \dots p_{r-1}$, and therefore $\Phi_m(x_r) \equiv \Phi_m(0) = 1 \bmod p_0 p_1 \dots p_{r-1}$, one gets p_r is a prime different from each of the p_i for all $i \in \{0, 1, \dots, r - 1\}$. This shows the proposition. □

For the proof in general one needs to use some character theory, and this character theory is mixed with arguments purely from complex analysis to conclude.

The way to do so goes via Dirichlet characters and L-functions. As usual call $R^\times$ the unit group of a ring R.

Definition 2.5.4. Let $m \in \mathbb{Z} \setminus \{0, 1, -1\}$. Then one defines for every irreducible character $\chi : (\mathbb{Z}/m\mathbb{Z})^\times \longrightarrow \mathbb{C}$ the associated *Dirichlet character* χ^D by

$$\mathbb{Z} \xrightarrow{\chi^D} \mathbb{C}$$
$$n \mapsto \begin{cases} \chi(n + m\mathbb{Z}) & \text{if } gcd(n, m) = 1 \\ 0 & \text{else} \end{cases}$$

Call χ^D a *Dirichlet character defined modulo m.*

We remark that if n divides m, then there is a group homomorphism $(\mathbb{Z}/m\mathbb{Z})^\times \longrightarrow (\mathbb{Z}/n\mathbb{Z})^\times$. Hence, every irreducible character χ of $(\mathbb{Z}/n\mathbb{Z})^\times$ gives rise to an irreducible character χ' of $(\mathbb{Z}/m\mathbb{Z})^\times$ by composing with this group homomorphism. If k is an integer such that $gcd(k, m) = 1$, then $\chi^D(k) = \chi'^D(k)$. Also if $gcd(k, n) \neq 1$, then $\chi^D(k) = \chi'^D(k) = 0$. But, if k is relatively prime to n but not relatively prime to m, then the value of the Dirichlet character, seen as character modulo m, evaluates k to 0: $\chi'^D(k) = 0$. Since necessarily k is in the kernel of the projection $(\mathbb{Z}/m\mathbb{Z})^\times \longrightarrow (\mathbb{Z}/n\mathbb{Z})^\times$ we get for the value of the Dirichlet character $\chi^D(k) = 1$.

Definition 2.5.5. Let χ be a Dirichlet character. Then, the *conductor* f_χ of χ is the smallest positive integer m such that χ is a Dirichlet character defined modulo m.

We are going to define complex valued functions by characters. Let χ be a Dirichlet character defined modulo m and an $s \in \mathbb{C}$ with real part $Re(s) > 1$. Then when $Re(s) \geq 1+\varepsilon$ for an $\varepsilon > 0$ one gets

$$|n^s| = |\exp(s\log(n))| = \exp(Re(s)\log(n)) \geq \exp(\log(n)\cdot(1+\varepsilon)).$$

Hence

$$\left|\sum_{n=k}^{m}\frac{\chi(n)}{n^s}\right| \leq \sum_{n=k}^{m}\left|\frac{\chi(n)}{n^s}\right| = \sum_{n=k}^{m}\frac{1}{|n^s|} = \sum_{n=k}^{m}\frac{1}{n^{Re(s)}} \leq \sum_{n=k}^{m}\frac{1}{n^{1+\varepsilon}}.$$

It is a classical fact that $\sum_{n=1}^{\infty}\frac{1}{n^s}$ converges for all real values $s > 1$. Hence, the series $\sum_{n=1}^{\infty}\frac{\chi(n)}{n^s}$ is a Cauchy series for those $s \in \mathbb{C}$ with real part $Re(s) > 1$, and converges therefore for every $s \in \mathbb{C}$ with real part $Re(s) > 1$.

By the same argument the Riemann ζ-function

$$\zeta(s) := \sum_{i=1}^{\infty}\frac{1}{n^s}$$

converges for $s \in \mathbb{C}$ with real part $Re(s) > 1$.

Definition 2.5.6. Let χ be a Dirichlet character defined modulo m. Then for every $s \in \mathbb{C}$ with $Re(s) > 1$ define

$$L(s,\chi) := \sum_{n=1}^{\infty}\frac{\chi(n)}{n^s}$$

the Dirichlet L-function.

It should be mentioned that one can define and study different types of generalised L-functions. We limit ourselves at the moment to Dirichlet L-functions.

The proof of Theorem 2.5.1 proceeds by analysing the degrees of the pôle at 1 of Dirichlet L-functions. It is therefore necessary to study the pôle in detail.

For this define the series $G(s) := \sum_{i=1}^{\infty}(-1)^n\frac{1}{n^s}$ which converges, by the theorem on alternating sums, as soon as the sequence $1/|n^s|$ is decreasing for fixed s. Hence, $G(s)$ converges for all $s \in \mathbb{C}$ with $Re(s) > 0$. Suppose now that $Re(s) > 1$. Then $G(s)$ converges absolutely and,

$$\begin{aligned} G(s) &= \sum_{n=1}^{\infty}(-1)^n\frac{1}{n^s} \\ &= \sum_{n=1}^{\infty}\frac{1}{(2n)^s} - \sum_{n=1}^{\infty}\left(\frac{1}{n^s} - \frac{1}{(2n)^s}\right) \end{aligned}$$

$$= \left(\frac{2}{2^s} - 1\right)\zeta(s)$$

Furthermore, $G(1) = \sum_{n=1}^{\infty}(-1)^n\frac{1}{n} = -\log(2)$ by the Taylor series of the function $\log(1+x)$.

Hence the Riemann ζ function has a pôle at $s = 1$ of order 1 and

$$\zeta(s) = \frac{1}{2^{(1-s)} - 1} \cdot G(s).$$

Let χ_0 be the principal Dirichlet character modulo m, that is $\chi_0(k) = 1$ if $gcd(k, m) = 1$ and $\chi_0(k) = 0$ when $gcd(k, m) > 1$.

We first need a product expression of the defining series of $L(\chi_0, s)$.

Recall that a number theoretically multiplicative function is a real valued function on natural numbers having the property that $f(nm) = f(n)f(m)$ whenever n and m do not have a non trivial common divisor.

Lemma 2.5.7. *Let $f : \mathbb{N} \longrightarrow \mathbb{R}$ be a non negative (in the number theoretical sense) multiplicative function and suppose that there is a real constant c such that for all primes p and all $k \in \mathbb{N}$ one has $f(p^k) < c$. Then $\sum_{n=1}^{\infty}\frac{f(n)}{n^s}$ converges for all s with $Re(s) > 1$ and*

$$\sum_{n=1}^{\infty}\frac{f(n)}{n^s} = \prod_{p\ prime}\left(1 + \sum_{k=1}^{\infty}\frac{f(p^k)}{p^{ks}}\right).$$

Proof. If all terms converge absolutely, the infinite product and the series are equal by the unique decomposition of integers into products of powers of primes.

We shall assume for a moment that s is real. The complex case is then a consequence, taking absolute values.

Put $a(p) := \sum_{k=1}^{\infty}\frac{f(p^k)}{p^{ks}}$. Then

$$a(p) = \frac{1}{p^s}\sum_{k=0}^{\infty}\frac{f(p^{k+1})}{p^{ks}} \le \frac{1}{p^s}\sum_{k=0}^{\infty}\frac{c}{p^{ks}} = \frac{c}{p^s(1-\frac{1}{p^s})} < \frac{2c}{p^s}$$

and therefore, using that $\sum_{p\text{ prime}}\frac{1}{p^s} < \sum_{n=1}^{\infty}\frac{1}{n^s}$, which converges as soon as $Re(s) > 1$,

$$\sum_{p\text{ prime},p\le N} a(p) < 2c\sum_{p\text{ prime}}\frac{1}{p^s} =: M < \infty.$$

Hence

$$\prod_{p\text{ prime},p\le N}(1 + a(p)) < \prod_{p\text{ prime},p\le N}\exp(a(p)) = \exp\left(\sum_{p\text{ prime},p\le N} a(p)\right)$$

Therefore,

$$\sum_{n=1}^{\infty} f(n)/n^s \le \exp(M)$$

and since f is non negative, $\sum_{n=1}^{\infty} f(n)/n^s$ converges. This shows the lemma. □

As an application of Lemma 2.5.7 one puts $f = \chi_0$, the principal Dirichlet character modulo m, and sees

$$\begin{aligned}
L(s,\chi_0) &= \prod_{p \text{ prime}} \frac{1}{1-\frac{\chi_0(p)}{p^s}} \\
&= \left(\prod_{p\nmid m} \frac{1}{1-\frac{\chi_0(p)}{p^s}}\right)\cdot\left(\prod_{p\mid m} \frac{1}{1-\frac{\chi_0(p)}{p^s}}\right) \\
&= \left(\prod_{p\nmid m} \frac{1}{1-\frac{\chi_0(p)}{p^s}}\right)\cdot\left(\prod_{p\mid m} \frac{1-\frac{1}{p^s}}{1-\frac{1}{p^s}}\right) \\
&= \left(\prod_{p\nmid m} \frac{1}{1-\frac{1}{p^s}}\right)\cdot\left(\prod_{p\mid m} \frac{1}{1-\frac{1}{p^s}}\right)\cdot\left(\prod_{p\mid m}\left(1-\frac{1}{p^s}\right)\right) \\
&= \left(\prod_{p\mid m}\left(1-\frac{1}{p^s}\right)\right)\zeta(s)
\end{aligned}$$

The term $(\prod_{p|m}(1-\frac{1}{p^s}))$ is a finite product and therefore everywhere convergent. The term $\zeta(s)$ has a simple pôle at $s = 1$ and defines a holomorphic function everywhere else on the complex half-plane $Re(s) > 0$.

For this we need a technical lemma.

Lemma 2.5.8. *Let $(a_n)_{n\in\mathbb{N}}$ and $(b_n)_{n\in\mathbb{N}}$ sequences of complex numbers and suppose that the series $\sum_{n=1}^{\infty} a_n b_n$ converges. Put $A_n := \sum_{i=1}^{n} a_i$ and suppose*

$$\lim_{n\to\infty} A_n b_n = 0.$$

Then

$$\sum_{n=1}^{\infty} a_n b_n = \sum_{n=1}^{\infty} A_n(b_n - b_{n+1}).$$

Proof. $S_N := \sum_{n=1}^{N} a_n b_n$. Put $A_0 = 0$.

$$\begin{aligned}
S_N &= \sum_{n=1}^{N} (A_n - A_{n-1}) b_n \\
&= \sum_{n=1}^{N} A_n b_n - \sum_{n=1}^{N} A_{n-1} b_n \\
&= \sum_{n=1}^{N} A_n b_n - \sum_{n=0}^{N-1} A_n b_{n+1} \\
&= A_N b_N + \sum_{n=0}^{N-1} A_n(b_n - b_{n+1})
\end{aligned}$$

This proves the lemma. ☐

Proposition 2.5.9. *The function $\zeta(s)-\frac{1}{(s-1)}$ has an analytic expansion into the half-plane* $\{s \in \mathbb{C} | \, Re(s) > 0\}$.

Proof.

$$\zeta(s) = \sum_{n=1}^{\infty} \frac{1}{n^s} = \sum_{n=1}^{\infty} n\left(\frac{1}{n^s} - \frac{1}{(n+1)^s}\right)$$

by Lemma 2.5.8.

For each $x \in \mathbb{R}$ put $[x] := max\{n \in \mathbb{Z} | \, n \leq x\}$ and $< x > := x - [x]$.

$$\begin{aligned}
\zeta(s) &= s \sum_{n=1}^{\infty} n \int_{x=n}^{n+1} \frac{1}{x^{1+s}} dx \\
&= s \sum_{n=1}^{\infty} \int_{x=n}^{n+1} [x] \frac{1}{x^{1+s}} dx \\
&= s \int_{x=1}^{\infty} [x] \frac{1}{x^{1+s}} dx \\
&= s \int_{x=1}^{\infty} \frac{1}{x^s} dx - s \int_{x=1}^{\infty} < x > \frac{1}{x^{1+s}} dx \\
&= \frac{s}{s-1} - s \int_{x=1}^{\infty} < x > \frac{1}{x^{1+s}} dx
\end{aligned}$$

As $| < x > | \leq 1$, the last integral converges and defines an analytic function as soon as $Re(s) > 0$. This shows the proposition. □

Lemma 2.5.10. *Let χ be a non trivial Dirichlet character modulo m. Then for all $N > 0$ one has $|\sum_{n=0}^{N} \chi(n)| \leq |(\mathbb{Z}/m\mathbb{Z})^{\times}| =: \varphi(m)$.*

Proof. Let $N = am + b$ for $0 \leq b < m$. Trivially $\chi(n + m) = \chi(n)$ for all $n \in \mathbb{Z}$. Hence

$$\sum_{n=1}^{N} \chi(n) = a \cdot \left(\sum_{n=0}^{m-1} \chi(n)\right) + \sum_{n=0}^{b} \chi(n).$$

Since χ is not the trivial character, its values sum up to 0, and so $\sum_{n=0}^{m-1} \chi(n) = 0$. Hence,

$$\left|\sum_{n=1}^{N} \chi(n)\right| = \left|\sum_{n=0}^{b} \chi(n)\right| \leq \sum_{n=0}^{b} |\chi(n)| \leq \varphi(m).$$

This shows the lemma. □

Proposition 2.5.11. *Let χ be a non trivial Dirichlet character modulo m. Then $L(s,\chi)$ has an analytic extension on $\{s \in \mathbb{C} | \, Re(s) > 0\}$.*

Proof. Define for every $x \in \mathbb{R}$ the function $S(x) := \sum_{n \le x} \chi(n)$. Lemma 2.5.8 shows that

$$\begin{aligned} L(s,\chi) &= \sum_{n=1}^{\infty} S(n)\left(\frac{1}{n^s} - \frac{1}{(n+1)^s}\right) \\ &= s \sum_{n=1}^{\infty} S(n) \int_{x=n}^{n+1} \frac{1}{x^{s+1}} dx \\ &= s \int_{x=1}^{\infty} S(x) \frac{1}{x^{s+1}} dx \end{aligned}$$

Since $|S(x)| \le \varphi(m)$ for all x, the integral converges for all s with $Re(s) > 0$. This proves the proposition. □

We define an auxiliary function

$$G(s,\chi) := \sum_{p \text{ prime}} \sum_{k=1}^{\infty} \frac{\chi(p^k)}{kp^{ks}}.$$

We remark that $|\chi(p^k)/(kp^{ks})| \le 1/p^{k \cdot Re(s)}$. Using that $\zeta(s)$ converges on the half-plane $Re(s) > 1$ and even uniformly on the half-plane $Re(s) > 1 + \delta$ for every positive δ, we obtain therefore that the series $G(s,\chi)$ converges on $\{s \in \mathbb{C} | \ Re(s) > 1\}$. Moreover, $G(s,\chi)$ is continuous in $\{s \in \mathbb{C} | \ Re(s) > 1\}$.

For all $z \in \mathbb{C}$ with $|z| \le 1$ one gets, using the complex Taylor series of the principal branch of $\log(1 - z)$,

$$\exp\left(\sum_{k=1}^{\infty} \frac{z^k}{k}\right) = \frac{1}{1-z}$$

and so, for $z = \chi(p)/p^s$,

$$\exp\left(\sum_{k=1}^{\infty} \frac{\chi(p^k)}{kp^{ks}}\right) = \frac{1}{1 - \frac{\chi(p)}{p^s}}$$

which implies

$$\begin{aligned} \exp(G(s,\chi)) &= \exp\left(\sum_{p \text{ prime}} \sum_{k=1}^{\infty} \frac{\chi(p^k)}{kp^{ks}}\right) \\ &= \prod_{p \text{ prime}} \exp\left(\sum_{k=1}^{\infty} \frac{\chi(p^k)}{kp^{ks}}\right) \\ &= \prod_{p \text{ prime}} \frac{1}{1 - \frac{\chi(p)}{p^s}} \\ &= L(s,\chi) \end{aligned}$$

Proposition 2.5.12. *Define the product* $F(s) := \prod_\chi L(s,\chi)$, *where* χ *runs through all Dirichlet characters modulo m. If* $s \in \mathbb{R}$ *and* $s > 1$, *then* $F(s)$ *is real and* $F(s) \geq 1$.

Proof.

$$\begin{aligned}\sum_\chi G(s,\chi) &= \sum_p \sum_{k=1}^{\infty} \sum_\chi \frac{\chi(p^k)}{kp^{ks}} \\ &= \sum_{p;p^k \equiv 1 \bmod m} \sum_{k=1}^{\infty} \frac{\varphi(m)}{kp^{ks}} \geq 0 \quad \text{since } s > 1\end{aligned}$$

Hence, $\prod_\chi L(s,\chi) = \exp(\sum_\chi G(s,\chi)) \geq e^0 = 1$. □

Proposition 2.5.13. *Let* χ_c *be a non trivial character modulo m and suppose* $\operatorname{im}(\chi_c) \not\subseteq \mathbb{R}$. *Then* $L(1,\chi_c) \neq 0$.

Proof. If s is real,

$$\overline{L(s,\chi_c)} = \sum_{n=1}^{\infty} \frac{\overline{\chi_c(n)}}{n^s} = L(s,\overline{\chi_c}).$$

Hence, if $L(1,\chi_c) = 0$, then $L(1,\overline{\chi_c}) = 0$. We know by Proposition 2.5.9 that $L(s,\chi_0)$ has a pôle of first order at $s = 1$ and by Proposition 2.5.11 all the other functions $L(s,\chi)$ for χ not the trivial character are analytic around $s = 1$. Hence, if $L(1,\chi_c) = 0$, also $L(1,\overline{\chi_c}) = 0$ and therefore two of the analytic functions in the product defining F vanish, whereas only one has a simple pôle, which implies $F(1) = 0$. But, $F(s) \geq 1$ for all $s > 1$ and F is continuous. This is a contradiction. □

We are left with real valued characters, i. e. $\operatorname{im}\chi \in \{0, 1, -1\}$.

Proposition 2.5.14. *Let* χ_r *be a real valued non trivial Dirichlet character modulo m. Then* $L(1,\chi_r) \neq 0$.

Proof. Suppose $L(1,\chi_r) = 0$. Put

$$\psi(s) := \frac{L(s,\chi_r)L(s,\chi_0)}{L(2s,\chi_0)}.$$

Since $L(s,\chi_0)$ has a simple pôle at $s = 1$ and since $L(1,\chi_r) = 0$ by hypothesis, $L(s,\chi_r)L(s,\chi_0)$ is bounded at $s = 1$, even analytic in $\{s \in \mathbb{C} |\ Re(s) > 0\}$. The function $L(2s,\chi_0)$ is analytic in $\{s \in \mathbb{C} |\ Re(s) > \frac{1}{2}\}$ and so $\psi(s)$ is analytic in $\{s \in \mathbb{C} |\ Re(s) > \frac{1}{2}\}$. Since $L(2s,\chi_0)$ has a pôle at $s = 1/2$, one gets $\lim_{s\to\frac{1}{2}} \psi(s) = 0$.

Let s be a real and $s > 1$.

$$\psi(s) = \prod_p \left(\frac{1}{1 - \frac{\chi_r(p)}{p^s}}\right) \cdot \left(\frac{1}{1 - \frac{\chi_0(p)}{p^s}}\right)\left(1 - \frac{\chi_0(p)}{p^{2s}}\right)$$

$$= \prod_{p\nmid m} \frac{(1-\frac{1}{p^{2s}})}{(1-\frac{\chi_r(p)}{p^s})(1-\frac{1}{p^s})}$$

Observe

$$\chi_r(p) = -1 \Rightarrow \frac{(1-\frac{1}{p^{2s}})}{(1-\frac{\chi_r(p)}{p^s})(1-\frac{1}{p^s})} = 1$$

and so

$$\begin{aligned}
\psi(s) &= \prod_{p;\chi(p)=1} \frac{1+\frac{1}{p^s}}{1-\frac{1}{p^s}} \\
&= \prod_{p;\chi(p)=1} \left(1+\frac{1}{p^s}\right)\sum_{k=0}^{\infty}\frac{1}{p^{ks}} \\
&= \prod_{p;\chi(p)=1} \left(1+\frac{2}{p^s}+\frac{2}{p^{2s}}+\frac{2}{p^{3s}}+\frac{2}{p^{4s}}+\cdots\right).
\end{aligned}$$

Lemma 2.5.7 shows that $\psi(s) = \sum_{n=1}^{\infty}\frac{a_n}{n^s}$ converges for the a_n obtained by the above formula.

Let $s \in \mathbb{C}$. Since $\psi(s)$ is analytic in $\{s \in \mathbb{C}|\ Re(s) > \frac{1}{2}\}$, and since the distance from 2 to the first possible singularity is 3/2, the function $\psi(s)$ can be expressed as $\psi(s) = \sum_{m=0}^{\infty} b_m(s-2)^m$ for s satisfying $|s-2| < \frac{3}{2}$.

We know by Taylor's theorem that $b_m = \frac{\psi^{(m)}(2)}{m!}$ where as usual $\psi^{(m)}$ denotes the m-th derivative. Since $\psi_n(s) = \sum_{n=1}^{\infty}\frac{a_n}{n^s}$ one gets

$$\psi^{(m)}(2) = \sum_{n=1}^{\infty}\frac{a_n}{n^2}(-\log_e(n))^m = (-1)^m c_m$$

for some $c_m \geq 0$.

Hence, $\psi(s) = \sum_{n=0}^{\infty} c_m(2-s)^m$ for $c_m \geq 0$ and

$$c_0 = \psi(2) = \sum_{n=1}^{\infty}\frac{a_n}{n^2} \geq a_1 = 1.$$

As a consequence, on $\{x \in \mathbb{R}|\ \frac{1}{2} < x < 2\}$ we have $\psi(s) \geq 1$ and on the other hand, $\lim_{s\to\frac{1}{2}} \psi(s) = 0$, as was shown at the beginning of the proof. This contradiction proves the proposition. □

We are now in the position to prove the Theorem 2.5.1.

We have

$$G(s,\chi) = \sum_{p \text{ prime}}\sum_{k=1}^{\infty}\frac{\chi(p^k)}{kp^{ks}}$$

$$= \sum_p \frac{\chi(p)}{p^s} + \sum_{p \text{ prime}} \sum_{k=2}^{\infty} \frac{\chi(p^k)}{kp^{ks}}$$
$$= \sum_p \frac{\chi(p)}{p^s} + g_\chi(s)$$

Now, $g_\chi(s)$ is absolutely convergent for all s with $Re(s) > \frac{1}{2}$; and uniformly convergent in $Re(s) > \frac{1}{2} + \delta$; since

$$\left|\frac{\chi(p^k)}{kp^{ks}}\right| = \left|\frac{1}{kp^{ks}}\right| \le \frac{1}{k}\left(\frac{1}{p^{Re(s)}}\right)^k$$

and the corresponding series converges as soon as $(\frac{1}{p^{Re(s)}}) < 1$ for all p, i. e. the range $Re(s) > \frac{1}{2}$ is sufficient.

Hence, for r fixed and relatively prime to m one gets

$$\sum_\chi \chi(r^{-1})G(s,\chi) = \left(\sum_p \sum_\chi \chi(r^{-1})\frac{\chi(p)}{p^s}\right) + \sum_\chi \chi(r^{-1})g_\chi(s).$$

Put $\sum_\chi \chi(r^{-1})g_\chi(s) =: g(s)$. The function $g(s)$ is again analytic in the half-plane $Re(s) > \frac{1}{2}$. Moreover, $\sum_\chi \chi(b) = 0$ if $b \not\equiv 1 \bmod m$ and $\sum_\chi \chi(b) = \varphi(m)$ if $b \equiv 1 \bmod m$. Hence

$$\sum_\chi \chi(r^{-1})G(s,\chi) = \sum_{p \equiv r \bmod m} \frac{\varphi(m)}{p^s} + g(s).$$

We consider the limit $s \longrightarrow 1$. As g is analytic in the half-plane $Re(s) > \frac{1}{2}$, $g(1)$ is a complex value. But we know that $L(s,\chi_0)$ has a simple pôle at $s = 1$, and therefore, since $\exp(G(s,\chi_0)) = L(s,\chi_0)$, also $G(s,\chi_0)$ has a simple pôle at $s = 1$. Moreover Proposition 2.5.13 and Proposition 2.5.14 show that $L(1,\chi) \ne 0$ whenever χ is not the principal character, and therefore $G(1,\chi)$ is a complex number for non principal characters χ.

Therefore, $\sum_\chi \chi(r^{-1})G(s,\chi)$ is unbounded when $s \to 1$. This means

$$\sum_{p \equiv r \bmod m} \frac{\varphi(m)}{p^s}$$

is unbounded when $s \to 1$. So, the sum must be an infinite series, and therefore there must be infinitely many primes p with $p \equiv r \bmod m$. This proves the theorem. □

2.6 Exercises

Exercise 2.1. Let G be a group which has the following character table, where $\zeta = e^{\frac{2\pi i}{7}}$.

	C_1	C_2	C_3	C_4	C_5	C_6
χ_1	1	1	1	1	1	1
χ_2	3	−1	1	$\zeta+\zeta^2+\zeta^4$	$\zeta^3+\zeta^5+\zeta^6$	0
χ_3	3	−1	1	$\zeta^3+\zeta^5+\zeta^6$	$\zeta+\zeta^2+\zeta^4$	0
χ_4	6	2	0	−1	−1	0
χ_5	7	−1	−1	0	0	1
χ_6	8	0	0	1	1	−1

a) Find the order of G and the order of each of the conjugacy classes C_i for $i \in \{1,\ldots,6\}$.
b) Show that G is simple.

Exercise 2.2. Let G be a finite group.
a) Show that the set $G =: X$ is a G-set if one defines an action of G on X by $g \cdot x := gxg^{-1}$ for each $g \in G$ and $x \in X$.
b) Let V be the $\mathbb{C}$-vector space of dimension $|X|$ with basis X. Show that V becomes a $\mathbb{C}G$-module by defining

$$\left(\sum_{g\in G} z_g g\right)\cdot\left(\sum_{x\in X} y_x x\right) := \sum_{g\in G}\sum_{x\in X} z_g y_x g\cdot x$$

for all $z_g, y_x \in \mathbb{C}$ and $x \in X$ and $g \in G$. (Hint: Take first $g \in G$ and extend then by linearity.)
c) Express the character value χ_V of this $\mathbb{C}G$-module V on each element $g \in G$ by a function which uses only the internal group law of G.
d) Let S be a simple $\mathbb{C}G$-module. Compute the scalar product (χ_S, χ_V) and show that the sum of the values of each row of the character table is a non-negative integer.

Exercise 2.3. Let $\mathfrak{S}_3$ be the symmetric group of order 6 and let C_3 be the cyclic group of order 3. Let $1, c, c^2$ be the elements of C_3 and let $G := \mathfrak{S}_3 \times C_3$.
a) Find the conjugacy classes of G.
b) Find the character table of G.
c) Find the centre of G.
d) Denote by $\overline{n}$ the class of $n \in \mathbb{Z}$ in $\mathbb{Z}/7\mathbb{Z}$. Let χ_- be the non trivial character of $\mathfrak{S}_3$ of degree 1. Show that G acts on $\mathbb{Z}/7\mathbb{Z}$ by $(\sigma, c^i)\cdot\overline{n} := \overline{\chi_-(\sigma)\cdot 2^i\cdot n}$.
Find the orbits of this action.
Denote by M_7 the $\mathbb{C}G$-permutation module associated with this action.
e) Find the character χ_{M_7} of M_7.

f) Show that χ_{M_7} is sum of 7 irreducible characters, and amongst them twice the trivial $\mathbb{C}G$-character.
g) Decompose χ_{M_7} in irreducible $\mathbb{C}G$-characters.

Exercise 2.4. Let ζ_8 be a primitive 8-th root of unity in $\mathbb{C}$. We find a character table of some group G.

	C_1	C_2	C_3	C_4	C_5	C_6	C_7	C_8
χ_1	1	1	1	1	1	1	1	1
χ_2	1	1	1	1	1	−1	−1	−1
χ_3	2	2	−1	−1	2	0	0	0
χ_4	2	−2	1	−1	0	$-\zeta_8-\zeta_8^3$	$\zeta_8+\zeta_8^3$	0
χ_5	2	−2	1	−1	0	$\zeta_8+\zeta_8^3$	$-\zeta_8-\zeta_8^3$	0
χ_6	3	3	0	0	−1	1	1	−1
χ_7	3	3	0	0	−1	−1	−1	1
χ_8	4	−4	−1	1	0	0	0	0

a) Find $|G|$.
b) Find the kernels of the irreducible characters of G in terms of the conjugacy classes of G.
c) Find the normal subgroups of G in terms of the conjugacy classes of G.
d) Find the centre $Z(G)$ of G in terms of the conjugacy classes of G.
e) Find the size of the conjugacy classes of G.
f) Find the centre of $G/Z(G)$ as well as the number of conjugacy classes of $G/Z(G)$.
g) Find the normal subgroups of $G/Z(G)$ in terms of the conjugacy classes of $G/Z(G)$.
h) Denote in the sequel $\overline{G} := G/Z(G)$. Let $\overline{N}$ be a normal subgroup of $\overline{G}$ different from $\{1\}$ and from $G/Z(G)$. Is the group $\overline{G}/\overline{N}$ abelien?
i) Find the character table of $\overline{G}/\overline{N}$.

Exercise 2.5. Let $x := \sqrt{2} \cdot i$, be a complex root of $X^2 + 2 = 0$. We find the following character table.

	C_1	C_2	C_3	C_4	C_5	C_6	C_7	C_8
χ_1	1	1	1	1	1	1	1	1
χ_2	1	−1	1	1	1	−1	−1	1
χ_3	9	−1	1	1	0	1	1	−1
χ_4	9	1	1	1	0	−1	−1	−1
χ_5	10	0	−2	2	1	0	0	0
χ_6	10	0	0	−2	1	$-x$	x	0
χ_7	10	0	0	−2	1	x	$-x$	0
χ_8	16	0	0	0	−2	0	0	1

a) Find the order of G.
b) Find the cardinality of each of the conjugacy classes of G.
c) Find the normal subgroups of G and determine the centre of G.
d) Is G isomorphic to the symmetric group of degree 6?

Exercise 2.6. We study three groups G_1, G_2 and G_3 with character tables CT_1, CT_2 and CT_3. Let ζ_n be a primitive n-th root of unity in $\mathbb{C}$.

CT_1:

	C_1^1	C_2^1	C_3^1
χ_1^1	1	1	1
χ_2^1	1	−1	1
χ_3^1	2	0	−1

CT_2:

	C_1^2	C_2^2	C_3^2	C_4^2	C_5^2
χ_1^2	1	1	1	1	1
χ_2^2	3	−1	$-\zeta_5-\zeta_5^{-1}$	$-\zeta_5^2-\zeta_5^3$	0
χ_3^2	3	−1	$-\zeta_5^2-\zeta_5^3$	$-\zeta_5-\zeta_5^{-1}$	0
χ_4^2	4	0	−1	−1	1
χ_5^2	5	1	0	0	−1

CT_3:

	C_1^3	C_2^3	C_3^3	C_4^3	C_5^3	C_6^3	C_7^3	C_8^3	C_9^3
χ_1^3	1	1	1	1	1	1	1	1	1
χ_2^3	7	−1	−2	1	1	1	0	0	0
χ_3^3	7	−1	1	$-\zeta_9^4-\zeta_9^5$	x	$-\zeta_9^2-\zeta_9^7$	0	0	0
χ_4^3	7	−1	1	$-\zeta_9^2-\zeta_9^7$	$-\zeta_9^4-\zeta_9^5$	x	0	0	0
χ_5^3	7	−1	1	x	$-\zeta_9^2-\zeta_9^7$	$-\zeta_9^4-\zeta_9^5$	0	0	0
χ_6^3	8	0	−1	−1	−1	−1	1	1	1
χ_7^3	9	1	0	0	0	0	$\zeta_7^3+\zeta_7^4$	$\zeta_7+\zeta_7^6$	$\zeta_7^2+\zeta_7^5$
χ_8^3	9	1	0	0	0	0	$\zeta_7^2+\zeta_7^5$	$\zeta_7^3+\zeta_7^4$	$\zeta_7+\zeta_7^6$
χ_9^3	9	1	0	0	0	0	$\zeta_7+\zeta_7^6$	$\zeta_7^2+\zeta_7^5$	$\zeta_7^3+\zeta_7^4$

for $x := \zeta_9^2+\zeta_9^4+\zeta_9^5+\zeta_9^7$.

a) Find the order of G_1, of G_2 and of G_3.
b) Show $G_1 \simeq \mathfrak{S}_3$.
c) Do the groups G_2 and G_3 possess a non-trivial normal subgroup?
d) Prove that each $g \in G_i$ is conjugate in G_i to its inverse g^{-1} for all $i \in \{1,2,3\}$.
e) What are the orders of the elements in C_1^2, C_3^2, C_4^2 and C_5^2?
f) It is known that there is a unique simple group of order 60. Show that G_2 is isomorphic to the alternating group of degree 5.
g) One knows that $G_1 < G_2 < G_3$. Find the restriction of the irreducible characters of G_2 to G_1 and decompose $\chi_i^2 \downarrow_{G_1}^{G_2}$ in a sum of irreducible characters.

Exercise 2.7 (with solution in Chapter 9). Consider the field $\mathbb{F}_7 = \mathbb{Z}/7\mathbb{Z}$ with 7 elements and the affine line $A^1(\mathbb{F}_7)$ on this field. The group $(\mathbb{F}_7 \setminus \{0\}, \cdot)$ acts on the set $\mathbb{F}_7$ by

$$\begin{aligned} \mathbb{F}_7 \setminus \{0\} \times \mathbb{F}_7 &\longrightarrow \mathbb{F}_7 \\ (a, v) &\mapsto av \end{aligned}$$

and the group $(\mathbb{F}_7, +)$ acts on the set $\mathbb{F}_7$ by

$$\begin{aligned} \mathbb{F}_7 \times \mathbb{F}_7 &\longrightarrow \mathbb{F}_7 \\ (x, v) &\mapsto x + v. \end{aligned}$$

Let $\delta_a(v) := av$ for $(a, v) \in \mathbb{F}_7 \setminus \{0\} \times \mathbb{F}_7$ and $\tau_x(v) := x + v$ for $(x, v) \in \mathbb{F}_7 \times \mathbb{F}_7$. Denote by $G := Aff^1(\mathbb{F}_7)$ the subgroup of $\mathfrak{S}_7$ generated by $\{\delta_a, \tau_x | \ a \in \mathbb{F}_7 \setminus \{0\}; x \in \mathbb{F}_7\}$. Let $\Delta := \{\delta_a | \ a \in \mathbb{F}_7 \setminus \{0\}\}$ and $T := \{\tau_x | \ x \in \mathbb{F}_7\}$.

a) Show $\delta_a \tau_x = \tau_{x \cdot a} \delta_a$. Deduce that each element of G can be written in the form $\tau_x \delta_a$ for an $(a, x) \in \mathbb{F}_7 \setminus \{0\} \times \mathbb{F}_7$. Show that T is a normal subgroup of G, that Δ is a group as well and that $\Delta \simeq G/T$.
b) Determine the order of G and show that G has a cyclic subgroup of order 6.
c) Show that T is the union of two conjugacy classes of G.
d) Find the conjugacy classes of G and their respective cardinality.
e) Find the degrees of the irreducible complex characters of G.
f) Find an element g of $\mathbb{F}_7$ such that $\mathbb{F}_7 \setminus \{0\} = \{g^n | \ n \in \mathbb{N}\}$.
g) Find the character table of G.

Exercise 2.8. Let G be a group which has the following character table:

	C_1	C_2	C_3	C_4	C_5	C_6	C_7
χ_1	1	1	1	1	1	1	1
χ_2	1	−1	1	1	−1	1	−1
χ_3	1	−1	j^2	1	$-j^2$	j	$-j$
χ_4	1	−1	j	1	$-j$	j^2	$-j^2$
χ_5	1	1	j^2	1	j^2	j	j
χ_6	1	1	j	1	j	j^2	j^2
χ_7	6	0	0	−1	0	0	0

where j is a primitive third root of unity.

a) Find the order of G.
b) For each $i \in \{1, \ldots, 7\}$ find the number of elements n_i in each conjugacy class C_i.
c) Is G abelian?
d) Show that $G/[G, G]$ is cyclic of order 6.
e) Show that G has a normal subgroup N which is cyclic of order 7.

f) For each $i \in \{1, \dots, 7\}$ find the order of the elements in C_i.
g) Show that G is isomorphic to the group of the affine line $\mathbb{F}_7$, i. e.

$$G \simeq (\mathbb{F}_7, +) \rtimes (\mathbb{F}_7^*, \cdot)$$

where $(x, \lambda) \cdot (y, \mu) := (x + \lambda y, \lambda\mu)$ for each $x, y \in \mathbb{F}_7$ and $\lambda, \mu \in \mathbb{F}_7^*$.

Exercise 2.9. Let D_4 be the dihedral group of order 8, let Q_8 be the quaternion group of order 8 and let C_3 be the cyclic group of order 3. Denote by $G_1 = D_4 \times C_3$ the direct product of D_4 and C_3 and by $G_2 = Q_8 \times C_3$ the direct product of Q_8 and C_3.
a) Find the conjugacy classes of G_1 and of G_2.
b) Find the degrees of the irreducible complex characters of G_1 and of G_2.
c) Find the character table of G_1 and of G_2.
d) Show that the centre of G_1 is isomorphic to the centre of G_2.
e) Let $\mathcal{N}(G_1)$ be the set of normal subgroups of G_1 and let $\mathcal{N}(G_2)$ be the set of normal subgroups of G_2. The sets $\mathcal{N}(G_1)$ and $\mathcal{N}(G_2)$ are partially ordered by inclusion of subgroups. Show that there is a bijection $\varphi : \mathcal{N}(G_1) \longrightarrow \mathcal{N}(G_2)$ which preserves the partial order:

$$\forall N, M \in \mathcal{N}(G_1) : \; N \subseteq M \Longrightarrow \varphi(N) \subseteq \varphi(M).$$

f) Show that the map φ preserves the order of the subgroups. Does it preserves the structure of the subgroups?

Exercise 2.10. Let ζ be a primitive 8-th root of unity in $\mathbb{C}$. We find the character table of some group G:

	C_1	C_2	C_3	C_4	C_5	C_6	C_7
χ_1	1	1	1	1	1	1	1
χ_2	1	-1	-1	1	1	1	1
χ_3	1	-1	1	1	1	-1	-1
χ_4	1	1	-1	1	1	-1	-1
χ_5	2	0	0	-2	2	0	0
χ_6	2	0	0	0	-2	$\zeta^3 - \zeta$	$\zeta - \zeta^3$
χ_7	2	0	0	0	-2	$\zeta - \zeta^3$	$\zeta^3 - \zeta$

a) Find the cardinality of the conjugacy classes of G.
b) Find the centre $Z(G)$ of G in terms of the conjugacy classes of G.
c) Show that $G/Z(G)$ is not abelian.
d) Find the partially ordered set of normal subgroups of G and the order of each of the normal subgroups.

e) Find the centre $Z(G/Z(G))$ of $G/Z(G)$ and show that $(G/Z(G))/(Z(G/Z(G)))$ is a group of exponent 2.
f) Find the partially ordered set of normal subgroups of $G/Z(G)$.
g) Let H be a group and let N be a normal subgroup of H. Let $\pi : H \longrightarrow H/N$ be the canonical homomorphism. If M is a normal subgroup of H/N show that $\pi^{-1}(M)$ is a normal subgroup of H.
h) Let $\pi : G \longrightarrow G/Z(G)$ and $\psi : G/Z(G) \longrightarrow (G/Z(G))/(Z(G/Z(G)))$ be the canonical homomorphisms. Find those normal subgroups of G which are of the form $\pi^{-1}(M)$ and those which are of the form $(\psi\circ\pi)^{-1}(L)$ for normal subgroups M of $G/Z(G)$ and normal subgroups L of $(G/Z(G))/(Z(G/Z(G)))$.

Exercise 2.11. Let G be a non-abelian group of order 12 and let S_2 be a Sylow 2-subgroup of G and let S_3 be a Sylow 3-subgroup of G.

a) Suppose for all questions ai) that $S_2 \trianglelefteq G$.
   a1) Considering the possible degrees of the irreducible complex characters of G show that G has 9 or 4 conjugacy classes.
   a2) If $S_3 \trianglelefteq G$, show that G is abelian. Deduce that there are four Sylow 3-subgroups of G and hence 8 elements of order 3 in G.
   a3) Use the orbit lemma and the fact that two elements are conjugate if and only if their inverse are conjugate as well to prove that a conjugacy class of elements of order 3 has either 2 or 4 elements.
   a4) Deduce that G has 4 conjugacy classes.
   a5) Find the degrees of the irreducible characters of G.
b) Suppose in all questions bi) that $S_3 \trianglelefteq G$.
   b1) Show that G has at least four irreducible complex characters of degree 1.
   b2) Deduce that G allows one or two irreducible complex characters of degree 2 and eight or four complex irreducible characters of degree 1.
   b3) Deduce from the fact that G is not abelian that S_3 is not in the centre of G.
   b4) Deduce that S_3 is the union of two conjugacy classes of G.
   b5) Use the commutator quotient to show that G cannot have 9 conjugacy classes.
   b6) Show that G has two irreducible complex characters of degree 2 and four irreducible complex characters of degree 1.
c) If S_3 is not normal in G, show that $S_2 \trianglelefteq G$.

Exercise 2.12. Let G be a group and let ζ be a root of $X^2 + X + 1$ in $\mathbb{C}$. The character table of G is

	C_1	C_2	C_3	C_4	C_5	C_6	C_7	C_8	C_9	C_{10}	C_{11}
χ_1	1	1	1	1	1	1	1	1	1	1	1
χ_2	1	1	ζ^2	1	1	ζ^2	ζ	1	ζ^2	ζ	ζ
χ_3	1	1	ζ	1	1	ζ	ζ^2	1	ζ	ζ^2	ζ^2
χ_4	1	ζ^2	1	1	ζ	ζ^2	1	1	ζ	ζ^2	ζ
χ_5	1	ζ	1	1	ζ^2	ζ	1	1	ζ^2	ζ	ζ^2
χ_6	1	ζ^2	ζ^2	1	ζ	ζ	ζ	1	1	1	ζ^2
χ_7	1	ζ	ζ	1	ζ^2	ζ^2	ζ^2	1	1	1	ζ
χ_8	1	ζ^2	ζ	1	ζ	1	ζ^2	1	ζ^2	ζ	1
χ_9	1	ζ	ζ^2	1	ζ^2	1	ζ	1	ζ	ζ^2	1
χ_{10}	3	0	0	$3\zeta^2$	0	0	0	3ζ	0	0	0
χ_{11}	3	0	0	3ζ	0	0	0	$3\zeta^2$	0	0	0

a) Find the order of G.
b) Show that $G/[G,G]$ is isomorphic to $C_3 \times C_3$.
c) Show that $C_1 \cup C_4 \cup C_8 =: N$ is a normal subgroup of G.
d) Find the centre of G.
e) Find the size of each of the conjugacy classes of G.
f) Let H be a group of order 27 and center $Z(H)$ of order 3. Find the character table of H.
g) Since the automorphism group of the cyclic group of order 9 is cyclic of order 6 there is a unique non abelian semidirect product $C_9 \rtimes C_3$. Denote by $UT(3,3)$ the subgroup

$$\left\{ \begin{pmatrix} 1 & x & y \\ 0 & 1 & z \\ 0 & 0 & 1 \end{pmatrix} \middle| x, y, z \in \mathbb{F}_3 \right\}$$

of $GL_3(\mathbb{F}_3)$. Show that $UT(3,3) \not\simeq C_9 \rtimes C_3$ and

$$Z(UT(3,3)) \simeq Z(C_9 \rtimes C_3) \simeq C_3.$$

Exercise 2.13. Let G be a finite group and let $C_1, C_2, \ldots, C_s$ be the conjugacy classes of G and for any $i \in \{1, \ldots, s\}$ let $K_i = \sum_{g \in C_i} g$. We convene that $C_1 = \{1\}$.

a) Recall that for any $k \in \{1, \ldots, s\}$ and any $g \in C_k$ the cardinal $c_{i,j}^k := |\{(x,y) \in C_i \times C_j \mid xy = g\}|$ satisfies $K_i \cdot K_j = \sum_{k=1}^s c_{i,j}^k K_k$.
b) Show that for any irreducible character χ of G we get

$$|C_i| \cdot |C_j| \cdot \chi(C_i) \cdot \chi(C_j) = \chi(1) \sum_{k=1}^{s} c_{i,j}^k |C_k| \cdot \chi(C_k)$$

c) Deduce from this, multiplying by $\overline{\chi(C_\ell)} \cdot \chi(1)^{-1}$ and summing over $\chi \in \mathrm{Irr}(G)$,

$$c_{i,j}^\ell = \frac{|C_i| \cdot |C_j|}{|G|} \sum_{\chi \in \mathrm{Irr}(G)} \frac{\chi(C_i) \cdot \chi(C_j) \cdot \overline{\chi(C_\ell)}}{\chi(1)}$$

d) Show that two groups with equal character tables have isomorphic centres, and two group algebras with isomorphic centres have equal character tables.
e) Recall from Exercise 2.6 the character table CT_2 of the alternating group $\mathfrak{A}_5$ of degree 5 and recall that C_3^2 and C_4^2 are conjugacy classes of elements of order 5. Posons $X = C_3^2$ et Calculer $c_{X,X}^X$.
f) Show that $s \cdot t \in X \Leftrightarrow t \cdot s \in X$. Define a graph $\Gamma_{\mathfrak{A}_5}(X)$ whose vertices are the elements of X, and two elements $s, t \in X$ are linked by an edge if $s \cdot t \in X$.
g) For each $s \in X$ show that precisely $c_{X,X}^X$ vertices have an edge with s.
h) Show that the group $\mathfrak{A}_5$ acts on the graph $\Gamma_{\mathfrak{A}_5}(X)$ by conjugation, and this action is transitive on the vertices. Show that $\Gamma_{\mathfrak{A}_5}(X)$ is connected.
i) If $\Gamma_{\mathfrak{A}_5}(X)$ is a planar graph, use Euler's polyhedral formula to show that $\Gamma_{\mathfrak{A}_5}(X)$ is the icosahedron graph.
j) (Radu Stancu) One can verify that $x \in X \Rightarrow x^2 \notin X$ and hence $x \in X \Leftrightarrow x^{-1} \in X$. Use this property in the sequel. Take a $g \in X$ and consider the adjacant elements $x_1, \ldots, x_5$. Show that all the $gx_1, \ldots, gx_5$ are in X, adjacent to g^{-1}, but not adjacent to g. Show that none of the $x_1, \ldots, x_5$ is adjacent to g^{-1}. Show that x_1 is adjacent to one of the $x_2, \ldots, x_5$, say x_2. Deduce that g is a vertex of $\Gamma_{\mathfrak{A}_5}$ which is adjacent to precisely 5 triangles. Similarly for g^{-1}. Show that x_i is adjacent to x_j^{-1} if and only if $i - j + 2$ or $i - j - 2$ is divisible by 5. We obtain the graph of an icosahedron.

NB: $\Gamma_{\mathfrak{A}_5}(X)$ is a graph with 12 vertices and each vertex is linked by an edge with 5 other vertices. Moreover, $\Gamma_{\mathfrak{A}_5}(X)$ is planar.

The automorphism group of an icosahedron is $\mathfrak{A}_5$ (cf. Exercise 8.6).

3 Tensor products, Mackey formulas and Clifford theory

Wenn es nur einmal so ganz stille wäre.
Wenn das Zufällige und Ungefähre
verstummte und das nachbarliche Lachen,
wenn das Geräusch, das meine Sinne machen,
mich nicht so sehr verhinderte am Wachen -:

Dann könnte ich in einem tausendfachen
Gedanken bis an deinen Rand dich denken

Rainer Maria Rilke

If only once it were totally silent.
If the random and the approximative
disappeared and the neighbour's laughing,
the sound my senses make,
would not get in the way of my awareness -:

then in one thousandfold
thought I could think you up to your very brink

Rainer Maria Rilke;[1]

3.1 The tensor product

We are going to introduce a fundamental construction which is absolutely necessary for higher algebra. We postponed this construction until now since by experience it poses difficulties to everyone learning it for the first time. Moreover, the tensor product provides many traps even for experienced users. Nevertheless, in order to be able to dive into more sophisticated properties of group rings, or algebras in general it is impossible to surround this construction. Tensor products are a central object in higher algebra.

3.1.1 The definition of a tensor product

The first concept is a free abelian group generated by a set. The models are the abelian groups $(\mathbb{Z}^n, +)$ for any integer n, which is a free abelian group on the set

$$\{(1,0,0,\dots,0),(0,1,0,\dots,0),\dots,(0,0,\dots,0,1)\}$$

1 in der Übersetzung von Susanne Brennecke, translation by Susanne Brennecke.

https://doi.org/10.1515/9783110702446-003

in the definition below Indeed, the above set is a basis of $\mathbb{Z}^n$ in the sense that for any abelian group B and any n elements $b_1, b_2, \ldots, b_n$ of B there is a unique homomorphism of abelian groups $\varphi : \mathbb{Z}^n \longrightarrow B$ such that

$$\begin{aligned}\varphi(1,0,0,\ldots,0) &= b_1,\\ \varphi(0,1,0,\ldots,0) &= b_2,\\ \ldots &\ldots \ldots\\ \varphi(0,0,\ldots,0,1) &= b_n.\end{aligned}$$

Indeed, $\varphi(a_1, a_2, \ldots, a_n) = \sum_{i=1}^n a_i b_i$ is a group homomorphism and is the unique one with the above properties.

Definition 3.1.1. An abelian group A is free on a subset S of A if for every abelian group B and every mapping $\varphi_S : S \longrightarrow B$ (as set!) there is a unique homomorphism $\varphi : A \longrightarrow B$ such that the restriction $\varphi|_S$ of φ to S equals φ_S.

Suppose A is a free abelian group on a set S_A, and suppose B is another free abelian group on a set S_B. If S_A and S_B are of the same cardinality (i. e. there is a bijection $\beta : S_A \longrightarrow S_B$) then A and B are isomorphic. Indeed, β defines a unique homomorphism of groups $\widehat{\beta} : A \longrightarrow B$ restricting to β on S_A, and β^{-1} defines a unique homomorphism of groups $\widehat{\beta^{-1}} : B \longrightarrow A$ restricting to β^{-1} on S_B. Now, the identity on S_A is the unique group homomorphism $A \longrightarrow A$ restricting to the identity on S_A. But, $\widehat{\beta^{-1}} \circ \widehat{\beta}$ is a group homomorphism restricting to $\beta^{-1} \circ \beta = id_{S_A}$ on S_A. Therefore, by the unicity, $\widehat{\beta^{-1}} \circ \widehat{\beta} = id_A$. Analogously, $\widehat{\beta} \circ \widehat{\beta^{-1}} = id_B$. Hence, $\widehat{\beta}$ is an isomorphism.

Up to now we did not show that free abelian groups exist, except for S being a finite set. Set

$$F_S := \{f \in Map(S, \mathbb{Z}) \mid f(s) = 0 \text{ for almost all } s \in S\}$$

the set of mappings (as set) of S to the abelian group $(\mathbb{Z}, +)$ being 0 for all but a finite number of points $s \in S$. This is clearly an abelian group by pointwise addition: $(f + g)(s) := f(s) + g(s)$ for all $f, g \in Map(S, \mathbb{Z})$ and $s \in S$. Define for every $s \in S$ the mapping $f_s \in Map(S, \mathbb{Z})$ by $f_s(t) := 0$ if $s \neq t$ and $f_s(s) = 1$. Let us prove that F_S is free on the set $\{f_s \mid s \in S\}$. Given an abelian group B and a family $(b_s)_{s\in S}$ of B. We want to define a homomorphism of abelian groups $\beta : F_S \longrightarrow B$ such that $\beta(f_s) = b_s$ for all $s \in S$. For all $s \in S$ put

$$\beta(f) := \sum_{f(s)\neq 0} f(s) b_s.$$

Since for every $f \in F_S$ there are only a finite number of $s \in S$ such that $f(s) \neq 0$, the sum is well defined. Let $s \in S$. Then

$$(\beta(f_s))(t) = \sum_{f_s(t)\neq 0} f_s(t) b_t = b_s.$$

The definition of β implies immediately that β is a group homomorphism and we just showed that $\beta(f_s) = b_s$ for all $s \in S$. Hence, F_S is free on $\{f_s | s \in S\}$; and this set is of the same cardinality as S.

We have just proved the following.

Lemma 3.1.2. *For every set S there is a free abelian group F that is free on a set F_S of the same cardinality as S.*

We are ready to define the tensor product.

Definition 3.1.3. Given a ring A, a left A-module M and a right A-module N, then take the free abelian group $F_{N\times M}$ on $N \times M$. In this abelian group we look for the smallest subgroup $R_{N\times M;A}$ which contains all the elements

$$\begin{array}{lrl}(R1) & (n_1+n_2,m)-(n_1,m)-(n_2,m); & \forall m\in M; n,n_1,n_2\in N\\ (R2) & (n,m_1+m_2)-(n,m_1)-(n,m_2); & \forall m_1,m_2\in M; n\in N\\ (R3) & (nr,m)-(n,rm); & \forall r\in A\end{array}$$

The *tensor product of N and M over A* is

$$N\otimes_A M := F_{N\times M}/R_{N\times M;A}.$$

A short look on the definition makes clear that the tensor product is a subtle thing which is not so easy to understand. Also, the tensor product $N \otimes_A M$ is an abelian group, and nothing more in general.

We call $n\otimes m$ for $n \in N$ and $m \in M$ the image of (n, m) in $N\otimes_A M$. The elements of $N\otimes_A M$ are sums of the form $\sum_{i=1}^{k} n_i \otimes m_i$ and there are in general many very different ways to present the same element by means of the relations in $R_{N\times M;A}$. Also the quotient $N \otimes_A M = F_{N\times M}/R_{N\times M;A}$ may become 0 even when neither of the modules N or M is 0.

Example 3.1.4. If p and q are relatively prime integers, then

$$\mathbb{Z}/p\mathbb{Z}\otimes_{\mathbb{Z}}\mathbb{Z}/q\mathbb{Z}=0.$$

Indeed, since $N \otimes_A M$ is generated by $n\otimes m$ for $n \in N$ and $m \in M$, which is the image of (n, m) in $N \times M$, we only need to show that $n \otimes m = 0$ in $\mathbb{Z}/p\mathbb{Z} \otimes_{\mathbb{Z}} \mathbb{Z}/q\mathbb{Z}$, for $n \in \mathbb{Z}/p\mathbb{Z}$ and $m \in \mathbb{Z}/q\mathbb{Z}$. Now, since p and q are relatively prime integers, there are integers x and y such that $1 = px + qy$ by Euclid's algorithm. Hence,

$$\begin{aligned} n\otimes m &= (n\cdot 1)\otimes m\\ &= n\cdot(px+qy)\otimes m\\ &= (npx\otimes m)+(nqy\otimes m) && \text{by Definition 3.1.3 (R1)}\\ &= (npx\otimes m)+(n\otimes qym) && \text{by Definition 3.1.3 (R3)}\\ &= (0\otimes m)+(n\otimes 0) && \text{since } n\in\mathbb{Z}/p\mathbb{Z} \text{ and } m\in\mathbb{Z}/q\mathbb{Z}\end{aligned}$$

Go back to the general case $N \otimes_A M$ for a moment. Then, $n \otimes 0 = 0 \in N \otimes_A M$ and $0 \otimes m = 0 \in N \otimes_R M$ for all $n \in N$ and $m \in M$. Indeed, denote by 0_A the 0 element in A and by 0_N resp. 0_M the zero element in M, and then

$$n \otimes 0_M = n \otimes 0_{\mathbb{Z}} \cdot 0_M = n \cdot 0_{\mathbb{Z}} \otimes 0_M = 0_N \otimes 0_M$$

and therefore, by the same argument

$$n \otimes 0_M = 0_N \otimes 0_M = 0_N \otimes m$$

for all n and m. Now,

$$(n \otimes m) + (0_N \otimes 0_M) = (n \otimes m) + (n \otimes 0_M) = (n \otimes (m + 0_M)) = (n \otimes m)$$

for all $n \in N$ and $m \in M$. Hence on the generators of $N \otimes_A M$ the element $0_N \otimes 0_M$ is neutral, and therefore it is neutral in all of $N \otimes_A M$.

3.1.2 Homomorphisms and tensor product

One would like to know now how it is possible to control these strange objects reasonably well. The main method is by a universal property.

Proposition 3.1.5. *Given a ring A, an A-left-module M and an A-right-module N, and an abelian group B. Then, any map (as sets!) $\varphi : M \times N \longrightarrow B$ satisfying*

$$\begin{aligned} \varphi((n_1 + n_2, m)) &= \varphi((n_1, m)) + \varphi((n_2, m)); \ \forall m \in M; n, n_1, n_2 \in N \\ \varphi((n, m_1 + m_2)) &= \varphi((n, m_1)) + \varphi((n, m_2)); \ \forall m_1, m_2 \in M; n \in N \\ \varphi((nr, m)) &= \varphi((n, rm)); \ \forall r \in A \end{aligned}$$

admits a unique extension $\hat{\varphi} : N \otimes_A M \longrightarrow B$ as homomorphism of abelian groups with $\hat{\varphi}(n \otimes m) = \varphi(n, m)$. A mapping φ satisfying the above properties is called A-balanced.

Proof. Since $F_{N \times M}$ is free there is a unique homomorphism of abelian groups $F_\varphi : F_{N \times M} \longrightarrow B$ extending φ. Since we supposed the above properties of φ, one gets that $F_\varphi(R_{N \times M;A}) = 0$. Therefore, $\hat{\varphi}$ exists with the required properties.

Uniqueness follows from the fact that the elements $n \otimes m$ generate $N \otimes_A M$ and so the images of all other elements in $N \otimes_A M$ are determined. □

A first application is the following Lemma.

Lemma 3.1.6. *Let N_1 and N_2 be R-right-modules, and let M_1 and M_2 be R-left-modules. Then, for every homomorphism $\alpha : N_1 \longrightarrow N_2$ of R-right-modules and every homomorphism $\beta : M_1 \longrightarrow M_2$ of R-left-modules there is a unique homomorphism of abelian groups $\gamma : N_1 \otimes_R M_1 \longrightarrow N_2 \otimes_R M_2$ satisfying $(\alpha \otimes \beta)(n \otimes m) = \alpha(n) \otimes \beta(m)$. Denote $\gamma =: \alpha \otimes \beta$.*

Proof. Since in the tensor product $N_2 \otimes_R M_2$ elements satisfy the required relations, Proposition 3.1.5 shows that $\alpha \otimes \beta$ exists and is unique. □

Recall the following concept from Definition 1.2.26:

Definition 3.1.7. Let $(R, +, \cdot)$ be a ring, then the *opposite ring* R^{op} is a ring $(R, +, \cdot^{op})$ being R as set, $(R, +)$ is the original structure as abelian group and for any $r_1, r_2 \in R$ define $r_1 \cdot^{op} r_2 := r_2 \cdot r_1$. If K is a commutative ring and if R is a K-algebra, then R^{op} becomes a K-algebra as well since the action of K commutes with the action of R; or otherwise said, the image of K in A as multiples of 1 is in the centre of R. It is clear that the centres of R and of R^{op} coincide.

Definition 3.1.8. Let R and S be two rings. Then an *R-S-bimodule* is an $R \times S^{op}$ left-module. More explicitly, one has an action of R on the left, of S on the right and that the actions commute in the sense that for every $r \in R$, $s \in S$, $m \in M$ one has $(rm)s = r(ms)$.

Lemma 3.1.9. *Let R, S and T be rings. Let M be an R-left-module and let N be an S-R-bimodule, and U be a T-R-bimodule. Then,*
1. *$N \otimes_R M$ is an S-left-module satisfying $s \cdot (n \otimes m) = (sn) \otimes m$ for all $s \in S$, $n \in N$ and $m \in M$.*
2. *$Hom_R(N, U)$ is a T-S-bimodule by an operation defined by $(t \cdot f \cdot s)(n) := tf(sn)$ for all $s \in S$, $n \in N$, $t \in T$ and $f \in Hom_R(N, U)$.*

Proof.
1. For every $s \in S$ the mapping $\lambda_s : N \times M \longrightarrow N \otimes_R M$ given by $\lambda_s(n, m) := (sn) \otimes m$ is obviously R-balanced just using the defining properties of a module and of the tensor product on the right of the mapping. By Proposition 3.1.5 there is a mapping $\hat{\lambda}_s : N \otimes_R M \longrightarrow N \otimes_R M$ satisfying the required property. Since N is a bimodule, $1_S \cdot (n \otimes m) = n \otimes m$, the operation is distributive and $(s_1 s_2) \cdot (n \otimes m) = s_1 \cdot (s_2 \cdot (n \otimes m))$ for all $s_1, s_2 \in S$ and $n \in N$, $m \in M$.
2. The fact that $(t \cdot f \cdot s)(n) := tf(sn)$ for all $s \in S$, $n \in N$, $t \in T$ and $f \in Hom_R(N, U)$ defines for all $s \in S$, $t \in T$, $f \in Hom_R(N, U)$ an element in $Hom_R(N, U)$ is clear using that f is R-linear and that N is a bimodule. The fact that this defines a bimodule structure on $Hom_R(N, U)$ is an easy exercise using the definitions of a bimodule and an R-linear homomorphism only.

This shows the lemma. □

Lemma 3.1.10. *Let R, S, T and V be four rings and let M be an T-V-bimodule, N be an S-T-bimodule, N' another S-T-bimodule, U be an R-S-bimodule, W an R-V-bimodule. Then,*
1. *$U \otimes_S (N \otimes_T M) \simeq (U \otimes_S N) \otimes_T M$ where the isomorphism maps $(u \otimes (n \otimes m))$ to $((u \otimes n) \otimes m)$ for all $u \in U$, $n \in N$, $m \in M$.*

2.

$$Hom_R(U \otimes_S N, W) \simeq Hom_S(N, Hom_R(U, W))$$

as T-V-bimodules. The isomorphism restricts to an isomorphism

$$Hom_{R-V}(U \otimes_S N, W) \simeq Hom_{S-V}(N, Hom_R(U, W))$$

and to an isomorphism

$$Hom_{T-R}(U \otimes_S N, W) \simeq Hom_{T-S}(N, Hom_R(U, W)).$$

3.

$$U \otimes_S (N \oplus N') \simeq (U \otimes_S N) \oplus (U \otimes_S N')$$

and

$$(N \oplus N') \otimes_T M \simeq (N \otimes_T M) \oplus (N' \otimes_T M).$$

Proof.

1. We first define for every $u \in U$ a mapping $\lambda_u : N \times M \longrightarrow (U \otimes_S N) \otimes_T M$ by $\lambda_u(n,m) := (u \otimes n) \otimes m$. It is immediate to verify that λ_u is T-balanced, and so it induces a group homomorphism $\lambda_u : N \otimes_T M \longrightarrow (U \otimes_S N) \otimes_T M$. Now, in a second step we define a mapping $U \times (N \otimes_T M) \longrightarrow (U \otimes_S N) \otimes_T M$ by $(u,x) \mapsto \lambda_u(x)$ for every $u \in U$ and $x \in N \otimes_T M$. It might be here a little less clear why this map is S-balanced. Here is a proof. Take $u_1, u_2 \in U$ and $x = \sum_{i=1}^k n_i \otimes m_i \in N \otimes_T M$. Then,

$$\begin{aligned}
\lambda_{u_1+u_2}(x) &= \sum_{i=1}^k ((u_1 + u_2) \otimes n_i) \otimes m_i \\
&= \sum_{i=1}^k ((u_1 \otimes n_i) \otimes m_i + (u_2 \otimes n_i) \otimes m_i) \\
&= \lambda_{u_1}\left(\sum_{i=1}^k n_i \otimes m_i\right) + \lambda_{u_2}\left(\sum_{i=1}^k n_i \otimes m_i\right) \\
&= \lambda_{u_1}(x) + \lambda_{u_2}(x)
\end{aligned}$$

   Similarly one shows the other conditions. This clearly defines an isomorphism.
2. Define

$$\begin{aligned}
\Psi : Hom_R(U \otimes_S N, W) &\longrightarrow Hom_S(N, Hom_R(U, W)) \\
f &\mapsto (n \mapsto (u \mapsto f(u \otimes n)))
\end{aligned}$$

First, for any $f \in Hom_R(U \otimes_S N, M)$ and any $n \in N$ the mapping $(u \mapsto f(u \otimes n))$ is R-linear. Indeed, $(ru \mapsto f(ru \otimes n))$, but $f(ru \otimes n) = rf(u \otimes n)$. Moreover, $\Psi(f)$ is S-linear, since

$$\begin{aligned}(\Psi(f)(sn))(u) &= f(u \otimes sn)\\ &= f(us \otimes n)\\ &= (\Psi(f)(n))(su)\\ &= (s \cdot \Psi(f)(n))(u)\end{aligned}$$

Define

$$\begin{aligned}\Phi : Hom_S(N, Hom_R(U, W)) &\longrightarrow Hom_R(U \otimes_S N, W)\\ f &\mapsto (u \otimes n \mapsto (f(n))(u))\end{aligned}$$

We need to show that this is well-defined, i. e. $\Phi(f)$ is S-balanced. But this follows immediately from the fact that f is S-linear. Moreover, Φ is inverse to Ψ.
Now, it is easy to show that Φ maps T-linear maps to T-linear maps, and likewise Ψ. Similarly, Φ maps V-linear maps to V-linear maps, and similarly Ψ.

3. $u\otimes(n+n') \mapsto (u\otimes n)\oplus(u\otimes n')$ defines a mapping $U\otimes_S(N\oplus N') \longrightarrow (U\otimes_S N)\oplus(U\otimes_S N')$ and $(u\otimes n)\oplus(u'\otimes n') \mapsto (u\otimes n)+(u'\otimes n')$ defines a mapping in the other direction. The fact that they are mutually inverse is clear.
The other associativity is done similarly.

This proves the lemma. □

3.1.3 The case of tensor products over fields

A somewhat simpler situation occurs if A is a K-algebra for a field K. Then tensor products over A become K-vector spaces and so one can count the dimension.

Lemma 3.1.11. *Let K be a commutative ring and let A be a K-algebra. If M is an A-right-module and N is an A left-module, then $M \otimes_A N$ is a K-module. If K is a field and M and N are finite dimensional over K, then $dim_K(M \otimes_K N) = dim_K(M) \cdot dim_K(N)$. Moreover, $A^s \otimes_A N \simeq N^s$ as A left-module. In particular $A \otimes_A N \simeq N$ by a mapping satisfying $a \otimes n \mapsto an$ for each $a \in A$ and $n \in N$.*

Proof. The fact that $M \otimes_A N$ is a K-module is a consequence of Lemma 3.1.9. Given a K-basis B_M of M and a K-basis B_N of N, it is clear that $\{b_M \otimes b_N |\ b_M \in B_M, b_N \in B_N\}$ is a generating set of $M \otimes_K N$. To show that this is a basis we shall prove the isomorphism $A^s \otimes_A N \simeq N^s$ first. Observe that $(a_1, a_2, \ldots, a_s) \times x \mapsto (a_1x, a_2x, \ldots, a_nx)$ is clearly balanced. Hence it defines a mapping $A^s \otimes N \xrightarrow{\alpha} N$.

Call for each $i \in \{1, 2, \ldots, s\}$ the element $e_i \in A^s$ which is 0 on each component except on the component i, where it is 1. We shall define an inverse mapping

$$\begin{aligned} N^s &\xrightarrow{\beta} A^s \otimes_A N \\ (n_1, n_2, \ldots, n_s) &\mapsto (e_1 \otimes n_1 + e_2 \otimes n_2 + \cdots + e_s \otimes n_s) \end{aligned}$$

We see that

$$\begin{aligned} (\beta \circ \alpha)((a_1, a_2, \ldots, a_s) \otimes x) &= (e_1 \otimes a_1 x + e_2 \otimes a_2 x + \cdots + e_s \otimes a_s x) \\ &= (a_1 \otimes x + a_2 \otimes x + \cdots + a_s \otimes x) \\ &= (a_1, a_2, \ldots, a_s) \otimes x \end{aligned}$$

Since the elements $(a_1, a_2, \ldots, a_s) \otimes x$ for $a_1, a_2, \ldots, a_s \in A$ and $x \in N$ is a generating set for $A^s \otimes_A N$, this implies that $\beta \circ \alpha = id_{A^s \otimes_A N}$. Moreover

$$(\alpha \circ \beta)(n_1, \ldots, n_s) = \alpha(e_1 \otimes n_1 + e_2 \otimes n_2 + \cdots + e_s \otimes n_s) = (n_1, \ldots, n_s)$$

as is seen directly. Clearly, α and β are A-linear.

As a consequence of the above one gets $K^n \otimes_K K^m \simeq K^{n \cdot m}$ as K-vector spaces for all $n, m \in \mathbb{N}$. □

Remark 3.1.12. By Lemma 3.1.10 (3) it is sufficient to show only $A \otimes_A M \simeq M$, i. e. the case $s = 1$.

Remark 3.1.13. The same proof yields actually that even for infinite direct sums $(\oplus_{i \in I} A) \otimes_A N \simeq \oplus_{i \in I} N$ for arbitrary index sets I. Here $\oplus_{i \in I} N$ is the set of those mappings $I \longrightarrow N$ such that the image of almost every $i \in I$ is 0.

Note that $(\prod_{i \in I} A) \otimes_A N \neq \prod_{i \in I} N$ in general. Indeed, the definition of the tensor product implies that an element of the left hand side is a finite linear combination of elementary tensors $x \otimes n$ for $x \in \prod_{i \in I} A$ and $n \in N$. The right hand side is a direct product of copies of N, whence an a priori infinite product.

3.1.4 Additional structure, algebras and modules

Lemma 3.1.14. *Given a commutative ring R and R-algebras A and B, then $A \otimes_R B$ becomes an R-algebra satisfying $(a_1 \otimes b_1) \cdot (a_2 \otimes b_2) = (a_1 a_2 \otimes b_1 b_2)$ for all $a_1, a_2 \in A$ and $b_1, b_2 \in B$.*

Proof. In Lemma 3.1.11 we have already seen that $A \otimes_R B$ is an R-module. The additive structure is therefore clear. We need to define the multiplicative structure.

For every $a_1 \in A$ and $b_1 \in B$ define a mapping $A \times B \longrightarrow A \otimes_R B$ by

$$(a_1, b_1) \cdot (a_2, b_2) := (a_1 a_2 \otimes b_1 b_2)$$

for all $a_2 \in A$ and $b_2 \in B$. It is immediate that this is R-balanced and so this defines a mapping

$$A \otimes_R B \longrightarrow A \otimes_R B$$

satisfying $(a_1, b_1)\cdot(a_2 \otimes b_2) = (a_1 a_2 \otimes b_1 b_2)$ for all $a_1, a_2 \in A$ and $b_1, b_2 \in B$. Moreover, it is clear that $(a_1 r, b_1)$ and (a_1, rb_1) induce the same mapping, and also that the mapping induced by $(a_1 + a_1', b_1)$ is the sum of the mappings induced by (a_1, b_1) and (a_1', b_1). Likewise the mapping induced by $(a_1, b_1 + b_1')$ is the sum of the mappings induced by (a_1, b_1) and (a_1, b_1').

Hence, $(a_1 \otimes b_1)\cdot(a_2 \otimes b_2) := (a_1 a_2 \otimes b_1 b_2)$ is well-defined. Now, one has to verify that this gives a structure of an R-algebra. But this is a straight forward verification. □

We come to an application for group rings.

Lemma 3.1.15. *Let R be a commutative ring and let G and H be two groups. Then $R(G \times H) \simeq RG \otimes_R RH$ as R-algebras.*

Proof. We define a mapping $\alpha : G \times H \longrightarrow RG \otimes_R RH$ by $\alpha(g, h) := g \otimes h$ for every $(g, h) \in G \times H$. Since $1_G \otimes 1_H$ is obviously the unit in $RG \otimes_R RH$, $\alpha(1_G, 1_H) = 1_{RG \otimes_R RH}$. Moreover,

$$\begin{aligned}\alpha((g_1, h_1) \cdot (g_2, h_2)) &= \alpha(g_1 g_2, h_1 h_2)\\ &= g_1 g_2 \otimes h_1 h_2\\ &= (g_1 \otimes h_1) \cdot (g_2 \otimes h_2)\\ &= \alpha(g_1, h_1) \cdot \alpha(g_2, h_2)\end{aligned}$$

and so α induces a homomorphism $R(G \times H) \longrightarrow RG \otimes_R RH$. Inversely, the elements $g \otimes h$ for $g \in G$ and $h \in H$ are a generating set for $RG \otimes_R RH$, since G is an R-basis for RG and H is an R-basis for RH. Then

$$\left(\sum_{g\in G} r_g g, \sum_{h\in H} r_h h\right) \mapsto \sum_{g\in G}\sum_{h\in H} r_g r_h (g, h)$$

is R-balanced and induces hence a mapping $\beta : RG \otimes_R RH \longrightarrow R(G \times H)$. It is clear that β is inverse to α. This proves the lemma. □

Given a commutative ring R, we have seen that for R-algebras A and B the space $A \otimes_R B$ is an R-algebra again.

Lemma 3.1.16. *Let R be a commutative ring and let A and B be R-algebras. Suppose that M is an A-module and N is a B-module. Then the space $M \otimes_R N$ is an $A \otimes_R B$-module.*

Proof. Lemma 3.1.9 implies that $M \otimes_R N$ is an A-module. Now, the B-left-module structure of N is a B^{op}-right-module structure. So, $M \otimes_R N$ is an $A \otimes_R B$-module if and only if

$M \otimes_R N$ is an A-B^{op}-bimodule. The analogue of Lemma 3.1.9 for right-modules implies that the only property to be verify is that the actions of A on the left commutes with the action of B^{op} on the right, and that the actions of R inside A and inside B coincide. But since this action is defined by the equation $a \cdot (m \otimes n) = am \otimes n$ and $(m \otimes n) \cdot b = m \otimes bn$, both of the statements are clear. □

We want to look for properties of this module.

Example 3.1.17. Recall Example 1.2.23. Let Q_8 be the quaternion group of order 8, that is the group generated by a and b subject to the relations $a^4 = 1$, $a^2 = b^2 = (ab)^2$ and $a^2b = ba^2$. This group has order 8. Its elements are $1, a, a^2, a^3, b, b^3, (ab), (ab)^3$. The elements a, b and (ab) are all of order 4, the element a^2 is central of order 2.

Let $\mathbb{H}$ be the real quaternion algebra. This is a 4-dimensional real vector space with basis 1, i, j and k. It is a skew-field when one defines 1 to be the neutral element of the multiplication, $i \cdot j = k$ and $i^2 = j^2 = k^2 = -1$.

Then $\mathbb{H}$ is an $\mathbb{R}Q_8$-module by identifying a with i and b with j. The relations in $\mathbb{H}$ are precisely the same as the relations in Q_8. Moreover, $\mathbb{H}$ is simple as $\mathbb{R}Q_8$-module. Indeed, since $\mathbb{H}$ is a skew-field and since Q_8 acts on by multiplication on the left, $End_{\mathbb{R}Q_8}(\mathbb{H}) \simeq \mathbb{H}$ is a skew-field. By Schur's Lemma, using that $\mathbb{R}Q_8$ is semisimple, $\mathbb{H}$ is simple.

As we have seen in Corollary 1.2.31 every skew-field that is finite dimensional over an algebraically closed field K is actually equal to K. This will be the key observation for the following construction.

The cyclic group C_4 of order 4 generated by c acts on $\mathbb{C}$ by multiplication by i, the imaginary unit. Then, $\mathbb{H} \otimes_{\mathbb{R}} \mathbb{C}$ is an $\mathbb{R}Q_8 \otimes_{\mathbb{R}} \mathbb{R}C_4$-module.

Actually, by Lemma 3.1.15 for two groups G and H and a commutative ring K we get $KG \otimes_K KH \simeq K(G \times H)$. Now, $\mathbb{R}C_4 \simeq \mathbb{R} \times \mathbb{R} \times \mathbb{C}$, and so

$$\begin{aligned}\mathbb{R}(Q_8 \times C_4) &\simeq (\mathbb{R} \otimes_{\mathbb{R}} \mathbb{R}Q_8) \times (\mathbb{R} \otimes_{\mathbb{R}} \mathbb{R}Q_8) \times (\mathbb{C} \otimes_{\mathbb{R}} \mathbb{R}Q_8)\\ &\simeq \mathbb{R}Q_8 \times \mathbb{R}Q_8 \times (\mathbb{C} \otimes_{\mathbb{R}} \mathbb{R}Q_8)\end{aligned}$$

as $\mathbb{R}$-algebra. Now,

$$\mathbb{R}Q_8 \simeq \mathbb{R} \times \mathbb{R} \times \mathbb{R} \times \mathbb{R} \times \mathbb{H}$$

and so we need to determine the 8-dimensional $\mathbb{R}$-algebra $\mathbb{C} \otimes_{\mathbb{R}} \mathbb{H}$ (for the statement that this is an $\mathbb{R}$-algebra see Lemma 3.1.14). Moreover, the algebra $\mathbb{C} \otimes_{\mathbb{R}} \mathbb{H}$ is actually a 4-dimensional $\mathbb{C}$-algebra, and moreover, the algebra is semisimple since $\mathbb{R}(Q_8 \times C_4)$ is semisimple. Since $\mathbb{C}$ is algebraically closed, and since $\mathbb{C} \otimes_{\mathbb{R}} \mathbb{H}$ is not commutative,

$$\mathbb{C} \otimes_{\mathbb{R}} \mathbb{H} \simeq Mat_{2\times2}(\mathbb{C}).$$

We may give an explicit isomorphism as follows. Define

$$\alpha : \mathbb{H} \otimes_{\mathbb{R}} \mathbb{C} \longrightarrow Mat_{2\times2}(\mathbb{C})$$

$$1 \otimes 1 \mapsto \begin{pmatrix} 1 & 0 \\ 0 & 1 \end{pmatrix}$$

$$1 \otimes i \mapsto \begin{pmatrix} i & 0 \\ 0 & i \end{pmatrix}$$

$$i \otimes 1 \mapsto \begin{pmatrix} i & 0 \\ 0 & -i \end{pmatrix}$$

$$j \otimes 1 \mapsto \begin{pmatrix} 0 & 1 \\ -1 & 0 \end{pmatrix}$$

and extend this multiplicatively and $\mathbb{R}$-bilinearly. It is straight forward to check that α is then a ring homomorphism. Moreover, α is injective since the 16 matrices corresponding to the image of the elements (g, c) for $g \in Q_8$ and $(g, 1)$ are $\mathbb{R}$-linearly independent as is easily checked. We observe now that via this isomorphism $\mathbb{H} \otimes_{\mathbb{R}} \mathbb{C}$ is isomorphic to two copies of a 2-dimensional simple $\mathbb{R}(Q_8 \times C_4)$-module.

This example shows that if A and B are K-algebras, if M is a simple A-module and if N is a simple B-module, then $M \otimes_K N$ is a not necessarily simple $A \otimes_K B$-module. Nevertheless, one can prove that it is always semisimple when M and N are finite dimensional over K.

We shall study the endomorphism algebra of such a tensor product $M \otimes_K N$ as $A \otimes_K B$-module for M an A-module and N a B-module.

Lemma 3.1.18. *Let K be a field and let A and B be K-algebras. Suppose M_1, M_2 are A-modules and N_1, N_2 are B-modules. Then there is a homomorphism of vector spaces*

$$Hom_A(M_1, M_2) \otimes_K Hom_B(N_1, N_2) \longrightarrow Hom_{A \otimes_K B}(M_1 \otimes_K N_1, M_2 \otimes_K N_2).$$

If $A = B = K$, then this homomorphism is an isomorphism.

Proof. First we get a homomorphism of K-modules

$$\Phi : Hom_A(M_1, M_2) \otimes_K Hom_B(N_1, N_2) \longrightarrow Hom_{A \otimes_K B}(M_1 \otimes_K N_1, M_2 \otimes_K N_2)$$

by Lemma 3.1.6, defining $\Phi(\alpha \otimes \beta) := \alpha \otimes \beta$.

Choosing K-bases B_{M_1} of M_1, B_{M_2} of M_2 and B_{N_1} of N_1, B_{N_2} of N_2, then $B_{M_1} \otimes B_{M_2} := \{b_1 \otimes b_2 | \; b_1 \in B_{M_1}; b_2 \in B_{M_2}\}$ is a basis for $M_1 \otimes_K M_2$ and likewise $B_{N_1} \otimes B_{N_2}$ is a basis for $N_1 \otimes_K N_2$. We obtain basis for $Hom_A(M_1, M_2)$ by putting for each $b_{M_1} \in B_{M_1}$ and each $b_{M_2} \in B_{M_2}$ the mappings $f_{b_{M_1}}^{b_{M_2}}$ defined by

$$f_{b_{M_1}}^{b_{M_2}}(b'_{M_1}) = \begin{cases} b_{M_2} & \text{if } b_{M_1} = b'_{M_1} \\ 0 & \text{if } b'_{M_1} \in B_{M_1} \setminus \{b_{M_1}\} \end{cases}$$

Similar constructions can be done for the vector space $Hom_B(N_1, N_2)$ and for the vector space $Hom_{A \otimes_K B}(M_1 \otimes_K N_1, M_2 \otimes_K N_2)$.

Hence, if

$$\Phi\Big(\sum_{b_{M_1}\in B_{M_1}, b_{N_1}\in B_{N_1}, b_{M_2}\in B_{M_2}, b_{N_2}\in B_{N_2}} \lambda_{b_{M_1},b_{N_1}}^{b_{M_2},b_{N_2}}(f_{b_{M_1}}^{b_{M_2}} \otimes f_{b_{N_1}}^{b_{N_2}})\Big) = 0,$$

then $\lambda_{b_{M_1},b_{N_1}}^{b_{M_2},b_{N_2}} = 0$ for all indices.

The surjectivity follows by the description of the basis elements. □

Lemma 3.1.19. *Let K be a field, let G_1 and G_2 be two finite groups and suppose that the orders of G_1 and G_2 are invertible in K. Suppose K is a splitting field for G_1 and for G_2. Then for each simple $K(G_1 \times G_2)$-module T there is a simple KG_1-module S_1 and a simple KG_2-module S_2 such that T is isomorphic to $S_1 \otimes_K S_2$. Moreover, for each simple KG_1-module S_1 and each simple KG_2-module S_2 the $K(G_1\times G_2)$-module $S_1 \otimes_K S_2$ is simple.*

Proof. Let T be a simple $K(G_1 \times G_2)$-module. We shall use $K(G_1 \times G_2) \simeq KG_1 \otimes_K KG_2$, which holds by Lemma 3.1.15.

We get

$$\begin{aligned} Hom_K(S_1 \otimes_K S_2, S_1 \otimes_K S_2) &= Hom_{K\otimes_K K}(S_1 \otimes_K S_2, S_1 \otimes_K S_2) \\ &= Hom_K(S_1, S_1) \otimes Hom_K(S_2, S_2). \end{aligned}$$

There is an injective mapping between these two spaces by Lemma 3.1.18. Moreover, the left hand side is the space of K-linear endomorphisms of the $\dim_K(S_1)\cdot\dim_K(S_2)$-dimensional vector space $S_1 \otimes_K S_2$, which is of dimension $(\dim_K(S_1)\cdot\dim_K(S_2))^2$. The dimension of the right hand side is $\dim_K(S_1)^2\cdot\dim_K(S_2)^2$. Now, $G_1\times G_2$ acts on the spaces on the left as a group acting on $S_1 \otimes_K S_2$, and on each of the spaces $End_K(S_1)$ and $End_K(S_2)$ on the right accordingly. The morphism of Lemma 3.1.18 commutes with this action and so one may take invariants on both sides, i. e. one considers the largest sub-module with trivial $G_1 \times G_2$-action. This are the $K(G_1 \times G_2)$ linear endomorphisms on the left and the tensor product of the KG_1-linear endomorphisms with the KG_2-linear endomorphisms on the right. Hence

$$Hom_{K(G_1\times G_2)}(S_1 \otimes_K S_2, S_1 \otimes_K S_2) = Hom_{KG_1}(S_1, S_1) \otimes Hom_{KG_2}(S_2, S_2).$$

But, on the right we get $Hom_{KG_1}(S_1, S_1) = K = Hom_{KG_2}(S_2, S_2)$ and so $S_1 \otimes_K S_2$ is simple.

The number of conjugacy classes of $G_1 \times G_2$ equals the product of the number of conjugacy classes of G_1 and of G_2. Hence each simple $K(G_1 \times G_2)$-module is of the form $S_1 \otimes_K S_2$ for simple KG_i-modules S_i for $i \in \{1, 2\}$. This proves the statement. □

3.2 Using tensor products, induced modules

We are going to use the tensor product mainly in two directions. First, if E is an extension field of K, A is a K-algebra and M is an A-module. Then by Lemma 3.1.14 $E \otimes_K A$ is an algebra again. Moreover, by Lemma 3.1.9 the tensor product $E \otimes_K A$ is an E-vector space. The multiplication inside $E \otimes_K A$ implies that the elements $e \otimes 1$ for $e \in E$ are all central, and hence $E \otimes_K A$ is an E-algebra. The module $E \otimes_K M$ is an $E \otimes_K A$-module by Lemma 3.1.16. This procedure is called change of rings, meaning change of base rings.

The second occasion we will use frequently is the induction of modules. We start with the observation that given a commutative ring R, an R-algebra A and an R-subalgebra B, the ring B acts on A by right multiplication. Otherwise said, restricting the action of A on the regular right-module of A to B gives A the structure of a right-B-module. Also, A can be seen as left-regular A-module. Both actions commute. Otherwise said, A is an A-B-bimodule by putting $a \cdot x \cdot b := axb$ for all $a, x \in A$ and $b \in B$.

Definition 3.2.1. Let G be a group, let R be a commutative ring and let H be a subgroup of G. Then for every RH-module M one defines the *induced module* $M \uparrow_H^G$ to be $RG \otimes_{RH} M$.

The induced module is of extreme importance in representation theory. A large part of representation theory is built to study various properties of this concept.

Remark 3.2.2. A third occasion to use the tensor product is specific to group rings, or more generally to so-called Hopf-algebras (for an introduction see Montgomery's book [Mon-93]), namely given a commutative ring K and a group G as well as two KG-modules M_1 and M_2, then $M_1 \otimes_K M_2$ is a KG-module as well by putting $g \cdot (m_1 \otimes m_2) := (gm_1 \otimes gm_2)$ for all $g \in G$ and $m_1 \in M_1$, $m_2 \in M_2$. Indeed, for all $g \in G$ the mapping

$$\begin{aligned} g\cdot \ : M_1 \times M_2 &\longrightarrow M_1 \otimes_K M_2 \\ (m_1, m_2) &\mapsto gm_1 \otimes gm_2 \end{aligned}$$

is K-balanced and so it induces a mapping

$$\begin{aligned} g\cdot \ : M_1 \otimes_K M_2 &\longrightarrow M_1 \otimes_K M_2 \\ m_1 \otimes m_2 &\mapsto gm_1 \otimes gm_2 \end{aligned}$$

which in turn induces a K-linear action of G on $M_1 \otimes_K M_2$, since

$$1_G\cdot \ = id_{M_1 \otimes_K M_2}$$

and

$$(g_1 g_2)\cdot \ = (g_1\cdot\ ) \circ (g_2\cdot\ )$$

for all $g_1, g_2 \in G$ as is immediately checked.

But, a warning should be given here. Elements $g_1 + g_2$ in KG *do not* act diagonally on $M_1 \otimes_K M_2$. In fact, for $m_1 \in M_1$ and $m_2 \in M_2$ one gets

$$\begin{aligned} (g_1 + g_2) \cdot (m_1 \otimes m_2) &= g_1 \cdot (m_1 \otimes m_2) + g_2 \cdot (m_1 \otimes m_2) \\ &= (g_1 m_1 \otimes g_1 m_2) + (g_2 m_1 \otimes g_2 m_2) \end{aligned}$$

whereas

$$(g_1+g_2)m_1\otimes(g_1+g_2)m_2=(g_1m_1\otimes g_1m_2)+(g_1m_1\otimes g_2m_2)\\ +(g_2m_1\otimes g_1m_2)+(g_2m_1\otimes g_2m_2)$$

Lemma 3.2.3. *Let G be a finite group and let K be a field such that the order of G is invertible in K. Then for two KG-modules M_1 and M_2 one has $(\chi_{M_1\otimes_K M_2})(g)=\chi_{M_1}(g)\cdot\chi_{M_2}(g)$ for each $g\in G$.*

Proof. By Lemma 1.3.6 one can choose for each $g\in G$ a K-basis B_1 of M_1 and a K-basis B_2 of M_2 such that g acts by diagonal matrices on M_1 and M_2. Since $B_1\otimes B_2=\{b_1\otimes b_2|\ b_1\in B_1;b_2\in B_2\}$ is a K-basis of $M_1\otimes M_2$ as in the proof of Lemma 3.1.18 one gets that on this basis as well g acts via a diagonal matrix and the trace of this matrix is the product of the traces of the matrices of g acting on M_1 and on M_2. This proves the lemma. □

We now continue to examine induced modules.

Remark 3.2.4. Let us give the action of G on $M\uparrow_H^G$ more explicitly. Let $G=\bigcup_{i\in I}g_iH$ be a disjoint union of cosets. Recall that the cardinality of I is the index of H in G. Then for all $g\in G$ and all $i\in I$ one has $gg_i\in g_{j(g,i)}H$. This means that there is an $h(g,i)\in H$ such that $gg_i=g_{j(g,i)}h(g,i)$ for all $i\in I$, $g\in G$. Now one gets for all $m\in M$

$$g\cdot(g_i\otimes m)=(gg_i)\otimes m=g_{j(g,i)}h(g,i)\otimes m=g_{j(g,i)}\otimes h(g,i)m.$$

Particular cases are simpler, for example whenever M is a trivial KH-module K. Then $g\cdot(g_i\otimes m)=g_{j(g,i)}\otimes m$ and so the action of G is by permuting the basis $g_i\otimes 1$ of $K\uparrow_H^G$.

Lemma 3.2.5. *If K is a field, if G is a group, if H is a subgroup of G and if M is a finite dimensional KG-module, then $M\uparrow_H^G$ is finite dimensional if and only if H is of finite index in G, that is there are only finitely many cosets $g_1H,g_2H,\dots,g_nH$ in G/H.*

Proof. If H is of finite index in G, then $G=\bigcup_{i\in I}^n g_iH$ for certain elements $g_i\in G;i\in I$ and such that $g_iH=g_jH\Rightarrow i=j$. But then KG as a KH-right module is just isomorphic to $\bigoplus_{i\in I}KH$. Hence, by Lemma 3.1.11 and Remark 3.1.13 we get the following isomorphisms of vector spaces:

$$(\dagger):KG\otimes_{KH}M\simeq\bigoplus_{i\in I}KH\otimes_{KH}M\simeq M^I.$$

So $M\uparrow_H^G$ is finite dimensional if and only if I is finite and M is finite dimensional. Hence, $M\uparrow_H^G$ is finite dimensional if and only if M is finite dimensional and H is of finite index in G. □

Corollary 3.2.6. *If K is a field, if G is a group, if H is a subgroup of G and if M is a finite dimensional KG-module, then*

$$dim_K(M\uparrow_H^G)=dim_K(M)\cdot|G:H|.$$

Proof. This follows immediately from the formula (†). □

We observe that in the extreme case of $H = G$ then $M \uparrow_G^G \simeq M$. This is clear since $M \uparrow_G^G = KG \otimes_{KG} M \simeq M$ by Lemma 3.1.11.

An important property of induced modules is that one can also induce morphisms.

Definition 3.2.7. Let G be a group, let H be a subgroup, let R be a commutative ring and let M and N be RH-modules. Define for every $\alpha \in Hom_{RH}(M,N)$ the morphism $id_{RG} \otimes \alpha =: \alpha \uparrow_H^G$.

We will see that $\alpha \uparrow_H^G$ is G-linear.

Lemma 3.2.8. *Let G be a group with subgroup H, let R be a commutative ring and let M and N be RH-modules. Then for every $\alpha \in Hom_{RH}(M,N)$ one gets $\alpha \uparrow_H^G \in Hom_{RG}(M \uparrow_H^G, N \uparrow_H^G)$.*

Proof. Let again $G = \bigcup_{i\in I}^n g_iH$ be a decomposition into cosets. We compute for all $g \in G$, $i \in I$ and $m \in M$

$$\begin{aligned} g\cdot(\alpha \uparrow_H^G (g_iH \otimes m)) &= g\cdot(g_iH \otimes \alpha(m)) \\ &= (gg_iH)\otimes\alpha(m) \\ &= \alpha\uparrow_H^G (gg_iH \otimes m) \\ &= \alpha\uparrow_H^G (g(g_iH\otimes m)) \end{aligned}$$

This proves the lemma. □

All KG-modules can be reached by induction. This is the statement of the next lemma.

Lemma 3.2.9. *Let G be a finite group, let H be a subgroup and let K be a field in which $|G|$ is invertible. Then for every simple KG-module N there is a simple KH-module M such that N is a direct factor of $M \uparrow_H^G$.*

Proof. First, if $M = M_1 \oplus M_2$ for KH-modules M, M_1 and M_2, then

$$M \uparrow_H^G = (M_1 \oplus M_2) \uparrow_H^G \simeq M_1 \uparrow_H^G \oplus M_2 \uparrow_H^G$$

by Lemma 3.1.10 and so, if N is an indecomposable KG-module and if N is a direct factor of $M \uparrow_H^G$ for some KH-module M, then there is an indecomposable KH-module M' such that N is a direct factor of $M' \uparrow_H^G$. Of course, simple modules are indecomposable.

Now, $KH \uparrow_H^G = KG \otimes_{KH} KH \simeq KG$ by Lemma 3.1.11 and so the regular H-module induced to G is the regular G-module. Since every simple KG-module is a direct factor of the regular one, we proved the lemma. □

Remark 3.2.10. It is important to note that this does not mean that every KG-module is isomorphic to a module of the form $M \uparrow_H^G$ for some subgroup H of G and a KH-module

M. One might need to go to quotients and submodules in order to get a specific KG-module M. This remark is crucial in case the order of the group is not invertible in K.

Lemma 3.2.11. *Let R be a commutative ring, let G be a group and let H be a subgroup of G and K be subgroup of H. Then for every two RK-modules M and N one gets*

$$(M \uparrow_K^H) \uparrow_H^G \simeq M \uparrow_K^G$$

and

$$(M \oplus N) \uparrow_H^G \simeq M \uparrow_H^G \oplus N \uparrow_H^G .$$

Proof. Using Lemma 3.1.10 one gets

$$M \uparrow_K^H \uparrow_H^G \simeq RG \otimes_{RH} (RH \otimes_{RK} M) \simeq (RG \otimes_{RH} RH) \otimes_{RK} M \simeq RG \otimes_{RK} M.$$

The second affirmation is proved in a similar way. This proves the lemma. □

Example 3.2.12. Let $\mathfrak{S}_n$ be the symmetric group on n letters $\{1, 2, \ldots, n\}$. The group $\mathfrak{S}_n$ acts on $\{1, 2, \ldots, n\}$ and one gets that $\mathfrak{S}_{n-1}$ is a subgroup of $\mathfrak{S}_n$ by just considering those permutations of $\{1, 2, \ldots, n\}$ which fix the last point n. In other words, $Stab_{\mathfrak{S}_n}(n) \simeq \mathfrak{S}_{n-1}$, where $Stab_G(S)$ denotes the pointwise stabiliser of the subset S of Ω in the group G acting on Ω. The case of a setwise stabiliser can be obtained by considering the induced action of $\mathfrak{S}_n$ on $\mathcal{P}(\Omega)$, the set of subsets of Ω. The setwise stabiliser is then the stabiliser of the point S of $\mathcal{P}(\Omega)$ under the action of $\mathfrak{S}_n$ on $\mathcal{P}(\Omega)$. Take K the trivial $\mathfrak{S}_n$-module, consider it as trivial $\mathfrak{S}_{n-1}$-module by just restricting the action of $\mathfrak{S}_n$ to $\mathfrak{S}_{n-1}$, and construct $M := K \uparrow_{\mathfrak{S}_{n-1}}^{\mathfrak{S}_n}$. Then M is of dimension n over K. We apply Lemma 3.1.10 to this situation and get

$$\begin{aligned} Hom_{K\mathfrak{S}_n}(M, K) &= Hom_{K\mathfrak{S}_n}(K\mathfrak{S}_n \otimes_{K\mathfrak{S}_{n-1}} K, K) \\ &\simeq Hom_{K\mathfrak{S}_{n-1}}(K, Hom_{K\mathfrak{S}_n}(K\mathfrak{S}_n, K)) \\ &\simeq Hom_{K\mathfrak{S}_{n-1}}(K, K) \\ &= K \end{aligned}$$

where the second isomorphism is Lemma 3.1.10 and the third isomorphism comes from the following fact. For every algebra A and every A-module V one has $Hom_A(A, V) \simeq V$ as A-module, using the A-A-bimodule structure of A by left and right multiplication. The isomorphism is given by $Hom_A(A, V) \ni f \mapsto f(1) \in V$. Hence, the trivial $K\mathfrak{S}_n$-module K is a quotient of M, and in case $n!$ is invertible in K, a direct factor of M. Suppose $n!$ is invertible in K. Then we have $M \simeq K \oplus N$ for another $K\mathfrak{S}_n$-module N of dimension $n - 1$. Is N simple? In order to decide this we shall use Lemma 3.1.10 again.

$$Hom_{K\mathfrak{S}_n}(M, M) = Hom_{K\mathfrak{S}_n}(K\mathfrak{S}_n \otimes_{K\mathfrak{S}_{n-1}} K, K\mathfrak{S}_n \otimes_{K\mathfrak{S}_{n-1}} K)$$

$$\simeq Hom_{K\mathfrak{S}_{n-1}}(K, Hom_{K\mathfrak{S}_n}(K\mathfrak{S}_n, K\mathfrak{S}_n \otimes_{K\mathfrak{S}_{n-1}} K))$$
$$\simeq Hom_{K\mathfrak{S}_{n-1}}(K, K\mathfrak{S}_n \otimes_{K\mathfrak{S}_{n-1}} K)$$

and we know that the dimension of this space is at least 2. The dimension is 2 if and only if N is simple and non isomorphic to K. Indeed,

$$\begin{aligned} Hom_{K\mathfrak{S}_n}(M,M) &= Hom_{K\mathfrak{S}_n}(K \oplus N, K \oplus N) \\ &= \begin{pmatrix} Hom_{K\mathfrak{S}_n}(K,K) & Hom_{K\mathfrak{S}_n}(N,K) \\ Hom_{K\mathfrak{S}_n}(K,N) & Hom_{K\mathfrak{S}_n}(N,N) \end{pmatrix} \\ &= \begin{pmatrix} K & Hom_{K\mathfrak{S}_n}(N,K) \\ Hom_{K\mathfrak{S}_n}(K,N) & Hom_{K\mathfrak{S}_n}(N,N) \end{pmatrix} \end{aligned}$$

and so, since $Hom_{K\mathfrak{S}_n}(N,N)$ contains at least the identity,

$$\begin{aligned} dim_K(Hom_{K\mathfrak{S}_n}(M,M)) = 2 \Rightarrow Hom_{K\mathfrak{S}_n}(N,K) = Hom_{K\mathfrak{S}_n}(K,N) = 0 \\ \text{and } N \text{ is simple.} \end{aligned}$$

We need to examine $Hom_{K\mathfrak{S}_{n-1}}(K, K\mathfrak{S}_n \otimes_{K\mathfrak{S}_{n-1}} K)$, or what is the same, how many copies of the trivial $K\mathfrak{S}_{n-1}$-module are a submodule of the induced module $K\mathfrak{S}_n \otimes_{K\mathfrak{S}_{n-1}} K$, considered as left $K\mathfrak{S}_{n-1}$-module. Now, call for every $m \in \{1, 2, \dots, n-1\}$ the 2-cycle $\sigma_m = (m\ n)$ the element in $\mathfrak{S}_n$ that interchanges m with n in $\{1, 2, \dots, n\}$. Then one computes $\sigma_m \sigma_k = (k\ n\ m) \notin \mathfrak{S}_{n-1}$. Hence, we can write $\mathfrak{S}_n$ as a union of left cosets modulo $\mathfrak{S}_{n-1}$:

$$\mathfrak{S}_n = \mathfrak{S}_{n-1} \cup \bigcup_{m=1}^{n-1} \sigma_m \mathfrak{S}_{n-1}.$$

By Remark 3.2.4 one gets that $\mathfrak{S}_{n-1}$ acts as permutation of the basis $\{\sigma_m \mathfrak{S}_{n-1} \otimes 1 \mid m \in \{1, 2, \dots, n-1\}\} \cup \{\mathfrak{S}_{n-1}\}$ by left multiplication with elements of $\mathfrak{S}_{n-1}$. In order to find the trivial submodule, just as in Remark 2.3.3, one needs to determine the $K\mathfrak{S}_{n-1}$-module

$$\left(\frac{1}{|\mathfrak{S}_{n-1}|} \sum_{\sigma \in \mathfrak{S}_{n-1}} \sigma\right) \cdot K\mathfrak{S}_n \otimes_{K\mathfrak{S}_{n-1}} K.$$

Put $e_1 := (\frac{1}{|\mathfrak{S}_{n-1}|} \sum_{\sigma \in \mathfrak{S}_{n-1}} \sigma)$. Since every element σ in $\mathfrak{S}_{n-1}$ has the property that $\sigma \mathfrak{S}_{n-1} = \mathfrak{S}_{n-1}$ it is clear that

$$e_1 \cdot (\mathfrak{S}_{n-1} \otimes_{K\mathfrak{S}_{n-1}} 1_K) = \mathfrak{S}_{n-1} \otimes_{K\mathfrak{S}_{n-1}} 1_K.$$

Moreover, since $(k\ m) \cdot (m\ n) \cdot (k\ m)^{-1} = (k\ n)$ for all $m, k \in \{1, 2, \dots, n-1\}$ one has

$$e_1 \cdot (\sigma_m \mathfrak{S}_{n-1} \otimes_{K\mathfrak{S}_{n-1}} 1_K) = \sum_{m=1}^{n-1} \sigma_m \mathfrak{S}_{n-1} \otimes_{K\mathfrak{S}_{n-1}} 1_K.$$

Therefore, $dim_K(Hom_{K\mathfrak{S}_n}(M,M)) = 2$, namely generated by the two elements $\mathfrak{S}_{n-1} \otimes_{K\mathfrak{S}_{n-1}} 1_K$ and $\sum_{m=1}^{n-1} \sigma_m \mathfrak{S}_{n-1} \otimes_{K\mathfrak{S}_{n-1}} 1_K$.

The methods of the example generalize to the class of all finite groups.

Definition 3.2.13. Let G be a group and let H be a subgroup. Suppose that R is a commutative ring. Then for any RG-module M denote by $M\downarrow_H^G$ the RH-module whose underlying R-module structure is the same as M, and H acts on M as G does. In other words, the group homomorphism $H \longrightarrow Aut_R(M)$ which gives M the structure of an RH-module is the composition of the embedding $H \longrightarrow G$ and the group homomorphism $G \longrightarrow Aut_R(M)$ which gives M the structure of an RG-module. Call $M\downarrow_H^G$ the *restriction of M to H*.

Using this we get the following statement.

Lemma 3.2.14 (Frobenius reciprocity). *Let G be a group and let H be a subgroup of G. Suppose R is a commutative ring. Then for every RG-module M and every RH-module N we get* $Hom_{RG}(N\uparrow_H^G, M) \simeq Hom_{RH}(N, M\downarrow_H^G)$, *natural in N and M.*

Proof. The main ingredient is Lemma 3.1.10.

$$\begin{aligned} Hom_{RG}(N\uparrow_H^G, M) &\simeq Hom_{RG}(RG \otimes_{RH} N, M)\\ &\simeq Hom_{RH}(N, Hom_{RG}(RG, M))\\ &\simeq Hom_{RH}(N, M\downarrow_H^G) \end{aligned}$$

This proves the lemma. □

Corollary 3.2.15. *Let G be a group and let H be a subgroup of G. Suppose K is a field such that* $|G|$ *is invertible in K. Then for every finite dimensional KG-module M and every finite dimensional KH-module N we get* $Hom_{KG}(M, N\uparrow_H^G) \simeq Hom_{KH}(M\downarrow_H^G, N)$.

Proof. This is a consequence of the fact that Wedderburn's theorem implies that for finite dimensional semisimple K-algebras A and finite dimensional A-modules M and N one has $dim_K(Hom_A(M,N)) = dim_K(Hom_A(N,M))$. □

Remark 3.2.16. If H is of finite index in G the result of Corollary 3.2.15 is true without the hypothesis that KG is semisimple. In fact, define

$$\begin{aligned} Hom_{RH}(M\downarrow_H^G, N) &\xrightarrow{\Phi} Hom_{RG}(M, N\uparrow_H^G)\\ \varphi &\mapsto \left(m \mapsto \sum_{gH\in G/H} g\otimes\varphi(g^{-1}m)\right) \end{aligned}$$

Then this formula does not depend on the representatives gH of G/H. Indeed,

$$\begin{aligned} (\Phi(\varphi))(m) &= \sum_{gH\in G/H} g\otimes\varphi(g^{-1}m)\\ &= \sum_{gH\in G/H} g\otimes\varphi(hh^{-1}g^{-1}m) \end{aligned}$$

$$= \sum_{gH\in G/H} g\otimes h\varphi(h^{-1}g^{-1}m)$$
$$= \sum_{gH\in G/H} gh\otimes \varphi(h^{-1}g^{-1}m)$$
$$= \sum_{ghH\in G/H} gh\otimes \varphi((gh)^{-1}m)$$

for all $m\in M$ and $h\in H$.

Moreover, $\Phi(\varphi)$ is G-linear. Indeed, for all $m\in M$ and $k\in G$ one has

$$\begin{aligned}(\Phi(\varphi))(km) &= \sum_{gH\in G/H} g\otimes\varphi(g^{-1}km)\\ &= \sum_{gH\in G/H} g\otimes\varphi((k^{-1}g)^{-1}m)\\ &= \sum_{gH\in G/H} kk^{-1}g\otimes\varphi((k^{-1}g)^{-1}m)\\ &= k\Big(\sum_{gH\in G/H}(k^{-1}g)\otimes\varphi((k^{-1}g)^{-1}m)\Big)\\ &= k\Big(\sum_{(k^{-1}g)H\in G/H}(k^{-1}g)\otimes\varphi((k^{-1}g)^{-1}m)\Big)\\ &= k(\Phi(\varphi)(m))\end{aligned}$$

We need to show that Φ is bijective. For this we define a mapping in the other direction. Since $RG\otimes_{RH}N=\bigoplus_{gH\in G/H}g\otimes N$ as RH-module, every $\varphi\in Hom_{RG}(M,RG\otimes_{RH}N)$ decomposes into $\varphi=\oplus_{gH\in G/H}\varphi_g$ with $\varphi_g: N\longrightarrow g\otimes M$. Define $\Psi(\varphi)=\varphi_1$ for every $\varphi\in Hom_{RG}(M,RG\otimes_{RH}N)$. It is clear by construction that φ_1 is RH-linear, and that Ψ is inverse to Φ.

We can give the character of an induced module.

Lemma 3.2.17. *Let G be a finite group, let K be a field such that the order of G is invertible in K, and let M be a KG-module. For every subgroup H of G and every element $g\in G$ put $\mathfrak{X}_{H,g}:=\{x\in G\mid g\in xHx^{-1}\}$. Then for all $g\in G$ one has*

$$\chi_{M\uparrow_H^G}(g)=\frac{1}{|H|}\sum_{x\in\mathfrak{X}_{H,g}}\chi_M(x^{-1}gx).$$

Proof. Let B be a K-basis of M and let $G=\bigcup_{i=1}^n g_iH$ such that $n=|G:H|$. Then $\{g_i\otimes b\mid i\in\{1,2,\dots,n\};b\in B\}$ is a K-basis of $M\uparrow_H^G=KG\otimes_{KH}M$. We need to sum the diagonal coefficients of the action of g on this basis.

Now,

$$g\cdot(g_i\otimes b)=(gg_i\otimes b)$$

and so one can get a non zero diagonal coefficient for this basis vector only if $gg_i \in g_iH$. This is equivalent to $g \in g_iHg_i^{-1}$, or otherwise said $g_i \in \mathfrak{X}_{H,g}$.

If $g \notin g_iHg_i^{-1}$, then the contribution of the basis vectors $g_i \otimes b$ to the character value is 0.

If $g \in g_iHg_i^{-1}$, then

$$g \cdot (g_i \otimes b) = (gg_i \otimes b) = (g_i \otimes g_i^{-1}gg_ib).$$

We consider hence $g_i^{-1}gg_i$ acting on all the basis vectors $b \in B$ of M, hence we consider $g_i^{-1}gg_i$ acting on M, and take the trace of this matrix then. By definition of a character the trace of the matrix corresponding to $g_i^{-1}gg_i$ acting on M is $\chi_M(g_i^{-1}gg_i)$. Therefore,

$$\begin{aligned}\chi_{M\uparrow_H^G}(g) &= \sum_{g_i \in \mathfrak{X}_{H,g};i=1}^{n} \chi_M(g_i^{-1}gg_i)\\ &= \frac{1}{|H|}\sum_{x \in \mathfrak{X}_{H,g}} \chi_M(x^{-1}gx)\end{aligned}$$

where in the last equation one sums over all of the coset g_iH for all $i \in \{1, 2, \ldots, n\}$ and gets the same result whenever one fixes i. Hence, if one divides by the size of the coset $|H|$, one gets the correct result. This finishes the proof. □

3.3 Mackey's formula

A very far reaching concept using induction from and restriction to subgroups is Mackey's formula. It should be observed that there is a very extensive literature on abstract versions of the Mackey formula. For a very complete account I recommend Bouc [Bouc-97].

The question there is what happens if one induces a module from a subgroup H of a group G and restricts the result to a possibly different subgroup K.

In order to get the right concept well suited to study this problem let G be a group and let H and K be two subgroups of G. Of course $H = K$ is allowed, but the subgroups may well differ. Define a relation ${}_K{\sim}_H$ on G by setting

$$g_1 \,{}_K{\sim}_H\, g_2 \Leftrightarrow g_1 \in \{hg_2k \mid h \in H; k \in K\}$$

for all $g_1, g_2 \in G$.

Lemma 3.3.1. *Let G be a group and H and K be two subgroups. Then ${}_K{\sim}_H$ is an equivalence relation.*

Proof. Of course,

$$g_1 \,{}_K{\sim}_H\, g_2 \Rightarrow \exists h \in H, k \in K : g_1 = hg_2k$$

$$\Rightarrow \exists h \in H, k \in K : h^{-1}g_1k^{-1} = g_2$$
$$\Rightarrow g_2 \,{}_K{\sim}_H\, g_1$$

and $g_1 \,{}_K{\sim}_H\, g_1$ for all $g_1 \in G$ taking $h = k = 1$.

Moreover

$$\begin{aligned} g_1 \,{}_K{\sim}_H\, g_2 \text{ and } g_2 \,{}_K{\sim}_H\, g_3 &\Rightarrow \exists h_1, h_2 \in H, k_1, k_2 \in K : g_1 = h_1g_2k_1 \\ &\qquad \text{and } g_2 = h_2g_3k_2 \\ &\Rightarrow \exists h_1, h_2 \in H, k_1, k_2 \in K : g_1 = (h_1h_2)g_2(k_2k_1) \end{aligned}$$

This shows the lemma. □

As a consequence one sees that two classes Kg_1H and Kg_2H are either identical or disjoint. Moreover, G is the disjoint union of double classes.

Definition 3.3.2. The set of equivalence classes of G under the relation ${}_K{\sim}_H$ is denoted by $K\backslash G/H$. The equivalence class of $g \in G$ is called double coset, or double class of g modulo K and H and is denoted by KgH.

Example 3.3.3. If $K = 1$ we get the usual left cosets of G/H as double cosets. There each coset gH for $g \in G$ is of cardinality $|H|$. This is not longer true if $K \neq 1$. The size of double cosets are not constant in general.

Let $D_4 = \langle a, b | \ a^4, b^2, baba\rangle$ be the dihedral group of order 8. Denote $B := \langle b\rangle$ the cyclic group of order 2 generated by b. Then D_4/B has 4 classes of length 2

$$\begin{aligned} D_4/B &= \{1, b\} \cup \{a, ab\} \cup \{a^2, a^2b\} \cup \{a^3, a^3b\} \\ &= (1 \cdot B) \cup (a \cdot B) \cdot (a^2 \cdot B) \cup (a^3 \cdot B). \end{aligned}$$

In $B\backslash D_4/B$ there are 2 classes $\{1, b\}$ and $\{a^2, a^2b\}$ of length 2 and one class $\{a, ab, a^3, a^3b\}$ of length 4.

$$\begin{aligned} B\backslash D_4/B &= \{1, b\} \cup \{a^2, a^2b\} \cup \{a, ab, a^3, a^3b\} \\ &= (B \cdot 1 \cdot B) \cup (B \cdot a \cdot B) \cup (B \cdot a^2 \cdot B). \end{aligned}$$

Lemma 3.3.4. *Let G be a group and let H and K be subgroups of G. If H or K is a normal subgroup of G, then the double cosets $K\backslash G/H$ are all of the same size.*

Proof. Suppose K is normal in G. Then $KgH = gKH$ and KH is a subgroup of G. Hence the double cosets of G modulo K and H are the same as the left cosets modulo the subgroup KH. The case of H being normal is similar. □

The reason why we need to study double cosets comes from the question to understand the group ring RG for G a group and R a commutative ring as $RK - RH$-bimodule, for K and H being subgroups of G. Once this is understood we are going to be able to deduce the important Mackey formula from there.

We need another rather general concept before. Let R be a commutative ring, let A be an R-algebra and let α be an automorphism of A as R-algebra. For any A-module M we form another A-module ${}^{\alpha}M$ by the following construction.

Since M is an A-module, the module structure on M is given by a homomorphism of R-algebras $A \xrightarrow{\mu} End_R(M)$. The composition

$$A \xrightarrow{\alpha} A \xrightarrow{\mu} End_R(M)$$

is a homomorphism of R-algebras again. Hence one gets a structure of an A-module on the *same set* M. This A-module is denote by ${}^{\alpha}M$.

Definition 3.3.5. Let R be a commutative ring, let A be an R-algebra and let M be an A-module. Given an automorphism α of A. Then the A-module ${}^{\alpha}M$ is called the twist of M by α.

Let us give a more explicit description of such an ${}^{\alpha}M$. Since α is R-linear, it is clear that *as R-module* one has that ${}^{\alpha}M$ is really equal to M. Denote by $\bullet$ the operation of A on ${}^{\alpha}M$ and by $\cdot$ the operation of A on M. Then for every $a \in A$ one has $a \bullet m := \alpha(a) \cdot m$.

Definition 3.3.6. An automorphism α of an R-algebra A is *inner* if there is an invertible element $u \in A$ such that $\alpha(a) = u \cdot a \cdot u^{-1}$ for all $a \in A$. Denote $Inn(A)$ the set of inner automorphisms of A.

Lemma 3.3.7. *Let A be an R-algebra and let M be an A-module. Then for all $\alpha \in Inn(A)$ one has ${}^{\alpha}M \simeq M$.*

Proof. Let u be an element of A such that $\alpha(a) = u \cdot a \cdot u^{-1}$ for all $a \in A$. Then

$$\begin{aligned} M &\xrightarrow{\varphi} {}^{\alpha}M \\ m &\mapsto u \cdot m \end{aligned}$$

is an A-module homomorphism. Indeed, denoting again by $\bullet$ the operation of A on ${}^{\alpha}M$, one gets

$$\varphi(a \cdot m) = u \cdot (a \cdot m) = (u \cdot a \cdot u^{-1}) \cdot (u \cdot m) = a \bullet \varphi(m)$$

for all $a \in A$ and $m \in M$. Of course, φ is a homomorphism of R-modules since the image of R in A is in the centre of A. Moreover, since u is invertible, φ is invertible as well. The inverse application is multiplication by u^{-1}. We proved the lemma. □

If α is an inner automorphism of A and $\alpha(a) = u \cdot a \cdot u^{-1}$ for all $a \in A$, then denote ${}^{\alpha}M = {}^{u}M$ for every A-module M. The proof of Lemma 3.3.7 suggests to denote ${}^{u}M$ by $u \cdot M$ as well.

Example 3.3.8. The inverse of Lemma 3.3.7 is not true in general. Let K be a field, let G be a group and let K be the trivial KG-module. Let α be an automorphism of G that is not inner in KG. This happens for example if G is abelian and $\alpha \in Aut(G)$ is not the identity. Then ${}^{\alpha}K \simeq K$ since every element of G acts as identity, and so $g \in G$ acts just the same way as $\alpha(g)$ does.

Lemma 3.3.9. *Let R be a commutative ring and let G be a group. Suppose H and K are subgroups of G. Then RG is an RK-RH-bimodule by multiplication of elements in RG. Moreover,*

$$RG \simeq \bigoplus_{(KgH \in K\backslash G/H)} RK \otimes_{R(K\cap gHg^{-1})} gRH$$

as RK-RH-bimodule.

Proof. Let

$$K\backslash G/H = \bigcup_{i \in I} Kg_iH$$

with elements $g_i \in G$ for all $i \in I$, an index set, such that $Kg_iH = Kg_jH \Leftrightarrow i = j$. Then for every $g \in G$ there is a unique $i(g) \in I$ and a unique $k(g) \in K$, $h(g) \in H$ such that $g = k(g)g_{i(g)}h(g)$. Define a mapping

$$\begin{aligned} RG &\xrightarrow{\alpha} \bigoplus_{i \in I} RK \otimes_{R(K\cap g_iHg_i^{-1})} g_iRH \\ g &\mapsto k(g) \otimes g_{i(g)}h(g) \end{aligned}$$

and an inverse mapping

$$\begin{aligned} \bigoplus_{i \in I} RK \otimes_{R(K\cap g_iHg_i^{-1})} g_iRH &\xrightarrow{\beta} RG \\ k \otimes gh &\mapsto kgh \end{aligned}$$

First, β is well defined. Actually gRH is an $R(gHg^{-1})$-module. More generally, if L is a subgroup of G and if α is an automorphism of G then for each RL-module M one has that ${}^{\alpha}M$ is an $R\alpha^{-1}(L)$-module. Indeed, $\ell \bullet m = \alpha(\ell)m$ and this is defined whenever $\ell \in \alpha^{-1}(L)$. Moreover as is immediately seen for every $x \in K \cap gHg^{-1}$ one has

$$\beta(kx \otimes gh) = kxgh = kg(g^{-1}xgh) = \beta(k \otimes g(g^{-1}xgh))$$

and so the operation of $K \cap gHg^{-1}$ on gRH is defined exactly in the way needed for giving a well-defined mapping.

It is clear by definition that α and β are mutually inverse mappings. Hence, β and α are bijective.

Moreover, β is trivially a homomorphism of RK-RH-bimodules. This shows the lemma. □

We come to the main result of Section 3.3.

Theorem 3.3.10 (Mackey). *Let R be a commutative ring, let G be a group with subgroups H and K and let M be an RH-module. Then*

$$M\uparrow_H^G\downarrow_K^G \simeq \bigoplus_{KgH\in K\backslash G/H} ({}^g(M\downarrow^H_{(H\cap gKg^{-1})}))\uparrow^K_{(gHg^{-1}\cap K)}$$

Proof.

$$\begin{aligned} M\uparrow_H^G\downarrow_K^G &\simeq \Big(\bigoplus_{KgH\in K\backslash G/H} RK\otimes_{R(K\cap gHg^{-1})} gRH\Big)\otimes_{RH} M \quad \text{(cf. Lemma 3.3.9)}\\ &\simeq \bigoplus_{KgH\in K\backslash G/H} RK\otimes_{R(K\cap gHg^{-1})} (gRH\otimes_{RH} M) \quad \text{(cf. Lemma 3.1.10)}\\ &\simeq \bigoplus_{KgH\in K\backslash G/H} RK\otimes_{R(K\cap gHg^{-1})} gM \quad \text{(cf. Lemma 3.1.11)}\\ &\simeq \bigoplus_{KgH\in K\backslash G/H} ({}^g(M\downarrow^H_{H\cap gKg^{-1}}))\uparrow^K_{gHg^{-1}\cap K} \end{aligned}$$

This proves the theorem. □

There are plenty of applications of this formula. Its force becomes evident when one discusses more sophisticated situations in particular when one studies representations over fields of positive characteristic dividing the group order.

Nevertheless, we give three applications right now.

Corollary 3.3.11. *Let G be a finite group, let H be a subgroup of G and let K be a field. Suppose M is an indecomposable KH-module with $End_{KH}(M) = K$. Then $End_{KG}(M\uparrow_H^G) = K$ if and only if for all $g\in G\setminus H$ one has*

$$Hom_{K(H\cap gHg^{-1})}(({}^gM\downarrow^{gHg^{-1}}_{(H\cap gHg^{-1})}), M\downarrow^H_{(H\cap gHg^{-1})}) = 0.$$

Proof.

$$\begin{aligned} K &= Hom_{KG}(M\uparrow_H^G, M\uparrow_H^G)\\ &= Hom_{KH}(M, M\uparrow_H^G\downarrow_H^G) \quad \text{(cf. Lemma 3.2.14)}\\ &= Hom_{KH}\Big(M, \bigoplus_{HgH\in K\backslash G/H} ({}^g(M\downarrow^H_{(H\cap gHg^{-1})}))\uparrow^H_{(gHg^{-1}\cap H)}\Big)\\ &= \bigoplus_{HgH\in H\backslash G/H} Hom_{KH}(M, ({}^g(M\downarrow^H_{(H\cap gHg^{-1})}))\uparrow^H_{(gHg^{-1}\cap H)}). \end{aligned}$$

Now, for $HgH = H$, the class with representative $g = 1$, one has

$$\begin{aligned} Hom_{KH}(M, ({}^1(M\downarrow^H_{(H\cap 1H1^{-1})}))\uparrow^H_{(1H1^{-1}\cap H)}) &= Hom_{KH}(M, M\downarrow^H_H\uparrow^H_H)\\ &= End_{KH}(M)\\ &= K \end{aligned}$$

This shows the Corollary. □

Of course, if the order of G is invertible in K and K is a splitting field for G, then $End_{KG}(M) = K$ if and only if M is simple. In case K is not a splitting field, similar arguments as in the proof of Corollary 3.3.11 may be applied as well. If the order of G is not invertible in K then the Maschke's theorem 1.2.9 does not apply anymore and in general there are modules with endomorphism algebra being K but which are not simple.

The method of proof of Corollary 3.3.11 gives another application which will be of importance later.

Corollary 3.3.12. *Let G be a finite group and, H be a subgroup and let R be a commutative ring. Then for every RH-module M one has that M is a direct factor of $M \uparrow_H^G\downarrow_H^G$.*

Proof.

$$M \uparrow_H^G\downarrow_H^G \simeq \bigoplus_{HgH\in H\backslash G/H} ({}^g(M \downarrow^H_{(H\cap gHg^{-1})})) \uparrow^H_{(gHg^{-1}\cap H)}$$

and in particular for the class $H1H = H$ one has

$$({}^1(M \downarrow^H_{(H\cap 1H1^{-1})})) \uparrow^H_{(1H1^{-1}\cap H)} = M \downarrow^H_H\uparrow^H_H = M$$

is a direct factor of $M \uparrow_H^G\downarrow_H^G$. This proves the corollary. □

Definition 3.3.13. Let G be a group and let R be a commutative ring. A finitely generated RG-module M is a *permutation module* if there is an R-base B of M such that for all $g \in G$ and all $b \in B$ one has $gb \in B$. Such a basis B is called *permutation basis*. A *p-permutation module* (or trivial source module) is a direct summand of a permutation module. A *transitive permutation module* is a permutation module with a permutation basis which is a transitive G-set.

Lemma 3.3.14. *Given a commutative ring R and a group G, then M is an RG-permutation module if and only if M is a direct sum of modules of the form $R \uparrow_H^G$ for some subgroups H of G.*

Proof. Let M be a permutation module and let B be a permutation basis. Let $B = B_1 \cup B_2 \cup \cdots \cup B_s$ be a disjoint composition into G-orbits. Then each B_i for $i \in \{1,\ldots,s\}$ is a transitive G-set and therefore for all $i \in \{1,\ldots,s\}$ one gets that $B_i \simeq G/H_i$ as G-set for subgroups $H_1, H_2, \ldots, H_s$ of G.

Moreover, denoting for all $i \in \{1,\ldots,s\}$ by RB_i the R-linear combinations inside M formes by elements in B_i one gets $M \simeq RB_1 \oplus RB_2 \oplus \cdots \oplus RB_s$ as RG-module.

Without loss of generality we may hence suppose $s = 1$ and we identify $B = B_1$ with G/H.

$$\begin{aligned} RG \otimes_{RH} R &\stackrel{\alpha}{\longrightarrow} M \\ \sum_{g\in G} r_g g \otimes 1 &\mapsto \sum_{g\in G} r_g \cdot (g \cdot H) \end{aligned}$$

Then α is well defined. Indeed,

$$\alpha\Big(\sum_{g\in G} r_g gh \otimes 1\Big) = \sum_{g\in G} r_g \cdot (gh \cdot H) = \sum_{g\in G} r_g \cdot (g \cdot hH)$$
$$= \sum_{g\in G} r_g \cdot (g \cdot H) = \alpha\Big(\sum_{g\in G} r_g g \otimes 1\Big)$$

Moreover α is RG-linear. We define a mapping in the other direction by

$$\begin{aligned} M &\xrightarrow{\beta} RG \otimes_{RH} R \\ \sum_{g\in G} r_g gH &\mapsto \sum_{g\in G} r_g g \otimes 1 \end{aligned}$$

Then again β is well-defined and obviously invers to α.

To show the other direction we first mention that direct sums of permutation modules are again permutation modules, the permutation basis of the sum being the union of the permutation bases of the summands. Hence we only need to show that $R \uparrow_H^G$ is a permutation module. But this is actually done in the first step, showing that $RG \otimes_{RH} R \simeq R(G/H)$, where $R(G/H)$ is $R^{G/H}$ as R-module and G acts on $R^{G/H}$ as G acts on G/H. This shows the lemma. □

Remark 3.3.15. In the literature sometimes p-permutation modules as defined in Definition 3.3.13 are called permutation modules as well. It will be of importance to distinguish genuine permutation modules from p-permutation modules.

Corollary 3.3.16. *Let K be a field and let G be a group. Then the G-fixed points of a finite dimensional transitive permutation module is one-dimensional K-space.*

Proof. Let $M = K(G/H)$ be a transitive G-permutation module. Then using Remark 3.2.16

$$\begin{aligned} M^G &= Hom_{KG}(K, M) \\ &= Hom_{KG}(K, K \uparrow_H^G) \\ &= Hom_{KH}(K \downarrow_H^G, K) \\ &= Hom_{KH}(K, K) \\ &= K \end{aligned}$$

This proves the lemma. □

We get an immediate consequence of Corollary 3.3.16. and Mackey's theorem 3.3.10.

Corollary 3.3.17. *Let G be a group, let H be a subgroup of finite index and let K be a field. Then $dim_K(End_{KG}(K(G/H))) = |H\backslash G/H|$.*

Proof.

$$\begin{aligned}
&End_{KG}(K(G/H)) \\
&\simeq Hom_{KG}(K\uparrow_H^G, K\uparrow_H^G) \\
&\simeq Hom_{KH}(K, K\uparrow_H^G\downarrow_H^G) \\
&\simeq Hom_{KH}\Big(K, \bigoplus_{HgH\in H\backslash G/H} ({}^g(K\downarrow^H_{(H\cap gHg^{-1})}))\uparrow^H_{(gHg^{-1}\cap H)}\Big) \\
&\simeq \bigoplus_{HgH\in H\backslash G/H} Hom_{KH}(K, ({}^g(K\downarrow^H_{(H\cap gHg^{-1})}))\uparrow^H_{(gHg^{-1}\cap H)}) \\
&\simeq \bigoplus_{HgH\in H\backslash G/H} Hom_{KH}(K, K\downarrow^H_{(H\cap gHg^{-1})}\uparrow^H_{(gHg^{-1}\cap H)})
\end{aligned}$$

and each of the modules

$$K\downarrow^H_{(H\cap gHg^{-1})}\uparrow^H_{(gHg^{-1}\cap H)} = K\uparrow^H_{(gHg^{-1}\cap H)}$$

is a transitive KH-permutation module. By Corollary 3.3.16 its fixed point space is one-dimensional. This shows the corollary. □

Example 3.3.18. Consider the subgroup $\mathfrak{S}_{n-1}$ in $\mathfrak{S}_n$ and a field K. The left cosets are represented by the cycles (i, n) for $i \in \{1, 2, \ldots, n-1\}$. Indeed, if $(i\ n)\mathfrak{S}_{n-1} = (i-1\ n)\mathfrak{S}_{n-1}$, then $(i-1\ n)\cdot(i\ n) = (i-1\ i\ n) \in \mathfrak{S}_{n-1}$, which is absurd. But, there are only two double cosets, namely $\mathfrak{S}_{n-1}$ and $\mathfrak{S}_{n-1}(1\ n)\mathfrak{S}_{n-1}$. Indeed,

$$(i\ j)\cdot(i\ n)\cdot(i\ j) = (j\ n)$$

for all $i \neq j$ with $i, j \in \{1, 2, \ldots, n-1\}$. Hence, if $n!$ is invertible in K, by Corollary 3.3.17 and by Corollary 3.3.16 the module

$$K\uparrow^{\mathfrak{S}_n}_{\mathfrak{S}_{n-1}} \simeq K \oplus L$$

is isomorphic to the direct sum of the trivial module K and a simple $n-1$-dimensional $K\mathfrak{S}_n$-module L. This completes Example 1.2.11.

3.4 Clifford theory

We come to the question if it is possible to link the representation theory of a group G over a commutative ring R to the representation theory of some normal subgroup of G. Of course in this generality the question cannot have a positive answer, since the trivial group with one element has a trivial representation theory and of course is a normal subgroup of every group.

3.4.1 Inertia group

In order to solve this problem one needs to study the concept of an inertia group. Given a group G and a normal subgroup N. Let R be a commutative ring and let M be an RN-module. Then for all $g \in G$ conjugation with g gives an automorphism γ_g on N, since N is normal in G. We may hence study the twisted KN-module ${}^{\gamma_g}M$ (cf. Definition 3.3.5) and denote it again, slightly abusing the notation, by ${}^gM$.

Definition 3.4.1. Let G be a group and let N be a normal subgroup of G. Let R be a commutative ring and let M be an RN-module.

$$I_G(M) := \{g \in G | \; {}^gM \simeq M \text{ as } RN\text{-module} \}$$

is the *inertia group* of M in G.

Lemma 3.4.2. *For G be a group, R a commutative ring, N a normal subgroup of G and M an RN-module, the inertia group $I_G(M)$ is a subgroup of G and N is a subgroup of $I_G(M)$. Moreover, N and $C_G(N) := \{g \in G| \; gn = ng \forall n \in N\}$ are subgroups of $I_G(M)$.*

Proof. Indeed, $\gamma(gh) = \gamma(g)\gamma(h)$ and so

$${}^{\gamma(gh)}M \simeq {}^{\gamma(g)}\left({}^{\gamma(h)}M\right) \simeq {}^{\gamma(g)}M \simeq M$$

whenever $g, h \in I_G(M)$. A similar proof shows $g \in I_G(M) \Rightarrow g^{-1} \in I_G(M)$.

The fact that N is a subgroup of $I_G(M)$ is a consequence of Lemma 3.3.7. The fact that $C_G(N)$ is a subgroup of $I_G(M)$ follows from the definition of a twisted module.

To show that $N \leq I_G(M)$ we may proceed just as above. However, we could argue more abstractly, which seems to be more instructive. For a K-algebra A and an A-module M we can twist M by any element α in $Aut_K(A)$ just the same way as we defined for conjugation $\gamma(g)$ by $g \in G$. We may define

$$I_{Aut_K(A)}(M) := \{\alpha \in Aut_K(A) \mid {}^{\alpha}M \simeq M\}.$$

Now,

$$Inn(A) \subseteq I_{Aut_K(A)}(M)$$

by Lemma 3.3.7 and since $N \trianglelefteq G$ we get a group homomorphism

$$\gamma : G \longrightarrow Aut_K(KN)$$

by sending $g \in G$ to conjugation $\gamma(g)$ by g, we obtain that the restriction of γ to N has image in $Inn_K(KN)$. This proves that $N \leq I_G(M)$ as well. This proves the lemma. □

In order to illustrate Clifford's theorem we examine Mackey's formula in case $K = H$ is normal in G.

We get

$$\begin{aligned} M \uparrow_H^G \downarrow_H^G &\simeq \bigoplus_{HgH \in H\backslash G/H} ({}^g(M \downarrow^H_{(H\cap gHg^{-1})})) \uparrow^H_{(gHg^{-1}\cap H)} \\ &\simeq \bigoplus_{gH \in G/H} ({}^g(M \downarrow^H_H)) \uparrow^H_H \\ &\simeq \bigoplus_{gH \in G/H} {}^gM. \end{aligned}$$

In particular we get that as RH-module, the induced module $M \uparrow_H^G$ is a direct sum of conjugates ${}^gM$ of M. Within the classes of G/H exactly the classes $I_G(M)/H$ lead to conjugates which are isomorphic to M.

3.4.2 Factor sets, second group cohomology

We first examine what happens when we consider $M \uparrow_H^{I_G(M)}$. We hence assume for the moment that $G = I_G(M)$. Then we know that ${}^gM \simeq M$, and let $\alpha_g : M \longrightarrow {}^gM$ be an isomorphism of KH-modules. For each $g \in I_G(M)$ choose such an isomorphism α_g, and we will choose $\alpha_1 = id_M$, of course. Now, ${}^gM \simeq {}^{hg}M$ for $g, h \in I_G(M)$, and we may take α_h as isomorphism ${}^gM \longrightarrow {}^{hg}M$. Let's check if this is KH-linear.

$$\alpha_h(x \cdot m) = \alpha_h(({}^gx)m) = ({}^gx) \cdot \alpha_h(m) = ({}^h({}^gx))\alpha_h(m) = ({}^{hg}x)m$$

for all $x \in H$ and $m \in M$, and where we denote by $\cdot$ the action of M on the various twisted versions of M to the contrast of the original action of H on M, which we denote by xm for $x \in H$ and $m \in M$. However, $\alpha_h \circ \alpha_g$ will differ in general from α_{hg} by an automorphism of the KH-module M. So,

$$\alpha_{hg}^{-1} \circ \alpha_h \circ \alpha_g =: f(h, g)$$

defines a mapping

$$f : I_G(M) \times I_G(M) \longrightarrow Aut_{KH}(M).$$

By definition, $f(1, g) = f(g, 1) = id_M$ for all $g \in I_G(M)$.

What about associativity in presence of $g, h, k \in I_G(M)$?

$$\begin{aligned} f(k, hg) \circ f(h, g) &= \alpha_{khg}^{-1} \circ \alpha_k \circ \alpha_{hg} \circ \alpha_{hg}^{-1} \circ \alpha_h \circ \alpha_g \\ &= \alpha_{khg}^{-1} \circ \alpha_k \circ \alpha_h \circ \alpha_g \\ &= \alpha_{khg}^{-1} \circ \alpha_{kh} \circ \alpha_g \circ \alpha_g^{-1} \circ \alpha_{kh}^{-1} \circ \alpha_k \circ \alpha_h \circ \alpha_g \end{aligned}$$

$$\begin{aligned}&= f(kh,g) \circ \alpha_g^{-1} \circ f(k,h) \circ \alpha_g \\ &= f(kh,g) \circ (f(k,h)^g)\end{aligned}$$

where we denote by $y^g := \alpha_g^{-1} \circ y \circ \alpha_g$ for any $y \in Aut_{KH}(M)$.

May we modify our choices α_g such that $\alpha_h \circ \alpha_g = \alpha_{hg}$ for all $g, h \in I_G(M)$? Hence $\beta_g = \alpha_g \circ c_g$ for some $c_g \in Aut_{KH}(M)$. this means

$$\begin{aligned}\beta_h \circ \beta_g &= \alpha_h \circ c_h \circ \alpha_g \circ c_g \\ &= \alpha_h \circ \alpha_g \circ (\alpha_g^{-1} \circ c_h \circ \alpha_g) \circ c_g \\ &= \alpha_{hg} \circ f(h,g) \circ (\alpha_g^{-1} \circ c_h \circ \alpha_g) \circ c_g \\ &\overset{!}{=} \alpha_{hg} \circ c_{hg} = \beta_{hg}\end{aligned}$$

This is equivalent to

$$f(h,g) = c_{hg} \circ c_g^{-1} \circ (\alpha_g^{-1} \circ c_h \circ \alpha_g)^{-1} = c_{hg} \circ c_g^{-1} \circ ((c_h)^g)^{-1}$$

where we abbreviate $(c_h)^g = (\alpha_g^{-1} \circ c_h \circ \alpha_g)$. Of course, $f(h,g) := c_{hg} \circ c_g^{-1} \circ ((c_h)^g)^{-1}$ satisfies $f(k,hg) \circ f(h,g) = f(kh,g) \circ (f(k,h)^g)$, as is readily verified.

Definition 3.4.3. Let G be a group, and let X be a group equipped with a mapping $\omega : G \to \mathfrak{S}_X$, denote by $g \mapsto (x \mapsto x^g)$. Then $f : G \times G \to X$ is a *factor set* if

$$f(k,hg) \cdot f(h,g) = f(kh,g) \cdot (f(k,h)^g).$$

The factor set is normalised if $f(1,g) = f(g,1) = 0$ for all $g \in G$.

If M is an abelian group and ω defines a $\mathbb{Z}G$-module structure on X, then a normalised factor set f is called a *2-cocycle*. A 2-cocycle f is a *2-coboundary* if there is a mapping $G \to X$, denoted by $h \mapsto c_h$, such that

$$f(h,g) = c_{hg} \circ c_g^{-1} \circ ((c_h)^g)^{-1}.$$

Denote by $Z^2(G,X)$ the set of normalised factor sets and $B^2(H,X)$ the set of 2-coboundaries.

If X is an abelian group, and ω defines a $\mathbb{Z}G$-module structure, then $Z^2(G,X)$ is an abelian group (as is seen almost by definition) and $B^2(G,X)$ is a subgroup of $Z^2(G,X)$.

Definition 3.4.4. Let G be a group and let X be a $\mathbb{Z}G$-right module. Then

$$H^2(G,X) = Z^2(G,X)/B^2(G,X)$$

is the *degree 2 group cohomology with values in X*.

Now, we consider the case when $Aut_{KH}(M) \simeq K^\times$ is given by the multiplicative group of K, or in other words all endomorphisms of K are just scalar multiplication. This happens for example in case M is a simple KH-module with endomorphism ring K, such as if $K = \mathbb{C}$ and M is a simple module, but there are other cases when this may occur. In this case we also have that the action of $I_G(M)$ on $Aut_K(M)$ is trivial, since this action is given by conjugation with an automorphism. Conjugation is trivial since $Aut_K(M)$ is commutative.

This situation gives a new structure.

Definition 3.4.5. Let Q be a group and let $f \in Z^2(Q, K^\times)$ be a 2-cocycle with a trivial action of Q on $K^\times$. Then we define K_fQ the K-vector space with basis $\{e_Q \mid q \in Q\}$ and a multiplication of the basis elements by $e_{q_1} \cdot e_{q_2} = f(q_1, q_2)e_{q_1q_2}$. Then $(K_fQ, +, \cdot)$ is a K-algebra, the *twisted group algebra*.

We need to verify that the multiplication defined in Definition 3.4.5 is associative.

$$\begin{aligned}(e_{q_1} \cdot e_{q_2}) \cdot e_{q_3} &= f(q_1, q_2)e_{q_1q_2}e_{q_3} \\ &= f(q_1, q_2)f(q_1q_2, q_3)e_{(q_1q_2)q_3} \\ &= f(q_1, q_2q_3)f(q_2, q_3)e_{q_1(q_2q_3)} \\ &= f(q_2, q_3)e_{q_1}e_{q_2q_3} \\ &= e_{q_1} \cdot (e_{q_2} \cdot e_{q_3})\end{aligned}$$

The second group cohomology occurs in another context as well We will need a concept of a new type of representations.

Definition 3.4.6. Let G be a group. A *projective representation* of dimension n over a field K is a homomorphism of groups $G \to PGL_n(K)$.

Recall that $PGL_n(K) := GL_n(K)/Z(GL_n(K))$, where it is elementary that the centre $Z(GL_n(K))$ of $GL_n(K)$ is the group of non zero scalar multiples of the identity matrix.

We consider a more general situation. Let H be a group and let $Z \leq Z(H)$ be a central subgroup of H. Denote $P := H/Z$. Since there is a canonical homomorphism of groups $H \overset{\pi}{\twoheadrightarrow} P$, for $H := GL_n(K)$ and $Z := Z(GL_n(K))$ a representation of G in the ordinary sense induces a projective representation. When is a projective representation induced from a representation in the ordinary sense? The answer will give an element in $H^2(G, K^\times)$.

Let $\alpha : G \to P$ be a group homomorphism, and choose for each $g \in G$ an element $\phi(g) \in H$ such that $\pi(\phi(g)) = \alpha(g)$. Now, since $\alpha(gh) = \alpha(g)\alpha(h)$, we get $f(g, h) := \phi(gh)^{-1}\phi(g)\phi(h) \in Z$ and we obtain again

$$\begin{aligned}f(k, hg) \cdot f(h, g) &= \phi(khg)^{-1}\phi(k)\phi(hg)\phi(hg)^{-1}\phi(h)\phi(g) \\ &= \phi(khg)^{-1}\phi(k)\phi(h)\phi(g)\end{aligned}$$

$$= \phi(khg)^{-1}\phi(kh)\phi(g) \cdot \phi(g)^{-1}\phi(kh)^{-1}\phi(k)\phi(h)\phi(g)$$
$$= f(kh, g) \cdot \phi(g)^{-1} f(k, h)\phi(g).$$

such that f is a 2-cocycle with values in Z. Moreover, since f has values in Z, conjugation of $f(k,h)$ by $\Phi(g)$ is trivial, and hence we have a trivial action of G. Therefore, we may simplify the definition of a 2-cocycle and of a 2-coboundary in case of a trivial action. Now, f is a 2-coboundary if and only if

$$\phi(gh)^{-1}\phi(g)\phi(h) = f(h, g) = c_{hg}c_g^{-1}(\phi(g)^{-1}c_h\phi(g))^{-1} = c_{hg}c_g^{-1}c_h^{-1},$$

and this is equivalent to

$$(\phi(gh)c_{gh})^{-1} = c_g^{-1}c_h^{-1}\phi(h)^{-1}\phi(g)^{-1}$$

or taking inverses, and using that c_g is in Z as well,

$$\phi(gh)c_{gh} = (\phi(h)c_h) \cdot (\phi(g)c_g).$$

This shows that f is a 2-coboundary if and only if

$$\psi : g \mapsto \phi(g)c_g =: \psi(g)$$

is a group homomorphism $G \to H$ with $\pi \circ \psi = \alpha$.

We obtained

Lemma 3.4.7. *Let H be a group, and let Z be a central subgroup of H. Denote $P := H/Z$, let $\pi : H \to P$ be the canonical morphism and let $\alpha : G \to P$ be a group homomorphism. Taking $\phi(g) \in \pi^{-1}(\alpha(g))$ the map $f : G \times G \to Z$ defined by $f(g,h) := \phi(gh)^{-1}\phi(g)\phi(h)$ is a 2-cocycle, and f is a 2-coboundary if and only if there is a group homomorphism $\psi : G \to H$ such that $\pi \circ \psi = \alpha$.*

A map ϕ as in Lemma 3.4.7 is called a lift of the projective representation α with cocycle f. The role of $H^2(G, Z)$ goes even further.

Lemma 3.4.8. *Let Z be a $\mathbb{Z}G$-right module, and let $f : G \times G \to Z$ be a 2-cocycle, then $G \times Z$ carries a group structure G_f given by $(g,z)\cdot(h,w) := (gh, f(g,h)+zh+w)$. Moreover, $f_1, f_2 \in Z^2(G,Z)$ represent the same element in $H^2(G,Z)$ if and only if there is an isomorphism $G_{f_1} \simeq G_{f_2}$ preserving the normal subgroup Z and inducing the identity on G.*

Proof. Associativity follows from the 2-cocycle condition:

$$\begin{aligned}
&((g_1,z_1)\cdot(g_2,z_2))\cdot(g_3,z_3)\\
&= ((g_1g_2, f(g_1,g_2) + z_1g_2 + z_2))\cdot(g_3,z_3)\\
&= ((g_1g_2)g_3, f(g_1g_2,g_3) + f(g_1,g_2)g_3 + (z_1g_2)g_3 + z_2g_3 + z_3))\\
&= ((g_1(g_2g_3)), f(g_1,g_2g_3) + f(g_2,g_3) + z_1(g_2g_3) + z_2g_3 + z_3))
\end{aligned}$$

$$= (g_1, z_1) \cdot (g_2 g_3, f(g_2, g_3) + z_2 g_3 + z_3)$$
$$= (g_1, z_1) \cdot ((g_2, z_2) \cdot (g_3, z_3))$$

The left neutral element is $(1, 0)$ since $(1, 0) \cdot (g, z) = (g, f(1, g) + z) = (g, z)$ since f is a normalised factor set. Now, $(g^{-1}, -zg^{-1}) \cdot (g, z) = (1, 0)$, which proves the first statement.

If $(f_2 - f_1)(g, h) = c_{gh} - c_g - c_h g$, then

$$\begin{array}{ccc} G_{f_1} & \xrightarrow{\rho} & G_{f_2} \\ (g, z) & \mapsto & (g, c_g + z) \end{array}$$

is an isomorphism preserving Z and inducing the identity on G. Indeed, bijectivity is clear. Moreover,

$$\begin{aligned} \rho((g_1, z_1)(g_2, z_2)) &= \rho(g_1 g_2, f_1(g_1, g_2) + z_1 g_2 + z_2) \\ &= (g_1 g_2, c_{g_1 g_2} + f_1(g_1, g_2) + z_1 g_2 + z_2) \\ &= (g_1 g_2, f_2(g_1, g_2) + c_{g_1} + c_{g_2} g_1) \\ &= (g_1, z_1 + c_{g_1}) \cdot (g_2, z_2 + c_{g_2}) \\ &= \rho(g_1, z_1) \cdot \rho(g_2, z_2) \end{aligned}$$

Conversely, an isomorphism sends $(g, z) \in G_{f_1}$ to $(g, z + c(g)) \in G_{f_2}$ for some $c(g) \in Z$. The fact that this map is a group homomorphism implies that $c(g)$ is a 2-coboundary such that the difference between f_1 and f_2 is c. □

Note that if $\hat{G}$ is a group with abelian normal subgroup N and $G := \hat{G}/N$, then N is actually a $\mathbb{Z}G$-module, where $g \in G$ acts on N by conjugation with an element $\phi(g)$. Here, we choose again $\phi(g) \in \pi^{-1}(g)$ where $\pi : \hat{G} \to G$ is the natural morphism. Lemma 3.4.7, applied to $\alpha = id$, shows that the (short exact) sequence of groups

$$1 \to N \to \hat{G} \xrightarrow{\pi} G \to 1$$

determines an element f in $H^2(G, N)$, and there is a group homomorphism $G \xrightarrow{\sigma} \hat{G}$ with $\pi \circ \sigma = id_G$ if and only if $f = 0$. Moreover, Lemma 3.4.8 shows that we may obtain any group E with an embedding $N \hookrightarrow E$ such that $G = E/N$ and the given action of G on N by conjugation via a choice of a preimage in E is obtained as some $G_f = E$ for some 2-cocycle f.

Lemma 3.4.9. *Let $\alpha : G \longrightarrow PGL_n(K)$ be a projective representation, and let f be the 2-cocycle given by the obstruction to lifting α to a linear representation $\psi : G \longrightarrow GL_n(K)$. Then there is a group homomorphism $\beta : G_f \longrightarrow GL_n(K)$ satisfying $\pi \circ \beta = \alpha \circ \nu$, where $\nu : G_f \longrightarrow G$ and $\pi : GL_n(K) \longrightarrow PGL_n(K)$ are the natural projections.*

Proof. The 2-cocycle f is obtained by a choice $\phi(g) \in \pi^{-1}(\alpha(g))$ for each $g \in G$, and then $\phi(gh) \cdot f(g, h) = \phi(g)\phi(h)$. Define $\beta(g, z) := \phi(g)z$ and compute

$$\beta((g_1, z_1)(g_2, z_2)) = \beta(g_1 g_2, f(g_1, g_2) z_1 z_2)$$

$$= \phi(g_1g_2)f(g_1,g_2)z_1z_2$$
$$= \phi(g_1)z_1 \cdot \phi(g_2)z_2$$
$$= \beta(g_1,z_1) \cdot \beta(g_2,z_2)$$

This shows the statement. □

Of course, it is most interesting to see when $H^2(G,N) = 0$. A first easy criterion is given in the next lemma.

Lemma 3.4.10. *Let K be a commutative ring and let G be a finite group of order $|G|$. Then $|G| \cdot H^2(G,M) = 0$ for every KG-module M.*

In particular, if $|G| \cdot M = M$, then $H^2(G,M) = 0$.

Proof. Let f be a 2-cocycle with values in M. By definition of a 2-cocycle we get

$$gf(k,h) - f(k,hg) + f(kh,g) = f(h,g)$$

for all $g,h,k \in G$. We sum these equations over $h \in G$. The right hand side is independent of k, and so, when we define $\sigma(x) := \sum_{k\in G} f(k,g)$, we obtain

$$|G|f(h,g) = \sum_{k\in G}(f(kh,g) + gf(k,h) - f(k,hg)) = \sigma(g) + g\sigma(h) - \sigma(hg).$$

This is a 2-coboundary, and hence $|G| \cdot H^2(G,M) = 0$.

If $|G| \cdot M = M$, then we get for all $f \in H^2(G,M)$ a $\hat{f} \in H^2(G,M)$ such that $|G| \cdot \hat{f} = f$ and therefore $f = 0 \in H^2(G,M)$. □

We see that a projective representation of G gives an ordinary representation of a central extension of G.

3.4.3 Clifford's main theorem

We first consider the case of modules with full inertia group.

Theorem 3.4.11 (Clifford [CuRe-82-86, Theorem 11.20]). *Let G be a group, let H be a normal subgroup of G. Let K be a field and let M be a simple KG-module. Let L be an indecomposable direct factor of $M \downarrow_H^G$ with $End_{KH}(L) = K$ and suppose that $I_G(L) = G$. Then L lifts to a projective representation with cocycle f, and then there is a projective representation U of KG such that*

$$M \simeq L \otimes_K U$$

as KG-module. The action of G is diagonal, in the sense that $g \cdot (\ell \otimes x) = (g\ell) \otimes (gx)$ for all $g \in G$, $\ell \in L$ and $x \in U$. Moreover, G acts on U as lift of a projective representation with cocycle $1/f$.

Proof. Consider $E := End_{KG}(L \uparrow_H^G)^{op}$ and define $Q := G/H$. Then $L \uparrow_H^G$ is a right E-module. We have $L \uparrow_H^G = \bigoplus_{gH \in Q} {}^gL$ as KH-module. Since $I_G(L) = G$, we have ${}^gL \simeq L$ as a KH-module for all $g \in G$. Choose an isomorphism $\varphi_g : L \to {}^gL$ for each $gH \in Q$, and obtain that $f(g_1, g_2) := \varphi_{g_1g_2}^{-1} \circ \varphi_{g_1} \circ \varphi_{g_2}$ is a KH-linear automorphism of L, and hence a multiplication by some scalar in $K^\times$. The discussion preceding Definition 3.4.3 shows that f is a 2-cocycle. This shows that $E = K_fQ$, the twisted group ring (cf. Definition 3.4.5), where we denoted by e_{gH} the basis element of Q representing gH.

Now, M is simple and

$$Hom_{KG}(M, L \uparrow_H^G) \simeq Hom_{KH}(M \downarrow_H^G, L) \neq 0$$

by Frobenius reciprocity Remark 3.2.16. Therefore M is a submodule of $L \uparrow_H^G$, using that M is simple. Let

$$U := \{\gamma \in E \mid \gamma(L \uparrow_H^G) \subseteq M\}.$$

Then, U is a left ideal of E. Indeed, if $\varphi \in E$, then interpret φ as an endomorphism of $L \uparrow_H^G$ and for all $\gamma \in U$, also interpreted as endomorphism of $L \uparrow_H^G$ we get

$$(\gamma \circ \varphi)(L \uparrow_H^G) \subseteq \gamma(L \uparrow_H^G) \subseteq M,$$

and use that E is the opposite of the endomorphism ring.

It is clear by definition of U that $M \supseteq L \uparrow_H^G \cdot U$. We shall show that we get equality here. Indeed, M is a simple KG-module and $L \uparrow_H^G \cdot U$ is a KG-submodule of M. We need to exclude $L \uparrow_H^G \cdot U = 0$. This is equivalent to $U = 0$. But L is a submodule of $M \downarrow_H^G$ and since $Hom_{KG}(L \uparrow_H^G, M) = Hom_{KH}(L, M \downarrow_H^G)$ we get $End_{KH}(L)$ is a subset of $Hom_{KG}(L \uparrow_H^G, M)$. But $End_{KH}(L)$ contains at least the identity, and since $M \leq L \uparrow_H^G$, any non zero morphism in $Hom_{KG}(L \uparrow_H^G, M)$ induces a non zero morphism in E. This shows that $M = L \uparrow_H^G \cdot U$.

We prove now that

$$\begin{aligned} L \otimes_K U &\xrightarrow{\alpha} L \uparrow_H^G \cdot U \\ \ell \otimes \gamma &\mapsto \ell \cdot \gamma = \gamma(\ell) \end{aligned}$$

with diagonal action of G on the left term is an isomorphism. First, it is clear that α is a well-defined K-linear mapping. Since by Frobenius reciprocity

$$Hom_{KG}(L \uparrow_H^G, M) \xrightarrow{\beta} Hom_{KH}(L, M \downarrow_H^G)$$

via $\varphi \mapsto (\ell \mapsto \varphi(1 \otimes_H \ell))$ is an isomorphism, we get

$$\beta(U) = V := \{\gamma \in Hom_{KH}(L, L \uparrow_H^G \downarrow_H^G) \mid \operatorname{im} \gamma \subseteq M \downarrow_H^G\}$$

and hence

$$L\uparrow_H^G \cdot U = L \cdot V.$$

By hypothesis, $K = End_{KH}(L)$, and so

$$L \otimes_K V = L \otimes_{End_{KH}(L)^{op}} V = L \cdot V$$

by the evaluation map, which proves the statement.

We consider the G-action. Recall that $E = K_f Q$ for some cocycle $f : Q \times Q \to K^\times$. Let $x \in G$. Then G acts on $L \otimes_{End_{KH}(L)^{op}} V$ as $x \cdot (\ell \otimes y) = x\ell e_{xH}^{-1} \otimes e_{xH} y$. The action of $x \in G$ on L given by $x\ell := x\ell e_{xH}^{-1}$ is a projective representation of G with 2-cocycle $1/f$. The action of $x \in G$ on U given by $x \cdot y := e_{xH} y$ is a projective representation of G on U with cocycle f. This shows the statement. □

The general case is given in the following result.

Theorem 3.4.12 (Clifford's theorem (1937)). *Let G be a finite group and let N be a normal subgroup of G. Let R be a commutative ring such that the Krull-Schmidt theorem is valid for RS-modules, for all subgroups S of G. Let M be an indecomposable RN-module and suppose*

$$M\uparrow_N^{I_G(M)} \simeq M_1 \oplus M_2 \oplus \cdots \oplus M_r$$

for indecomposable $RI_G(M)$-modules $M_1, M_2, \ldots, M_r$.

Then for all $i, j \in \{1, 2, \ldots, r\}$ one gets $M_i\uparrow_{I_G(M)}^G$ is indecomposable and $M_i\uparrow_{I_G(M)}^G \simeq M_j\uparrow_{I_G(M)}^G$ implies $M_i \simeq M_j$.

Proof. As in Section 3.4.1 we get

$$M\uparrow_N^G\downarrow_N^G \simeq \bigoplus_{gN\in G/N} {}^gM.$$

and since for every $g_1, g_2 \in G$ one has ${}^{g_1}M \simeq {}^{g_2}M$ whenever $g_1 I_G(M) = g_2 I_G(M)$, we get

$$M\uparrow_N^G\downarrow_N^G \simeq \bigoplus_{gI_G(M)\in G/I_G(M)} ({}^gM)^{|I_G(M):N|}.$$

But, by Lemma 3.2.11 one gets

$$M\uparrow_N^G \simeq (M\uparrow_N^{I_G(M)})\uparrow_{I_G(M)}^G \simeq M_1\uparrow_{I_G(M)}^G \oplus M_2\uparrow_{I_G(M)}^G \oplus \cdots \oplus M_r\uparrow_{I_G(M)}^G .$$

Hence there are integers $n_1, n_2, \ldots, n_r$ with $\sum_{i=1}^r n_i = |I_G(M) : N|$ such that

$$M_i\uparrow_{I_G(M)}^G\downarrow_N^G \simeq \bigoplus_{gI_G(M)\in G/I_G(M)} ({}^gM)^{n_i}.$$

Since, as before, for every $g_1, g_2 \in G$ one has ${}^{g_1}M \simeq {}^{g_2}M$ if and only if $g_1I_G(M) = g_2I_G(M)$, we get $M_i \downarrow_N^{I_G(M)} \simeq M^{n_i}$.

Corollary 3.3.12 implies that for each $i \in \{1, 2, \ldots, r\}$ the $RI_G(M)$-module M_i is a direct factor of $M_i \uparrow_{I_G(M)}^G \downarrow_{I_G(M)}^G$. Hence for all $i \in \{1, 2, \ldots, r\}$ there is an indecomposable RG-module F_i such that F_i is a direct factor of $M_i \uparrow_{I_G(M)}^G$ and such that M_i is a direct factor of $F_i \downarrow_{I_G(M)}^G$. Since $M_i \downarrow_N^{I_G(M)} \simeq M^{n_i}$ one has that M^{n_i} is a direct factor of $(F_i \downarrow_{I_G(M)}^G) \downarrow_N^{I_G(M)} = F_i \downarrow_N^G$. But for each $i \in \{1, 2, \ldots, r\}$ one has that F_i is an RG-module, and so M^{n_i} is a direct factor of $F_i \downarrow_N^G$ implies ${}^gM^{n_i}$ is a direct factor of ${}^gF_i \downarrow_N^G \simeq F_i \downarrow_N^G$ for all $g \in G$. We use again that for every $g_1, g_2 \in G$ one has ${}^{g_1}M \simeq {}^{g_2}M$ if and only if $g_1I_G(M) = g_2I_G(M)$, such that whenever we sum over cosets modulo $I_G(M)$ we get different isomorphism classes of modules, and so, $\bigoplus_{gI_G(M)\in G/I_G(M)}({}^gM)^{n_i}$ is a direct factor of $F_i \downarrow_N^G$. But

$$M_i \uparrow_{I_G(M)}^G \downarrow_N^G \simeq \bigoplus_{gI_G(M)\in G/I_G(M)} ({}^gM)^{n_i}$$

and F_i is a direct factor of $M_i \uparrow_{I_G(M)}^G$ with

$$F_i \downarrow_N^G \simeq \bigoplus_{gI_G(M)\in G/I_G(M)} ({}^gM)^{n_i}$$

as well. Therefore, $F_i \simeq M_i \uparrow_{I_G(M)}^G$ for all $i \in \{1, 2, \ldots, r\}$. This shows that $M_i \uparrow_{I_G(M)}^G$ is indecomposable for all $i \in \{1, 2, \ldots, r\}$.

We need to show that $M_i \uparrow_{I_G(M)}^G \simeq M_j \uparrow_{I_G(M)}^G$ implies $M_i \simeq M_j$.

One gets, using Mackey's theorem and that N is normal in G and contained in $I_G(M)$,

$$(\dagger): \quad M_i \uparrow_{I_G(M)}^G \downarrow_N^G \simeq \bigoplus_{gI_G(M)\in G/I_G(M)} {}^g(M_i \downarrow_N^{I_G(M)}).$$

Suppose M_j is a direct summand of $M_i \uparrow_{I_G(M)}^G \downarrow_{I_G(M)}^G$. Since M_i is a direct factor of $M_i \uparrow_{I_G(M)}^G \downarrow_{I_G(M)}^G$, and since its restriction to N corresponds to the coset $I_G(M)$ in $G/I_G(M)$ in the decomposition $(\dagger)$, the restriction to M_j to N is a direct sum of copies of ${}^gM$ for $g \in G \setminus I_G(M)$. But since ${}^gM \not\simeq M$ for exactly the $g \in G \setminus I_G(M)$ we get $M_i \uparrow_{I_G(M)}^G \simeq M_j \uparrow_{I_G(M)}^G$ implies $M_i \simeq M_j$. □

3.4.4 Small and big inertia groups: Blichfeldt's and Ito's theorem

Given a field K, a group G and a normal subgroup N, and suppose that the order of G is invertible in K. We know that all simple KG-modules are direct summands of the induced modules of simple KN-modules. In order to find the decomposition of these inductions we use Clifford's theorem. Hence, Clifford's theorem reduces in some sense the question to find simple KG-modules to the question of finding for each simple

KN-module S its inertia group $I_G(S)$ and then the decomposition of S induced to its inertia group. Often $I_G(S)$ is smaller than G, but not always.

There is an immediate consequence of the case when the inertia group is the whole group.

Corollary 3.4.13. *Let R be a commutative ring and G be a finite group such that the Krull-Schmidt property holds for all RL-modules, for all subgroups L of G. Let H be a normal subgroup of G and let S be a simple RG-module. Let T be a simple submodule of $S\downarrow_H^G$. If $I_G(T) = G$, then $S\downarrow_H^G\simeq T^n$ for some integer n.*

Proof. Indeed, $\sum_{g\in G} gT$ is a KG-submodule of S. Since S is simple, $\sum_{g\in G} gT = S$. But, $gT \simeq {}^gT$ as KH-module, and so gT is a simple KH-module as well. Since $I_G(T) = G$ we get $T \simeq {}^gT \simeq gT$ for all $g \in G$ and so $S\downarrow_A^G\simeq T^n$, for some integer n. □

Example 3.4.14.

1. Suppose G is a group and N is a subgroup of the centre of G. Then for all commutative rings R and all RG-modules M one has $I_G(M) = G$. Indeed, Lemma 3.4.2 shows that $I_G(M)$ contains $C_G(N)$ but since $N \subseteq Z(G)$ one gets
$$G \supseteq I_G(M) \supseteq C_G(N) \supseteq C_G(Z(G)) = G.$$
2. Recall the complex character table of the alternating group $\mathfrak{A}_4$ of degree four from Example 2.3.7.

$\mathfrak{A}_4$ size of class	$\{1\}$ 1	$C_{(1\ 2)(3\ 4)}$ 3	$C_{(1\ 2\ 3)}$ 4	$C_{(1\ 3\ 2)}$ 4
χ_1	1	1	1	1
χ_ζ	1	1	ζ	ζ^2
χ_{ζ^2}	1	1	ζ^2	ζ
χ_3	3	−1	0	0

Since $\mathfrak{A}_4$ is a normal subgroup of $\mathfrak{S}_4$, we try to look for the inertia groups of the four simple $\mathbb{C}\mathfrak{A}_4$-modules.

Since a twist by a group automorphism of the trivial module is again the trivial module, the inertia group of the trivial module is the whole group: $I_{\mathfrak{S}_4}(\mathbb{C}) = \mathfrak{S}_4$. The twist of a simple module by any group automorphism is again a simple module. Moreover, its dimension stays the same by definition. Now, since there is only one irreducible character of degree 3 of $\mathfrak{A}_4$ there is only one three-dimensional simple $\mathbb{C}\mathfrak{A}_4$-module M_3. Therefore $I_{\mathfrak{S}_4}(M_3) = \mathfrak{S}_4$.

There are two remaining simple one-dimensional non trivial $\mathbb{C}\mathfrak{A}_4$-modules M_ζ and M_{ζ^2}. They differ by the action of $t := (1\ 2\ 3)$ on it. The element t acts on M_ζ by multiplication with ζ and t acts on M_{ζ^2} by multiplication with ζ^2. Now,
$$\mathfrak{S}_4 = \mathfrak{A}_4 \cup (2\ 3)\mathfrak{A}_4$$

and so we need to see what is the module ${}^{(2\,3)}(M_\zeta)$. The element $t = (1\,2\,3)$ acts on ${}^{(2\,3)}(M_\zeta)$ as $(2\,3)\cdot(1\,2\,3)\cdot(2\,3)^{-1}$ acts on M_ζ. But $(2\,3)\cdot(1\,2\,3)\cdot(2\,3)^{-1} = (1\,3\,2) = (1\,2\,3)^2$ and so ${}^{(2\,3)}(M_\zeta) \simeq M_{\zeta^2}$. Therefore,

$$I_{\mathfrak{S}_4}(M_\zeta) = \mathfrak{A}_4 = I_{\mathfrak{S}_4}(M_{\zeta^2}).$$

Clifford's theorem implies that $M_\zeta \uparrow_{\mathfrak{A}_4}^{\mathfrak{S}_4}$ is a two-dimensional simple $\mathbb{C}\mathfrak{S}_4$-module.

3. Let $G = \mathfrak{S}_3$ be the symmetric group of degree 3 and let $H = \mathfrak{A}_3$ be the cyclic group of order 3. Then $\mathbb{C}\mathfrak{A}_3$ has three simple modules up to isomorphism, whereas $\mathbb{Q}\mathfrak{A}_3$ has only 2 simple modules up to isomorphism, one is the trivial module $\mathbb{Q}_+$, the other is the module $\mathbb{Q}(\zeta)$ for ζ being a third root of unity. In Example 1.2.43 we have determined the $\mathbb{Q}\mathfrak{S}_3$-modules. Recall that there are three simple modules, one, $S_\mathbb{Q}$ of dimension 2, and two, $\mathbb{Q}_+$ the trivial module and $\mathbb{Q}_-$ the signature, of dimension 1 coming from the isomorphism $\mathfrak{S}_3/\mathfrak{A}_3 \simeq C_2$. All three simple modules stay simple when we tensor with $\mathbb{C}$ over $\mathbb{Q}$. Let S be the simple $\mathbb{C}\mathfrak{S}_3$-module of dimension two. Then $S \downarrow_{\mathfrak{A}_3}^{\mathfrak{S}_3}$ is a direct sum of two simple $\mathbb{C}\mathfrak{A}_3$-modules. We want to determine which modules appear here. For this we pass to $\mathbb{Q}$ and tensor afterwards with $\mathbb{C}$ over $\mathbb{Q}$. Let $S_\mathbb{Q}$ be the two-dimensional simple $\mathbb{Q}\mathfrak{S}_3$-module for which $\mathbb{C}\otimes_\mathbb{Q} S_\mathbb{Q} \simeq S$. Then

$$\begin{aligned} Hom_{\mathbb{Q}C_3}(S_\mathbb{Q} \downarrow_{C_3}^{\mathfrak{S}_3}, \mathbb{Q}_+) &\simeq Hom_{\mathbb{Q}\mathfrak{S}_3}(S_\mathbb{Q}, \mathbb{Q}_+ \uparrow_{C_3}^{\mathfrak{S}_3}) \\ &\simeq Hom_{\mathbb{Q}\mathfrak{S}_3}(S_\mathbb{Q}, \mathbb{Q}(\mathfrak{S}_3/C_3)) \\ &\simeq Hom_{\mathbb{Q}\mathfrak{S}_3}(S_\mathbb{Q}, \mathbb{Q}_+ \oplus \mathbb{Q}_-) \\ &= 0 \end{aligned}$$

Therefore $S_\mathbb{Q} \downarrow_{C_3}^{\mathfrak{S}_3} \simeq \mathbb{Q}(\zeta)$. Call α this isomorphism. Then $id_\mathbb{C} \otimes \alpha$ gives an isomorphism

$$S \downarrow_{C_3}^{\mathfrak{S}_3} \simeq \mathbb{C} \otimes_\mathbb{Q} \mathbb{Q}(\zeta) \simeq \mathbb{C}_\zeta \oplus \mathbb{C}_{\zeta^2}$$

Now $I_{\mathfrak{S}_3}(\mathbb{C}_\zeta) = C_3$, and so $S \simeq \mathbb{C}_\zeta \uparrow_{C_3}^{\mathfrak{S}_3}$. Hence S is induced from $\mathfrak{A}_3$, whereas $S_\mathbb{Q} \downarrow_{C_3}^{\mathfrak{S}_3} \simeq \mathbb{Q}(\zeta)$ is simple and so $I_{\mathfrak{S}_3}(\mathbb{Q}(\zeta)) = \mathfrak{S}_3$. Therefore $S_\mathbb{Q}$ is not induced from $\mathfrak{A}_3$.

Recall Definition 1.1.22 on faithful modules:

Definition 3.4.15. Let G be a group and let R be a commutative ring. An RG-module S is *faithful* if whenever for $g \in G$ one has $g \cdot s = s$ for all $s \in S$, implies $g = 1$.

Examples for faithful modules are the regular module, the two-dimensional simple $\mathbb{R}Q_8$-module, for Q_8 being the quaternion group of order 8 (cf. Example 1.2.23), and also the 2-dimensional simple $\mathbb{Q}\mathfrak{S}_3$-module of Example 1.2.43.

Proposition 3.4.16 (Blichfeld's theorem). *Let G be a finite group. Let K be a field such that the order of G is invertible in K. Let S be a simple faithful KG-module and let A be an abelian normal subgroup of G not contained in the centre of G. Suppose that K is a splitting field for A. Let T be a simple submodule of $S\downarrow_A^G$. Then $I_G(T) \neq G$.*

Proof. Suppose $I_G(T) = G$. We know that T is a direct summand of $S\downarrow_A^G$. By Corollary 3.4.13 all simple direct summands of $S\downarrow_A^G$ are isomorphic to T. Since K is a splitting field for A and therefore each simple KA-module is one-dimensional $S\downarrow_A^G\simeq T^{dim_K(S)}$.

Since each simple KA-module is one-dimensional, there is a homomorphism of groups $\alpha : A \longrightarrow K^\times$ such that for all $t \in T$ one gets $a \cdot t = \alpha(a)t$. Since $S\downarrow_A^G\simeq T^{dim_K(S)}$, one gets for all $a \in A$ and all $s \in S$ that $a \cdot s = \alpha(a)s$. The KG-module S is faithful, and so $\sigma|_A = \alpha$, denoting by $\sigma : G \longrightarrow Gl_{dim_K(S)}(K)$ the group homomorphism which defines the KG-module structure on S. But the image A' of α in $Gl_{dim_K(S)}(K)$ are diagonal matrices and so A' is central in $Gl_{dim_K(S)}(K)$. The KG-module S is faithful if and only if φ is injective. Hence $G \simeq \varphi(G) \subseteq Gl_{dim_K(S)}(K)$. Since

$$A' = \varphi(A) \subseteq Z(Gl_{dim_K(S)}(K)) \subseteq Z(\varphi(G))$$

we get that A is contained in the centre of G. This contradiction proves the proposition. □

Remark 3.4.17. Let G be a finite group, let K be a field and suppose that $|G|$ is invertible in K. What happens if one restricts a simple KG-module S to a normal subgroup N. Clifford's theorem 3.4.12 implies that if one restricts S to N, taking any simple KN-submodule T of $S\downarrow_N^G$ then for any simple submodule T' of $T\uparrow_N^{I_G(T)}$ the module $T'\uparrow_{I_G(T)}^G$ is simple. Different simple direct submodules T' and T'' of $T\uparrow_N^{I_G(T)}$ yield non isomorphic induced modules $T'\uparrow_N^{I_G(T)}$ and $T''\uparrow_N^{I_G(T)}$. But Frobenius reciprocity shows that

$$Hom_{KG}(S\downarrow_N^G\uparrow_N^G, S) \simeq Hom_{KN}(S\downarrow_N^G, S\downarrow_N^G) \ni id_S \neq 0$$

and so by the semisimplicity of KG the simple module S is a direct factor of $S\downarrow_N^G\uparrow_N^G$. Therefore there is a simple KN-submodule T_S of $S\downarrow_N^G$ and a unique simple $KI_G(T_S)$-submodule T_S' of $T_S\uparrow_N^{I_G(T_S)}$ such that $S \simeq T_S'\uparrow_{I_G(T_S)}^G$.

Lemma 3.4.18. *Let G be a finite group and let N be a normal subgroup of G. Let K be a field such that $|G|$ is invertible in K, and such that K is a splitting field for G and all of its subgroups (in particular contains $|G|$-th roots of unity), and let S be a simple KG-module. Then there is a simple KN-module T_S such that when one takes n_S the maximal integer such that there is a monomorphism $T_S^{n_S} \hookrightarrow S\downarrow_N^G$ the image T_S' of this monomorphism carries a natural structure of a $KI_G(T_S)$-module and $S \simeq T_S'\uparrow_{I_G(T_S)}^G$.*

Proof. Remark 3.4.17 shows everything except the characterisation of T_S' as image of a monomorphism as claimed.

Let $e^2 = e$ be the central primitive idempotent in KN with $eT_S = T_S$. Since N is normal in G, for all $g \in G$ the idempotent $g{\cdot}e{\cdot}g^{-1} =: {}^g e$ is a central primitive idempotent of KN as well. We claim that

$$(\dagger)\quad {}^g e = e \Leftrightarrow g \in I_G(T_S).$$

Indeed, if $g \in I_G(T_S)$, then

$${}^g T_S = {}^g(eT_S) = ({}^g e)\,{}^g T_S = ({}^g e)T_S$$

and so, since T_S is simple and since ${}^g e$ is a primitive central idempotent, ${}^g e = e$. If ${}^g e = e$, then

$$eKN = ({}^g e)KN = {}^g(eKN)$$

and so ${}^g T_S \simeq T_S$. Hence $g \in I_G(T_S)$.

Therefore $T' := eS$ is a $K(I_G(T_S))$-module. It is clear that T' is the sum of all KN-submodules of S isomorphic to T_S. We still need to show that $T' = T'_S$.

We need to understand $T_S \uparrow_N^{I_G(T_S)}$ in order to be able to detect the direct factor T'_S.

We get

$$\begin{aligned}
T_S \uparrow_N^{I_G(T_S)} \downarrow_N^{I_G(T_S)} &= (K(I_G(T_S)) \otimes_{KN} T_S) \downarrow_N^{I_G(T_S)} \\
&= \Bigl(\bigoplus_{gN \in I_G(T_S)/N} gKN \otimes_{KN} T_S \Bigr) \downarrow_N^{I_G(T_S)} \\
&= \bigoplus_{gN \in I_G(T_S)/N} (gKN \otimes_{KN} T_S) \\
&= \bigoplus_{gN \in I_G(T_S)/N} {}^g T_S \\
&\simeq T_S^{|I_G(T_S):N|}
\end{aligned}$$

since $I_G(T_S)$ is the subgroup of those elements $g \in G$ with ${}^g T_S \simeq T_S$ as KN-module.

Therefore, by the multiplication map

$$\begin{aligned}
K(I_G(T_S)) \otimes_{KN} T_S &\longrightarrow S \\
x \otimes t &\mapsto xt
\end{aligned}$$

the module $T_S \uparrow_N^{I_G(T_S)}$ maps to a submodule of T'.

Now,

$$\sum_{gI_G(T_S) \in G/T_G(T_S)} gT' =: S'$$

is a KG-submodule of S. Since S is simple,

$$\sum_{gI_G(T_S) \in G/I_G(T_S)} gT' = S.$$

But then

$$\begin{aligned} KG \otimes_{KI_G(T_S)} T' &\longrightarrow S \\ x \otimes t &\mapsto xt \end{aligned}$$

is an epimorphism of KG-modules.

We shall show that T' is actually a simple $K(I_G(T_S))$-module. Let U be a simple $K(I_G(T_S))$-submodule of T'. Since

$$\sum_{gI_G(T_S)\in G/I_G(T_S)} g \otimes T' \simeq \bigoplus_{gI_G(T_S)\in G/I_G(T_S)} {}^gT'$$

and since we get

$${}^{g_1}T' \simeq {}^{g_2}T' \Leftrightarrow g_1 I_G(T_S) = g_2 I_G(T_S)$$

by (†), one has that

$$\sum_{gI_G(T_S)\in G/I_G(T_S)} g \otimes U \simeq \bigoplus_{gI_G(T_S)\in G/I_G(T_S)} {}^gU$$

is a KG-submodule mapping to the direct sum decomposition

$$\sum_{gI_G(T_S)\in G/I_G(T_S)} g \cdot U = S = \sum_{gI_G(T_S)\in G/I_G(T_S)} g \cdot T'.$$

Therefore, for $g = 1$ in particular, $U = T'$. This shows the statement. □

As promised in the remarks after Theorem 2.4.9 we are now giving a more precise statement on the dimension of simple modules using Clifford theory. The proof of Ito's theorem which we are going to give here is due to Carlson and Roggenkamp and uses in an essential way Blichfeld's theorem Proposition 3.4.16.

Theorem 3.4.19 (Ito's theorem [Ito-51]; Carlson-Roggenkamp [CarRog-88]). *Let G be a finite group and H an abelian normal subgroup of G. Let K be a field in which the order of G is invertible containing all $|G|$-th roots of unity. Then for every simple KG-module S one gets that $\dim_K(S)$ divides $|G : H|$.*

Remark 3.4.20. Theorem 2.4.9 was the special case when $H = 1$. The method there was number theoretical, namely to study integral elements in cyclotomic fields. The proof here is going to be representation theoretic, using module theory.

Proof of Theorem 3.4.19. We prove the theorem by induction on $|G|$. Let S be a simple KG-module and let T_S be a simple submodule of $S \downarrow_H^G$ and T'_S be the simple $KI_G(T_S)$-submodule of S as in Lemma 3.4.18.

Suppose first that $I_G(T_S) \neq G$. Then, by the induction hypothesis there is an integer t such that $dim_K(T_S') \cdot t = |I_G(T_S) : H|$. By Clifford's theorem 3.4.12

$$\begin{aligned} t \cdot dim_K(S) &= t \cdot dim_K(T_S' \uparrow_{I_G(T_S)}^G) \\ &= t \cdot |G : I_G(T_S)| \cdot dim_K(T_S') \\ &= |G : I_G(T_S)| \cdot |I_G(T_S) : H| \\ &= |G : H| \end{aligned}$$

The statement follows in this case.

Suppose now $I_G(T_S) = G$. Since H is abelian and since K is a splitting field for H we get $dim_K(T_S) = 1$. Then by Corollary 3.4.13 we get

$$S \downarrow_H^G \simeq T_S^{dim_K(S)},$$

using that K is a splitting field for H and so each simple KH-module is one-dimensional. Let $\varphi : G \longrightarrow Gl_{dim_K(S)}(K)$ be the representation of G which affords the KG-module S. Then, since T_S is one-dimensional, as in the proof of Blichfeld's theorem Proposition 3.4.16 the image $\varphi(H)$ consists of multiples of the identity matrix. Hence, as in the proof of Proposition 3.4.16 we get $H/(\ker(\varphi) \cap H)$ is a subgroup of $Z(G/\ker(\varphi))$. As $\ker(\varphi)$ acts as id_S on S, the module S is actually a simple $G/\ker(\varphi)$-module. Put $\overline{G} := G/\ker(\varphi)$ and $\overline{H} := H/(\ker(\varphi) \cap H)$. Proposition 3.4.16 implies that $\ker(\varphi) \neq 1$. Hence $|\overline{G}| < |G|$. The induction hypothesis implies now that $dim_K(S)$ divides $|\overline{G} : \overline{H}|$. But $|\overline{G} : \overline{H}|$ divides $|G : H|$. This proves the Theorem. □

Just as for Mackey's formula there are plenty of applications of Clifford theory. A systematic account is given in Andrei Marcus [Marc-99].

3.5 Exercises

Exercise 3.1. Let K be a field and let D be a skew field containing K as sub-algebra. Let A be a simple finite dimensional K-algebra, let V be a simple A-module and suppose $D = End_A(V)$.

a) Let $Z(D) := \{d \in D \mid d \cdot e = e \cdot d \ \forall e \in D\}$. Show that $Z(D) =: E$ is a field containing K.
b) Show that left multiplication with elements in A induce an injective algebra homomorphism $A \longrightarrow End_D(V)$. Use Proposition 1.2.39 and Wedderburn's theorem to show that $A \simeq End_D(V)$.
c) Suppose that B is a K-subalgebra of A and suppose that B is a simple K-algebra. Show that V is a $B \otimes_K D$-module if one puts $(b \otimes \delta) \cdot v := \delta(bv)$ for each $b \in B$, $\delta \in D$ et $v \in V$.
d) Find a K-subalgebra of the algebra of 2×2 matrices over K which is not semisimple.

e) Show that the dimension of the finite dimensional simple algebra A over $D = Z(A)$ is n^2 where $n \in \mathbb{N}$. Find an example of a semisimple K-algebra A such that $K = Z(A)$ and $dim_K(A)$ is not a square.

Exercise 3.2. Consider the group $GL_2(3)$ of invertible matrices of size 2×2 with entries in the field $\mathbb{F}_3$ with 3 elements.

a) Show that the group $SL_2(3)$ of matrices of size 2×2 with entries in $\mathbb{F}_3$ and determinant 1 is a normal subgroup of index 2 of $GL_2(3)$.
b) Find the centre $Z(SL_2(3))$ of $SL_2(3)$ and determine the order of the quotient $PSL_2(3) = SL_2(3)/Z(SL_2(3))$.
c) Show that $PSL_2(3)$ acts transitively on the set of one-dimensional sub vector spaces of $\mathbb{F}_3^2$ (i. e. the points of the projective line over $\mathbb{F}_3$) and deduce that $PSL_2(3)$ is isomorphic to the alternating group of order 12.
d) Show that $Z(SL_2(3))$ is also a normal subgroup of $GL_2(3)$ and find the index of $PSL_2(3)$ in $PGL_2(3) := GL_2(3)/Z(SL_2(3))$.
e) Find the inertia group of each irreducible character of $PSL_2(3)$ in $PGL_2(3)$.

Exercise 3.3. Using the notations from Exercise 2.1, decompose $\chi_i \cdot \chi_j$ into indecomposable characters for each $i, j \in \{1, \dots, 6\}$

Exercise 3.4. Let G be a finite group with subgroup H. For any $\mathbb{C}G$-module M inducing the character χ_M and any $\mathbb{C}H$-module N inducing the character χ_N let $res_H^G\chi_M = \chi_{M\downarrow_H^G}$ and $ind_H^G\chi_N = \chi_{N\uparrow_H^G}$. Show for any $\chi \in Irr(G)$ and $\psi \in Irr(H)$

1. $\langle ind_H^G\chi, \psi\rangle = \langle \chi, res_H^G(\psi)\rangle$
2. $ind_H^G(\psi \cdot res_H^G\chi) = ind_H^G\psi \cdot \chi$.

Exercise 3.5. Let G be a finite group and let H_1 and H_2 be two subgroups of G. Let M_1 be a simple $\mathbb{C}H_1$-module and let M_2 be a simple $\mathbb{C}H_2$-module. For $x, y \in G$ let ${}^xH_1 \cap {}^yH_2 := xH_1x^{-1} \cap yH_2y^{-1}$.

a) Let $x, y, z, w \in G$. If xy^{-1} and zw^{-1} belong to the same double class C of $H_1\backslash G/H_2$ then show that
$$Hom_{\mathbb{C}({}^xH_1\cap{}^yH_2)}({}^xM_1, {}^yM_2) \simeq Hom_{\mathbb{C}({}^zH_1\cap{}^wH_2)}({}^zM_1, {}^wM_2)$$
b) Show that $M\uparrow_{H_1}^G$ is a simple $\mathbb{C}G$-module if and only if
$$Hom_{\mathbb{C}(H_1\cap{}^xH_1)}(M_1, {}^xM_1) = 0$$
for all $x \notin H_1$.
c) Let D_n be the dihedral group of order $2n$ for an integer $n \geq 3$ and let $\zeta_n := e^{2\pi i/n} \in \mathbb{C}$. Let C_n be the cyclic group of order n, and suppose that C_n is generated by $c \in C_n$.
   c1) Show that for all $i \in \{0, 1, \dots, n-1\}$ one defines a simple one-dimensional $\mathbb{C}C_n$-module T_i with the action $c \cdot m := \zeta_n^i m$ for each $m \in T_i$.

c2) Find the inertia group $I_{D_n}(T_i)$ for each $i \in \{0, 1, \ldots, n-1\}$.
c3) Use part b) to find the $i \in \{0, 1, \ldots, n-1\}$ for which the $\mathbb{C}D_n$-module $T_i \uparrow_{C_n}^{D_n}$ is simple.
c4) Use Clifford's theorem to decompose the $\mathbb{C}I_{D_n}(T_i)$-module $T_i \uparrow_{C_n}^{I_{D_n}(T_i)}$ in a direct sum of simple $\mathbb{C}I_{D_n}(T_i)$-modules.

Exercise 3.6 (with solution in Chapter 9). Let D_n be the dihedral group of order $2n$. Recall that D_n is generated by an element a of order n and b of order 2 satisfying $baba = 1$ and the subgroup C_n generated by a is normal and cyclic of order n.
a) If n is odd, and if M is an indecomposable $\mathbb{C}C_n$-module such that the inertia group $I_{D_n}(M)$ contains b, show that M is the trivial module.
b) If n is even, show that there are exactly two indecomposable $\mathbb{C}C_n$-modules M_0 and M_1 not isomorphic to each other, and such that $b \in I_{D_n}(M_0)$ and $b \in I_{D_n}(M_1)$.
c) Let $n \in \mathbb{N}$ and $n \geq 3$. Find for each indecomposable $\mathbb{C}C_n$-module M the inertia subgroup $I_{D_n}(M)$.
d) Let n be odd. Decompose for each indecomposable $\mathbb{C}C_n$-module M the $\mathbb{C}D_n$-module $M \uparrow_{C_n}^{D_n}$ in a direct sum of indecomposable $\mathbb{C}D_n$-modules.

Exercise 3.7. Let G be a finite group and let N be a normal subgroup of G such that G/N is of prime order p.
a) If M is a simple $\mathbb{C}N$-module, show that either $M \uparrow_N^G$ is a simple $\mathbb{C}G$-module, or else $End_{\mathbb{C}G}(M \uparrow_N^G)$ is a p-dimensional $\mathbb{C}$-algebra.
b) Study the number of pairwise non isomorphic direct summands of $M \uparrow_N^G$ in case $p = 2$ et $p = 3$.

Exercise 3.8. Let $Q_8 = \langle i, j \mid i^2 = j^2 = (ij)^2 = z; z^2 = 1; iz = zi; jz = zj \rangle$ be the quaternion group of order 8. Recall $Z(Q_8) = \{1, i^2\}$.
a) Let K be a field of characteristic different from 2 which contains an element x with $x^2 + 1 = 0$. Show that

$$i \mapsto \begin{pmatrix} x & 0 \\ 0 & \frac{1}{x} \end{pmatrix} \qquad j \mapsto \begin{pmatrix} 0 & 1 \\ -1 & 0 \end{pmatrix}$$

defines a representation ρ of Q_8 over K. Let M be the KQ_8-module defined by ρ.
b) Show that M is simple.
c) Show that the equation $X^2 + 1 = 0$ has a root in $K = \mathbb{F}_5$.
Suppose in the sequel that $K = \mathbb{F}_5$.
d) Let $G := M \rtimes Q_8$, which is a group of order 200. Find the character table of M.
e) Show that if U is a subgroup of Q_8 with $|Z(Q_8) \cap U| = 1$, then $|U| = 1$.
f) If V is a simple non trivial $\mathbb{C}M$-module, determine $I_G(V)$.
g) Show there is a simple $\mathbb{C}G$-module of dimension 8.
h) How many orbits does Q_8 have with respect to the action on M?
i) Find the degrees of the irreducible characters of G.

Exercise 3.9. Let $\mathfrak{S}_6$ be the symmetric group of degree 6 on the symbols $\{1,2,3,4,5,6\}$.

a) Find the order of $\mathfrak{S}_6$, the conjugacy classes of $\mathfrak{S}_6$, and the cardinality of each class.

b) How many irreducible characters of $\mathfrak{S}_6$ are there? How many irreducible characters of $\mathfrak{S}_6$ are of degree 1.

c) It is known that the symmetric group $\mathfrak{S}_5$ of degree 5 is naturally a subgroup of $\mathfrak{S}_6$, in that this is the stabiliser of the symbol 6. Denote by $\mathbb{C}_{\mathfrak{S}_5}$ the trivial $\mathbb{C}\mathfrak{S}_5$-module and by $\mathbb{C}_{\mathfrak{S}_6}$ the trivial $\mathbb{C}\mathfrak{S}_6$-module. Show $\mathbb{C}_{\mathfrak{S}_5}\uparrow_{\mathfrak{S}_5}^{\mathfrak{S}_6}\simeq \mathbb{C}_{\mathfrak{S}_6}\oplus L$ for a simple $\mathbb{C}\mathfrak{S}_6$-module L. Show that $\{id; (1\ 6), (2\ 6), (3\ 6), (4\ 6), (5\ 6)\}$ is a system of representants of the classes $\mathfrak{S}_6/\mathfrak{S}_5$. Compute the character χ_5^+ of L.

d) Denote by $\mathbb{C}_{\mathfrak{S}_6}^-$ the non-trivial $\mathbb{C}\mathfrak{S}_6$-module of dimension 1. Let M be a simple $\mathbb{C}\mathfrak{S}_6$-module. Show that $\mathbb{C}_{\mathfrak{S}_6}^-\otimes_{\mathbb{C}} M$ is a $\mathbb{C}\mathfrak{S}_6$-module if one defines the action of $\mathfrak{S}_6$ by $g\cdot(x\otimes m) := gx\otimes gm$ for all $x\in\mathbb{C}_{\mathfrak{S}_6}^-$, all $m\in M$ and all $g\in\mathfrak{S}_6$. Let χ_M be the character of M and χ_M^- be the character of $\mathbb{C}_{\mathfrak{S}_6}^-\otimes_{\mathbb{C}} M$. Show $\chi_M^-(g)=\chi_M(g)$ for each $g\in\mathfrak{S}_6$ if and only if $\chi_M(g)=0$ for each g of signature -1. Deduce $M\not\simeq\mathbb{C}_{\mathfrak{S}_6}^-\otimes_{\mathbb{C}} M$ if and only if there is a $g\in\mathfrak{S}_6$ of signature -1 with $\chi_M(g)\neq 0$.

e) The group $\mathfrak{S}_6$ admits an automorphism σ of order 2 with the property

$$\sigma(((1\ 2\ 3)(5\ 6))^{\mathfrak{S}_6}) = (((1\ 2\ 3\ 4\ 5\ 6))^{\mathfrak{S}_6}.$$

Deduce that $\mathfrak{S}_6$ admits four pairwise different irreducible characters of degree 5.

f) Show that σ changes the class of $(1\ 2)(3\ 4)(5\ 6)$ with the class of $(1\ 2)$, and the class of $(1\ 2\ 3)$ with the class of $(1\ 2\ 3)(4\ 5\ 6)$. We know that σ fixes all the other conjugacy classes. Determine the values of the irreducible characters of $\mathfrak{S}_6$ of degree 5.

g) Let E_2 be the set of subsets of cardinality 2 of $\{1,2,3,4,5,6\}$; (e. g. $\{1,2\}\in E_2$, $\{5,6\}\in E_2$ etc.). Show that the action of $\mathfrak{S}_6$ on $\{1,2,3,4,5,6\}$ induces an action of $\mathfrak{S}_6$ on E_2.

h) Denote by W_2 the permutation module associated to this set. Compute the character of W_2 and show that W_2 contains a simple submodule of degree 9. This 9-dimensional simple module is denoted L_2. Compute the character χ_9 of L_2. Deduce that there is another 9-dimensional irreducible character χ_9^- of $\mathfrak{S}_6$.

i) Use Ito's theorem to see that there is an irreducible character χ_{16} of degree 16.

j) Show that $\chi_{16}(g)=0$ for each element $g\in\mathfrak{S}_6$ with signature -1. Prove $\chi_{16}((1\ 2\ 3)) = \chi_{16}((1\ 2\ 3)(4\ 5\ 6))$.

k) Is there in irreducible character of degree 14? Prove that there are two irreducible characters χ_{10} et χ_{10}^- of degree 10.

l) Show $\chi_{10}((1\ 2))=2$ and $\chi_{10}^-((1\ 2))=-2$. Deduce that for all $g\in\mathfrak{S}_6$ we get $|\chi_{10}(g)| = |\chi_{10}^-(g)|$.

m) Use the fact that all character values of a symmetric group are integers to determine the character table of $\mathfrak{S}_6$.

4 Bilinear forms on modules

Und wenn es uns glückt,
Und wenn es sich schickt,
So sind es Gedanken!

Johann Wolfgang von Goethe, Faust Der Tragödie Erster Teil

and if we succeed,
and if things suit,
then they are thoughts!

Johann Wolfgang von Goethe, Faust first part of the tragedy;[1]

One might consider representation theory of a finite group G over a field K as linear algebra of K-vector spaces in presence of additional symmetries, namely those of the group G. Seen this way, it is tempting to try to see what happens with bilinear forms, scalar products and similar structures in presence of this additional symmetry G. As one knows in many disciplines, such as functional analysis, the structure of spaces in presence of a scalar product is much richer and also much more rigid than just the linear structure.

4.1 Invariant bilinear forms

Bilinear forms are classical objects in linear algebra. We recall some of the basic definitions.

Definition 4.1.1. Let K be a field and let V be a K-vector space. A *bilinear form* b on V is an element of $Hom_K(V \otimes_K V, K)$. A bilinear form b is
- *symmetric* if $b(v,w) = b(w,v)$ for all $v, w \in V$
- *antisymmetric* if $b(v,w) = -b(w,v)$ for all $v, w \in V$
- *alternating* if $b(v,v) = 0$ for all $v \in V$.

Remark 4.1.2. Let b be an alternating bilinear form. Then

$$0 = b(v+w, v+w) = b(v,v) + b(v,w) + b(w,v) + b(w,w) = b(v,w) + b(w,v)$$

and hence an alternating bilinear form is anti-symmetric. If the characteristic of the base field is different from 2, then the converse is true as well since if b is skew-symmetric, then $b(v,w) = -b(w,v)$ and for $v = w$ this implies that b is alternating. This is false over fields of characteristic 2. Indeed, if V is 2-dimensional with basis v_1, v_2,

1 in der Übersetzung von Susanne Brennecke, translation by Susanne Brennecke.

https://doi.org/10.1515/9783110702446-004

then define a bilinear form b by $b(v_i, v_j) = 1$ for all $i, j \in \{1, 2\}$. Then $b(v_1, v_1) = 1 \neq 0$ and

$$b(\lambda_1 v_1 + \lambda_2 v_2, \mu_1 v_1 + \mu_2 v_2) = \lambda_1\mu_1 + \lambda_2\mu_2 + \lambda_1\mu_2 + \mu_1\lambda_2 = b(\mu_1 v_1 + \mu_2 v_2, \lambda_1 v_1 + \lambda_2 v_2)$$

and hence b is anti-symmetric.

It is clear that a symmetric or antisymmetric bilinear form b on V has the property that

$$b(v, w) = 0 \Leftrightarrow b(w, v) = 0$$

for all $v, w \in V$. Hence the following definition makes sense.

Definition 4.1.3. Let V be a K-vector space and let b be a symmetric or antisymmetric bilinear form on B. Then

- the *kernel of b* is defined to be $\ker(b) := \{v \in V \mid b(v, -) = 0\}$, where we denote by $b(v, -)$ the K-linear map $V \ni w \mapsto b(v, w) \in K$.
- A symmetric or antisymmetric bilinear form b is *non degenerate* if $\ker(b) = 0$.
- A vector space V with non degenerate alternating bilinear form b is *symplectic*.

Let K be a field, let G be a finite group and let V be a KG-module.

Definition 4.1.4. A bilinear form $b : V \times V \longrightarrow K$ on V is said to be G-invariant if $b(gv, gw) = b(v, w)$ for all $g \in G$ and $v, w \in V$.

Remark 4.1.5. It is not possible to give a KG-invariant bilinear form. Indeed, if b is G-invariant and bilinear, one has for all $v, w \in V$ and $g_1, g_2 \in G$

$$\begin{aligned} b((g_1 + g_2)v, (g_1 + g_2)w) &= b(g_1 v, g_1 w) + b(g_1 v, g_2 w) \\ &\quad + b(g_2 v, g_1 w) + b(g_2 v, g_2 w) \\ &= b(v, w) + b(g_2^{-1} g_1 v, w) + b(g_1^{-1} g_2 v, w) + b(v, w) \\ &= 2b(v, w) + b((g_2^{-1} g_1 + g_1^{-1} g_2)v, w) \end{aligned}$$

and there is no reason why this should be equal to $b(v, w)$. In particular for $g_1 = g_2$ one gets $b((g_1 + g_2)v, (g_1 + g_2)w) = 4b(v, w)$. It is more reasonable to ask that $b(xv, xw) = \varepsilon(x)^2 b(v, x)$, where $\varepsilon : KG \to K$ is the augmentation map $\varepsilon(\sum_{g \in G} k_g g) = \sum_{g \in G} k_g$. But even this will not work since in this case $b((g_2^{-1} g_1 + g_1^{-1} g_2)v, w)$ should be $2b(v, w)$, and there is no reason why this should hold. Actually simple examples show that this will not be true in general. Hence, the invariance of the form is a property of the group rather than of the group ring.

Let $b : V \times V \longrightarrow K$ be a G-invariant bilinear form. Then b induces a K-linear mapping

$$\begin{aligned} \varphi_b : V &\longrightarrow V^* \\ v &\mapsto (w \mapsto b(v, w)) \end{aligned}$$

Indeed, for all $v_1, v_2 \in V$ and $g \in G$ one gets

$$\begin{aligned}\varphi_b(\lambda_1 v_1 + \lambda_2 v_2) &= b(\lambda_1 v_1 + \lambda_2 v_2, -)\\ &= \lambda_1 b(v_1, -) + \lambda_2 b_2(v_2, -)\\ &= \lambda_1 \varphi_b(v_1) + \lambda_2 \varphi_b(v_2)\end{aligned}$$

Moreover, since b is G-invariant, one gets

$$\begin{aligned}\varphi_b(gv) &= b(gv, -)\\ &= b(g^{-1}gv, g^{-1}-)\\ &= b(v, g^{-1}-)\\ &= g \cdot (\varphi_b(v))\end{aligned}$$

when one remembers that V^* is naturally a right kG-module and using the mapping $G \ni g \mapsto g^{-1} \in G$ it becomes a left KG-module.

Conversely given a KG-linear homomorphism $\varphi : V \longrightarrow V^*$, one defines $b_\varphi : V \times V \longrightarrow K$ by $b_\varphi(v, w) := \varphi(v)(w)$. We remark that b_φ is KG-linear. Indeed, for all $v, w \in V$ and $g \in G$ one has

$$\begin{aligned}b_\varphi(gv, gw) &= (\varphi(gv))(gw)\\ &= (g\varphi(v))(gw)\\ &= \varphi(v)(g^{-1}gw)\\ &= \varphi(v)(w)\\ &= b_\varphi(v, w)\end{aligned}$$

The fact that b_φ is bilinear follows immediately by the fact that φ is linear.

We proved the following lemma.

Lemma 4.1.6. *Giving a G-invariant bilinear form $b : V \times V \longrightarrow K$ is equivalent to giving a homomorphism φ_b in $Hom_{KG}(V, V^*)$. The form b is non degenerate if and only if the homomorphism φ_b is an isomorphism.*

Remark 4.1.7. Instead of the elementary verification one might alternatively apply Frobenius reciprocity in a slightly more sophisticated version observing carefully the ways the group is acting. More precisely

$$(Hom_K(V \otimes_K V, K))^G \simeq Hom_K(V \otimes_{KG} V, K) \simeq Hom_{KG}(V, Hom_K(V, K))$$

where the G-invariant bilinear forms are just the dual of the G-coinvariant tensors, and hence the elements in the dual of the elements in $V \otimes_{KG} V$, taking into account that V becomes right KG-module by letting act g via g^{-1}, and hence $Hom_K(V, K)$ is a left KG-module this way.

Lemma 4.1.6 shows that the most interesting case is the case of self-dual modules, in which case there is a non degenerate invariant bilinear form.

Lemma 4.1.8. *Let G be a group, let K be a field and let V be a KG-module such that $End_{KG}(V) \simeq K$. If $V \simeq V^*$ as KG-modules, then there is an up to scalar unique non zero G-invariant bilinear form b on V. Moreover, b is non degenerate.*

Proof. Lemma 4.1.6 shows that there exists a G-invariant bilinear form b on V. Let b' be another G-invariant bilinear form on V. Then $\varphi_{b'}^{-1} \circ \varphi_b \in End_{KG}(V) \simeq K$. Hence there is a $\lambda_{b',b} \in K^\times$ such that $\varphi_{b'} = \lambda_{b',b} \cdot \varphi_b$. Translating this equality back to bilinear forms gives $b' = \lambda_{b',b} \cdot b$. Since $End_{KG}(V) \simeq K$ and since $V \simeq V^*$, any homomorphism $V \longrightarrow V^*$ is either 0 or an automorphism. This proves the lemma. □

The theory of bilinear and quadratic forms differs very much, if the base field is of characteristic 2 or not. We will first deal with the easier case of fields of characteristic different from 2.

Remark 4.1.9. Let K be a field of characteristic different from 2. Given a bilinear form b on V, then we get

$$b(v,w) = \frac{1}{2}(b(v,w) + b(w,v)) + \frac{1}{2}(b(v,w) - b(w,v))$$

for all $v, w \in V$. If we put

$$b_s(v,w) := \frac{1}{2}(b(v,w) + b(w,v)) \quad \text{and} \quad b_a(v,w) := \frac{1}{2}(b(v,w) - b(w,v))$$

for all $v, w \in V$ we obtain that every bilinear form on V is a sum of a symmetric bilinear form b_s on V and an antisymmetric bilinear form b_a on V. It is clear by definition that if b is invariant, also b_s is G-invariant and b_a is G-invariant.

Corollary 4.1.10. *Let G be a finite group, let K be a field of characteristic different from 2 and let V be a KG-module such that $End_{KG}(V) \simeq K$. If $V \simeq V^*$ as KG-modules, then there is either a symmetric G-invariant non degenerate bilinear form or an antisymmetric G-invariant non degenerate bilinear form on V. If V is simple and $V \not\simeq V^*$ then there is no G-invariant non degenerate bilinear form on V.*

Proof. Lemma 4.1.6 shows existence of a G-invariant bilinear form on V. Remark 4.1.9 shows that b is a sum of a symmetric G-invariant bilinear form b_s on V and an antisymmetric G-invariant bilinear form b_a on V. Lemma 4.1.8 shows that the G-invariant bilinear form on V is either 0 or non-degenerate, and then unique up to a scalar. Hence $b = b_s$ or $b = b_a$. A non zero G-invariant bilinear form on V is equivalent to a homomorphism $V \longrightarrow V^*$. If V is simple, any such morphism is an isomorphism. □

Definition 4.1.11. Let K be a field of characteristic different from 2 and let G be a finite group. Let V be a KG-module with $End_{KG}(V) = K$ and $V \simeq V^*$. Then V is called of *symmetric type* if there is a G-invariant symmetric bilinear form on V and of *alternating type* otherwise.

Corollary 4.1.10 shows that modules V of alternating type admit a G-invariant non degenerate antisymmetric bilinear form on V.

4.2 The Frobenius Schur indicator

Is it possible to see from the character table if a simple module is self-dual, symmetric or antisymmetric?

Definition 4.2.1. Let χ be a complex character of a finite group. The *Frobenius Schur indicator* of χ is the value

$$FS(\chi) := \frac{1}{|G|}\sum_{g\in G}\chi(g^2).$$

Theorem 4.2.2. *Let χ_V be an irreducible complex character of the simple $\mathbb{C}G$-module V of the finite group G. Then the Frobenius Schur indicator of χ_V has value* 0, 1 *or* −1. *Moreover*

- *the module V is of symmetric type if and only if $FS(\chi_V) = 1$,*
- *the module V is of antisymmetric type if and only if $FS(\chi_V) = -1$,*
- *the module V is not self-dual if and only if $FS(\chi_V) = 0$.*

Proof. Bilinear forms on V are parameterised by $V^*\otimes V^*$. Indeed, any element $\sum_i f_i\otimes g_i$ for $f_i, g_i \in V^*$ give a bilinear form by

$$(v,w)\mapsto \sum_i f_i(v)g_i(w).$$

Conversely, a bilinear form b on V defines a homomorphism $V \longrightarrow V^*$ and so an element in

$$Hom_K(V,V^*) \simeq Hom_K(V\otimes_K V,K) = (V\otimes_K V)^* \simeq V^*\otimes V^*$$

since for all vector spaces W_1, W_2 one gets

$$(W_1\otimes W_2)^* \simeq W_2^*\otimes W_1^*.$$

Now, the cyclic group C_2 of order 2 acts on $W\otimes_K W$ for each vector space W by interchanging the two components. Symmetric bilinear forms are therefore the dual of those elements in $(V\otimes_K V)$ which are invariant under this action, whence the coinvariants, that is the elements in

$$((V\otimes_K V)^{C_2})^* = ((V\otimes_K V)^*)_{C_2} = (V^*\otimes_K V^*)_{C_2}.$$

Here we denote for every vector space W by

$$S^2(W) := (W \otimes_K W)_{C_2}$$

the quotient of $W \otimes_K W$ by the sub-vector space generated by elements of the form $w_1 \otimes w_2 - w_2 \otimes w_1$, for $w_1, w_2 \in W$. Likewise the alternating bilinear forms are parameterised by

$$\Lambda^2(V^*) := (V^* \otimes_K V^*)/\langle f \otimes f \mid f \in V^* \rangle.$$

Now, Remark 4.1.9 shows that

$$V^* \otimes_K V^* \simeq \Lambda^2(V^*) \oplus S^2(V^*).$$

G acts on each of these spaces by diagonal action:

$$g \cdot (f_1 \otimes f_2) = (gf_1) \otimes (gf_2)$$

for every $f_1, f_2 \in V^*$ and $g \in G$, and the G-invariant subspaces are the G-invariant forms of the specific types. We obtain the G-invariant subspace of $V^* \otimes_K V^*$ by

$$\langle \chi_1, \chi_{V^* \otimes V^*} \rangle = \frac{1}{|G|} \sum_{g \in G} \chi_V(g^{-1})^2 = \frac{1}{|G|} \sum_{g \in G} \chi_V(g)^2.$$

The character of $\Lambda^2(V^*)$ is

$$\chi_{\Lambda^2(V^*)}(g) = \frac{1}{2}(\chi_V(g)^2 - \chi_V(g^2)).$$

Indeed, we take for the K-linear dual of $\Lambda^2(V^*)$ as basis the elements $w_{i,j} := v_i \otimes v_j - v_j \otimes v_i$ for $i < j$ and $v_i; 1 \le i \le n$ a basis of V^*. Then $g \in G$ acts on $w_{i,j}$ as

$$g \cdot w_{i,j} = \sum_{r,s} (a_{i,r}a_{j,s} - a_{j,r}a_{i,s}) v_r \otimes v_s = \sum_{r<s} (a_{i,r}a_{j,s} - a_{j,r}a_{i,s}) w_{r,s}$$

if $g \cdot v_i = \sum_r a_{i,r} v_r$. Hence

$$\chi_{\Lambda^2(V^*)}(g) = \sum_{i<j} a_{ii}a_{jj} - a_{ij}a_{ji}$$

and

$$\begin{aligned} 2\chi_{\Lambda^2(V^*)}(g) &= \sum_{i \ne j} a_{ii}a_{jj} - a_{ij}a_{ji} \\ &= \Big(\sum_i a_{ii}\Big)\Big(\sum_j a_{jj}\Big) - \sum_{i,j} a_{ij}a_{ji} \\ &= \chi_{V^*}(g)^2 - \chi_{V^*}(g^2). \end{aligned}$$

As $V^* \otimes_K V^* \simeq \Lambda^2(V^*) \oplus S^2(V^*)$ we obtain immediately the formula for the character of $S^2(V^*)$:

$$\chi_{S^2(V^*)}(g) = \chi_{V^*\otimes V^*}(g) - \chi_{\Lambda^2(V^*)}(g) = \frac{1}{2}(\chi_{V^*}(g)^2 + \chi_{V^*}(g^2))$$

Hence

$$FS(\chi) = \frac{1}{|G|}\sum_{g\in G}\chi(g^2) = \langle \chi_1, \chi_{S^2(V^*)} - \chi_{\Lambda^2(V^*)}\rangle.$$

If V is simple, there is a either no non zero invariant bilinear form, in which case

$$\langle \chi_1, \chi_{S^2(V^*)}\rangle = \langle \chi_1, \chi_{\Lambda^2(V^*)}\rangle = 0$$

and $SF(\chi_V) = 0$, or a symmetric invariant non degenerate bilinear form, in which case $\langle \chi_1, \chi_{S^2(V^*)}\rangle = 1$ and $\langle \chi_1, \chi_{\Lambda^2(V^*)}\rangle = 0$, or an antisymmetric invariant non degenerate bilinear form, in which case $\langle \chi_1, \chi_{S^2(V^*)}\rangle = 0$ and $\langle \chi_1, \chi_{\Lambda^2(V^*)}\rangle = 1$. This proves the theorem. □

4.3 Quadratic modules in characteristic 2

Recall the definition of a quadratic form on a vector space.

Definition 4.3.1. Let K be a field and let V be a K-vector space. A mapping $Q : V \to K$ is a *quadratic form on V* if
- $Q(\lambda v) = \lambda^2 Q(v)$ for all $v \in V$ and $\lambda \in K$.
- The map $b : V \times V \to K$ defined by $b_Q(v, w) := Q(v+w) - Q(v) - Q(w)$ for all $v, w \in V$ is a bilinear form on V.

If V is a KG-module, then a quadratic form Q on V is G-invariant if $Q(gv) = Q(v)$ for all $v \in V$ and $g \in G$.

Remark 4.3.2.
- By definition a quadratic form Q on V induces a bilinear form b_Q, called its *polarisation*. If V is a KG-module and if Q is G-invariant, then also its polarisation b_Q is G-invariant.
- Moreover, if b is a bilinear form on V, then $Q_b(v) := b(v, v)$ is a quadratic form. If V is a KG-module, and if b is a G-invariant bilinear form on V, then Q_b is a G-invariant quadratic form on V.
- If b is a bilinear form, then

$$\begin{aligned} b_{Q_b}(v, w) &= Q_b(v + w) - Q_b(v) - Q_b(w) \\ &= b(v + w, v + w) - b(v, v) - b(w, w) \end{aligned}$$

$$= b(v,w) + b(w,v)$$
$$= 2b_s(v,w) \quad \text{(cf. Remark 4.1.9)}$$

and hence $b_{Q_b} = 2b_s$, the double of the symmetrisation of b.
- If Q is a quadratic form, then

$$Q_{b_Q}(v) = b_Q(v,v) = Q(v+v) - Q(v) - Q(v) = 2Q(v).$$

The last two items imply that if K is of characteristic different from 2, then (G-invariant) symmetric bilinear forms are in bijection with (G-invariant) quadratic forms via the correspondence $b \mapsto \frac{1}{2}Q_b$.

The situation is different in characteristic 2. Still, by definition, a quadratic form induces a symmetric bilinear form, but in general not all symmetric bilinear forms are polarisations of quadratic forms.

Definition 4.3.3. Let K be a field. A self-dual simple KG-module V with endomorphism ring K is called *quadratic* if there is a quadratic form Q on V such that the non-zero G-invariant bilinear form on V is the polarisation of a G-invariant quadratic form on V.

As we have seen, if the characteristic of K is different from 2, then all self-dual simple KG-modules with trivial endomorphism algebra are either quadratic or admit a non degenerate antisymmetric invariant form. We shall study the question when this is the case over fields of characteristic 2.

In order to do so we need to introduce a new concept which we will detail in the next section.

4.3.1 First group cohomology

In Definition 3.4.4 we obtained an abelian group from the question of extending an action of a normal subgroup of G on a module M to all of G in case $I_G(M) = G$. In this section we shall define the first group cohomology by means of considerations of G-fixed points.

Let M and N be $\mathbb{Z}G$-modules and let $\alpha : M \to N$ be a homomorphism. Then $\alpha(M^G) \subseteq N^G$, where we denote by $M^G := \{m \in M \mid gm = m \ \forall g \in G\}$ the G-fixed points on M, and likewise for N. Indeed, if $m \in M^G$, then

$$\alpha(m) = \alpha(gm) = g\alpha(m)$$

for all $g \in G$. Hence α induces a homomorphism $\alpha^G : M^G \to N^G$ of abelian groups. If α is injective, then α^G is injective as well. Indeed, we have more generally

$$\ker(\alpha^G) = M^G \cap \ker \alpha.$$

Example 4.3.4. Let p be a prime, let K be a field of characteristic $p > 0$ and let G be the cyclic group of order p. Since the trivial KG-module K is simple there is a surjective KG-module homomorphism

$$\pi : KG \twoheadrightarrow K$$

of the regular module of G to the trivial module of G. Explicitly π is the augmentation mapping, defined by $\pi(g) = 1$ for all $g \in G$. The G-fixed points of the regular KG-module is given by $KG^G = K \cdot \sum_{g \in G} g$. But this shows that $\pi^G = 0$.

Let K be a commutative ring. For any KG-module M we shall define a K-module $H^1(G, M)$ such that if $\pi : M \twoheadrightarrow N$ is an epimorphism of KG-modules, then there is a canonical K-module homomorphism

$$\operatorname{coker} \pi \to H^1(G, \ker \pi).$$

So, let M be a KG-module and consider the *derivations of G with values in M*

$$Der(G, M) := \{f : G \to M \mid f(gh) = f(g) + gf(h)\ \forall g, h \in G\}.$$

This is a K-module, since M is a KG-module. A special kind of derivations are called *inner derivations*, namely

$$IDer(G, M) := \{f : G \to M \mid \exists m_0 \in M \forall\ g \in G : f(g) = gm_0 - m_0\}.$$

Now, $IDer(G, M)$ is a K-submodule of $Der(G, M)$. Indeed, let f_m and f_n be two inner derivations given by $f_m(g) = gm - m$ and $f_n(g) = gn - n$. Then $f_m - f_n = f_{m-n}$ with the obvious notation. Moreover, since

$$f_m(gh) - f_m(g) - gf_m(h) = (ghm - m) - (gm - m) - g(hm - m) = 0$$

an inner derivation is a derivation.

Definition 4.3.5. Let G be a group, let K be a commutative ring and let M be a KG-module. Then $Der(G, M) := \{f : G \to M \mid f(gh) = f(g) + gf(h)\ \forall g, h \in G\}$ are the *derivations of G with values in M* and $IDer(G, M) := \{f : G \to M \mid \exists m_0 \in M \forall\ g \in G : f(g) = gm_0 - m_0\}$ are the *inner derivations of G with values in M*. The K-module

$$H^1(G, M) := Der(G, M)/IDer(G, M)$$

is the *degree* 1 *group cohomology*. An element in $Der^1(G, M)$ is called a 1-*cocycle* with values in M, and an element in $IDer(G, M)$ is also called a 1-*coboundary*.

Remark 4.3.6. Recall the definition of a 2-coboundary. A map $f : G \times G \to M$ is a 2-coboundary if there is $\sigma : G \to M$ such that $f(g, h) = \sigma(gh) - \sigma(g) - g\sigma(h)$. The similarity with the defining equation of a derivation should be noted. This is not an accident, but rather a very shadow of a vast theory behind the scenes. For more details the reader may consult the monograph by Adem-Milgram [AdMi-04] for example.

Proposition 4.3.7. *Let K be a commutative ring, let G be a group and let M and N be KG-modules. Let $\alpha : N \twoheadrightarrow M$ be a KG-module epimorphism and let α^G be the induced mapping on the fixed point spaces. Then there is a K-module monomorphism $\beta : \operatorname{coker}(\alpha^G) \hookrightarrow H^1(G, \ker\alpha)$.*

Proof. Let now K be a commutative ring and G be a group, and let M and N be KG-modules, and let $\alpha : N \twoheadrightarrow M$ be an epimorphism of KG-modules. Let $C := \ker\alpha$. We need to define a map $M^G \to H^1(G, C)$. For this purpose we need to define for each $m \in M^G$ a map $\beta_m : G \to C$ satisfying the defining condition to be a derivation. In this case we shall define $\beta(m) := \beta_m$. Let $m \in M^G$ and let $n_m \in N$ with $\alpha(n_m) = m$. For any $g \in G$ we get $gm = m$, and so $\alpha(gn_m) = \alpha(n_m)$. This shows that $gn_m - n_m \in C$. Let $\beta_m(g) := gn_m - n_m$. Then

$$\beta_m(gh) = ghn_m - n_m = g(hn_m - n_m) + (gn_m - n_m) = g\beta_m(h) + \beta_m(g).$$

We need to show that the class of β_m in $H^1(G, C)$ does not depend on the choice of n_m. If $\alpha(n_m) = \alpha(n'_m)$, then

$$(gn_m - n_m) - (gn'_m - n'_m) = g(n_m - n'_m) - (n_m - n'_m)$$

such that we may suppose that $m = 0$. But in this case $n_0 \in C$ and since $\beta_{n_0}(g) = gn_0 - n_0$ we have that β_{n_0} is an inner derivation. Further, if $m \in \alpha(N^G)$, then there is $n_m \in N^G$ with $m = \alpha(n_m)$. In this case we may take this element n_m for the definition of β_m, and in this case $\beta_m(g) = gn_m - n_m = n_m - n_m = 0$. Now, since α is K-linear, β is K-linear as well.

We still need to show that β is injective. Let $m \in M$ with $\beta_m = 0 \in H^1(G, C)$. This is equivalent with $\beta_m \in IDer(G, C)$, and therefore there is $c_m \in C$ such that $\beta_m(g) = gc_m - c_m$ for all $g \in G$. Therefore

$$\beta_m(g) = gn_m - n_m = gc_m - c_m$$

for all $g \in G$. This shows that $g(n_m - c_m) = (n_m - c_m)$ for all $g \in G$. Moreover,

$$m = \alpha(n_m) = \alpha(n_m - c_m),$$

and therefore $m \in \alpha(N^G)$ and this shows in turn that β is injective. □

Proposition 4.3.7 now gives a criterion of when α^G is surjective. Indeed, denoting by $V^{[1]}$ the inverse Frobenius twist of V, if $H^1(G, V^{[1]}) = 0$, then necessarily $\operatorname{coker}(\alpha^G) = 0$, and hence α^G is surjective. How can we give such a statement on $H^1(G, L)$ for a KG-module L? An easy criterion is the subject of the following statement.

Lemma 4.3.8. *Let K be a commutative ring, let G be a finite group of order $|G|$ and let L be a KG-module. Then $|G| \cdot H^1(G, L) = 0$. In particular, if $|G| \cdot L = L$, then $H^1(G, L) = 0$.*

Proof. Once we proved that $|G| \cdot H^1(G, L) = 0$, we get the statement. Indeed, for every $f' \in H^1(G, L)$, we have $|G| \cdot f' = 0$, but since $|G|L = L$, we have that for each $f \in H^1(G, L)$ there is $\hat{f} \in H^1(G, L)$ such that $|G| \cdot \hat{f} = f$. This is zero by what we said before.

Let $f : G \to L$ be a derivation. We have to show that f is inner. Let $\ell := -\sum_{g \in G} f(g)$. Then, by the defining equation of a derivation we get

$$f(g) = f(gh) - gf(h)$$

for all $g, h \in G$. The left hand side is independent of h, and summing over $h \in G$ we get

$$|G|f(g) = \sum_{h \in G} (f(gh) - gf(h)) = \sum_{h \in G} f(gh) - g \sum_{h \in G} f(h) = g\ell - \ell,$$

which is hence an inner derivation. This shows the statement. □

Remark 4.3.9. Lemma 4.3.8 shows that if $|G|K = K$, then $H^1(G, L) = 0$.

4.3.2 Properties of symmetric and exterior products

We recall from Section 4.2 the interpretation of bilinear forms by elements of the tensor product. Let K be a field and let V be a finite dimensional K-vector space. Denote $V^* := Hom_K(V, K)$. A bilinear form on V is by definition a K-bilinear map $V \times V \to K$, and this is by definition an element in $(V \otimes_K V)^* \simeq V^* \otimes_K V^*$. Now,

$$\Lambda^2(V) := V \otimes_K V / \langle v \otimes v \mid v \in V \rangle_K$$

is the exterior product and taking co-invariants of the "switching the position action of the cyclic group of order 2", we get the symmetric product

$$S^2(V) := (V \otimes_K V)_{\mathfrak{S}_2}.$$

This may be defined by

$$S^2(V) := (V \otimes_K V) / \langle v_1 \otimes v_2 - v_2 \otimes v_1 \mid v_1, v_2 \in V \rangle$$

Moreover, let the symmetric powers

$$\Gamma^2(V) := (V \otimes_K V)^{\mathfrak{S}_2}$$

the fixed point space under the "switching the position action of the cyclic group of order 2". Now, we get a natural isomorphism

$$(\Gamma^2(V))^* \simeq S^2(V^*).$$

By definition we have natural monomorphisms

$$\Gamma^2(V) \hookrightarrow V \otimes_K V$$

which gives by dualising a natural epimorphism

$$V^* \otimes_K V^* \overset{\zeta}{\twoheadrightarrow} S^2(V^*).$$

We have $\ker(\zeta) = \Lambda^2(V^*)$ as is immediately verified. Moreover, by definition we obtain immediately

$$V^* \otimes_K V^* \twoheadrightarrow \Lambda^2(V^*).$$

Suppose now that K is a field of characteristic 2. Then we compute in $\Lambda^2(V^*)$

$$f_1 \otimes f_2 + f_2 \otimes f_1 = f_1 \otimes f_2 + f_2 \otimes f_1 + f_1 \otimes f_1 + f_2 \otimes f_2 = (f_1 + f_2) \otimes (f_1 + f_2) = 0$$

and hence the natural map $V^* \otimes_K V^* \to \Lambda^2(V^*)$ induces a natural map

$$\begin{aligned} S^2(V^*) &\overset{\pi}{\to} \Lambda^2(V^*) \\ f_1 \otimes f_2 &\mapsto f_1 \otimes f_2 \end{aligned}$$

on the quotient. Indeed, we showed that $f_1 \otimes f_2$ and $f_2 \otimes f_1$ have the same image. Hence there is a natural epimorphism

$$S^2(V^*) \overset{\pi}{\twoheadrightarrow} \Lambda^2(V^*)$$

such that the composition

$$V^* \otimes_K V^* \twoheadrightarrow S^2(V^*) \overset{\pi}{\twoheadrightarrow} \Lambda^2(V^*)$$

is the natural mapping

$$V^* \otimes_K V^* \twoheadrightarrow \Lambda^2(V^*).$$

We shall see later that the map π is the polarisation map. Since

$$\Lambda^2(V^*) = V^* \otimes V^* / \langle f \otimes f \mid f \in V^* \rangle_K,$$

it is clear that the image of $\langle f \otimes f \mid f \in V^* \rangle_K$ in $S^2(V^*)$ is in the kernel of π. Since the dimension of $\langle f \otimes f \mid f \in V^* \rangle_K$ in $V^* \otimes V^*$ equals the dimension of $\langle f \otimes f \mid f \in V^* \rangle_K$ in $S^2(V^*)$ we see that $\ker \pi = \langle f \otimes f \mid f \in V^* \rangle_K$. There is a bijection $V^* \to \ker(\pi)$ given by $V \ni f \mapsto f \otimes f \in \ker \pi$. However,

$$\lambda \cdot f \mapsto (\lambda f) \otimes (\lambda f) = \lambda^2 (f \otimes f) \in \ker \pi.$$

Hence, the action of $\lambda \in K$ on V^* is mapped to the action of λ^2 on $\ker \pi$. Suppose now that K is a perfect field of characteristic 2. Therefore

$$(V^*)^{[1]} \simeq \ker \pi$$

where for each K-vector space W we denote by $W^{[1]}$ the $\mathbb{F}_2$ vector space W, but the action of K on W is twisted by the inverse of the Frobenius automorphism of K.

Proposition 4.3.10. *Let G be a group, let K be a perfect field of characteristic 2 and let V be a finite dimensional K-vector space. Let $D := \operatorname{coker}((S^2(V^*))^G \longrightarrow (\Lambda^2(V^*))^G)$. Then there is a monomorphism*

$$D \hookrightarrow H^1(G, V^{[1]}).$$

Proof. This is an immediate consequence of the fact that

$$(V^*)^{[1]} \simeq \ker(S^2(V^*) \xrightarrow{\pi} \Lambda^2(V^*)),$$

which was shown just above, and Proposition 4.3.7. □

4.3.3 Quadratic modules and group cohomology

The results of this section are mainly taken from [SinWi-91]. Let G be a group, let K be a field and let V be a KG-module. Then, V^* is a KG-module, and also $V^* \otimes_K V^*$ is a KG-module as well, where $g \in G$ acts on $f_1 \otimes f_2 \in V^* \otimes_K V^*$ as $g \cdot (f_1 \otimes f_2) = (gf_1) \otimes (gf_2)$.

Then, the K-vector space of bilinear forms on V is $V^* \otimes_K V^* = (V \otimes_K V)^*$, and the space of G-invariant bilinear forms on V is given by $(V^* \otimes_K V^*)^G$. Now, $\Lambda^2(V^*)$ is the K-vector space of symplectic forms over V with values in K. Recall that a bilinear form $f : V \times V \to K$ is *symplectic* if $f(v, v) = 0$ for all $v \in V$, and if in addition f is *non-degenerate* in the sense that $f(v, w) = 0$ for all $w \in V$ implies $v = 0$.

For fields of characteristic 2 the space of quadratic K-valued forms on V is given by

$$((V \otimes_K V)^{\mathfrak{S}_2})^* = (V^* \otimes_K V^*)_{\mathfrak{S}_2} = S^2(V^*).$$

This follows as in the proof of Proposition 4.3.10 and by the elementary approach in Exercise 4.10.

We obtain the following first result.

Proposition 4.3.11 (Fong). *Let K be a field of characteristic 2, let G be a group and let V be a KG-module. If $V \simeq V^*$ as KG-module, and if $V^G = 0$, then V admits a nonzero G-invariant symplectic form.*

Proof. Since $V \simeq V^*$, using Lemma 4.1.8 we have $(V^* \otimes_K V^*)^G \neq 0$. Since $V^G = 0$, and since V is self-dual, we also get $(V^*)^G = 0$ and hence

$$\ker(\pi^G) = \ker(\pi)^G = ((V^*)^{[1]})^G = 0.$$

Therefore π^G is injective. But then, $\pi^G \circ \zeta^G$ either has a nonzero image or a nonzero kernel. Since

$$\ker(\pi^G \circ \zeta^G) = \ker(\zeta^G) = (\Lambda^2(V^*))^G,$$

this shows the statement. □

Theorem 4.3.12. *Let K be a field of characteristic 2 and let V be a self-dual simple non-trivial KG-module with $\mathrm{End}_{KG}(V) = K$. Then there is an up to scalar unique non-degenerate G-invariant bilinear symplectic form on V. If $H^1(G, V^{[1]}) = 0$, then this form is of quadratic type, i. e. is the polarisation of a G-invariant quadratic form.*

Proof. The existence of the symplectic form is Proposition 4.3.11. Let Q be a quadratic form on V, and let b_Q be its polarisation. Then $b_Q(v_1, v_2) = Q(v_1 + v_2) - Q(v_1) - Q(v_2)$ and

$$b_Q(v, v) = 4Q(v) - 2Q(v) = 0$$

since Q is of characteristic 2. Hence b_Q is symplectic. Symplectic bilinear forms are precisely the elements of $\Lambda^2(V^*)$. Hence we get a map

$$S^2(V^*) \to \Lambda^2(V^*).$$

This is precisely the map π from Section 4.3.2. Proposition 4.3.10 shows the statement. □

4.4 Exercises

Exercise 4.1. Show that a simple $\mathbb{C}G$-module V admits a G-invariant bilinear form if and only if its character has values in $\mathbb{R}$.

Exercise 4.2. Let G be a finite group of odd order n. For any non trivial irreducible character χ of G compare the Frobenius Schur indicator of χ with the scalar product of χ with the trivial character. Deduce that G does not allow any non trivial self-dual simple module, and that any non trivial simple $\mathbb{C}G$-module does not admit any G-invariant non degenerate bilinear form.

Exercise 4.3 (with solution in Chapter 9). Let ζ_7 be primitive 7-root of unity in the complex numbers. We find the character table of a group G as follows:

	C_1	C_2	C_3	C_4	C_5	C_6	C_7	C_8	C_9
χ_1	1	1	1	1	1	1	1	1	1
χ_2	6	2	3	−1	0	0	1	−1	−1
χ_3	10	−2	1	1	1	0	0	$\zeta_7^3+\zeta_7^5+\zeta_7^6$	$\zeta_7+\zeta_7^2+\zeta_7^4$
χ_4	10	−2	1	1	1	0	0	$\zeta_7+\zeta_7^2+\zeta_7^4$	$\zeta_7^3+\zeta_7^5+\zeta_7^6$
χ_5	14	2	2	2	−1	0	−1	0	0
χ_6	14	2	−1	−1	2	0	−1	0	0
χ_7	15	−1	3	−1	0	−1	0	1	1
χ_8	21	1	−3	1	0	−1	1	0	0
χ_9	35	−1	−1	−1	−1	1	0	0	0

a) Find the order of G?
b) Show that G is simple.
c) Find the size of the conjugacy classes of G.
d) Use the orthogonality relations to decompose

$$\chi := (\ 27\quad -1\quad 6\quad 2\quad 3\quad 1\quad 2\quad -1\quad -1\ )$$

as sum of irreducible characters.
e) Note by M_i the simple $\mathbb{C}G$-module corresponding to the irreducible character χ_i. Which are the modules M_i allowing a non degenerate invariant bilinear form?
f) We know that the map $G \ni x \mapsto x^2 \in G$ fixes C_1, C_5 et C_7, sends C_2 to C_1, C_3 to C_2, C_4 to C_3, C_6 to C_2, C_8 to C_9, and C_9 to C_8. If M_i allows an invariant non degenerate bilinear form, find its type (symmetric or antisymmetric).
g) The module $M_3 \oplus M_4$, does it allow a non degenerate G-invariant bilinear form?

Exercise 4.4 (M. Isaacs). Let G be a finite group and let $n \in \mathbb{N} \setminus \{1\}$. For each $g \in G$ consider

$$\theta_n(g) := |\{h \in G|\ h^n = g\}|.$$

a) Prove that $\theta_n(g) \geq 2$ implies that n divides the order of G.
b) Show that $\theta_n : G \to \mathbb{N}$ is a class function (cf. Definition 2.2.3).
c) Denote by $Irr(G)$ the set of irreducible complex characters of G. Deduce that there are complex valued functions $\mu_n : Irr(G) \to \mathbb{C}$ such that

$$\theta_n = \sum_{\chi \in Irr(G)} \mu_n(\chi)\chi$$

d) For each irreducible character χ of G show that

$$\theta_n(g)\overline{\chi}(g) = \sum_{h\in G;h^n=g} \overline{\chi(h^n)}$$

and use the orthogonality relations to see that

$$\mu_n(\chi) = \frac{1}{|G|}\sum_{g\in G}\chi(g^n)$$

and in particular $FS(\chi) = \mu_2(\chi)$.

e) Let C_2 be the number of elements of order 2 in G. Prove that

$$1 + C_2 = \sum_{\chi\in Irr(G)} FS(\chi)\chi(1)$$

f) Recall that a group is perfect if $G = G'$, i. e. the largest abelian quotient is the trivial group 1. Let G be a perfect group of odd order. Show that if G admits a simple non trivial module with an invariant symmetric bilinear form, then G also admits a simple non trivial module with an invariant antisymmetric bilinear form, and vice versa.

Exercise 4.5. Let G be a cyclic group of order n, generated by g and let M be a $\mathbb{Z}G$-module. Let ${}_nM := \{m \in M|\ \sum_{g\in G} g\cdot m = 0\}$. Show that $H^1(G,M) \simeq {}_nM/(g-1)M$.

Exercise 4.6. Let G be a group, and let K be a commutative ring. Let M_1 and M_2 be KG-modules. Show that $H^1(G, M_1 \oplus M_2) \simeq H^1(G, M_1) \oplus H^1(G, M_2)$.

Exercise 4.7. Let G be a group, let K be a commutative ring, and let M and M' be KG-modules, let N be a KG-submodule of M and $C := M/N$. Denote by $\iota : N \to M$ the embedding and by $\pi : M \to C$ the natural quotient mapping. Let furthermore $\alpha : M \to M'$ be a homomorphism of KG-modules.

a) Show that $H^1(G,M) \ni f \mapsto \alpha\circ f =: H^1(G,\alpha)(f) \in H^1(G,M')$ is a well-defined homomorphism of K-modules.
b) Show that $H^1(G,\pi)\circ H^1(G,\iota) = 0$.
c) Let $\beta : C^G \to H^1(G,N)$ be the homomorphism defined in Proposition 4.3.7. Show that $H^1(G,\iota)\circ\beta = 0$.
d) Prove that $\ker(H^1(G,\pi)) = \operatorname{im}(H^1(G,\iota))$.
e) Prove that $\ker(H^1(G,\iota)) = \operatorname{im}(\beta)$.
f) For any derivation $f : G \longrightarrow C$ let $\hat{f} : G \longrightarrow M$ be a map with $\pi\circ\hat{f} = f$. Then define $\gamma(f) : G\times G \longrightarrow M$ by $\gamma(f)(g_1, g_2) := g_1\hat{f}(g_2) - \hat{f}(g_1g_2) + \hat{f}(g_2)$. Show that γ is a K-module homomorphism $H^1(G,C) \xrightarrow{\gamma} H^2(G,N)$ such that $\ker(\gamma) = \operatorname{im}(H^1(G,\pi))$.

Exercise 4.8. Let G be a group and let H be a normal subgroup of G. Let M be a $\mathbb{Z}G$-module, and denote by M^H the $\mathbb{Z}G/H$-module formed by the H-fixed points of M.

a) Let $f : G/H \longrightarrow M^H$ be a 1-cocycle. Prove that the composition $\hat{f}$ of the natural maps $G \longrightarrow\!\!\!\!\!\rightarrow G/H \xrightarrow{f} M^H \hookrightarrow M$ is a 1-cocycle of G with values in M.
b) Show that if f is a 1-coboundary on G/H, then $\hat{f}$ is a 1-coboundary on G with values in M.
c) Prove that $H^1(G/H, M^H) \ni f \mapsto \hat{f} \in H^1(G, M)$ is a well-defined homomorphism $inf^G_{G/H}$ of abelian groups.
d) Show that if f is a 1-cocycle on G/H such that $\hat{f}$ is a 1-coboundary, then f is a 1-coboundary as well.
e) Let f be a 1-cocycle of G with values in M. Show that the restriction $f|_H$ of f to H is a 1-cocycle of H with values in M.
f) Show that restriction to elements in H induces a well-defined group homomorphism $res^G_H : H^1(G, M) \to H^1(H, M)$.
g) Let f be a 1-cocycle with values in M^H, show that $\hat{f}|_H$ is constant and equal to $f(1) = 0$.
h) Let f be a 1-cocycle of G with value in M, and suppose that $f|_H$ is a 1-coboundary of H with values in M. Show that one can modify f by a 1-coboundary on G with values in M, such that $f|_H = 0$. Show further that in this case f is constant on the cosets of G modulo H, and that $f = inf^G_{G/H}(\check{f})$ for some 1-cocycle $\check{f}$ on G/H with values in M^H.
i) Show that $inf^G_{G/H}$ is a monomorphism and that $\ker(res^G_H) = \mathrm{im}(inf^G_{G/H})$.

NB: This construction is part of what is called the five term sequence.

Exercise 4.9. This exercise is an application of the first group cohomology to a nice problem in number theory.
a) Let F be a field and let $\sigma_1, \ldots, \sigma_n$ be pairwise distinct automorphisms of F. We will show that the $\mathbb{Z}$-linear endomorphism $\sum_{i=1}^n c_i\sigma_i$ of F is 0 if and only if $c_i = 0$ for all i.
   a1) Let r be minimal such that $\lambda := \sum_{j=1}^r c_{i_j}\sigma_{i_j}$ is 0 with all c_{i_j} non zero and $i_j \neq i_k$ for all $j \neq k$. Show that $r > 1$.
   a2) Consider for an $a \in F$ the endomorphism $\lambda(ax) - \sigma_{i_r}(a)\lambda(x)$ and obtain a contradiction to the minimality of r.
b) Let L be a finite Galois extension of a field K with Galois group G. Then $(L, +)$ is a KG-module, but also the multiplicative group $(L^\times, \cdot)$ is a KG-module. In this part of the exercise we show the Hilbert 90 theorem, which says that $H^1(G, L^\times) = 0$. Let $f : G \to L^\times$ be a 1-cocycle and define a K-linear endomorphism $\varphi := \sum_{g\in G} f(g)g$ of L.
   b1) Use a) to show that there is $x \in L^\times$ such that $\beta := \varphi(x) \neq 0$.
   b2) Show that $g \cdot \beta = f(g)\beta$ for all $g \in G$.
   b3) Deduce that f is a 1-coboundary and $H^1(G, L^\times) = 0$.
c) We follow here basically N.Elkies (American Mathematical Monthly 110 (2003) 678). Let $K = \mathbb{Q}$ and $L = \mathbb{Q}(i)$. We will show in this exercise that the only in-

teger solutions of $x^2 + y^2 = z^2$ are proportional of those of the form $(x, y, z) = (m^2 - n^2, 2mn, m^2 + n^2)$ for integers m, n. Recall that triples of integers x, y, z with $x^2 + y^2 = z^2$ are called pythagorean triples.

c1) For such a pythagorean triple (x, y, z) as above consider $u := \frac{x+iy}{z}$. Show that $u \cdot \overline{u} = 1$.

c2) Use Hilbert 90 to get an element $a \in L$ such that $u = a \cdot \overline{a}^{-1}$, and show that one may assume that a is an element of $\mathbb{Z}[i]$, the Gaussian integers.

c3) Conclude.

Exercise 4.10. Recall that a field K of characteristic 2 is perfect if $K \ni u \mapsto u^2 \in K$ is surjective. A bilinear form g is the polar form of a quadratic form q if $g(x, y) = q(x + y) + q(x) + q(y)$ for all $x, y \in V$. We denote in this case $g = f_q$. Let K be a perfect field of characteristic 2, let V be a K-vector space of finite dimension and let f be a bilinear form on V. Suppose that f is the polar form of the quadratic forms q_1 and q_2 on V.

a) Show $q_1 + q_2$ is additive, i. e. $q_1 + q_2$ is a homomorphisms of abelian groups.
b) Show that there is a linear form $\varphi \in Hom_K(V, K)$ such that $(q_1 + q_2)(x) = \varphi(x)^2$ for all $x \in V$.
c) If $\varphi \in Hom_K(V, K)$ is a linear form on V, show that $V \ni x \mapsto \varphi(x)^2 \in K$ is a quadratic form q_φ on V.
d) Find the polar form of q_φ.
e) Denote by $Bil(V)$ the K-vector space consisting of all bilinear forms on V and by $Q(V)$ the K-vector space of all quadratic forms on V. Show that the map $Q(V) \ni q \mapsto f_q \in Bil(V)$, denoted by Ψ, is a homomorphism of vector spaces. Compute its kernel.
f) Show that the image of Ψ is in the subspace of symplectic bilinear forms on V.
g) Use Gram matrices to determine the dimension of $Q(V)$ and the dimension of the space of symplectic bilinear forms on V. Deduce that Ψ is surjective.
h) If V is a KG-module, show that the restriction of Ψ to the space $Q(V)^G$ of G-invariant quadratic forms has image in the space of symplectic G-invariant bilinear forms on V. Describe the kernel in this case.
i) If V is a KG-module, and suppose that there is no non trivial KG-linear homomorphism from V to the trivial KG-module, what can we say about the restriction of Ψ to the space of G-invariant quadratic forms $Q(V)^G$?

Exercise 4.11. (Sin-Willems) Let G be a finite group which contains a non trivial normal subgroup N of odd order and let K be a field of characteristic 2. Let V be a faithful, self-dual, indecomposable KG-module.

a) Let $f_N := \frac{1}{|N|} \sum_{g \in N} g$. Show that $f_N^2 = f_N$ and that for all KG-module M the module $f_N \cdot M$ is a KG-module, and

$$M = f_N \cdot M \oplus (1 - f_N) \cdot M.$$

Prove that N acts trivially on $f_N \cdot M$.

b) Let K be the trivial KG-module. Deduce that $f_N \cdot V = 0$ and that $f_N \cdot K = K$.
c) Show that $V^G \simeq Hom_{KG}(K, V)$ and deduce that $V^G = 0$.
d) Use the decomposition into two central idempotents $1 = f_N + (1 - f_N)$ of KN to show that there is an isomorphism of algebras $KN = K \times X$ for some K-algebra X. Show that $H^1(N, V) = 0$.
e) Use Exercise 4.8 to show that $H^1(G, V) = 0$.
f) Prove that V admits a non zero G-invariant symplectic form which is the polarisation of a G-invariant quadratic form.

5 Brauer induction, Brauer's splitting field theorem

Mit der Wahrheit hadern wir nicht,
weil wir uns irrten, sondern stets,
weil sie ist wie sie ist.

Reiner Kunze, Gedichte @ 2001, S. Fischer Verlag Frankfurt am Main

We don't bemoan truth
because we are wrong, but ever
because it is the way it is.

Reiner Kunze, poetry;[1]

We have seen in Lemma 3.2.9 that all KG-modules are direct summands of modules induced from subgroups, whenever the order of G is invertible in K. But, in order to get this, it happens that it is impossible to get a specific KG-module M as a module $N\uparrow_H^G$ for some KH-module N, H being a subgroup of G. It might happen that only a number n_M of copies of M can be achieved this way. It is desirable to be able to get a single module M as induced modules, in some way. The way this is done is by considering so-called virtual modules, that is formal $\mathbb{Z}$-linear combinations of isomorphism classes of modules, and moreover by not only considering one specific subgroup from where one would like to induce, but by considering a class of very easily controllable subgroups. Then the important induction theorem of Brauer says that all modules can be achieved as $\mathbb{Z}$-linear combinations of modules induced from modules over this class of groups. A proof of this result is given in Section 5.2.

A consequence will be the most important theorem due to Brauer saying that if G is a group of exponent n, then fields containing a primitive n-th root of unity are splitting fields for G. This statement is suggested by the fact that the character values, which are the traces of the matrices generating the group ring in the Wedderburn decomposition are in such fields. One would like to have that actually the representations should be realisable over the integers of this field. In general this is not true, but for solvable groups this is true again, as is proved by Cliff, Ritter and Weiss [ClRiWe-92]. A much more natural property holds for odd order groups, as is shown by Cram and Neiße [CrNe-96]. For a group of odd order take the smallest cyclotomic field containing all character values of the group. Then every simple representation is already realisable over the algebraic integers over this field.

5.1 Grothendieck and character ring

We start to generalise characters to so-called virtual characters.

1 in der Übersetzung von Susanne Brennecke, translation by Susanne Brennecke.

https://doi.org/10.1515/9783110702446-005

Definition 5.1.1. Let G be a finite group and let K be a field such that the order of G is invertible in K. Let $Irr_K(G)$ be the set of irreducible K-valued characters of G. Then the smallest abelian subgroup of $CF_K(G)$ containing $Irr_K(G)$ is the set of *virtual characters* of G.

Recall that Proposition 2.2.4 assumes that K is a splitting field for G. Here we are dealing with a more general case.

Observe that a virtual character of G is an object of the form $\chi_M - \chi_N$ for two KG-modules M and N. Indeed, ψ is a virtual character if and only if there are integers $x_\chi \in \mathbb{Z}$ for $\chi \in Irr_K(G)$ such that $\psi = \sum_{\chi \in Irr_K(G)} x_\chi \chi$. Hence

$$\begin{aligned}\psi &= \sum_{\chi \in Irr_K(G)} x_\chi \chi \\ &= \sum_{\chi \in Irr_K(G); x_\chi > 0} |x_\chi| \chi - \sum_{\chi \in Irr_K(G); x_\chi < 0} |x_\chi| \chi \\ &= \chi_M - \chi_N\end{aligned}$$

where $M = \oplus_{\chi \in Irr_K(G); x_\chi > 0} S_\chi^{x_\chi}$ and $N = \oplus_{\chi \in Irr_K(G); x_\chi < 0} S_\chi^{-x_\chi}$, denoting as usual S_χ the simple module whose character is χ. If the characteristic of K is $p > 0$, then χ_S represents the same element as $\chi_{S^{1+p}} = (1+p) \cdot \chi_S$. Hence, usually it is common to work over fields K of characteristic 0 whenever one looks at characters.

There is a module theoretic counterpart, the Grothendieck group, which we are going to use. Actually the Grothendieck group is a much more general concept which is of tremendous importance in modern algebra. It seems to be a good occasion to introduce it here.

Let A be a ring and let M be an A-module. Denote by $\{M\}$ the isomorphism class of M.

Definition 5.1.2. Let A be an R-algebra for R being a commutative ring. Then let F_{A-mod} be the free abelian group on the set of isomorphism classes of finitely generated A-modules. Let R_{A-mod} be the subgroup of F_{A-mod} generated by the objects $\{M\} - \{N\} - \{L\}$ where L is a finitely generated submodule of M and $N \simeq M/L$. Then the *Grothendieck group* of finitely generated A-modules is

$$G_0(A - mod) = F_{A-mod}/R_{A-mod}.$$

Denote by $[M]$ the image of $\{M\} \in F_{A-mod}$ in $G_0(A - mod)$.

Remark 5.1.3.

1. In case of a field R and finite dimensional R-algebras A we usually consider only finitely generated A-modules. The problem when working over general commutative rings is that if M is a finitely generated A-module and L is a submodule, then L does not need to be finitely generated. In case R is a field or more generally R and A are Noetherian, then finitely generated modules are Noetherian again and

sub-objects of Noetherian objects are finitely generated. We shall elaborate on the properties of being Noetherian and its definition in Section 6.1 when it is going to become necessary.

2. The reason of considering finitely generated modules only is the following easy observation. $\prod_{i=1}^{\infty} M \simeq \prod_{i=0}^{\infty} M$, and $M\times\prod_{i=1}^{\infty} M = \prod_{i=0}^{\infty} M$. Hence, if $M^{\infty} := \prod_{i=1}^{\infty} M$, we get M is a submodule of M^{∞} with quotient being isomorphic to M^{∞}. Hence, if we would consider all A-modules in the definition of the Grothendieck group, we would end up with the group with only one element.
3. The remarks preceding Definition 5.1.2 show that taking characters is a mapping from $G_0(KG)$ to the abelian group of virtual characters of G over K for a finite group G and a field K in which the order of G is invertible. More precisely, map $\{M\} \in F_{KG-Mod}$ to χ_M in the group of virtual characters of G over K. Then, since every KG-module is semisimple, the map factors through $G_0(KG)$, using Lemma 2.1.5. If moreover K is a field of characteristic 0, then for every virtual character χ one gets two KG-modules M and N with $\chi = \chi_M - \chi_N$. We may assume that $\langle\chi_M,\chi_N\rangle = 0$ since KG is semisimple, and then the decomposition $\chi = \chi_M - \chi_N$ is unique.
Hence if K is a field of characteristic 0 and G is a finite group then $G_0(KG)$ is isomorphic to the group of virtual characters of G over K.

There are several operations on $G_0(KG)$. For the moment suppose only that G is a finite group and that K is a field.

- First $G_0(KG)$ is trivially an abelian group.
- Remark 3.2.2 gives a multiplicative structure on $G_0(KG)$: If L is a KG-submodule of M, and if N is another KG-module, then $N\otimes_K L$ is a submodule of $N\otimes_K M$. Indeed, if ι is the embedding of L into M, then $id_N\otimes\iota$ is a mapping of $N\otimes_K L$ to $N\otimes_K M$. Since there exists a K-basis of N, Lemma 3.1.11 shows that $id_N\otimes\iota$ is injective. Moreover, again by Lemma 3.1.11 and since N has a K-basis,

$$(N\otimes_K M)/(N\otimes_K L) \simeq N\otimes_K (M/L).$$

Therefore N defines an operation of F_{KG-mod} on $G_0(KG)$ by

$$\{N\}\otimes_K [M] := [N\otimes_K M].$$

By the same argument the operation of F_{KG-mod} on $G_0(KG)$ factors through an operation of $G_0(KG)$ on $G_0(KG)$. Hence we get a well-defined mapping

$$\begin{aligned} -\otimes_K - : G_0(KG)\times G_0(KG) &\longrightarrow G_0(KG)\\ ([M],[N]) &\mapsto [M\otimes_K N] \end{aligned}$$

Lemma 5.1.4. *Let K be a field and let G be a group. Then $G_0(KG)$ is a commutative ring with unit by setting $[M]\cdot[N] := [M\otimes_K N]$ and $[M]+[N] := [M\oplus N]$ for all KG-modules M and N. The unit element in $G_0(KG)$ is the trivial KG-module.*

Proof.

- Recall that $g \in G$ acts on $n \otimes m \in N \otimes M$ by $g \cdot (n \otimes m) = (gn) \otimes (gm)$. The product $- \otimes_K - : G_0(KG) \times G_0(KG) \longrightarrow G_0(KG)$ is commutative since

$$\begin{aligned} M \otimes_K N &\simeq N \otimes_K M \\ m \otimes n &\mapsto n \otimes m \end{aligned}$$

  is obviously well-defined, bijective and a homomorphism of KG-modules.
- The product $- \otimes_K -$ is associative, since tensor products are associative up to isomorphism by Lemma 3.1.10 and since the associativity isomorphism in Lemma 3.1.10 is clearly linear with respect to the diagonal G-action.
- We get that the trivial KG-module K acts as identity on $G_0(KG)$. Indeed, $K \otimes_K M \simeq M$ by Lemma 3.1.11. Here we use again the action of G on the tensor product. Hence the product $- \otimes_K - : G_0(KG) \times G_0(KG) \longrightarrow G_0(KG)$ has a unit element, the trivial module.
- The product $- \otimes_K - : G_0(KG) \times G_0(KG) \longrightarrow G_0(KG)$ is distributive since

$$M \otimes_K (N \oplus L) \simeq (M \otimes_K N) \oplus (M \otimes_K L)$$

  by Lemma 3.1.10.

We showed the Lemma. □

Remark 5.1.5. The ring structure on $G_0(KG)$ uses the fact that if M and N are KG-modules, then $M \otimes_K N$ is a KG-module again by the action of G on $m \otimes n \in M \otimes_K N$ as $g \cdot (m \otimes n) := (gm) \otimes (gn)$. This actually is a special case of a so-called bialgebra structure, namely an algebra homomorphism $A \longrightarrow A \otimes_K A$ satisfying a number of properties. This is given by

$$KG \ni g \mapsto g \otimes g \in KG \otimes_K KG$$

in case of group rings KG. The structure theory of bialgebras or more restrictively of Hopf algebras is a well-established field in mathematics.

Let G be a finite group and let K be a field. Given a subgroup H of G the group ring KG is isomorphic to $(KH)^{|G:H|}$ as KH-right module. Hence, by Lemma 3.1.11 one gets a mapping

$$\begin{array}{ccc} G_0(KH) & \overset{(\ )\uparrow_H^G}{\longrightarrow} & G_0(KG) \\ [M] & \mapsto & [M \uparrow_H^G] \end{array}$$

Lemma 5.1.6. *Let G be a group, let H be a subgroup and let K be a commutative ring. Then the image $G_0(KH) \uparrow_H^G$ of $G_0(KH) \overset{(\ )\uparrow_H^G}{\longrightarrow} G_0(KG)$ is an ideal in $G_0(KG)$. Moreover, for* *every subgroup H of G and every KH-module N and KG-module M one gets*

$$(N \uparrow_H^G) \otimes_K M \simeq (N \otimes_K M \downarrow_H^G) \uparrow_H^G$$

as KG-modules.

Proof. Since

$$M \downarrow_H^G \uparrow_H^G \oplus N \uparrow_H^G \simeq (M \downarrow_H^G \oplus N) \uparrow_H^G$$

for KH-modules M and N we get that $G_0(KH) \uparrow_H^G$ is a subgroup of the additive group of $G_0(KG)$. We need to show that for every KG-module M and every KH-module N one has

$$M \otimes_K (N \uparrow_H^G) \in G_0(KH) \uparrow_H^G .$$

Let $g \in G$ be a fixed element. For this g define a mapping

$$\begin{aligned} N \times M &\longrightarrow (KG \otimes_{KH} N) \otimes_K M \\ (n, m) &\mapsto (g \otimes n) \otimes gm \end{aligned}$$

which is clearly K-balanced. Hence it induces a mapping

$$\begin{aligned} N \otimes_K M &\xrightarrow{\mu_g} (KG \otimes_{KH} N) \otimes_K M \\ (n \otimes m) &\mapsto (g \otimes n) \otimes gm \end{aligned}$$

of K-vector spaces. Then we define a mapping

$$\begin{aligned} KG \times (N \otimes_K M) &\xrightarrow{\mu} (KG \otimes_{KH} N) \otimes_K M \\ \left(\sum_{g \in G} x_g g, y\right) &\mapsto \sum_{g \in G} x_g \mu_g(y) \end{aligned}$$

This is KH-balanced since for every $h \in H$

$$\begin{aligned} \left(\sum_{g \in G} x_g gh, (n \otimes m)\right) \mapsto \sum_{g \in G} x_g \mu_{gh}(n \otimes m) &= \sum_{g \in G} x_g (gh \otimes n) \otimes ghm \\ &= \sum_{g \in G} x_g (g \otimes (hn)) \otimes g(hm) \\ &= \sum_{g \in G} x_g \mu_g(h \cdot (n \otimes m)) \end{aligned}$$

and the latter expression is the image of $(\sum_{g\in G} x_g g, h\cdot(n\otimes m))$. Hence we get an induced mapping

$$\begin{aligned} KG\otimes_{KH}(N\otimes_K M) &\stackrel{\mu}{\longrightarrow} (KG\otimes_{KH}N)\otimes_K M\\ \Big(\sum_{g\in G} x_g g\Big)\otimes y &\mapsto \sum_{g\in G} x_g\mu_g(y). \end{aligned}$$

Observe that the isomorphism is defined by

$$g\otimes(n\otimes m)\mapsto(g\otimes n)\otimes gm$$

and hence is different from the isomorphism following from the associativity of the tensor product in Lemma 3.1.10.

Completely analogously one defines an inverse mapping

$$\begin{aligned} (KG\otimes_{KH}N)\otimes_K M &\stackrel{\nu}{\longrightarrow} KG\otimes_{KH}(N\otimes_K M)\\ (g\otimes n)\otimes m &\mapsto g\otimes(n\otimes g^{-1}m) \end{aligned}$$

by first fixing $m\in M$ and then varying $m\in M$ as well. Then one observes that $\mu\circ\nu = id_{(KG\otimes_{KH}N)\otimes_K M}$ and $\nu\circ\mu = id_{KG\otimes_{KH}(N\otimes M)}$ to see that μ is an isomorphism of K-modules. It is actually an isomorphism of KG-modules. Indeed, for any $g_1, g_2\in G$, $m\in M$, $n\in N$ one gets

$$\begin{aligned} \mu(g_1(g_2\otimes(n\otimes m))) &= \mu((g_1g_2)\otimes(n\otimes m))\\ &= ((g_1g_2)\otimes n)\otimes g_1g_2m\\ &= g_1\cdot((g_2\otimes n)\otimes g_2m)\\ &= g_1\cdot\mu(g_2\otimes(n\otimes m)) \end{aligned}$$

This proves the lemma. □

Example 5.1.7.

1. Let $K=\mathbb{Q}$ and $G=\mathfrak{S}_3$. Then there are three simple $\mathbb{Q}\mathfrak{S}_3$-modules; the trivial module $\mathbb{Q}_+$, the sign representation $\mathbb{Q}_-$ and the two-dimensional simple module S_2. Using the characters of these modules and Lemma 3.2.3, as well as the orthogonality relations for characters one gets that

   $$S_2\otimes_{\mathbb{Q}} S_2\simeq\mathbb{Q}_+\oplus\mathbb{Q}_-\oplus S_2.$$

   Moreover,

   $$\mathbb{Q}_-\otimes_{\mathbb{Q}}\mathbb{Q}_-\simeq\mathbb{Q}_+$$

   and

   $$\mathbb{Q}_-\otimes_{\mathbb{Q}} S_2\simeq S_2.$$

Remark 5.1.3 and Proposition 2.2.4 show that $G_0(\mathbb{Q}\mathfrak{S}_3)$ is isomorphic to $\mathbb{Z}\oplus\mathbb{Z}\oplus\mathbb{Z}$ as abelian group. So,

$$G_0(\mathbb{Q}\mathfrak{S}_3) \simeq \mathbb{Z}[X,Y]/(Y^2-1, X^2-X-Y-1, XY-X)$$

when we identify X with $[S_2]$ and Y with $[\mathbb{Q}_-]$.

2. Recall the character table of the alternating group of degree 4 from Example 2.3.7:

$\mathfrak{A}_4$ size of class	$\{1\}$ 1	$C_{(1\ 2)(3\ 4)}$ 3	$C_{(1\ 2\ 3)}$ 4	$C_{(1\ 3\ 2)}$ 4
χ_1	1	1	1	1
χ_ζ	1	1	ζ	ζ^2
χ_{ζ^2}	1	1	ζ^2	ζ
χ_3	3	−1	0	0

where ζ is a primitive third root of unity. Then $(\chi_\zeta)^3 = \chi_1$ and $\chi_\zeta^2 = \chi_{\zeta^2}$. Moreover $\chi_3\chi_\zeta = \chi_3$ and $\chi_3^2 = 2\cdot\chi_3 + \chi_1 + \chi_\zeta + \chi_\zeta^2$. Hence

$$G_0(\mathbb{C}\mathfrak{A}_4) \simeq \mathbb{Z}[X,Y]/(Y^3-1, XY-X, X^2-2X-Y^2-Y-1)$$

identifying X with χ_ζ and Y with χ_3.

3. The ring $\mathbb{Q}\mathfrak{A}_4$ has three simple modules; the trivial module $\mathbb{Q}_+$, a three dimensional simple module S_3 coming from the complex character χ_3 and a two-dimensional simple module $S_2 = \mathbb{Q}[\zeta]$ coming from the quotient $\mathfrak{A}_4 \longrightarrow C_3$. Then

$$S_3 \otimes_{\mathbb{Q}} S_3 \simeq S_3^2 \oplus S_2 \oplus \mathbb{Q}_+$$

by the above computations,

$$S_3 \otimes_{\mathbb{Q}} S_2 \simeq S_3 \oplus S_3$$

since over $\mathbb{C}$ one gets $\chi_3\chi_\zeta = \chi_3$ and

$$S_2 \otimes_{\mathbb{Q}} S_2 \simeq \mathbb{Q}_+^2 \oplus S_2$$

since

$$(\chi_\zeta + \chi_{\zeta^2})^2 = 2\cdot\chi_1 + \chi_\zeta + \chi_{\zeta^2}.$$

Therefore,

$$G_0(\mathbb{Q}\mathfrak{A}_4) \simeq \mathbb{Z}[X,U]/(U^2-U-2, XU-2X, X^2-2X-U-1)$$

identifying X with $[S_3]$ and U with $[S_2]$.

The fact that $G_0(A-mod)$ is free abelian for semisimple algebras A is almost trivial but useful to mention.

Lemma 5.1.8. *Let K be a field and let A be a finite dimensional semisimple K-algebra. Then $G_0(A)$ is a free abelian group with basis being the set $\{[S] \mid S$ simple A module$\}$.*

Proof. Every A-module M is a direct sum of uniquely determined indecomposable modules by the Krull-Schmidt theorem. Remark that here we use that A is finite dimensional. Whenever A is semisimple every indecomposable module is simple. Hence $\{[S] \mid S$ simple A module$\}$ is a basis of $G_0(A)$. □

5.2 Brauer induction formula

The main theorem in this section is due to Richard Brauer [Bra-47] saying basically that every complex valued character can be decomposed as a difference of characters induced from a special class of subgroups.

There are various proofs and versions of this quite deep statement. The classical one uses character theory and some number theoretic congruences on character values in an essential way. This approach gives a pure existence theorem. In particular the coefficients of the linear combinations are largely unclear. Another approach is given by Victor Snaith [Sna-88] who uses topological invariants, Euler classes for unitary groups to give an explicit formula for the coefficients. A third approach is given by Robert Boltje [Bol-88] who uses Mackey functor techniques to give a purely algebraic and explicit formula for the linear coefficients. It is interesting to note that the formulas are different for these two approaches. Finally a third approach is given by Peter Symonds [Sym-91]. Symonds uses geometric arguments, such as line bundles over projective spaces associated to representations, to obtain another explicit formula. We refer to Snaith [Sna-94a] for a detailed discussion of the theory.

Since Snaith's approach is elegant but theoretically sophisticated we feel that it is not appropriate for the purpose of this book. Boltje's formula could be presented right now, but the proof is highly complicate and a presentation would produce quite a large entire chapter which seems to be inappropriate as well. Symonds' method is quite far from our spirit at this point, though most interesting, and we refrain from giving this theory. We instead give a version of Brauer's original proof, which is displayed in different manners and modifications in Curtis-Reiner [CuRe-62, § 40] or Müller [Mül-80, § 5.7].

First we need to define the specific class of subgroups from where we want to induce.

Definition 5.2.1. Let p be a prime. A *p-elementary group* E is a direct product $E = C \times P$ where P is a p-group and C is a cyclic group of order prime to p. A group is *elementary* if it is p-elementary for some p. Let $\mathfrak{H}_p(G)$ be the set of p-elementary subgroups of G.

Given a finite group G every element $g \in G$ is an element of a p-elementary subgroup of G. Indeed, let $g \in G$ and let n be the order of g in G. Then $n = p^{n_p} \cdot n_{p'}$ where n_p is an integer and $gcd(p, n_{p'}) = 1$. Then the cyclic group $\tilde{C}$ generated by g is actually the direct product of the cyclic group of order $n_{p'}$ generated by $g^{p^{n_p}}$ and the cyclic group of order p^{n_p} generated by $g^{n_{p'}}$. The cyclic group generated by g is therefore p-elementary. We proved actually the following lemma.

Lemma 5.2.2. *Every finite cyclic group is p-elementary for every prime p.*

Proof. We may take G a cyclic group generated by g and the proof preceding the statement implies that this cyclic group is p-elementary. □

Theorem 5.2.3 (R.Brauer [Bra-47]). *Let G be a finite group and let K be a field of characteristic 0 and suppose that K is a splitting field for G and all of is subgroups. Then*

$$\sum_{p\ prime} \sum_{H \in \mathfrak{H}_p(G)} (G_0(KH) \uparrow_H^G) = G_0(KG).$$

As a consequence every complex valued character of G is a $\mathbb{Z}$-linear combination of complex valued characters induced from p-elementary abelian subgroups. The important statement is that the linear combination is a $\mathbb{Z}$-linear combination and no fraction is needed.

The proof of this theorem will cover the present Subsection 5.2 entirely.

We divide the proof of Theorem 5.2.3 into several lemmas.

We first start with a number theoretic result which can be found in almost any text in algebraic number theory. Let R be the ring of algebraic integers in $\mathbb{Q}(\zeta)$ where ζ is a primitive n-th root of unity and where n is the exponent of G.

Lemma 5.2.4. *Let ζ be a primitive n-th root of unity in $\mathbb{C}$. Then*

$$R = \text{algint}_{\mathbb{Z}}(\mathbb{Q}(\zeta)) = \mathbb{Z}[\zeta].$$

Proof. We postpone the proof of this fact until Lemma 7.1.5. The proof can also be found in e. g. [Was-97, Exercise 1.1, Theorem 2.6]. □

Let $\kappa_p(G) := \sum_{H \in \mathfrak{H}_p(G)} (G_0(KH) \uparrow_H^G)$ and $\kappa(G) := \sum_{p\ \text{prime}} \kappa_p(G)$.

Lemma 5.2.5. *If the trivial KG-module K_+ is in $R \otimes_{\mathbb{Z}} \kappa(G)$, then $\kappa(G) = G_0(KG)$.*

Proof. By hypothesis there are elements $\alpha_1, \alpha_2, \dots, \alpha_\ell$ in R and there are KG-modules $M_1, M_2, \dots, M_\ell \in \kappa(G)$ such that

$$[K_+] = \sum_{j=1}^{\ell} \alpha_j [M_j].$$

By Lemma 5.1.8 the group $G_0(KG)$ has a $\mathbb{Z}$-basis

$$\mathcal{S} := \{[S] \mid S \text{ simple } KG-\text{ module}\}.$$

Hence there are integers $\gamma_{[S],j} \in \mathbb{Z}$ for $[S] \in \mathcal{S}$ and $j \in \{1, 2, \ldots, \ell\}$ such that

$$[M_j] = \sum_{[S]\in\mathcal{S}} \gamma_{[S],j}[S].$$

By Lemma 5.2.4 we know that R has a $\mathbb{Z}$-basis $\{1, \zeta, \zeta^2, \ldots, \zeta^{\varphi(n)-1}\}$ for φ being the Euler φ-function, i. e. $\varphi(n) = |(\mathbb{Z}/n\mathbb{Z})^\times|$. Hence there are integers $\beta_{i,j}$ for $i \in \{0, 1, \ldots, \varphi(n) - 1\}$ and $j \in \{1, 2, \ldots, \ell\}$ such that

$$\alpha_j = \sum_{i=0}^{\varphi(n)-1} \beta_{i,j}\zeta^i$$

and so

$$[K_+] = \sum_{j=1}^{\ell} \sum_{i=0}^{\varphi(n)-1} \beta_{i,j}\zeta^i \sum_{[S]\in\mathcal{S}} \gamma_{[S],j}[S] = \sum_{[S]\in\mathcal{S}} \left(\sum_{j=1}^{\ell} \sum_{i=0}^{\varphi(n)-1} \beta_{i,j}\zeta^i\gamma_{[S],j} \right)[S].$$

Since $\mathcal{S}$ is a $\mathbb{Z}$-basis of $G_0(KG)$, and since $[K_+] \in \mathcal{S}$,

$$\sum_{j=1}^{\ell} \sum_{i=0}^{\varphi(n)-1} \beta_{i,j}\zeta^i\gamma_{[S],j} = \delta_{[S],[K_+]},$$

where $\delta_{x,y}$ is the Kronecker symbol, being 1 if the symbols x and y are equal and 0 else. Since $\{1, \zeta, \zeta^2, \ldots, \zeta^{\varphi(n)-1}\}$ is a $\mathbb{Z}$-basis of R, this implies

$$\sum_{j=1}^{\ell} \beta_{i,j}\gamma_{[S],j} = 0 \quad \text{whenever } [S] \neq [K_+] \text{ or } i \neq 0$$

and

$$\sum_{j=1}^{\ell} \beta_{0,j}\gamma_{[S],j} = 1 \quad \text{whenever } [S] = [K_+].$$

Therefore,

$$\sum_{j=1}^{\ell} \beta_{0,j}\gamma_{[S],j} = \delta_{[S],[K_+]}$$

and hence

$$\begin{aligned} \sum_{j=1}^{\ell} \beta_{0,j}[M_j] &= \sum_{j=1}^{\ell} \beta_{0,j} \sum_{[S]\in\mathcal{S}} \gamma_{[S],j}[S] \\ &= \sum_{[S]\in\mathcal{S}} \sum_{j=1}^{\ell} \beta_{0,j}\gamma_{[S],j}[S] \end{aligned}$$

$$= \sum_{[S]\in\mathcal{S}} \delta_{[S],[K_+]}[S]$$
$$= [K_+]$$

Therefore $[K_+] \in \kappa(G)$.

Now, $[K_+]$ is the unit in the ring $G_0(KG)$ by Lemma 5.1.4 and $\kappa(G)$ is an ideal in $G_0(KG)$ by Lemma 5.1.6 as a sum of ideals. As a consequence one gets $\kappa(G) = G_0(KG)$ which proves the lemma. □

Let E be an elementary subgroup of G. We may suppose that $E = A \times B$ for an abelian group A and a group B, and we may suppose moreover $gcd(|A|, |B|) = 1$. We are going to construct some specific element in $\kappa(G)$. Since K is a splitting field for all subgroups of G, the field K is a splitting field for A as well. Therefore there are $|A|$ distinct simple KA-modules up to isomorphism. All of them are of dimension 1 over K. Hence, each $a \in A$ acts on such a simple KA-module S by multiplication by $\zeta^{n_a(S)}$ for some integer n_a. Since $E = A \times B \longrightarrow A$ by the usual quotient mapping, each of the simple KA-modules is a simple KE-module as well. For S a KA-module call $inf_A^E(S)$ the KE-module which is produced by this quotient mapping $\pi : E \twoheadrightarrow A$.

Denote by $\mathcal{R}_A$ a set of representatives of the isomorphism classes of simple KA-modules. For all $a \in A$ let

$$[M_a(E)] := \sum_{S\in\mathcal{R}} \zeta^{-n_a(S)} \cdot [inf_A^E(S)] \in R \otimes_{\mathbb{Z}} G_0(KE)$$

and

$$[N_a(E)] := \sum_{S\in\mathcal{R}} \zeta^{-n_a(S)} \cdot [inf_A^E(S) \uparrow_E^G] \in R \otimes_{\mathbb{Z}} G_0(KG).$$

In order to simplify notations we shall use the notation S also for the module $inf_A^E(S)$.

Lemma 5.2.6. *Let $\chi_{N_a(E)}$ be the character of $N_a(E)$. Then*

$$\chi_{N_a(E)}(g) = \begin{cases} |C_G(a) : B| & \text{if } g \text{ is conjugate to } a \text{ in } G \\ 0 & \text{else} \end{cases}$$

Proof. Each of the KE-modules S constructed above as inflation from a simple KA-modules is one-dimensional. Let $g \in E$. Then

$$\chi_{[N_a(E)]}(g) = \sum_{S\in\mathcal{R}_A} \zeta^{-n_a(S)} \cdot \chi_{[S]}(g)$$
$$= \sum_{S\in\mathcal{R}_A} \chi_{[S]}(a^{-1}) \cdot \chi_{[S]}(g)$$
$$= \begin{cases} 0 & \text{if } a \text{ is not conjugate to } g \text{ in } A \\ |A| & \text{else} \end{cases}$$

by the orthogonality relations for conjugacy classes. Now, the structure of E being a direct product $A \times B$ and A being abelian implies that $g \in E$ is conjugate to $a \in A$ if and only if $g = a$. Therefore

$$\chi_{[N_a(E)]}(g) = \delta_{a,g}|A|$$

where δ denotes the Kronecker symbol.

We deduce the corresponding value for $\chi_{[N_a(E)]}(g)$ for $g \in G$ using Lemma 3.2.17. In fact this lemma shows that the value $\chi_{[N_a(E)]}(g)$ will be non zero only if there is an $x \in G$ such that $xgx^{-1} \in E$. The character value $\chi_{[N_a(E)]}(g)$ will then be a sum of the evaluations of $\chi_{[M_a(E)]}(xgx^{-1})$ divided by $|E|$. But $\chi_{[M_a(E)]}(xgx^{-1})$ is non zero exactly when $xgx^{-1} \in aB$ and then its value is $|A|$. Hence, one gets constant value $|A|$ on exactly the conjugacy class of a and 0 elsewhere. Therefore for g conjugate to an element in aB one has

$$\begin{aligned}
\chi_{[N_a(E)]}(g) &= \frac{1}{|E|}\left|\{x \in G \mid xgx^{-1} \in E\}\right| \cdot \chi_{[M_a(E)]}(xgx^{-1}) \\
&= \frac{1}{|E|}\left|\{x \in G \mid xgx^{-1} \in aB\}\right| \cdot |A| \\
&= \frac{1}{|B|} \cdot |B| \cdot \left|C_G(a) : B\right|
\end{aligned}$$

since $\{x \in G \mid xgx^{-1} \in aB\}$ is a set of left cosets of B, and modulo B the set is by definition the centralizer of a in G.

If g is not conjugate to an element in aB, none of the elements xgx^{-1} is ever in aB, and so $\chi_{[N_a(E)]}(g) = 0$. This shows the lemma. □

The group G decomposes into a disjoint union of conjugacy classes

$$G = C_{g_1} \cup C_{g_2} \cup \cdots \cup C_{g_m}$$

where $C_g = \{hgh^{-1} | \ h \in G\}$. Put $R_G := \{g_1, g_2, \ldots, g_m\}$ and fix a prime p. If p does not divide the order of a $g \in R_G$, then let B_g be a Sylow p-subgroup of the centraliser $C_G(g)$ of g in G. Let A_g be the cyclic group generated by g. Then $E_g := A_g \times B_g$ is a p-elementary subgroup of G. Since two Sylow p-subgroups of C_g are conjugate, the element $[N_g(E_g)]$ is independent of the choice of the representative g of the conjugacy class C_g and the choice of the Sylow subgroup.

By Lemma 5.2.6 and by construction, $\chi_{[N_g(E_g)]}(g) \notin p\mathbb{Z}$ and so there is an integer $m_g \in \mathbb{Z}$ such that the character value of $m_g \cdot [N_g(E_g)] = [(N_g(E_g))^{m_g}]$ on g is in $1 + p\mathbb{Z}$. Put

$$[F_p] := \sum_{g \in R_G;\ p \nmid |g|} m_g \cdot [N_g(E_g)] \in R \otimes_{\mathbb{Z}} G_0(KG).$$

Lemma 5.2.6 implies that $\chi_{[F_p]}(g) \in \mathbb{Z}$ for all $g \in G$.

Lemma 5.2.7. *For all $g \in G$ one gets*

$$\chi_{[F_p]}(g) \equiv 1 \mod p.$$

Proof. Let $g \in G$. Then the cyclic group $\langle g\rangle$ generated by g is abelian and hence has a normal Sylow p-subgroup. Therefore $\langle g\rangle$ is a direct product of this Sylow p-subgroup and another group whose order is prime to p. As a consequence there are uniquely determined elements g_p and $g_{p'}$ of G such that $g = g_p \cdot g_{p'}$.

We call the element g_p the *p-singular part* of g and the element $g_{p'}$ the *p-regular part* of g. An element $g \in G$ is called *p-regular* if its p-singular part is 1. Likewise an element $g \in G$ is called *p-singular* if its p-regular part is 1.

Let now $h \in G$ and $a \in G$ be a fixed p-regular element of G and $E_a = A_a \times B_a$ as above. Abbreviate $B = B_a$ and $N_a(E_a)$ by N_a. If $h_{p'}$ is not in aB, then no conjugate of h is in aB. Indeed, let

$$(xhx^{-1})_{p'} = ab \in aB.$$

Then

$$h = (x^{-1}ax) \cdot (x^{-1}bx) = h_{p'}h_p$$

and, since a and b commute, the same is true for the elements $(x^{-1}ax)$ and $(x^{-1}bx)$. By unicity of the p-singular and p-regular part of an element one gets that the p-regular element $x^{-1}ax$ equals $h_{p'}$ and $x^{-1}bx = h_p$. Hence $h_{p'}$ is conjugate to a, contradiction. We get that if no conjugate of h lies in aB, then $\chi_{[N_a]}(h) = 0$.

We now come to the proof of the fact that $\chi_{[F_p]}(h) \equiv 1 \mod p$. In order to see this we remark that the p-regular part of h is contained in exactly one p-regular conjugacy class. Moreover when the p-regular part of h is not conjugate to a regular element a, then $\chi_{[N_a]}(h) = 0$ by the previous argument. Hence it is sufficient to show that when h is conjugate to an element in aB, then

$$m_a \cdot \chi_{[N_a]}(h) \equiv 1 \mod p.$$

Suppose hence for a moment that h is conjugate to an element in aB for a p-regular element a. Let $\langle h\rangle$ be the subgroup of G generated by h and let

$$\chi_{N_a} = c_1 \cdot \chi_{X_1} + c_2 \cdot \chi_{X_2} + \cdots + c_u \cdot \chi_{X_u}$$

be a decomposition of the character of N_a into irreducible characters of $\langle h\rangle$, for $c_1, \ldots, c_u \in R$. Then each irreducible $K\langle h\rangle$-character χ_{X_j} is one-dimensional since K is a splitting field for all subgroups of G, including $\langle h\rangle$. Let d be such that $h_p^{p^d} = 1$. Then p is in the ideal generated by $1 - \zeta$ where ζ is a p^d-th root of unity, and since

$(x+y)^p - x^p - y^p \in pR$ for all $x, y \in R$,

$$\begin{aligned}
\chi_{[N_a]}(h) &\equiv \chi_{[N_a]}(h)^{p^d} (\operatorname{mod} p)\\
&\equiv \left(\sum_{j=1}^{u} c_j \cdot \chi_{X_j}(h)\right)^{p^d} (\operatorname{mod} p)\\
&\equiv \left(\sum_{j=1}^{u} c_j^{p^d} \cdot \chi_{X_j}(h^{p^d})\right) (\operatorname{mod} p)\\
&\equiv \left(\sum_{j=1}^{u} c_j^{p^d} \cdot \chi_{X_j}(h_{p'}^{p^d})\right) (\operatorname{mod} p)\\
&\equiv \left(\sum_{j=1}^{u} c_j \cdot \chi_{X_j}(h_{p'})\right)^{p^d} (\operatorname{mod} p)\\
&\equiv \left(\sum_{j=1}^{u} c_j \cdot \chi_{X_j}(h_{p'})\right) (\operatorname{mod} p)\\
&= \chi_{[N_a]}(h_{p'})\\
&= \chi_{[N_a]}(a)
\end{aligned}$$

Since $m_a \cdot \chi_{[N_a]}(a) \equiv 1 \bmod p$ this shows that $\chi_{[F_p]}(h) \equiv 1 \bmod p$. □

Let $|G| = p^{e_p} \cdot t_p$ for some integers e_p and t_p, and suppose that t_p is relatively prime to p. By construction

$$[F_p] \in R \otimes_{\mathbb{Z}} \sum_{H \in \mathfrak{H}_p} G_0(KH) \uparrow_H^G$$

Lemma 5.2.8.

$$V := t_p \cdot ([\underbrace{F_p \otimes_K \cdots \otimes_K F_p}_{p^{e_p} \text{ times}}] - [K_+]) \in R \otimes_{\mathbb{Z}} \sum_{H \in \mathfrak{H}_p} G_0(KH) \uparrow_H^G .$$

Proof. Let $C_{g_1}, C_{g_2}, \ldots, C_{g_s}$ be the p-regular conjugacy classes of G. Then, denoting by Γ_i a set of representatives of the isomorphism classes of simple $K\langle g_i\rangle$-modules,

$$[U_{g_i}] := \sum_{S \in \Gamma_i} \zeta^{-n_i(S)} \cdot [S \uparrow_E^G] \in R \otimes_{\mathbb{Z}} \sum_{H \in \mathfrak{H}_p} G_0(KH) \uparrow_H^G$$

for $n_i(S)$ being the integer such that g_i acts on S by $\zeta^{n_i(S)}$. As in the proof of Lemma 5.2.6 one gets that $\chi_{U_{g_i}}(h) = 0$ if h is not conjugate to g_i and $\chi_{U_{g_i}}(h) = |C_G(g_i)|$ if h is conjugate to g_i.

Recall that by Lemma 5.2.7 we have $\chi_{[F_p]}(h) \equiv 1 \bmod p$ for all $h \in G$. One gets

$$\begin{aligned}
\chi_{[\underbrace{F_p \otimes_K \cdots \otimes_K F_p}_{p^{e_p} \text{ times}}]}(h) &= \chi_{[F_p]}^{p^{e_p}}(h)\\
&\equiv 1 \bmod p^{e_p} \quad \text{for all } h \in G
\end{aligned}$$

and so the character of

$$t_p \cdot ([\underbrace{F_p \otimes_K \cdots \otimes_K F_p}_{p^{e_p} \text{ times}}] - [K_+])$$

on each $h \in G$ is in $|G| \cdot \mathbb{Z}$. We get

$$[V] = \sum_{j=1}^{s} \frac{1}{|G|} \chi_V(g_j) \frac{|G|}{|C_G(g_j)|} [U_{g_j}]$$

in the Grothendieck group $G_0(KG)$. Indeed, the left hand side and the right hand side have the same character values. But,

$$\frac{1}{|G|} \chi_V(g_j) \frac{|G|}{|C_G(g_j)|} \in \mathbb{Z}$$

and so we proved the statement. □

Lemma 5.2.8 implies that

$$t_p \cdot [K_+] \in R \otimes_{\mathbb{Z}} \sum_{H \in \mathfrak{H}_p} G_0(KH) \uparrow_H^G .$$

Let y be the greatest common divisor of all those t_p for which p divides $|G|$. Moreover, suppose a prime q divides y. Then q divides $|G|$ and we get $|G| = q^{e_q} \cdot t_q$ where t_q is not divisible by q. Hence q does not divide the greatest common divisor of all t_p for primes p dividing $|G|$, since t_q is one of them. We therefore get that $y = 1$.

Then

$$[K_+] = y \cdot [K_+] \in R \otimes_{\mathbb{Z}} \left(\sum_{p \text{ prime}} \sum_{H \in \mathfrak{H}_p} G_0(KH) \uparrow_H^G \right).$$

Using Lemma 5.2.5 this shows Theorem 5.2.3. □

5.3 Brauer's splitting field theorem

In this section we shall prove the most famous Brauer's theorem on splitting fields, saying that whenever K is a field containing an n-th root of unity with n being the exponent of G, then K is a splitting field of G. The example of a cyclic group of order n shows that this is best possible for general finite groups. However, on special classes of groups one can obtain much better results. For example for finite symmetric groups *every field* is a splitting field. This particular result is proved in a completely different way. Classical references on representations of symmetric groups are James [Jam-78] and Green [Gre-80], or a more recent and very original approach by Kleshchev [Kles-05].

Since on every KG-module M every group element $g \in G$ acts by a diagonalisable matrix with eigenvalues being $n(g)$-th roots of unity for $n(g)$ being the order of g, the statement of Brauer's theorem does not seem to be very hard. Actually, it is hard and not at all trivial, since the group elements are *not simultaneously diagonalisable* and a priori there is no reason why the way two elements commute does not involve matrices with strange coefficients.

Actually the most standard proof of Brauer's splitting field theorem is via Brauer's induction Theorem 5.2.3.

We first need a preparation.

Proposition 5.3.1. *Let G be a finite group and let K be a field such that the order of G is invertible in K. Suppose K is a splitting field for each abelian normal subgroup of G.*

- *If G is a p-group, then for each simple KG-module S there is a subgroup $H(S)$ of G and a one-dimensional $KH(S)$-modules $T(S)$ such that*

$$S \simeq T(S)\uparrow_{H(S)}^{G}.$$

- *Let $G = G_1 \times G_2$ and suppose that for each $i \in \{1,2\}$ for every simple G_i-module S there is a subgroup $H_i(S)$ and a one-dimensional $KH_i(S)$-module $T(S)$ such that $S \simeq T(S)\uparrow_{H_i(S)}^{G_i}$. Then for every simple G-module S there is a subgroup $H(S)$ and a one-dimensional $KH(S)$-module $T(S)$ such that*

$$S \simeq T(S)\uparrow_{H(S)}^{G}.$$

Proof. Let G be a p-group and let S be a simple KG-module. We shall show the statement by induction on the order of G. We know that $\ker(S) := \{g \in G \mid gs = s\ \forall s \in S\}$ is a normal subgroup of G and S is a faithful $K(G/\ker(S))$-module.

If $\ker(S) \neq \{1\}$, that is if S is not faithful, then by induction the proposition is true.

If $\ker(S) = \{1\}$, then S is faithful. If G has an abelian normal subgroup which is not in the centre of G, then Blichfeld's Theorem 3.4.16 shows that S is induced from the inertia group $I(S) \neq G$. By induction, since the inertia group is a proper subgroup of a p-group, and therefore a p-group as well, the proposition is true.

We need to show that p-groups have abelian normal subgroups not contained in their centre.

Lemma 5.3.2. *Let G be a non abelian p-group. Then there is an abelian subgroup A not contained in the centre of G.*

Proof. By Lemma 2.4.13 the centre of a p-group is non trivial. Moreover, it is immediate to see that any subgroup of the centre of a group is normal as well. The group $G/Z(G)$ is a p-group again, and so the centre of $G/Z(G)$ is non trivial. Hence there exists an element $d \cdot Z(G) \in G/Z(G)$ of order p in $Z(G/Z(G))$. We claim that the smallest group A containing d and $Z(G)$ is normal, abelian and not central.

First, A is not central since d is not in the centre of G, the element $d \cdot Z(G)$ being non trivial in $G/Z(G)$.

Second A is abelian since d, and all powers of d commute with all elements in $Z(G)$ by definition of the centre of G.

Third, A is normal. Indeed let $g \in G$. Since $d \cdot Z(G) \in Z(G/Z(G))$,

$$gdg^{-1} \cdot Z(G) = g \cdot (d \cdot Z(G)) \cdot g^{-1} = d \cdot Z(G).$$

Hence

$$gdg^{-1} \in d \cdot Z(G) \subseteq A$$

and this shows that A is normal. This proves the lemma. □

The first part of the proposition follows for non abelian groups.

If G is abelian, since K is a splitting field for G, all simple KG-modules are one-dimensional. Therefore the proposition is true in this case as well.

This shows at once the first part of the proposition.

For the second part let T be a simple $K(G_1 \times G_2)$-module. By Lemma 3.1.19 there is a simple KG_1-module S_1 and a simple KG_2-module S_2 such that $T \simeq S_1 \otimes_K S_2$.

We know that $S_1 \simeq T(S_1) \uparrow_{H(S_1)}^{G_1}$ for a one-dimensional simple $KH(S_1)$-module $T(S_1)$ and $S_2 \simeq T(S_2) \uparrow_{H(S_2)}^{G_2}$ for a one-dimensional simple $KH(S_2)$-module $T(S_2)$. Therefore

$$\begin{aligned} T \simeq S_1 \otimes_K S_2 &\simeq T(S_1) \uparrow_{H(S_1)}^{G_1} \otimes_K T(S_2) \uparrow_{H(S_2)}^{G_2} \\ &\simeq (T(S_1) \otimes_K T(S_2)) \uparrow_{H(S_1)\times H(S_2)}^{G_1 \times G_2} \end{aligned}$$

and by Lemma 3.1.11 one has

$$dim_K(T(S_1) \otimes_K T(S_2)) = dim_K(T(S_1)) \cdot dim_K(T(S_2)) = 1.$$

This proves the proposition. □

Corollary 5.3.3. *Let G be an elementary group and let K be a splitting field for G such that the order of G is invertible in K. Then for each simple KG-module S there is a subgroup $H(S)$ of G and a one-dimensional $KH(S)$-modules $T(S)$ such that*

$$S \simeq T(S) \uparrow_{H(S)}^{G} .$$

Proof. Let G be p-elementary. Then $G \simeq C \times P$ where P is a p-group and C is cyclic. Since cyclic groups are direct products of q-groups for different primes q, Proposition 5.3.1 shows directly the statement. □

We come to Brauer's splitting field Theorem.

Theorem 5.3.4. *Let G be a finite group of exponent n and let K be a field of characteristic 0 such that $X^n - 1 \in K[X]$ has n roots in K. Then K is a splitting field for G.*

Remark 5.3.5. Before starting the proof we note that K is a splitting field for G if and only if the endomorphism algebra of any simple KG-module is K. Further, Wedderburn's structure theorem implies that if for a simple KG-module S we have $End_{KG}(S) = K$, if and only if the representation S of G can be realised as a group homomorphism $G \longrightarrow GL_{\dim_K(S)}(K)$.

Proof. Let S be a simple KG-module. Brauer's induction theorem 5.2.3 shows that

$$\sum_{p \text{ prime}} \sum_{H \in \mathfrak{H}_p(G)} (G_0(KH) \uparrow_H^G) = G_0(KG)$$

for $\mathfrak{H}_p(G)$ being the class of p-elementary subgroups of G.

Since K contains all n-th roots of unity for n being the exponent of G, K is a splitting field for all abelian subgroups of G. Here one uses that trivially for cyclic groups C a splitting field is given by fields containing an $|C|$-th root of unity (cf. Example 1.1.21; (1)), and that abelian groups are direct products of cyclic groups. Moreover, subgroups of elementary groups are elementary, and induction from subgroups is transitive. Hence, Proposition 5.3.1 implies that

- there are elementary subgroups $H_1^+, H_2^+, \dots, H_n^+$ and $H_1^-, \dots, H_m^-$ of G
- and one dimensional KH_i^+-modules T_i^+ and one dimensional KH_j^--modules T_j^- for each $(i,j) \in \{1,\dots,n\} \times \{1,\dots,m\}$
- and integers $r_1^+, r_2^+, \dots, r_n^+ \in \mathbb{N}$ and integers $r_1^-, r_2^-, \dots, r_m^- \in \mathbb{N}$

such that

$$[S] = \left(\sum_{i=1}^{n} r_i^+ [T_i^+ \uparrow_{H_i^+}^G]\right) - \left(\sum_{j=1}^{m} r_j^- [T_j^- \uparrow_{H_j^-}^G]\right).$$

But

$$r_i^+ [T_i^+ \uparrow_{H_i^+}^G] = [(T_i^+ \uparrow_{H_i^+}^G)^{r_i^+}].$$

Put

$$P_S := \bigoplus_{i=1}^{n} (T_i^+ \uparrow_{H_i^+}^G)^{r_i^+}$$

and likewise

$$N_S := \bigoplus_{j=1}^{m} (T_j^- \uparrow_{H_j^-}^G)^{r_j^-}.$$

Then $[S] = [P_S] - [N_S]$ and so we have defined KG-modules P_S and N_S with

$$S \oplus N_S \simeq P_S.$$

Moreover, we know that the modules T_i^+ and T_j^- are simple one-dimensional modules. Hence, their endomorphism ring is K. By Mackey's Theorem 3.3.10 one gets that also the endomorphism ring of each $T_i^+ \uparrow_{H_i^+}^G$ and of each $T_j^- \uparrow_{H_j^-}^G$ are matrix rings over K. Hence, every simple direct factor of P_S as KG-module has endomorphism ring K, and likewise for N_S.

The following argument basically is due to Feit as is reported in [CuRe-82-86]. We need to show that we may choose $N_S = 0$. By the above we can write

$$S \oplus X \simeq Y$$

for two modules X and Y whose representation matrices for any indecomposable direct factor can be realised over K. We may reduce indecomposable direct factors so that the dimensions of X and Y are minimal with this property.

If $X \neq 0$, let U be a simple KG-module not isomorphic to S, being a direct factor of Y. Since U is not isomorphic to S, since $[S] = [X] - [Y]$ in $G_0(KG)$, and the fact that the simple KG-modules form a basis for $G_0(KG)$ (cf. Lemma 5.1.8) implies that U is also a direct factor of X. This contradicts the minimality of X and Y. Hence by minimality $S \simeq Y$. Since all simple modules which are indecomposable direct factors of X and of Y have endomorphism ring K, one sees that S as well has endomorphism ring K. This proves the Theorem. □

5.4 The semisimple algebra case

In this section we shall show that for a field K and semisimple finite dimensional K-algebras A there is always a splitting field L which is finite dimensional over K. In general it is however not so easy to find explicitly. As we have seen, group rings are therefore particular with this respect.

We first observe that if L is a splitting field of A and if E is an extension of L, then E is a splitting field as well. Indeed,

$$E \otimes_K A \simeq E \otimes_L L \otimes_K A \simeq E \otimes_L \prod_{i=1}^{k} Mat_n(L) = \prod_{i=1}^{k} Mat_n(E).$$

By Wedderburn's theorem 1.2.30 it is clear that we may suppose that A is actually simple, since finite dimensional semisimple algebras are finite direct products of simple algebras. If we manage to get a splitting field for every such algebra we take the field which is generated by all these finite extensions, which in turn is a finite extension of K and is itself a splitting field by the above argument. We may hence assume that $A = Mat_n(D)$ for some skew field D, which is finite dimensional over K.

If we can prove the result under the additional hypothesis that $Z(A) \simeq Z(D) \simeq K$, then we also get the general case. Indeed, $Z(D)$ is a finite extension of K, and a finite extension of $Z(D)$ is also a finite extension of K.

We may suppose therefore that A is a finite dimensional central simple K-algebra. Further, since $A \simeq Mat_n(D)$ and since $L \otimes_K Mat_n(D) \simeq Mat_n(L \otimes_K D)$ it is necessary and sufficient to show the result in case $A = D$ is a skew field, finite dimensional over K, with $Z(D) = K$.

Proposition 5.4.1. *Let D be a skew field with center K, and finite dimensional over K. If B is a skew-field containing K. Then $D \otimes_K B$ is artinian simple with centre $K \otimes_K Z(B) \simeq Z(B)$.*

Proof. Let $(b_i)_{i \in I}$ be a K-basis of B and let $u = \sum_{i \in I} d_i \otimes b_i \in Z(D \otimes_K B)$ with all but a finite number of d_i non zero. For any $d \in D$ we have $(d \otimes 1)u = u(d \otimes 1)$ and therefore $\sum_{i \in I}(dd_i - d_i d) \otimes b_i = 0$. Since $(b_i)_{i \in I}$ be a K-basis of B, we get $(dd_i - d_i d) = 0$ for all $d \in D$, and hence $d_i \in Z(D) = K$. This shows that $Z(D \otimes_K B) \subseteq K \otimes_K B \simeq B$. We now use that any such $u \in B$ has to be in $Z(B)$. On the other hand, it is trivial to see that $K \otimes_K Z(B) \subseteq Z(D \otimes_K B)$. Skew-fields are artinian, and therefore $D \otimes_K B$ is artinian. Let $J \neq 0$ be a two-sided ideal of $D \otimes_K B$. Let $0 \neq x = \sum_{i \in I} \delta_i \otimes b_i \in J$ such that the cardinal of $\{i \in I \mid \delta_i \neq 0\} =: N_x$ is minimal. Then for a $\delta_i \neq 0$ also $(\delta_i^{-1} \otimes 1) \cdot x =: y \in J$ with $|N_x| = |N_y| =: m$. We may hence suppose that $x = 1 \otimes b_1 + \sum_{i=2}^m \delta_{j_i} \otimes b_{j_i}$ for $\delta_{i_2}, \dots \delta_{i_m}$ non zero. But then for any $\delta \in D$

$$z := x - (\delta \otimes 1) \cdot x \cdot (\delta \otimes 1)^{-1} = \sum_{i=2}^{m} \delta \delta_{j_i} \delta^{-1} \otimes b_{j_i} \in J$$

with $|N_x| = |N_z| + 1$. By minimality of $|N_x|$ we have $z = 0$ and since $(b_i)_{i \in I}$ is a K-basis, $\delta_{i_1}, \dots, \delta_{i_m} \in Z(D) = K$. But then $x \in B$ is a unit, and therefore $J = D \otimes_K B$. □

Let A be a K-algebra and let M be an A-module. We say that the A-module M has the *double centralizer property* if for $D := End_A(M)$ we get $A = End_D(M)$. Note that M is a right D-module.

Lemma 5.4.2. *Let A be a K-algebra and let N be an A-module. Then the A-module $A \oplus N$ has the double centralizer property. Moreover, if for an A-module M and the positive integer k the module M^k has the double centralizer property, then also M has the double centralizer property.*

Proof.

$$D := End_A(A \oplus N) = \begin{pmatrix} End_A(A) & Hom_A(N,A) \\ Hom_A(A,N) & End_A(N) \end{pmatrix}$$

Let $\varphi \in End_D(A \oplus N)$. Since $End_D(A \oplus N) \subseteq End_K(A \oplus N)$, we may write

$$\varphi = \begin{pmatrix} f_{11} & f_{12} \\ f_{21} & f_{22} \end{pmatrix} \in \begin{pmatrix} End_K(A) & Hom_K(N,A) \\ Hom_K(A,N) & End_K(N) \end{pmatrix}.$$

Since $\begin{pmatrix} id_A & 0 \\ 0 & 0 \end{pmatrix} \circ \varphi = \varphi \circ \begin{pmatrix} id_A & 0 \\ 0 & 0 \end{pmatrix}$ we get $f_{12} = 0$ and $f_{21} = 0$. For any $a \in A$ we denote by μ_a right multiplication by $a \in A$, and we have $\begin{pmatrix} \mu_a & 0 \\ 0 & 0 \end{pmatrix} \circ \varphi = \varphi \circ \begin{pmatrix} \mu_a & 0 \\ 0 & 0 \end{pmatrix}$. Hence $f_{11} \in End_A(A_A) = A$, i. e. f_{11} is an A-endomorphism of the right regular module, which is isomorphic to A, where the isomorphism is given by $End_A(A_A) \ni f_{11} \mapsto f_{11}(1) \in A$. For any $n_0 \in N$ we have a map $\delta_{n_0} \in D$ given by $\delta_{n_0}(a,n) = (0, an_0)$. Then

$$0 = (\varphi \circ \delta_{n_0} - \delta_{n_0} \circ \varphi)(a,n) = \big(0, f_{22}(an_0)\big) - \big(0, f_{11}(an_0)\big)$$

shows $f_{22}(an_0) = f_{11}(a)n_0$ for all $a \in A$ and $n_0 \in N$. In particular, for $a = 1$ we get that $f_{22}(n_0) = f_{11}(1)n_0$, and hence φ is completely determined by f_{11}. Hence $End_D(A \oplus N) = A$.

Let $D = End_A(M)$ and $V = M^k$. Then M is an $A - D$-bimodule and with $E := End_A(V) = Mat_{k\times k}(D)$ the module V is an $A - E$-bimodule. For $f \in End_D(M)$ let

$$f^{(k)} := \begin{pmatrix} f & & \\ & \ddots & \\ & & f \end{pmatrix} \in End_E(V) = A$$

Hence $f^{(k)}$ is given by left multiplication by $a \in A$, and therefore so is f. □

Corollary 5.4.3. *Let A be a semisimple finite dimensional K-algebra over some field K, then any finitely generated faithful A-module M has the double centralizer property.*

Proof. Indeed, any simple A-module S has to be isomorphic to a direct factor of M. Hence, if d_S is the multiplicity of S as a direct factor of A, then putting $d := max(d_S \mid S \text{ simple } A\text{-module})$ we have that A is a direct factor of M^d. Hence $M^d = A \oplus T$ for some A-module T, has the double centralizer property by the first statement of Proposition 5.4.2 and therefore M has the double centralizer property by the second statement of Proposition 5.4.2. □

Proposition 5.4.4. *Let A be a skew-field with $Z(A) = K$ and let B be a simple subring of A, such that*

$$K \subseteq B \subseteq A.$$

Then $B' := \{x \in A \mid xb = bx \ \forall b \in B\}$ is a simple artinian ring and $B = \{x \in A \mid xb' = b'x \ \forall b' \in B'\}$.

Proof. Let S be a simple A-module and let $D = End_A(S)$. Then S is an $A - D$-bimodule and $A = End_D(S_D)$ by Corollary 5.4.3. For any $a \in A$, $d \in D$ denote by λ_a the D-module endomorphism of S given by left multiplication by a and by ρ_d the A-module endomorphism of S given by right multiplication with $d \in D$. By Proposition 5.4.1 $D^{op} \otimes_K B$ is a simple artinian algebra with centre $Z(B)$. Put $D_R := \{\rho_d \mid d \in D\}$ and $B_L := \{\lambda_b \mid b \in B\}$. Then D_R and B_L are both subalgebras of the finite dimensional K-algebra $End_K(S)$, and moreover, $\rho_d \circ \lambda_b = \lambda_b \circ \rho_d$ for any $b \in B$ and $d \in D$ each $d \in D$ being an A-linear endomorphism, whence also B-linear. Consider the natural map

$$D^{op} \otimes B \ni d \otimes b \mapsto \rho_d \circ \lambda_b \in D_R \cdot B_L.$$

The kernel of this map is a two-sided ideal and hence 0, $D^{op} \otimes_K B$ being simple. By definition the map is surjective, and hence is an isomorphism. Therefore $C := D_R \cdot B_L$ is a finite dimensional simple K-algebra, and S is a faithful C-module. Hence the C-module S has the double centralizer property by Lemma 5.4.2. Any $\varphi \in End_C(S) \subseteq End_D(S) \simeq A$ is given by left multiplication by $a_\varphi \in A$, which in addition commutes with any element in B_L. Since A and B act faithfully on S, we get $a_\varphi \in B'$ and therefore $B' = B'_L = End_C(S)$. As in Section 1.2.3 and the proof of Theorem 1.2.30 we see that B' is simple again. Since the C-module S has the double centralizer property, as seen above, $C = End_{B'}(S)$. If $a \in A$ commutes with any element in B', then λ_a commutes with any element in B'_L and therefore

$$\lambda_a \in C = D_R \cdot B_L \quad (\dagger).$$

Since $\lambda_a \in A_L$, also λ_a commuted with any $\rho_d \in C$ for all $d \in C$. Since $C = D^{op} \otimes_K B$ and the elements in $D^{op} \otimes_K B$ which commute with all elements of $D^{op} \otimes_K K$ are in $Z(D) \otimes_K B$ by the proof of Proposition 5.4.1. Since $Z(D) = K$ we have $\lambda_a \in B_L$, whence $a \in B$. □

Remark 5.4.5. Equation (†) in the proof of Proposition 5.4.4 shows that $D \otimes_K B \simeq End_{B'}(S)$, using the notation introduced in the proof there.

Theorem 5.4.6. *Let D be a skew-field and let $K = Z(D)$. Suppose that $\dim_K D < \infty$. Then, any maximal subfield E of D contains K and is a splitting field for D.*

Proof. Since $\dim_K D < \infty$ there are maximal subfields of D. Let E be a maximal subfield of D. Since $K = Z(D)$, $E \cdot K$ is a subfield of D. By Maximality of E we see $K \subseteq E$. Apply Proposition 5.4.4 to the situation $K \subseteq E \subseteq D$. Since E is a field,

$$E \subseteq \{x \in D \mid xb = bx \ \forall b \in E\}.$$

If $x \in \{y \in D \mid yb = by \ \forall b \in E\}$ then $E(x)$ is a subfield of D. By maximality of E, we have $x \in E$. By Remark 5.4.5 we see that $D \otimes_K E \simeq End_E(D)$. Since D is an E-vector space of dimension r, say,

$$End_E(D) = End_E(E^r) = Mat_r(E).$$

This proves the statement. □

Remark 5.4.7. It is not very difficult to show that $\dim_E(D) = \dim_K(E)$ and that $\dim_K(D) = \dim_K(E)^2$. Indeed

$$\dim_E(D)^2 = r^2 = \dim_E(Mat_r(E)) = \dim_E(D \otimes_K E) = \dim_K(D).$$

Since $\dim_K(D) = \dim_E(D) \cdot \dim_K(E)$, by the usual result on dimensions of towers of field extensions, we get $\dim_E(D) = \dim_K(E)$.

5.5 Exercises

Exercise 5.1.

a) Find all elementary subgroups of $\mathfrak{A}_4$ and find all elementary subgroups of $\mathfrak{A}_5$. Find all 5-elementary subgroups of $\mathfrak{A}_8$.
b) Show that subgroups of p-elementary groups are p-elementary.
c) Let $p \neq q$ be two distinct primes. Show that each p-elementary group which is also q-elementary is actually cyclic.

Exercise 5.2 (Artin). Let G be a finite group and let $\mathcal{X}$ be set of subgroups of G. Let K_G be the set of conjugacy classes of G.

a) Show that every cyclic subgroup of G is conjugate to a subgroup of an element of $\mathcal{X}$ if and only if $\bigcup_{H\in\mathcal{X}}\bigcup_{x\in G} x^{-1}Hx = G$.
b) If $\bigcup_{H\in\mathcal{X}}\bigcup_{x\in G} x^{-1}Hx = G$, show that for every conjugacy class C of G there is $H(C) \in \mathcal{X}$ with $C \cap H(C) \neq \emptyset$. Let $g_C \in C \cap H(C)$ be an element of order n_C and Z_C be the cyclic group generated by g_C. Let $\zeta_C \in CF_{\mathbb{C}}(Z_C)$ defined by

$$\zeta_C(h) = \begin{cases} n_C & \text{if } h = g_C \\ 0 & \text{else} \end{cases} .$$

Show that for all complex valued class functions $\eta \in CF_{\mathbb{C}}(G)$

$$\eta = \sum_{C\in K_G} ind_{H(C)}^G\left(\frac{\eta(g_C)}{|C_G(g_C)|} ind_{Z_C}^{H(C)}(\zeta_C)\right).$$

Show that $ind_{\mathcal{X}} := \sum_{H\in\mathcal{X}} ind_H^G$ is an epimorphism of $\mathbb{C}$-vector spaces

$$ind_{\mathcal{X}} : \bigoplus_{H\in\mathcal{X}} CF_{\mathbb{C}}(H) \longrightarrow CF_{\mathbb{C}}(G).$$

c) Suppose that $ind_{\mathcal{X}}$ is an epimorphism of vector spaces, and let χ be a complex valued character of G. Show that for all $\alpha \in CF_{\mathbb{C}}(G)$, all $H \in \mathcal{X}$, all $\zeta_H \in Irr(H)$ there are coefficients $c_{H,\zeta_H}(\alpha) \in \mathbb{C}$ such that

$$\chi = \sum_{H\in\mathcal{X}}\sum_{\zeta_H\in Irr(H)} c_{H,\zeta_H}(\alpha) ind_H^G(\zeta_H).$$

Deduce that for each $\alpha \in CF_{\mathbb{C}}(G)$ and each $\chi \in Irr(G)$ we have

$$\langle\alpha,\chi\rangle = \sum_{H\in\mathcal{X}}\sum_{\zeta_H\in Irr(H)} c_{H,\zeta_H}(\alpha)\cdot\langle ind_H^G(\zeta_H),\chi\rangle.$$

Consider the system of equations

$$\langle\alpha,\chi\rangle = \sum_{H\in\mathcal{X}}\sum_{\zeta_H\in Irr(H)} X_{H,\zeta_H}\cdot\langle ind_H^G(\zeta_H),\chi\rangle$$

in the variables X_{H,ζ_H} to show that we may choose $c_{H,\zeta_H}(\alpha) \in \mathbb{Q}$ whenever α is a character of G.

d) Show that if for all $H \in \mathcal{X}$, all $\zeta_H \in Irr(H)$ and all characters α of G there are $q_{H,\zeta_H}(\alpha) \in \mathbb{Q}$ such that

$$\alpha = \sum_{H \in \mathcal{X}} \sum_{\zeta_H \in Irr(H)} q_{H,\zeta_H} ind_H^G \zeta_H,$$

then $\bigcup_{H \in \mathcal{X}} \bigcup_{x \in G} x^{-1}Hx = G$.

NB: Use for this the trivial character χ_1 as α and Lemma 3.2.17.

Exercise 5.3 (with solution in Chapter 9). Let K be a field and let $A = \begin{pmatrix} K & K \\ 0 & K \end{pmatrix}$ be the algebra of upper 3×3 triangular matrices. Then any indecomposable A-module is isomorphic to one of the three modules

$$P_1 := \begin{pmatrix} K \\ 0 \end{pmatrix}, \quad P_2 := \begin{pmatrix} K \\ K \end{pmatrix}, \quad S_2 := P_2/P_1.$$

a) Give a list of pairwise non isomorphic A-left modules such that any A-left module of dimension at most 5 is isomorphic to one of the modules in the list.
b) Find in this list all A-modules having the double centralizer property.
c) Find all finite dimensional A-modules with double centralizer property.

Exercise 5.4. Recall from Exercise 1.12 the algebra $(\frac{a,b}{K})$.
a) Find the maximal subfield of $(\frac{-1,-1}{K})$ for any subfield K of $\mathbb{R}$.
b) If none of the elements a, b, $-ab$ is a square in K, find a maximal subfield of $(\frac{a,b}{K})$ for any subfield K of $\mathbb{R}$.

6 Some homological algebra methods in ring theory

Notre tête est ronde pour permettre à la pensée de changer de direction

Francis Picabia; La pomme des pins

Our head is round to allow thoughts to change direction

Francis Picabia; La pomme des pins[1]

In order to continue with more advanced topics in representation theory we need to use some classical methods from homological algebra. Homological algebra is a main tool and provides a notational and conceptual framework for all of modern algebra, and in particular for representation theory. Nevertheless, homological algebra is vast, and we only mention some topics as far as we shall use them in the sequel. For a systematic introduction the existing literature offers a large variety of textbooks and monographs.

6.1 Some facts about Noetherian and artinian modules

As long as we deal with finite dimensional algebras A over a field K we did not need to consider finiteness conditions for modules. We just consider finitely generated A-modules M, and automatically M is a finite dimensional K-vector space. Now in Section 8 we shall try to understand systematically RG-modules for certain integral domains R, and the situation becomes weird in this generality. A very successful concept for an appropriate finiteness condition is the concept of Noetherian and of artinian modules.

6.1.1 Elementary properties and definitions

Definition 6.1.1. Let R be a commutative ring and let A be an R-algebra. Then an A-module M is
- *Noetherian* if for any sequence of submodules
$$S_1 \leq S_2 \leq S_3 \leq S_4 \leq \cdots \leq M$$
there is n_0 such that $S_k = S_{n_0}$ for all $k \geq n_0$.
- *artinian* if for any sequence of submodules
$$M \geq S_1 \geq S_2 \geq S_3 \geq S_4 \geq \ldots$$
there is n_0 such that $S_k = S_{n_0}$ for all $k \geq n_0$.

1 translation by the author.

https://doi.org/10.1515/9783110702446-006

An algebra A is Noetherian if the regular module is Noetherian. An algebra A is artinian if the regular module is artinian.

Note that we are dealing with left modules and left ideals in Definition 6.1.1. An algebra is in this case better noted 'left Noetherian', respectively 'left artinian'. In case we are dealing with right ideals we get an a priori different concept of 'right Noetherian', respectively 'right artinian'.

Example 6.1.2.
1. If A is a finite dimensional K-algebra over a field K, then any finitely generated A-module is Noetherian and artinian. Indeed, any such module is a finite dimensional vector space and hence the modules in any increasing (respectively decreasing) sequence of submodules have increasing (respectively decreasing) dimension.
2. $\mathbb{Z}$ is Noetherian but not artinian. Indeed, any ideal of $\mathbb{Z}$ is principal, say $I = n\mathbb{Z}$. If $n\mathbb{Z} \leq m\mathbb{Z}$, then m divides n. Since any integer only has finitely many divisors, $\mathbb{Z}$ is Noetherian. However, $\mathbb{Z} \geq 2\mathbb{Z} \geq 4\mathbb{Z} \geq \dots$ is an infinite decreasing sequence.
3. For any fixed non zero integer n the ring $\mathbb{Z}/n\mathbb{Z}$ is artinian. Indeed, $\mathbb{Z}/n\mathbb{Z}$ is a finite set. The statement is clear from there.

Proposition 6.1.3. *Let R be a commutative ring and let A be an R-algebra. Then*
1. *any submodule of a Noetherian A-module is a Noetherian A-module,*
2. *any quotient of a Noetherian A-module is a Noetherian A-module.*
3. *If L is a Noetherian A-submodule of an A-module M, and if M/L is a Noetherian A-module, then M is Noetherian as well.*

Proof. We need to show 1. Let M be Noetherian and let L be a submodule of M. Let

$$S_1 \leq S_2 \leq S_3 \leq \cdots \leq L$$

be an increasing sequence of submodules of L. Since $L \leq M$, this is also a sequence of increasing submodules of M. Since M is Noetherian, there is n_0 such that $S_k = S_{n_0}$ for all $k \geq n_0$. This shows 1.

We need to show 2. Let M be Noetherian and $Q = M/L$ for some submodule L of M. Denote by $\pi : M \longrightarrow Q$ the canonical epimorphism. Let

$$S_1 \leq S_2 \leq S_3 \leq \cdots \leq Q$$

be an increasing sequence of submodules of Q. Then for the submodules $\pi^{-1}(S_i) =: T_i$ of M we get

$$T_1 \leq T_2 \leq T_3 \leq \cdots \leq M.$$

Hence, since M is Noetherian, there is n_0 such that $T_k = T_{n_0}$ for all $k \geq n_0$. But this shows

$$\pi(T_k) = \pi(T_{n_0})$$

for all $k \geq n_0$. Since π is an epimorphism, $\pi(T_n) = S_n$ for all $n \in \mathbb{N}$. This shows 2.

We need to show 3. Let M be an A-module and let L be a Noetherian submodule of M, and suppose that $Q := M/L$ is Noetherian. Denote by $\pi : M \longrightarrow Q$ the canonical epimorphism. Let

$$T_1 \leq T_2 \leq T_3 \leq \cdots \leq M$$

be an increasing sequence of submodules of M. Then denoting $S_n := \pi(T_n)$ for all $n \in \mathbb{N}$,

$$S_1 \leq S_2 \leq S_3 \leq \cdots \leq Q$$

is an increasing sequence of submodules of Q. Since Q is Noetherian, there is n_0 such that $S_k = S_{n_0}$ for all $k \geq n_0$. Similarly,

$$L \cap T_1 \leq L \cap T_2 \leq L \cap T_3 \leq \cdots \leq M$$

is an increasing sequence of submodules of the Noetherian module L, and hence there is $n_1 \in \mathbb{N}$ such that $L \cap T_k = L \cap T_{n_1}$ for all $k \geq n_1$. Let $n_2 := max(n_0, n_1)$ and choose $k \geq n_2$. Let $v \in T_k$. Since $\pi(T_{n_2}) = \pi(T_k)$ there is $u \in T_{n_2}$ with $v - u \in \ker(\pi) = L$. Hence there is $x \in L$ with $v - u = x$. Since $T_{n_2} \leq T_k$ we see that $x \in T_k \cap L = T_{n_2} \cap L$. Hence $v = x + u \in T_{n_2}$. This shows that $T_k = T_{n_2}$. Item 3 follows. □

Proposition 6.1.4. *Let R be a commutative ring and let A be an R-algebra. Then*
1. *any submodule of an artinian A-module is an artinian A-module,*
2. *any quotient of an artinian A-module is an artinian A-module.*
3. *If L is an artinian A-submodule of an A-module M, and if M/L is an artinian A-module, then M is artinian as well.*

Proof. The arguments of Proposition 6.1.3 apply mutatis mutandis to this situation as well. □

Proposition 6.1.5. *Let R be a commutative ring and let A be an R-algebra. Let M be an A-module. Then the following statements are equivalent.*
1. *M is Noetherian*
2. *every submodule of M is finitely generated*
3. *every non empty family of submodules of M has a maximal element.*

Proof. 1 implies 2: If M is Noetherian, let L be a submodule of M. If L is not finitely generated, let G be a minimal generating system of L in the sense that no element in G is superfluous. By hypothesis, G is infinite, and contains hence an infinite sequence $\{g_1, g_2, g_3, \dots\}$ of elements. Since G is minimal, $g_n \notin \sum_{i=1}^{n-1} Ag_i =: S_{n-1}$. Hence

$$S_1 \leq S_2 \leq S_3 \leq \cdots \leq L \leq M$$

is a strictly increasing sequence of submodules of M. This is a contradiction to M being Noetherian. Note that a very similar argument also shows 3 implies 2.

1 implies 3: Let $\mathcal{X}$ be a non empty family of submodules of M. This is partially ordered by inclusion. Let $\mathcal{Y}$ be a totally ordered subset of $\mathcal{X}$. Since M is Noetherian, the subset $\mathcal{Y}$ has to be finite. $\mathcal{Y} = (M_0 < M_1 < \cdots < M_s)$ and then M_s is a maximal element of $\mathcal{Y}$. By Zorn's lemma $\mathcal{X}$ has maximal elements.

3 implies 1: Let $M_0 < M_1 < M_2 < \cdots \leq M$ be an infinite strictly increasing sequence of submodules of M. If this sequence is not finite, then $\{M_n \mid n \in \mathbb{N}\}$ is a family of submodules without a maximal element.

2 implies 3: Let $\mathcal{X}$ be a non empty family of submodules of M, partially ordered by inclusion. We shall use Zorn's lemma again. Let $\mathcal{Y}$ be a totally ordered subset of $\mathcal{X}$. Then

$$L := \bigcup_{Y \in \mathcal{Y}} Y$$

is a submodule of M. Hence by 2, L is finitely generated. Therefore, there is a finite number of elements $Y_1, \dots, Y_n$ containing the generators of L. Since $\mathcal{Y}$ is totally ordered, one of these modules $Y_1, \dots, Y_n$ is maximal, say Y_n, and hence $L = Y_n \in \mathcal{Y}$. By Zorn's lemma $\mathcal{X}$ contains maximal elements. □

Proposition 6.1.6. *Let R be a commutative ring and let A be an R-algebra. Then an A-module M is artinian if and only if every non empty family of submodules of M contains a minimal element.*

Proof. Suppose that M is artinian and suppose that $\mathcal{X}$ is a family of submodules of M. This family is partially ordered by *inverse* inclusion:

$$X_1 \preceq X_2 :\Leftrightarrow X_2 \leq X_1.$$

Since M is artinian, any totally ordered subfamily $\mathcal{Y}$ has to be finite, and by Zorn's Lemma we obtain that $\mathcal{X}$ contains minimal elements.

Suppose that any non empty family of submodules of M contains a minimal element. The proof of Proposition 6.1.5, step 3 implies 1 applies analogously. □

Corollary 6.1.7. *Let R be a commutative ring and let A be an R-algebra. If $M \neq 0$ is an artinian A-module, then M contains a simple submodule S.*

Proof. The family of non zero submodules of M contains M, is hence non empty, and contains a minimal element by Proposition 6.1.6. This minimal element has to be simple since else it would contain an even smaller non zero submodule. □

Remark 6.1.8. The *socle* of a module M is the sum of all simple submodules of M. The socle may well be 0, as for example there is no simple $\mathbb{Z}$-submodule of $\mathbb{Z}$. Corollary 6.1.7 shows that if M is artinian, then the socle of M is non zero.

6.1.2 Composition series

Modules which are at once Noetherian and artinian have a very important property. They are glued together by simple modules, and the pieces are uniquely determined. For this very important result we show the most elegant following tool.

Lemma 6.1.9 (Zassenhaus' butterfly lemma). *Let M_1 and M_2 be submodules of M, let H_1 be a submodule of M_1 and let H_2 be a submodule of M_2. Then*

$$\begin{aligned}(H_1 + (M_1 \cap M_2)) \;/\; (H_1 + (M_1 \cap H_2)) &\\ \simeq (M_1 \cap M_2)/((H_1 \cap M_2) + (M_1 \cap H_2)) &\\ \simeq (H_2 + (M_1 \cap M_2))/(H_2 + (M_2 \cap H_1)). &\end{aligned}$$

The situation is illustrated by the following "butterfly diagram" (head down tail up)

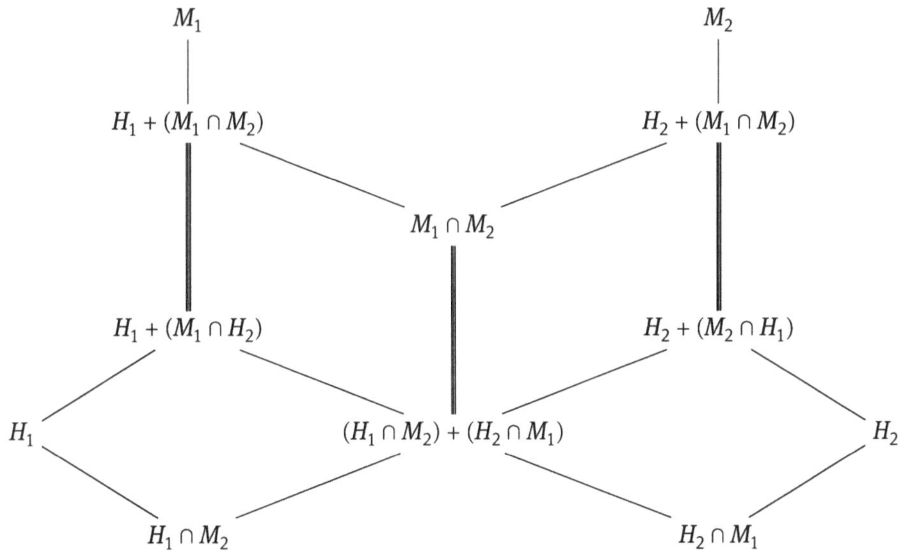

where the bold lines indicate isomorphic quotients.

Proof. Since

$$(H_1 + (M_1 \cap H_2)) \cap (M_1 \cap M_2) \simeq (H_1 \cap M_2) + (H_2 \cap M_1)$$

and likewise

$$(H_2 + (M_2 \cap H_1)) \cap (M_1 \cap M_2) \simeq (H_1 \cap M_2) + (H_2 \cap M_1)$$

the result is a direct consequence of the isomorphism theorem. □

Theorem 6.1.10. *Let R be a commutative ring and let A be an R-algebra. Let M be an artinian and Noetherian A-module. Then there is a finite sequence of submodules*

$$0 = L_0 < L_1 < L_2 < \cdots < L_\ell < L_{\ell+1} = M$$

such that L_i/L_{i-1} is simple for each $i \in \{1, \ldots, \ell + 1\}$. Moreover, if there is another such sequence

$$0 = N_0 < N_1 < N_2 < \cdots < N_n < N_{n+1} = M$$

such that N_i/N_{i-1} is simple for each $i \in \{1, \ldots, n + 1\}$ we get that $n = \ell$ and there is an element $\sigma \in \mathfrak{S}_n$ and A-module isomorphisms $N_i/N_{i-1} \simeq L_{\sigma(i)}/L_{\sigma(i)-1}$ for each $i \in \{1, \ldots, n + 1\}$.

Proof. By Corollary 6.1.7 there is a simple submodule L_1 of M. Then, consider M/L_1. By Proposition 6.1.4 this module is artinian again and using Corollary 6.1.7 contains a simple submodule S_2. Let $\pi_1 : M \longrightarrow M/L_1$ the natural projection. Then $\pi_1^{-1}(S_2) =: S_2$ is a submodule of M with $L_2/L_1 \simeq S_2$. Let $\pi_2 : M \longrightarrow M/L_2$ and let S_3 be a simple submodule of M/L_2. Then $L_3 = \pi_2^{-1}(S_3)$ is a submodule of M with $L_3/L_2 \simeq S_3$. Continuing this way we obtain inductively an increasing sequence

$$0 < L_1 < L_2 < L_3 < \cdots < M.$$

Since M is Noetherian, this sequence is finite. Let $L = \bigcup_{i \in \mathbb{N}} L_i$. By construction, $L = M$, since else there is a simple submodule S of M/L and the preimage of S in M extends the sequence, contrary to our assumption. Hence there is ℓ such that $L_{\ell+1} = M$.

Suppose that we have another such sequence

$$0 = N_0 < N_1 < N_2 < \cdots < N_n < N_{n+1} = M$$

such that N_i/N_{i-1} is simple for each $i \in \{1, \ldots, n + 1\}$. Refine this sequence as follows:

$$N_{i,j} := N_{i-1} + (L_j \cap N_i).$$

Then

$$\begin{aligned} M = N_{n+1} = N_{n+1,\ell+1} &\geq N_{n+1,\ell} \geq N_{n+1,\ell-1} \geq \ldots \\ \cdots \geq N_{n+1,1} \geq N_n \geq N_{n,\ell+1} &\geq \cdots \geq N_{n,1} \geq N_{n-1} \geq N_{n-1,\ell+1} \geq \cdots \geq 1 \end{aligned}$$

Likewise we put

$$L_{j,i} := L_{j-1} + (N_i \cap L_j).$$

By the Butterfly Lemma 6.1.9

$$N_{i,j}/N_{i,j-1} \simeq L_{j,i}/L_{j,i-1}.$$

Hence the two refined sequences of submodules are equivalent in the sense that they have the same length and there is a bijection between the (non zero) successive quotients. This shows the theorem. □

Definition 6.1.11. The sequence of successive quotients are the *composition factors* of M. A finite sequence

$$0 < L_1 < L_2 < \cdots < L_\ell < M$$

with simple successive quotients is a *composition series* of M.

Remark 6.1.12. Let R be a commutative ring and let A be an R-algebra. If M is an A-module having a composition series, then M is Noetherian and artinian. Indeed, any ascending sequence can be refined as in the proof of Theorem 6.1.10 and then the sequence is finite by the Butterfly Lemma 6.1.9. Similarly, any descending sequence of submodules can be refined as in the proof of Theorem 6.1.10 and then the sequence is finite by the Butterfly Lemma 6.1.9.

Remark 6.1.13. We note that with slightly different methods further properties of Noetherian modules over commutative rings are shown in Section 7.5 below.

6.2 On projective modules

Let A be an R-algebra for a commutative ring R, and let L, M and N be A-modules. If

$$M \xrightarrow{\varphi} N$$

is an epimorphism, then φ induces a homomorphism of R-modules

$$\begin{array}{ccc} Hom_A(L,M) & \xrightarrow{Hom_A(L,\varphi)} & Hom_A(L,N) \\ \psi & \longmapsto & \varphi \circ \psi \end{array}$$

However, $Hom_A(L, \varphi)$ is not surjective in general. Actually, $Hom_A(L, \varphi)$ is surjective if, and only if for any $\chi : L \to N$ there is $\psi : L \to M$ making the diagram

$$\begin{array}{ccc} M & \overset{\varphi}{\twoheadrightarrow} & N \\ & \nwarrow^{\psi} & \uparrow \chi \\ & & L \end{array}$$

commutative. Consider here the following Example.

Example 6.2.1. Let $A = R = \mathbb{Z}$ and consider the natural epimorphism

$$\mathbb{Z} \overset{\varphi}{\twoheadrightarrow} \mathbb{Z}/2\mathbb{Z}.$$

Then $Hom_{\mathbb{Z}}(\mathbb{Z}/4\mathbb{Z}, \mathbb{Z}) = 0$ but there is a natural epimorphism $\chi : \mathbb{Z}/4\mathbb{Z} \to \mathbb{Z}/2\mathbb{Z}$. Hence, there is no ψ

$$\begin{array}{ccc} \mathbb{Z} & \overset{\text{mod } 2}{\twoheadrightarrow} & \mathbb{Z}/2\mathbb{Z} \\ & \nwarrow^{?\psi?} & \uparrow \text{mod } 2 \\ & & \mathbb{Z}/4\mathbb{Z} \end{array}$$

as required.

Definition 6.2.2. Let R be a commutative ring, and let A be an R-algebra. An A-module P is called

- *projective* if for any A-modules M and N and any epimorphism

$$M \overset{\varphi}{\twoheadrightarrow} N$$

  the induced morphism $Hom_A(P, \varphi)$ is surjective.
- *free* if M contains an A-basis, i. e. there is a subset S of M such that for any A-module N and any set theoretic map $\alpha_S : S \longrightarrow N$ there is a unique A-module homomorphism $\alpha : M \longrightarrow N$ with $\alpha|_S = \alpha_S$. We say that F is free on S.

Example 6.2.3. By the above, $\mathbb{Z}/4\mathbb{Z}$ is not a projective $\mathbb{Z}$-module. A similar argument shows that for any integer $n > 1$ the abelian group $\mathbb{Z}/n\mathbb{Z}$ is not a projective $\mathbb{Z}$-module.

Example 6.2.4. A free A-module F on a finite set S of cardinal n is isomorphic to A^n. Indeed, if $S = \{s_1, \ldots, s_n\}$ then the set theoretical map

$$\begin{array}{l} s_1 \mapsto (1, 0, 0 \ldots, 0, 0) \in A^n \\ s_2 \mapsto (0, 1, 0 \ldots, 0, 0) \in A^n \\ \vdots \;\vdots\; \vdots \\ s_n \mapsto (0, 0, 0 \ldots, 0, 1) \in A^n \end{array}$$

induces an A-module homomorphism $F \xrightarrow{\mu} A^n$. Inversely $A^n \xrightarrow{\nu} F$ given by $\nu(a_1, \dots, a_n) = \sum_{i=1}^n a_i s_i$ is an A-module homomorphism as well. Further, $\nu \circ \mu(s_i) = s_i$ for all $i \in \{1, \dots, n\}$. Hence $\nu \circ \mu$ and id_F are identical on each element of S. By the unicity of the extension of the set theoretical map, $\nu \circ \mu = \mathrm{id}_F$. The fact that $\mu \circ \nu = \mathrm{id}_{A^n}$ is verified by inspection.

There is a quite complete characterisation of projective modules.

Proposition 6.2.5. *Let A be an R-algebra over some commutative ring R and let M be an A-module. Then the following are equivalent.*
1. *M is a projective A-module.*
2. *Every epimorphism $N \xrightarrow{\pi} M$ of A-modules splits (i. e. there is $\psi : M \longrightarrow N$ such that $\pi \circ \psi = \mathrm{id}_M$).*
3. *M is a direct summand of a free module.*

Proof. Suppose that M is projective. Then consider the identity map on M in the diagram

$$\begin{array}{ccc} N & \xrightarrow{\pi} & M \\ & & \uparrow \mathrm{id} \\ & & M \end{array}$$

and by definition of M being projective there is $\psi : M \longrightarrow N$ making the diagram

$$\begin{array}{ccc} N & \xrightarrow{\pi} & M \\ & \nwarrow^{\psi} & \uparrow \mathrm{id} \\ & & M \end{array}$$

commutative. Since $\pi \circ \psi = \mathrm{id}_M$, the epimorphism π splits.

Suppose that every epimorphism $\pi : N \longrightarrow M$ splits. Then M is the epimorphic image of the free module $F := \bigoplus_{m \in M} A$ with epimorphism $\pi((a_m)_{m \in M}) := \sum_{m \in M} a_m m$. Clearly π is an epimorphism. Hence there is $\psi : M \longrightarrow F$ with $\pi \circ \psi = \mathrm{id}_M$. Then $\psi \circ \pi$ is an idempotent endomorphism of F. Indeed,

$$(\psi \circ \pi)^2 = (\psi \circ \pi) \circ (\psi \circ \pi) = \psi \circ (\pi \circ \psi) \circ \pi = \psi \circ \pi$$

and hence

$$F = \mathrm{im}(\psi \circ \pi) \oplus \ker(\psi \circ \pi)$$

by Lemma 1.2.8. Since π is an epimorphism and ψ is a monomorphism, $\mathrm{im}(\psi \circ \pi) = \mathrm{im}\, \psi \simeq M$.

Suppose that M is a direct summand of a free module F. Then let $F \overset{\mu}{\twoheadrightarrow} M$ be the natural projection onto M, and let $M \overset{\nu}{\hookrightarrow} F$ be the natural embedding. Consider a diagram

$$\begin{array}{ccc} L & \overset{\varphi}{\twoheadrightarrow} & N \\ & & \uparrow{\scriptstyle\chi} \\ & & M \end{array}$$

and we need to lift χ to L, i. e. we need to find ψ making the diagram

$$\begin{array}{ccc} L & \overset{\varphi}{\twoheadrightarrow} & N \\ & {\scriptstyle\psi}\nwarrow & \uparrow{\scriptstyle\chi} \\ & & M \end{array}$$

commutative. However, this is part of a bigger diagram, where a γ

$$\begin{array}{ccc} L & \overset{\varphi}{\twoheadrightarrow} & N \\ & & \uparrow{\scriptstyle\chi} \\ & {\scriptstyle\gamma}\nwarrow & M \\ & & \uparrow{\scriptstyle\mu} \\ & & F \end{array}$$

with $\varphi \circ \gamma = \chi \circ \mu$ is easily found. Indeed, using that φ is surjective, for each $n \in N$ let $\ell_n \in L$ be an element with $\varphi(\ell_n) = n$. If F is free on a subset S, say, then let $\gamma(s) := \ell_{\chi\circ\mu(s)}$ for each $s \in S$. Then γ is actually the restriction to S of an A-module homomorphism $\gamma : F \to L$ by the property of F being free on S. Then, $\psi := \gamma \circ \nu$ satisfies

$$\varphi \circ \psi = \varphi \circ \gamma \circ \nu = \chi \circ \mu \circ \nu = \chi$$

has therefore the required property. □

Corollary 6.2.6. *In particular, the regular module is free and therefore projective.*

We need a new concept.

Definition 6.2.7. Let A be a ring and let M, N, Q be A-modules with A-module morphisms $\mu : M \longrightarrow Q$ and $\nu : N \longrightarrow Q$. The *pullback* of

$$M \xrightarrow{\mu} Q \xleftarrow{\nu} N$$

is the A-module

$$P := \{(m, n) \in M \times N \mid \mu(m) = \nu(n)\}.$$

From a structural point of view it is more interesting to observe that pullbacks have a 'universal' property and are characterised by this property.

Lemma 6.2.8. *Let A be a ring and let M, N, Q be A-modules with A-module morphisms $\mu : M \longrightarrow Q$ and $\nu : N \longrightarrow Q$. Then the pullback P of the diagram $M \xrightarrow{\mu} Q \xleftarrow{\nu} N$ gives a commutative diagram*

$$\begin{array}{ccc} P & \overset{\sigma}{\dashrightarrow} & M \\ {\scriptstyle\tau}\downarrow & & \downarrow{\scriptstyle\mu} \\ N & \underset{\nu}{\longrightarrow} & Q \end{array}$$

such that whenever there is an A-module X and morphisms $\xi : X \longrightarrow M$ and $\eta : X \longrightarrow N$ with $\mu \circ \xi = \nu \circ \eta$, then there is a unique homomorphism of A-modules $\zeta : X \longrightarrow P$ with $\sigma \circ \zeta = \xi$ and $\tau \circ \zeta = \eta$:

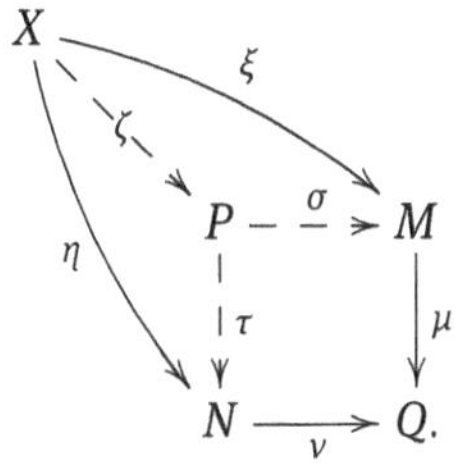

If there is an A-module P' and morphisms σ', τ' making the diagram below commutative and having the property that whenever there is an A-module X and morphisms $\xi : X \longrightarrow M$ and $\eta : X \longrightarrow N$ with $\mu \circ \xi = \nu \circ \eta$, then there is a unique homomorphism of A-modules $\zeta' : X \longrightarrow P'$ with $\sigma' \circ \zeta' = \xi$ and $\tau' \circ \zeta' = \eta$

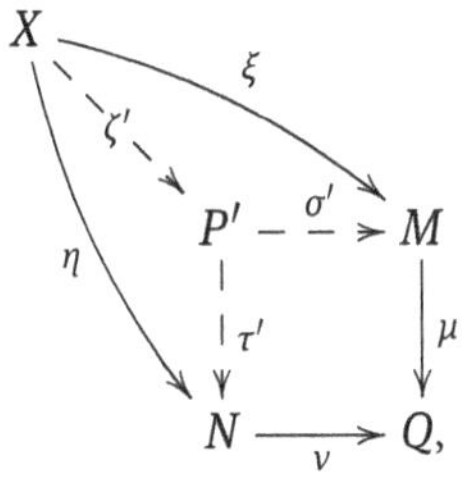

then $P \simeq P'$.

Proof. With the definition of $P := \{(m, n) \in M \times N \mid \mu(m) = \nu(n)\}$ we just let σ be the projection onto the first component and τ be the projection onto the second component. Moreover, if there is X and morphisms ξ and η as in the statement, then we just put $\zeta(x) = (\xi(x), \eta(x))$, and it is obvious that this is the only possibility to make the diagram commutative. Moreover, ζ satisfies the requirements of the statement.

Finally, let P' be another such object. By hypothesis P and P' both satisfy the universal properties, and hence we may use P, respectively P' as X. Therefore there is $\zeta : P' \to P$ and $\zeta : P \to P'$ such that $\sigma' \circ \zeta' = \xi$ and $\tau' \circ \zeta' = \eta$ respectively $\sigma \circ \zeta = \xi$ and $\tau \circ \zeta = \eta$. However, $\zeta \circ \zeta'$ and $\mathrm{id}_{P'}$ satisfy both the universal property, as well as $\zeta' \circ \zeta$ and id_P. By the uniqueness of the map satisfying the commutativity of the diagram, $\zeta \circ \zeta' = \mathrm{id}_{P'}$ and $\zeta' \circ \zeta = \mathrm{id}_P$. This shows the statement. □

Lemma 6.2.9. *Let A be an algebra over some commutative ring and let*

$$\begin{array}{ccc} P & \xrightarrow{\sigma} & M \\ {\scriptstyle\tau}\downarrow & & \downarrow{\scriptstyle\mu} \\ N & \xrightarrow[\nu]{} & Q \end{array}$$

be a pullback diagram. Then there is an isomorphism $\ker(\sigma) \xrightarrow{\rho} \ker(\nu)$ making the diagram

$$\begin{array}{ccccc} \ker(\sigma) & \hookrightarrow & P & \xrightarrow{\sigma} & M \\ {\scriptstyle\rho}\downarrow & & {\scriptstyle\tau}\downarrow & & \downarrow{\scriptstyle\mu} \\ \ker(\nu) & \hookrightarrow & N & \xrightarrow[\nu]{} & Q \end{array}$$

commutative. Moreover, if ν is an epimorphism, then also σ is an epimorphism.

Proof. The easiest proof in this case works with Definition 6.2.7. Indeed,

$$P = \{(m, n) \in M \times N \mid \mu(m) = \nu(n)\}.$$

Moreover, $\sigma(m, n) = m$. Now

$$\begin{aligned} \ker(\sigma) &= P \cap (\{0\} \times N) \\ &= \{0\} \times \ker(\nu) \end{aligned}$$

and ρ, as well as τ are just the projection onto the second component. If ν is surjective, then for any $m \in M$ there is $n(m) \in N$ with $\nu(n(m)) = \mu(m)$. Hence $(m, n(m)) \in P$ and σ is an epimorphism. □

An alternative and much more general proof of the first part of Lemma 6.2.9 uses Lemma 6.2.8. The property of the kernel gives immediately a unique map ρ making the diagram commutative. Apply now Lemma 6.2.8 to $X = \ker(\nu)$, with the natural map η, and $\xi = 0$. Then by Lemma 6.2.8 there is a unique map $\widehat{\rho} : \ker(\nu) \longrightarrow \ker(\sigma)$ making the diagrams commutative. Then it is not hard to show that $\widehat{\rho}$ is an inverse to ρ, using the unicity properties. The second part is a lot more complicated if one wants to avoid computing with elements. However, it is still possible to give an element-free proof of this fact as well, using what is generally called the snake lemma (cf. Exercise 6.8). We refer to e. g. [Zim-14, Chapter 3] for this more general theory.

An amusing and very useful application is Schanuel's lemma.

Lemma 6.2.10 (Schanuel's lemma). *Let A be an algebra over some commutative ring, and let M be an A-module. Suppose there are two projective A-modules P_1 and P_2 and two epimorphisms*

$$P_1 \xrightarrow{\pi_1} M \xleftarrow{\pi_2} P_2.$$

Then

$$\ker(\pi_1) \oplus P_2 \simeq \ker(\pi_2) \oplus P_1.$$

Proof. We form the pullback

$$\begin{array}{ccc} X & \xrightarrow{\sigma} & P_1 \\ {\scriptstyle\tau}\downarrow & & \downarrow{\scriptstyle\pi_1} \\ P_2 & \xrightarrow[\pi_2]{} & M \end{array}$$

By Lemma 6.2.9 since π_2 is surjective, also σ is surjective, and since P_1 is projective,

$$X \simeq P_1 \oplus \ker(\sigma) \simeq P_1 \oplus \ker(\pi_2).$$

But the situation is symmetrical with respect to P_1 and P_2, and hence

$$X \simeq P_2 \oplus \ker(\tau) \simeq P_2 \oplus \ker(\pi_1).$$

This shows the statement. □

6.3 Extension groups

Given a Noetherian and artinian module M over an algebra A by Theorem 6.1.10 it has a composition series and composition factors. How many modules M share the same composition factors with M? This number can be counted by extension groups. Their definition is slightly involved, and gives en passant a highly useful construction and invariant, namely the groups $Ext_A^n(S,T)$ for any positive integer n and any two A-modules S and T.

6.3.1 Degree 1 via syzygies

Projective modules are a central object in representation theory. We shall use them in the sequel in particular within a concept evoked earlier in Definition 3.4.4 and Definition 4.3.5 for the case of groups. This is actually a very special case of a much vaster concept, called derived functors. We refer to e. g. [Zim-14, Chapter 3] for a more exhaustive treatment of the subject.

We come to the very important object of extension group. Let A be an algebra over some commutative ring. Note that for each A-module M we may choose a generating set S of M as an A-module, and then there is an epimorphism

$$\bigoplus_{s\in S} A \xrightarrow{\pi} M$$

$$(a_s)_{s\in S} \longmapsto \sum_{s\in S} a_s s$$

where we observe that since we are dealing with the direct sum, all but a finite number of values a_s are 0. Since free modules are projective (cf. Proposition 6.2.5), we obtain that for any A-module M there is a projective A-module P and an epimorphism $P \longrightarrow M$.

Definition 6.3.1. Let A be an algebra over some commutative ring R, and let M and N be two A-modules. Then we choose a projective A-module P and an epimorphism $P \xrightarrow{\pi} M$. The *syzygy* $\Omega_\pi M$ *of* M is defined to be $\ker(\pi)$.

By definition, the syzygy of M depends on the choice of $P \xrightarrow{\pi} M$. If we choose another projective module Q with an epimorphism $Q \xrightarrow{\rho} M$, then by Schanuel's Lemma 6.2.10 we get

$$\Omega_\pi(M) \oplus Q = \ker(\pi) \oplus P \simeq \ker(\rho) \oplus P = \Omega_\rho(M) \oplus P.$$

Hence the syzygy is well-defined up to projective direct factors. Moreover, for each choice

$$P \xrightarrow{\pi} M$$

we get an induced monomorphism

$$\Omega_\pi(M) \xhookrightarrow{\iota_\pi} P.$$

This then induces a homomorphism of abelian groups

$$Hom_A(P, N) \xrightarrow{h_\pi(N)} Hom_A(\Omega_\pi(M), N)$$

$$f \longmapsto f \circ \iota_\pi$$

which of course depends on the choice of π. However, $\mathrm{coker}(h_\pi(N))$ does not depend on the choice of π, as we shall show below, and which merits a definition.

Definition 6.3.2. Let A be an algebra over some commutative ring R, and let M and N be two A-modules. With the notations above, let

$$Ext_A^1(M,N) := \mathrm{coker}(h_\pi(N)).$$

Lemma 6.3.3. *Let A be an algebra over some commutative ring R, and let M and N be two A-modules. Let $P \xrightarrow{\pi} M$ and $Q \xrightarrow{\rho} M$ be two epimorphisms with projective A-modules P and Q. Then*

$$\mathrm{coker}(h_\pi(N)) \simeq \mathrm{coker}(h_\rho(N)).$$

Proof. We have monomorphisms

$$\Omega_\pi(M) \overset{\iota_\pi}{\hookrightarrow} P \qquad\qquad \Omega_\rho(M) \overset{\iota_\rho}{\hookrightarrow} Q$$

As in the proof of Schanuel's lemma we form the pullback diagram

$$\begin{array}{ccccc}
 & & \Omega_\pi(M) & = & \Omega_\pi(M) \\
 & & \downarrow{\scriptstyle \hat\iota_\pi} & & \downarrow{\scriptstyle \iota_\pi} \\
\Omega_\rho(M) & \overset{\hat\iota_\rho}{\hookrightarrow} & X & \overset{\hat\rho}{\twoheadrightarrow} & P \\
\| & & \downarrow{\scriptstyle \hat\pi} & & \downarrow{\scriptstyle \pi} \\
\Omega_\rho(M) & \overset{\iota_\rho}{\hookrightarrow} & Q & \overset{\rho}{\twoheadrightarrow} & M
\end{array}$$

Here we adopt a practical convention (cf. Definition 6.3.10 below). We call a sequence of morphisms

$$S \overset{\sigma}{\hookrightarrow} T \overset{\tau}{\twoheadrightarrow} U$$

a *short exact sequence* if σ is a monomorphism, τ is an epimorphism, and $\ker(\tau) = \mathrm{im}(\sigma)$. We now consider the spaces $Hom_A(Y,N)$ for all Y occurring in the diagram. Since P is projective, and since Q is projective, the morphisms $\hat\pi$ and $\hat\rho$ split, and hence projecting on the other direct factor, also $\hat\iota_\pi$ and $\hat\iota_\rho$ split, i. e. have a left inverse, and hence the induced morphisms $Hom_A(X,N) \longrightarrow Hom_A(\Omega_\pi,N)$ and $Hom_A(X,N) \longrightarrow Hom_A(\Omega_\rho,N)$ are surjective. By precomposition this gives again a commutative diagram with exact lines and columns as follows:

$$\begin{array}{ccccc}
Hom_A(M,N) & \hookrightarrow & Hom_A(Q,N) & \xrightarrow{h_p(N)} & Hom_A(\Omega_p(M),N) \\
\downarrow & & \downarrow & & \| \\
Hom_A(P,N) & \hookrightarrow & Hom_A(X,N) & \twoheadrightarrow & Hom_A(\Omega_p(M),N) \\
\downarrow{\scriptstyle h_\pi(N)} & & \downarrow & & \\
Hom_A(\Omega_\pi(M),N) & = & Hom_A(\Omega_\pi(M),N) & & \\
\downarrow & & \downarrow & & \\
\text{coker}(h_\pi) & \longrightarrow & 0 & &
\end{array}$$

and by the snake lemma (cf. Exercise 6.8), applied to the diagram in the second and third row, there is an isomorphism $\text{coker}(h_\pi(N)) \simeq \text{coker}(h_p(N))$ as required. □

A direct consequence of the definition is the following lemma.

Lemma 6.3.4. *Let A be an algebra over some commutative ring R, and let M, M_1, M_2 and N, N_1, N_2 be A-modules. Then $Ext^1_A(M,N) = 0$ whenever M is projective. Moreover,*

$$Ext^1_A(M_1,N) \oplus Ext^1_A(M_2,N) \simeq Ext^1_A(M_1 \oplus M_2,N)$$

and

$$Ext^1_A(M,N_1) \oplus Ext^1_A(M,N_2) \simeq Ext^1_A(M,N_1 \oplus N_2).$$

Proof. If M is projective, then we may choose $P = M$ and $\pi = \text{id}_M$ so that $\Omega(M) = 0$. Hence $Hom_A(\Omega(M),N) = 0$. We have seen that $Ext^1_A(M,N)$ does not depend on the choice of π.

If $M = M_1 \oplus M_2$, then we may choose $P_1 \xrightarrow{\pi_1} M_1$ and $P_2 \xrightarrow{\pi_2} M_2$ for projective modules P_1 and P_2, and epimorphisms π_1 and π_2. Then

$$\begin{pmatrix} \pi_1 & 0 \\ 0 & \pi_2 \end{pmatrix} : P_1 \oplus P_2 \longrightarrow M_1 \oplus M_2$$

is an epimorphism. Hence

$$\Omega_{\begin{pmatrix} \pi_1 & 0 \\ 0 & \pi_2 \end{pmatrix}}(M_1 \oplus M_2) = \Omega_{\pi_1}(M_1) \oplus \Omega_{\pi_2}(M_2)$$

and

$$\iota_{\begin{pmatrix} \pi_1 & 0 \\ 0 & \pi_2 \end{pmatrix}} = \begin{pmatrix} \iota_{\pi_1} & 0 \\ 0 & \iota_{\pi_2} \end{pmatrix}.$$

This, together with $Hom_A(X_1 \oplus X_2,N) \simeq Hom_A(X_1,N) \oplus Hom_A(X_2,N)$, shows

$$Ext^1_A(M_1,N) \oplus Ext^1_A(M_2,N) \simeq Ext^1_A(M_1 \oplus M_2,N)$$

The fact that

$$Ext^1_A(M,N_1) \oplus Ext^1_A(M,N_2) \simeq Ext^1_A(M,N_1 \oplus N_2)$$

is clear from the additivity of taking homomorphisms. □

Lemma 6.3.5. *Let X be an A-module and let M be an A-submodule of X. If $X/M \simeq N$, and if $Ext^1_A(N,M) = 0$, then M has a supplement in X, i. e. there is an A-submodule N' of X such that $N' \simeq N$ and such that $M \oplus N' = X$.*

Proof. Indeed, let $P \xrightarrow{\pi} N$ be an epimorphism with kernel $\Omega^1_\pi(N)$. Denote by ι_π the inclusion of $\Omega^1_\pi(N)$ into P and by μ the inclusion of M into X. Since P is projective and since the quotient map $X \xrightarrow{\nu} N$ is surjective, there is an A-module homomorphism $P \xrightarrow{\lambda} X$ such that $\pi = \nu \circ \lambda$. Let τ be the restriction of λ to $\Omega^1_\pi(N)$. The image of λ is in M. Since $Ext^1_A(N,M) = 0$, there is an A-module homomorphism $P \xrightarrow{\sigma} M$ with $\sigma \circ \iota_\pi = \tau$. Hence,

$$(\lambda - \mu \circ \sigma) \circ \iota_\pi = \lambda \circ \iota_\pi - \mu \circ \sigma \circ \iota_\pi = \mu \circ \tau - \mu \circ \tau = 0$$

and by the universal property of the cokernel, the morphism $(\lambda - \mu \circ \sigma)$ factors through the cokernel of ι_π. Hence there is a unique A-module homomorphism $N \xrightarrow{\gamma} X$ with

$$\lambda = \mu \circ \sigma + \gamma \circ \pi.$$

But then $\nu \circ \gamma = \mathrm{id}_N$, which shows that $\gamma \circ \nu$ is an idempotent endomorphism of X giving by Lemma 1.2.8 the direct sum decomposition as required. □

Remark 6.3.6. Actually, $Ext^1_A(N,M)$ parameterises in a precise meaning the ways how M can be embedded into an A-module X such that the quotient is isomorphic to N. The very developed theory of this property may be found in e. g. [Zim-14, Section 1.8.2].

6.3.2 Higher *Ext*-groups and applications

Definition 6.3.7. Let A be an algebra over some commutative ring R, and let M and N be two A-modules. Then we define

$$Ext^i_A(M,N) := Ext^{i-1}_A(\Omega(M),N)$$

for each $i \geq 2$.

Note that Schanuel's Lemma 6.2.10 and Lemma 6.3.4 show together that the definition of $Ext^i_A(M,N)$ is well-defined for all $i \geq 1$. Note moreover that $Ext^i_A(M,N)$ is an R-module for any i and any M,N since *Hom*-spaces are R-modules. Further, up to a projective direct factor $\Omega_\pi(M)$ does not depend on π. Hence, we may note $\Omega(M)$ instead, keeping in mind that this is well-defined up to projective direct factors only. Using this notion, we may put

$$\Omega^n(M) := \Omega(\Omega^{n-1}(M))$$

for all $n \geq 2$, and $\Omega^1(M) = \Omega(M)$. Using this notation we see immediately that the recursive definition implies

$$Ext_A^i(M,N) = Ext_A^1(\Omega^{i-1}(M),N).$$

We obtain the following important consequence. We will come back to this result in Section 6.4.

Proposition 6.3.8. *Let A be an algebra over some commutative ring R. Then any A-submodule of any projective A-module is itself projective if and only if $Ext_A^2(M,N) = 0$ for all A-modules M and N.*

Proof. Let P be a projective A-module and let S be an A-submodule of P. Define $M := P/S$ and let $\pi : P \longrightarrow M$ be the canonical projection. Then $\Omega_\pi(M) = S$ and $\iota_\pi : S \longrightarrow P$ be the canonical inclusion. Let Q be a projective A-module and $\kappa : Q \longrightarrow S$ be an epimorphism. Denote by $T := \ker(\kappa)$ and $\lambda : T \longrightarrow Q$ the inclusion. Then $\Omega_\kappa(\Omega_\pi(M)) = T$. If $Ext_A^2(U,V) = 0$ for all A-modules U and V, in particular $Ext_A^2(M,T) = 0$. Consider $\mathrm{id}_T : \Omega_\kappa(S) \longrightarrow T$, which is in the image of $h_\kappa(T)$:

$$\begin{aligned}0 &= Ext_A^2(M,T)\\ &= Ext_A^1(S,T)\\ &= \mathrm{coker}(h_\kappa(T))\\ &= \mathrm{coker}(Hom_A(Q,T) \stackrel{h_\kappa(T)}{\longrightarrow} Hom_A(T,T)).\end{aligned}$$

Hence there is $\alpha \in Hom_A(Q,T)$ such that

$$\mathrm{id}_T = h_\kappa(T)(\alpha) = \alpha \circ \kappa.$$

This shows that

$$(\kappa \circ \alpha)^2 = \kappa \circ \alpha \circ \kappa \circ \alpha = \kappa \circ \alpha$$

and then, using Lemma 1.2.8 and the fact that $\mathrm{id}_T = \alpha \circ \kappa$ implies that κ is a monomorphism, and α is an epimorphism, $Q \simeq T \oplus S$. Hence, since Q is projective by construction, also S is projective.

Suppose that all submodules of projective modules are projective. Let M be an A-module. Then for any choice of an epimorphism $\pi : P \longrightarrow M$ for a projective A-module P we have $\Omega_\pi(M) = \ker(\pi)$ is a submodule of P, and therefore projective. We may choose the projective module $\Omega_\pi(M)$ and the epimorphism $id_{\Omega_\pi(M)}$ which then implies

$$\Omega_{id_{\Omega_\pi(M)}} = 0.$$

Then clearly $Ext_A^2(M,N) = 0$ for any A-module N. This shows the statement. □

Lemma 6.3.9. *Let R be a commutative ring and let A be an R-algebra. Then for all A-modules M, M_1, M_2, N, N_1, N_2, we get*

$$Ext_A^i(M_1,N) \oplus Ext_A^i(M_2,N) \simeq Ext_A^i(M_1 \oplus M_2, N)$$

and

$$Ext_A^i(M,N_1) \oplus Ext_A^i(M,N_2) \simeq Ext_A^i(M,N_1 \oplus N_2)$$

for all $i \geq 1$.

Proof. Since $\Omega(M_1 \oplus M_2) \simeq \Omega(M_1) \oplus \Omega(M_2)$ as in the proof of Lemma 6.3.4 we get the result by induction on i and the statement of Lemma 6.3.4. □

We have seen that we deal very often with sequences of homomorphisms such that the kernel of one is the image of the previous one. We shall use a specific language for this situation.

Definition 6.3.10. Let A be an algebra over a commutative ring R and let L, M, N be A-modules. Let $\alpha \in Hom_A(L,M)$ and $\beta \in Hom_A(M,N)$. Then

$$L \xrightarrow{\alpha} M \xrightarrow{\beta} N$$

is said to be *exact* at M if $\ker(\beta) = \operatorname{im}(\alpha)$. A diagram

$$0 \longrightarrow L \xrightarrow{\alpha} M \xrightarrow{\beta} N \longrightarrow 0$$

is called a *short exact sequence* if it is exact at L, at M, and at N.

We shall need a particular case of a well-known and very useful lemma.

Lemma 6.3.11 (Horseshoe lemma). *Let A be an algebra over a commutative ring R and let L, M, N be A-modules. Suppose that*

$$0 \longrightarrow L \xrightarrow{\alpha} M \xrightarrow{\beta} N \longrightarrow 0$$

is a short exact sequence, let $P_L \xrightarrow{\pi_L} L$ be an epimorphism with a projective A-module P_L, and let $P_N \xrightarrow{\pi_N} N$ be an epimorphism with a projective A-module P_N. Then there is *an epimorphism $P_L \oplus P_N \xrightarrow{\pi_M} M$ making the diagram*

$$\begin{array}{ccccc}
L & \xrightarrow{\alpha} & M & \xrightarrow{\beta} & N \\
\uparrow{\scriptstyle \pi_L} & & \uparrow{\scriptstyle \pi_M} & & \uparrow{\scriptstyle \pi_N} \\
P_L & \xrightarrow{\binom{1}{0}} & P_L \oplus P_N & \xrightarrow{(0,1)} & P_N \\
\uparrow{\scriptstyle \iota_L} & & \uparrow{\scriptstyle \iota_M} & & \uparrow{\scriptstyle \iota_N} \\
\ker \pi_L & \xrightarrow{\Omega_\alpha} & \ker \pi_M & \xrightarrow{\Omega_\beta} & \ker \pi_N
\end{array}$$

commutative, for ι_L, ι_M and ι_N being the canonical inclusions. Furthermore,

$$0 \longrightarrow \ker \pi_L \xrightarrow{\Omega_\alpha} \ker \pi_M \xrightarrow{\Omega_\beta} \ker \pi_N \longrightarrow 0$$

is a short exact sequence.

Proof. Since P_N is a projective module, and since β is an epimorphism, by the defining property of a projective module (respectively Proposition 6.2.5) there is a $\sigma \in Hom_A(P_N, M)$ with $\pi_N = \beta \circ \sigma$. If we define $\pi_M := (\alpha \circ \pi_L, \sigma)$, i. e. $\pi_M(u, v) = \alpha \circ \pi_L(u) + \sigma(v)$ and

$$P_L \oplus P_N \xrightarrow{(\alpha \circ \pi_L, \sigma)} M$$

then

$$\beta \circ (\alpha \circ \pi_L, \sigma) = (0, \pi_N) = \pi_N \circ (0, 1).$$

The fact that π_M makes the upper two squares commutative is equivalent to the existence of a homomorphism $\sigma : P_N \longrightarrow M$ with $\beta \circ \sigma = 0$ and π_M takes the form

$$P_L \oplus P_N \xrightarrow{(\alpha \circ \pi_L, \sigma)} M.$$

Hence, the above square is commutative. Note that we did not use the fact that α is injective. The homomorphisms Ω_α and Ω_β are defined by restriction to the kernels. Since the left upper square is commutative, the image of Ω_α is in $\ker(\pi_M)$, and since the right upper square is commutative, the image of Ω_β is in $\ker(\pi_N)$.

If $x \in \ker(\pi_N)$, then $\pi_M(0, x) = \sigma(x)$. Since $\beta \circ \pi_M(0, x) = \pi_N(x) = 0$, we get $\sigma(x) \in \ker \beta = \mathrm{im}(\alpha)$. Hence there is a $y \in L$ with $\alpha(y) = \sigma(x)$. Since π_L is surjective, there is $z \in P_L$ with $\pi_L(z) = y$, and therefore $\alpha \circ \pi_L(z) = \sigma(x)$. Hence

$$\pi_M(-z, x) = -\alpha \circ \pi_L(z) + \sigma(x) = -\alpha(y) + \sigma(x) = 0.$$

This shows that $(-z, x) \in \ker \pi_M$ and Ω_β is an epimorphism.

Let $y = (y_L, y_N) \in \ker(\Omega_\beta) \subseteq P_L \oplus P_N$. Then $y_N = 0$, and since $y \in \ker(\pi_M)$, we get $0 = \pi_M(y_L, 0) = \alpha \circ \pi_L(y_L)$. Since α is a monomorphism, $y_L \in \ker(\pi_L)$ and hence $y_L \in \ker(\pi_L)$. Hence $\Omega_\alpha(y_L) = (y_L, 0) = y$. This shows that the sequence

$$\ker \pi_L \xrightarrow{\Omega_\alpha} \ker \pi_M \xrightarrow{\Omega_\beta} \ker \pi_N \longrightarrow 0$$

is exact at $\ker \pi_N$ and at $\ker \pi_M$.

Finally, since $\binom{\mathrm{id}_{P_L}}{0}$ is an injective morphism $P_L \longrightarrow P_L \oplus P_N$, its restriction to the kernel $\ker \pi_L$ is injective as well. This shows the statement. □

Remark 6.3.12. Observe that we may iterate the horseshoe Lemma 6.3.11. The conclusion

$$0 \longrightarrow \ker \pi_L \xrightarrow{\Omega_\alpha} \ker \pi_M \xrightarrow{\Omega_\beta} \ker \pi_N \longrightarrow 0$$

is a short exact sequence may serve as ingredient again, replacing the assumption that $0 \longrightarrow L \longrightarrow M \longrightarrow N \longrightarrow 0$ is a short exact sequence.

Remark 6.3.13. A much simpler proof of the horseshoe lemma can be given using derived categories. We refer to [Zim-14, Lemma 3.5.50]. However, derived categories is a very heavy framework which is largely above the objective of our text.

In order to prepare the next statement we shall prove the following technical but important lemma.

Lemma 6.3.14. *Let A be an algebra over a commutative ring R and let L, M, N be A-modules*

- *If*

$$L \xrightarrow{\alpha} M \xrightarrow{\beta} N \longrightarrow 0$$

is a sequence which is exact at M and N. Then for all A-modules K the sequence

$$0 \longrightarrow Hom_A(N, K) \xrightarrow{\beta^*} Hom_A(M, K) \xrightarrow{\alpha^*} Hom_A(L, K),$$

with $\alpha^*(f) = f \circ \alpha$ and $\beta^*(f) = f \circ \beta$, is exact at $Hom_A(N, K)$ and at $Hom_A(M, K)$.

- *If*

$$0 \longrightarrow L \xrightarrow{\alpha} M \xrightarrow{\beta} N$$

is a sequence which is exact at M and L. Then for all A-modules K the sequence

$$0 \longrightarrow Hom_A(K, L) \xrightarrow{\alpha_*} Hom_A(K, M) \xrightarrow{\beta_*} Hom_A(K, N),$$

with $\alpha_*(g) = \alpha \circ g$ and $\beta_*(g) = \beta \circ g$ is exact at $Hom_A(K, L)$ and at $Hom_A(K, M)$.

Proof. Since $\beta \circ \alpha = 0$ and hence

$$0 = (\beta \circ \alpha)^* = \alpha^* \circ \beta^*$$

and

$$0 = (\beta \circ \alpha)_* = \beta_* \circ \alpha_*$$

we get

$$\mathrm{im}(\beta^*) \subseteq \ker(\alpha^*)$$

and

$$\mathrm{im}(\alpha_*) \subseteq \ker(\beta_*).$$

Let $f : K \longrightarrow M$ with $\beta_*(f) = \beta \circ f = 0$. Then, by the property of the kernel, there is a unique $g \in Hom_A(K, L)$ with $\alpha_*(g) = \alpha \circ g = f$. Hence in the second case the sequence is exact at $Hom_A(K, M)$. If for some $f : K \longrightarrow L$ we get $\alpha_*(f) = \alpha \circ f = 0$, then $f = 0$ since α is injective. Hence in the second case the sequence is exact at $Hom_A(K, L)$.

Let $f : M \longrightarrow K$ with $f \circ \alpha = \alpha^*(f) = 0$. Then by the property of being a cokernel, there is a unique $g : N \longrightarrow K$ with $f = g \circ \beta = \beta^*(g)$. Hence the upper sequence is exact at $Hom_A(M, K)$. If $f : N \longrightarrow K$ with $\beta^*(f) = f \circ \beta = 0$. Since β is surjective, $f = 0$ and the sequence is exact at $Hom_A(N, K)$. □

In order to be able to work with the groups $Ext^i_A(M, N)$ it is particularly useful to have the following

Proposition 6.3.15. *Let A be an algebra over a commutative ring R. Let K, L, M, N be A-modules.*

1. *If $\alpha : M \longrightarrow N$ and $\beta : N \longrightarrow L$ are A-module homomorphisms, then we get induced R-module homomorphisms*

$$Ext^i_A(K, M) \xrightarrow{\alpha^{(i)}_*} Ext^i_A(K, N) \quad \text{and} \quad Ext^i_A(K, N) \xrightarrow{\beta^{(i)}_*} Ext^i_A(K, L)$$

   as well as

$$Ext^i_A(N, K) \xrightarrow{\alpha^*_{(i)}} Ext^i_A(M, K) \quad \text{and} \quad Ext^i_A(L, K) \xrightarrow{\beta^*_{(i)}} Ext^i_A(N, K).$$

2. *If $0 \longrightarrow M \xrightarrow{\alpha} N \xrightarrow{\beta} L \longrightarrow 0$ is a short exact sequence, then*
   (a) $\mathrm{im}(\alpha^{(i)}_*) = \ker(\beta^{(i)}_*)$
   (b) $\mathrm{im}(\beta^*_{(i)}) = \ker(\alpha^*_{(i)})$.

Proof. The morphisms α and β induce R-linear homomorphisms

$$\begin{aligned} Hom_A(K,M) &\xrightarrow{\alpha_*} Hom_A(K,N) \\ \gamma &\mapsto \alpha\circ\gamma \end{aligned}$$

respectively

$$\begin{aligned} Hom_A(K,N) &\xrightarrow{\beta_*} Hom_A(K,L) \\ \gamma &\mapsto \beta\circ\gamma. \end{aligned}$$

Hence, $\beta_*\circ\alpha_* = (\beta\circ\alpha)_*$. Similarly, α and β induce R-linear homomorphisms

$$\begin{aligned} Hom_A(N,K) &\xrightarrow{\alpha^*} Hom_A(M,K) \\ \gamma &\mapsto \gamma\circ\alpha \end{aligned}$$

respectively

$$\begin{aligned} Hom_A(L,K) &\xrightarrow{\beta^*} Hom_A(N,K) \\ \gamma &\mapsto \gamma\circ\beta \end{aligned}$$

satisfying therefore $(\beta\circ\alpha)^* = \alpha^*\circ\beta^*$. The construction used to define $Ext_A^i(M,N)$ is trivially compatible with the first construction, showing hence the existence of $\alpha_*^{(i)}$ and $\beta_*^{(i)}$. Moreover, as $\beta_*\circ\alpha_* = (\beta\circ\alpha)_*$ we also get $\beta_*^{(i)}\circ\alpha_*^{(i)} = (\beta\circ\alpha)_*^{(i)}$ for all i. In order to define $\alpha^{(1)}$ we choose epimorphisms $P_M \xrightarrow{\pi_M} M$ and $P_N \xrightarrow{\pi_N} N$ for projective modules P_M and P_N. Then consider the diagram

$$\begin{array}{ccc} P_M & \xrightarrow{\pi_M} & M \\ \downarrow{\scriptstyle \exists\alpha_P} & & \downarrow{\scriptstyle \alpha} \\ P_N & \xrightarrow[\pi_N]{} & N \end{array}$$

where π_N is an epimorphism, and since P_M is projective, there is a (non unique) morphism α_P making this diagram commutative. Now, consider the kernels Ω_{π_M} and Ω_{π_N}. By the property of being a kernel there is a unique morphism $\Omega(\alpha)$ such that the diagram

$$\begin{array}{ccccc} \Omega_{\pi_M} & \xrightarrow{\iota_M} & P_M & \xrightarrow{\pi_M} & M \\ \downarrow{\scriptstyle \exists!\Omega(\alpha)} & & \downarrow{\scriptstyle \exists\alpha_P} & & \downarrow{\scriptstyle \alpha} \\ \Omega_{\pi_N} & \xrightarrow{\iota_N} & P_N & \xrightarrow[\pi_N]{} & N \end{array}$$

is commutative. Note however, that α_P is not uniquely defined. Suppose we have chosen another α'_P giving then $\Omega(\alpha)'$. What is the difference $\Omega(\alpha)' - \Omega(\alpha)$? We obtain this difference if we subtract the two results, fitting therefore in the diagram for $\alpha = 0$:

$$\begin{array}{ccccc} \Omega_{\pi_M} & \xrightarrow{\iota_M} & P_M & \xrightarrow{\pi_M} & M \\ \downarrow {\scriptstyle \exists!\Omega(0)} & & \downarrow {\scriptstyle \exists\nu} & & \downarrow {\scriptstyle 0} \\ \Omega_{\pi_N} & \xrightarrow{\iota_N} & P_N & \xrightarrow[\pi_N]{} & N \end{array}$$

But since the right square of the diagram is commutative, $\pi_N \circ \nu = 0$, and therefore there is a morphism $h : P_M \longrightarrow \Omega_{\pi_N}$ with $\iota_{\pi_N} \circ h = \nu$. But this shows

$$\Omega(\alpha)' - \Omega(\alpha) = \Omega(0) = h \circ \iota_{\pi_M}.$$

Therefore $\Omega(\alpha)' - \Omega(\alpha)$ induce the morphism

$$Ext^1_A(N,K) \xrightarrow{0} Ext^1_A(M,K),$$

or what is the same, the morphisms $\Omega(\alpha)'$ and $\Omega(\alpha)$ coincide as morphisms

$$Ext^1_A(N,K) \longrightarrow Ext^1_A(M,K).$$

By induction on i we get well-defined morphisms

$$Ext^i_A(N,K) \xrightarrow{\alpha^*_{(i)}} Ext^i_A(M,K)$$

for all i. Since $(\beta \circ \alpha)^* = \alpha^* \circ \beta^*$ and since the induced map on the R-modules $Ext^{(i)}_A(-,-)$ is well-defined, we obtain that $(\beta \circ \alpha)^*_{(i)} = \alpha^*_{(i)} \circ \beta^*_{(i)}$ for all i. This shows item 1.

We will show 2a. By the horseshoe Lemma 6.3.11 we choose projectives and epimorphisms $P_M \xrightarrow{\pi_M} M$, $P_L \xrightarrow{\pi_L} L$, and $P_N = P_M \oplus P_L$ with an epimorphism π_M according to Lemma 6.3.11. Consider the sequence

$$Ext^1_A(K,M) \xrightarrow{\alpha^{(1)}_*} Ext^1_A(K,N) \xrightarrow{\beta^{(1)}_*} Ext^1_A(K,L).$$

If we prove the statement in this case the general case follows since

$$Ext^i_A(K,X) = Ext^1_A(\Omega^{i-1}(K),X)$$

for all $i > 1$, and replacing K by $\Omega^{i-1}(K)$ gives the statement. Since $\beta \circ \alpha = 0$, as

$$\beta^{(1)}_* \circ \alpha^{(1)}_* = (\beta \circ \alpha)^{(1)}_* = 0$$

we have

$$\mathrm{im}(\alpha^{(1)}_*) \subseteq \ker(\beta^{(1)}_*).$$

Let $f \in Hom_A(\Omega_\pi(K), N)$, giving hence $f \in Ext_A^1(K, N)$, if now $\beta_*^{(1)}(f) = 0$, then $\beta \circ$ $f \in Hom_A(\Omega_\pi(K), M)$ factors through the monomorphism $\Omega_\pi(K) \xrightarrow{\iota_\pi} P$. More precisely there is $\sigma \in Hom_A(P, L)$ such that

$$\beta \circ f = \sigma \circ \iota_\pi.$$

$$
\begin{array}{ccc}
\Omega_\pi(K) & \xrightarrow{\iota_\pi} & P \\
\downarrow{\scriptstyle f} & & \downarrow{\scriptstyle \sigma} \\
M \xrightarrow{\alpha} N & \xrightarrow{\beta} & L
\end{array}
$$

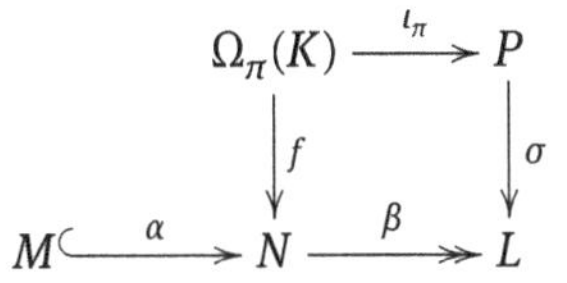

As β is an epimorphism and as P is projective, there is $\tau \in Hom_A(P, N)$ such that $\sigma = \beta \circ \tau$. Since f is defined up to morphisms factoring through ι_π only, we may replace f by $\tilde{f} = f - \tau \circ \iota_\pi$. Then, $\beta \circ \tilde{f} = 0$ and there is a unique $\lambda \in Hom_A(\Omega_\pi(K), M)$ such that

$$\tilde{f} = f - \tau \circ \iota_\pi = \alpha\lambda.$$

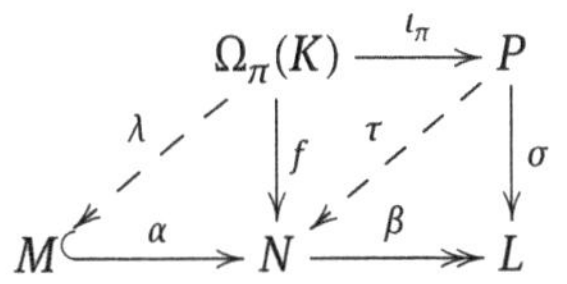

$$
\begin{array}{ccc}
\Omega_\pi(K) & \xrightarrow{\iota_\pi} & P \\
{\scriptstyle \lambda} \swarrow \quad \downarrow{\scriptstyle f} \quad {\scriptstyle \tau} \nearrow & & \downarrow{\scriptstyle \sigma} \\
M \xrightarrow{\alpha} N & \xrightarrow{\beta} & L
\end{array}
$$

Hence $\tilde{f} \in im(\alpha_*^{(1)})$.

We now will show 2b. Consider the sequence

$$Ext_A^1(L, K) \xrightarrow{\beta_{(1)}^*} Ext_A^1(N, K) \xrightarrow{\alpha_{(1)}^*} Ext_A^1(M, K).$$

We can write the groups Ext^1 as quotients

$$Ext_A^1(X, K) = coker(Hom_A(P_X, K) \longrightarrow Hom_A(\Omega(X), K))$$

and so

$$
\begin{array}{ccccc}
Hom_A(P_L, K) & \xrightarrow{\beta_P^*} & Hom_A(P_N, K) & \xrightarrow{\alpha_P^*} & Hom_A(P_M, K) \\
\downarrow{\scriptstyle \iota_L^*} & & \downarrow{\scriptstyle \iota_N^*} & & \downarrow{\scriptstyle \iota_M^*} \\
Hom_A(\Omega(L), K) & \xrightarrow{\Omega_\beta^*} & Hom_A(\Omega(N), K) & \xrightarrow{\Omega_\alpha^*} & Hom_A(\Omega(M), K) \\
\downarrow{\scriptstyle q_L} & & \downarrow{\scriptstyle q_N} & & \downarrow{\scriptstyle q_M} \\
Ext_A^1(L, K) & \xrightarrow{\beta_{(1)}^*} & Ext_A^1(N, K) & \xrightarrow{\alpha_{(1)}^*} & Ext_A^1(M, K)
\end{array}
$$

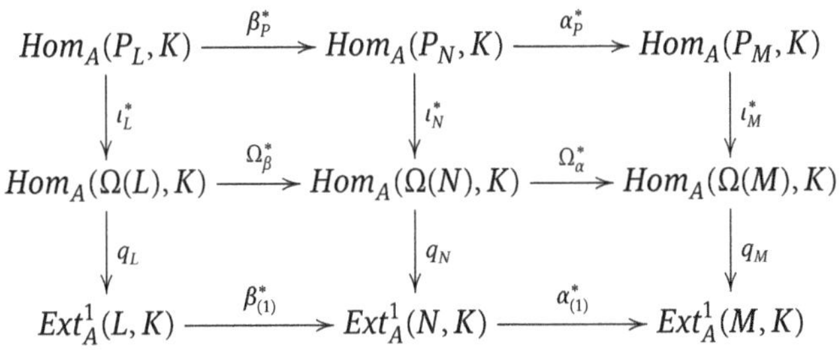

where the morphisms Ω_α, Ω_β, α_P, β_P are induced by α, β, the fact that P_L, P_M, P_N are all projective, where by the Horseshoe Lemma 6.3.11 we may suppose that $P_N = P_L \oplus$

P_M, and the morphisms π_M, π_N, π_L being surjective, as indicated in the commutative diagram

$$\begin{array}{ccccc}
M & \xrightarrow{\alpha} & N & \xrightarrow{\beta} & L \\
{\scriptstyle \pi_M}\uparrow & & {\scriptstyle \pi_N}\uparrow & & {\scriptstyle \pi_L}\uparrow \\
P_M & \xrightarrow{\alpha_P} & P_N & \xrightarrow{\beta_P} & P_L \\
{\scriptstyle \iota_M}\uparrow & & {\scriptstyle \iota_N}\uparrow & & {\scriptstyle \iota_L}\uparrow \\
\Omega(M) & \xrightarrow{\Omega_\alpha} & \Omega(N) & \xrightarrow{\Omega_\beta} & \Omega(L).
\end{array}$$

Again, since

$$\alpha^*_{(1)} \circ \beta^*_{(1)} = (\beta \circ \alpha)^*_{(1)}$$

we get

$$\mathrm{im}(\beta^*_{(1)}) \subseteq \ker(\alpha^*_{(1)}).$$

Let $g \in Hom_A(\Omega(N), K)$, representing an element $q_N(g)$ in $Ext^1_A(N, K)$ with

$$q_M(g \circ \Omega_\alpha) = \alpha^*_{(1)}(q_N(g)) = 0 \quad \text{in } Ext^1_A(M, K).$$

Hence, there is $h \in Hom_A(P_M, K)$ with

$$g \circ \Omega_\alpha = h \circ \iota_M = \iota^*_M(h).$$

Since α is a monomorphism, α^*_P is surjective. Hence, there is $\widetilde{h} \in Hom_A(P_N, K)$ with

$$h = \widetilde{h} \circ \alpha_P = \alpha^*_P(\widetilde{h}).$$

Therefore

$$\iota^*_N(\widetilde{h}) = \iota^*_N(\widetilde{h} \circ \alpha_P) = \widetilde{h} \circ \alpha_P \circ \iota_N = \widetilde{h} \circ \iota_L \circ \Omega_\beta = \Omega^*_\beta(\widetilde{h} \circ \iota_L).$$

Then we may replace g by $g - \iota^*_N(\widetilde{h})$ and get

$$\Omega^*_\alpha(g - \iota^*_N(\widetilde{h})) = g \circ \Omega_\alpha - \iota^*_M \circ \alpha^*_P(\widetilde{h}) = 0$$

Since by the Horseshoe Lemma 6.3.11 we get $\ker(\Omega_\beta) = \mathrm{im}(\Omega_\alpha)$, and Ω_β surjective, then Ω^*_β is injective and $\mathrm{im}(\Omega^*_\beta) = \ker(\Omega^*_\alpha)$. Therefore, $q_N(g) = \beta^*_{(1)}(q_L(\widehat{g}))$ for some $\widehat{g} \in Hom_A(\Omega(L), K)$. This shows then the proposition. □

6.4 Hereditary algebras

Some particular algebras will be of a special importance.

Definition 6.4.1. Let A be an algebra over some commutative ring. Then A is called *hereditary* if each submodule of any projective module is itself projective.

Example 6.4.2.
1. Since for semisimple algebras A each module is projective, semisimple algebras are hereditary.
2. For any field K the algebra $A = \begin{pmatrix} K & K \\ 0 & K \end{pmatrix}$ is hereditary, as was shown in Exercise 1.4.

Remark 6.4.3. Note that we are dealing with left modules, and hence the concept introduced above should be called left hereditary. Analogously there is the concept of right hereditary rings, namely those for which submodules of projective right modules are always projective right modules. We mention that the concepts of left and right hereditary rings do not coincide in general.

The following result is very important. For the proof we follow the monograph of Auslander-Buchsbaum [AuBu-74].

Theorem 6.4.4. *Let A be an algebra over a commutative ring. Then A is hereditary if and only if every ideal of A is a projective A-module.*

Moreover, if A is hereditary, then every projective A-module is a direct sum of ideals of A.

Proof. First, suppose that A is hereditary. Then the regular A-module is projective, and if I is an ideal of A, then I is an A-submodule of the regular module. Since A is hereditary, I is a projective module.

Conversely suppose that all ideals of A are projective A-modules and let P be a projective module. Then P is a direct summand of a free module by Proposition 6.2.5. Hence, if M is a submodule of P, then M is also a submodule of the free module. We may hence suppose that P is free and that X is a basis of P. For any subset Y of X let $\langle Y\rangle_A$ denote the A-submodule of P generated by Y.

We prove the result using Zorn's lemma. Let $\mathcal{G}$ be the set of couples (Y, S) where $Y \subseteq X$ and $S = \{S_\lambda \mid \lambda \in \Lambda\}$ is a set of submodules S_λ of $\langle Y\rangle_A \cap M$ such that each S_λ is isomorphic to an ideal of A and such that

$$\langle Y\rangle_A \cap M = \bigoplus_{\lambda\in\Lambda} S_\lambda.$$

This set $\mathcal{G}$ is partially ordered by $(Y^{(1)}, S^{(1)}) \leq (Y^{(2)}, S^{(2)})$ if and only if $Y^{(1)} \subseteq Y^{(2)}$ and $S^{(1)} \subseteq S^{(2)}$. The set is non empty since (Y, S) may be $(\emptyset, \{0\})$. Let then $(Y_\gamma, S_\gamma)_{\gamma\in\Gamma}$ be a totally ordered subset of $\mathcal{G}$. Put

$$\widehat{Y} := \bigcup_{\gamma\in\Gamma} Y_\gamma \quad\text{and}\quad \widehat{S} := \bigcup_{\gamma\in\Gamma} S_\gamma.$$

Then we get

$$\langle \widehat{Y}\rangle_A = \bigcup_{y\in\Gamma} \langle Y_y\rangle_A$$

and

$$\langle \widehat{Y}\rangle_A \cap M = \bigcup_{y\in\Gamma} \langle Y_y\rangle_A \cap M = \bigcup_{y\in\Gamma} \langle Y_y \cap M\rangle_A.$$

Since $\langle Y_y \cap M\rangle_A = \bigoplus_{S\in S_y} S$, and since the family $(Y_y, S_y)_{y\in\Gamma}$ is a totally ordered subset of $\mathcal{G}$,

$$\langle \widehat{Y}\rangle_A \cap M = \bigoplus_{S\in\widehat{S}} S$$

which means that $(\widehat{Y}, \widehat{S}) \in \mathcal{G}$. Zorn's lemma then implies that there is a maximal element (Y^0, S^0) in $\mathcal{G}$. We first show that $Y^0 = X$. Else there is $x \in X \setminus Y^0$ and put $Y^1 := Y^0 \cup \{x\}$. Then put

$$I := \{a \in A \mid ax \in M + \langle Y^0\rangle_A\}.$$

This is a left ideal of A.

We now show that for any $m \in \langle Y^1\rangle_A \cap M$ there is a unique $a \in I$ with $m = v + ax$ for some $v \in \langle Y^0\rangle_A$. Indeed, since $m \in \langle Y^1\rangle_A$ there is $v \in \langle Y^0\rangle_A$ and $a \in A$ with $m = v + ax$. Since $m \in M$, we get $ax = m - v \in M + \langle Y^0\rangle_A$, and therefore $a \in I$. This shows the existence. For unicity we suppose

$$v' + a'x = m = v + ax$$

for some $v, v' \in \langle Y^0\rangle_A$ and $a, a' \in I$. Then

$$v - v' = (a' - a)x \in \langle Y^0\rangle_A \cap Ax = \{0\}$$

Hence $v = v'$ and $a = a'$.

We now define a map f by

$$\begin{aligned} M \cap \langle Y^1\rangle_A &\xrightarrow{f} I \\ m = v + ax &\mapsto a \end{aligned}$$

Now,

$$\begin{aligned} f(am_1 + m_2) &= f(a(v_1 + a_1x) + (v_2 + a_2x)) \\ &= f((av_1 + v_2) + (aa_1 + a_2)x) \\ &= aa_1 + a_2 = af(m_1) + f(m_2) \end{aligned}$$

for all $m_1, m_2 \in M \cap \langle Y^1 \rangle_A$ and $a, a_1, a_2 \in A$. f is surjective, since for any $a \in I$ we have $ax \in M + \langle Y^0 \rangle_A$. Hence there is $m \in M$ and $v \in \langle Y^0 \rangle_A$ such that $ax = m + v$. Therefore $m \in M \cap \langle Y^1 \rangle_A$, and $f(m) = f(ax - v) = a$. Since I is an ideal of A, it is a projective A-module by hypothesis. Since $\ker(f) = M \cap \langle Y^0 \rangle_A$ by Proposition 6.2.5 we get that

$$M \cap \langle Y^1 \rangle_A \simeq I \oplus (M \cap \langle Y^0 \rangle_A).$$

But this contradicts the maximality of (Y^0, S^0) since $(Y^1, S^0 \cup \{I\}) \in \mathcal{G}$ and $(Y^0, S^0) < (Y^1, S^0 \cup \{I\})$.

Therefore $Y^0 = X$ and M is the direct sum of left ideals of A, whence direct sum of projective modules. A direct sum of projective modules is projective itself. □

Remark 6.4.5. Note the following very useful consequence. If A is an algebra over some commutative ring. By Theorem 6.4.4 the algebra A is hereditary if and only if every ideal is projective. By Proposition 6.3.8 we need to show that $Ext_A^2(M, N) = 0$ for all A-modules M, N. Since we only need to consider ideals of A we only need to verify the property for cyclic modules M. If A is now Noetherian, then any ideal of A is finitely generated, and hence the second syzygy of M is finitely generated again. Hence for Noetherian A we only need to verify the vanishing of $Ext_A^2(M, N)$ for cyclic M and finitely generated N only.

Theorem 6.4.6. *An algebra A is hereditary if and only if for all A-modules M and all A-modules N we have $Ext_A^2(M, N) = 0$. Moreover, in case of an artinian algebra A it is sufficient to verify $Ext_A^2(M, N) = 0$ for all simple A-modules M and N.*

Proof. Proposition 6.3.8 shows that A is hereditary if and only if

$$Ext_A^2(M, N) = 0$$

for all A-modules M and all A-modules N.

We shall show that we only need to verify this for simple A-modules in case of artinian algebras A. By Hopkins Theorem (cf. Exercise 6.4) we know that A is artinian and Noetherian. Hence any finitely generated A-module has a composition series. By Remark 6.4.5 we may assume that M is cyclic and N is finitely generated.

We first show that we may assume M simple by induction on the composition length of M. Let hence M be a finitely generated A-module, and let S be a simple submodule. Then $\overline{M} := M/S$ has smaller composition length and therefore by Proposition 6.3.15 we get that $Ext_A^2(\overline{M}, N) = 0$ for all N and $Ext_A^2(S, N) = 0$ for each fixed N implies that $Ext_A^2(M, N) = 0$ for each fixed N.

Since A is artinian, N has a composition series. By induction on the composition length of N, just as for M, we show that we may assume that N is simple. Let T be a simple submodule of N. Then for $\overline{N} := N/T$ we get by Proposition 6.3.15 that $Ext_A^2(M, \overline{N}) = 0$ for each fixed M and $Ext_A^2(M, T) = 0$ for each fixed M implies that $Ext_A^2(M, N) = 0$ for each fixed M. This proved the result. □

Remark 6.4.7. For non Noetherian algebras A there is a weaker concept. An algebra A is called *semihereditary* if all finitely generated ideals of A are projective. There is an extensive literature on semihereditary algebras. We will not go into details here.

6.5 Flatness

There is a somewhat dual concept to extension groups replacing $Hom_A(X, Y)$ by tensor products.

If R is a commutative ring and A is an R-algebra, and if $f : M \longrightarrow N$ is an epimorphism of A-modules, then for all A-right modules L the homomorphism

$$\mathrm{id}_L \otimes f : L \otimes_A M \longrightarrow L \otimes_A N$$

is an epimorphism as well. Indeed, the R-module $L \otimes_A N$ is generated by elements of the form $\ell \otimes n$ for $\ell \in L$ and $n \in N$. Since f is an epimorphism, for all $n \in N$ there is an $m \in M$ such that $f(m) = n$. Hence

$$\ell \otimes n = (\mathrm{id} \otimes f)(\ell \otimes m).$$

However, injectivity is not preserved in general as shows the following example.

Example 6.5.1. Multiplication by 2 on $\mathbb{Z}$ is an injective endomorphism:

$$\mathbb{Z} \overset{\cdot 2}{\hookrightarrow} \mathbb{Z}$$

Consider the induced morphism $\mathrm{id}_{2\mathbb{Z}} \otimes (\cdot 2)$:

$$\begin{array}{ccc} \mathbb{Z}/2\mathbb{Z} \otimes_{\mathbb{Z}} \mathbb{Z} & \xrightarrow{\mathrm{id}_{2\mathbb{Z}} \otimes (\cdot 2)} & \mathbb{Z}/2\mathbb{Z} \otimes_{\mathbb{Z}} \mathbb{Z} \\ \downarrow \simeq & & \downarrow \simeq \\ \mathbb{Z}/2\mathbb{Z} & \xrightarrow{0} & \mathbb{Z}/2\mathbb{Z} \end{array}$$

is commutative, since multiplication by 2 on $\mathbb{Z}/2\mathbb{Z}$ is just the 0 map. Of course, the 0 map is not injective.

Definition 6.5.2. Let R be a commutative ring and let A be an R-algebra. A right A-module L is called *flat* if for any monomorphism $f : M \longrightarrow N$ of A-modules the induced morphism

$$\mathrm{id} \otimes f : L \otimes_A M \longrightarrow L \otimes_A N$$

is a monomorphism of R-modules. Similarly, a left A-module L is called flat if for any monomorphism $f : M \longrightarrow N$ of A-right modules the induced morphism

$$f \otimes \mathrm{id} : M \otimes_A L \longrightarrow N \otimes_A L$$

is a monomorphism of R-modules.

An immediate consequence is the following

Example 6.5.3. Free modules are flat. Indeed,

$$\left(\bigoplus_{i\in I} A\right) \otimes_A X \simeq \bigoplus_{i\in I}(A \otimes_A X) \simeq \bigoplus_{i\in I} X$$

and if $f : M \longrightarrow Y$ is a homomorphism, then

$$\begin{array}{ccc} (\bigoplus_{i\in I} A) \otimes_A M & \xrightarrow{\mathrm{id}\otimes f} & (\bigoplus_{i\in I} A) \otimes_A N \\ \downarrow \simeq & & \downarrow \simeq \\ \bigoplus_{i\in I} M & \xrightarrow{(\bigoplus_{i\in I} f)} & \bigoplus_{i\in I} N \end{array}$$

is commutative. Since f is a monomorphism, $(\bigoplus_{i\in I} f)$ is a monomorphism as well and hence any free module is flat.

By a very similar argument also direct factors of free modules are flat. Hence projective modules are flat. The converse is not true in general.

Proposition 6.5.4. *Let R be a commutative ring, and let A be a Noetherian R-algebra. Then any finitely generated flat module is projective.*

The proof is quite involved. We refer to e. g. [Lam-98, 2. Theorem 4.30].

This motivates the following definition, which is somehow analogous to the definition of *Ext*-groups.

Definition 6.5.5. Let R be a commutative ring and let A be an R-algebra. For a left A-module M let P_M be a projective A-module and an epimorphism $P_M \xrightarrow{\pi} M$ with $\Omega_\pi(M) := \ker(\pi)$. Then let $\Omega_\pi(M) \xrightarrow{\iota} P_M$ be the natural inclusion. Define for any right A-module N the kernel of $N \otimes_A \Omega_\pi(M) \longrightarrow N \otimes_A P_M$ to be

$$Tor_1^A(N, M) := \ker(\mathrm{id}_N \otimes \iota)$$

the torsion group of N with M.

Note that Lemma 3.1.10 implies that

$$Tor_1^A(N, M_1 \oplus M_2) \simeq Tor_1^A(N, M_1) \oplus Tor_1^A(N, M_2)$$

and

$$Tor_1^A(N_1 \oplus N_2, M) \simeq Tor_1^A(N_1, M) \oplus Tor_1^A(N_2, M)$$

for any right A-modules N, N_1, N_2 and any left A-modules M, M_1, M_2. Since for projective A-modules P we have $Tor_1^A(N, P) = 0$, Schanuel's Lemma 6.2.10 shows then that $Tor_1^A(N, M)$ is independent of the choice of the projective module P_M and the epimorphism π.

Definition 6.5.6. Let R be a commutative ring and let A be an R-algebra. The *R-torsion submodule* M_t of an A-module M is formed by the set of $m \in M$ such that there is a non zero divisor $r \in R$ with $rm = 0$. An A-module M is said to be without R-torsion if $M_t = 0$.

Since R is commutative, the torsion submodule M_t is actually an A-submodule of M. Note that $A = R$ is explicitly allowed.

For a commutative ring R, a multiplicative subset S of R and an R-module M we let M_S be the R_S-module given by equivalence classes of $(x, s) \in M \times S$ where (x, s) is equivalent to (y, t) if and only if there is $u \in S$ with $utx = usy$. Now, $R_S \otimes_R M \simeq M_S$ by the map

$$M_S \ni (m, s) \mapsto (1, s) \otimes m \in R_S \otimes_R M$$

and in the other direction by

$$R_S \otimes_R M \ni (r, s) \otimes m \mapsto (rm, s) \in M_S.$$

Both maps are well-defined and inverse one to the other. As for localisation of rings, by abuse of language, if $\wp$ is a prime ideal of R, then we denote by $M_\wp$ the localisation of M at the multiplicative set $R \setminus \wp$. This will not give a notational problem since $\wp$ is not a multiplicative subset.

Lemma 6.5.7. *Let R be an integral domain and let S be a multiplicative subset of R. Then R_S is a flat R-module.*

Proof. Let $L \hookrightarrow M$ be an injective R-module homomorphism. We need to show that

$$R_S \otimes_R L \longrightarrow R_S \otimes_R M$$

is injective. We may suppose that L is actually a submodule of M. But if $(x, s) \in L \times S$ is a representative which is $0 = (0, 1)$ in $R_S \otimes M$. Then there is $u \in S$ such that $ux = 0$. But this means that $(x, s) = 0$ in $R_S \otimes L$ □

Lemma 6.5.8. *Let R be an integral domain and let M be an R-module. Then M is flat if and only if for every prime ideal $\wp$ of R the localisation $M_\wp$ is flat as $R_\wp$-module.*

Proof. Let M be a flat R-module. Suppose that $L(\wp) \hookrightarrow N(\wp)$ is a monomorphism of $R_\wp$-modules. Then consider

$$M_\wp \otimes_{R_\wp} L(\wp) \longrightarrow M_\wp \otimes_{R_\wp} N(\wp)$$

Since

$$M_\wp \otimes_{R_\wp} L(\wp) = (M \otimes_R R_\wp) \otimes_{R_\wp} L(\wp) = M \otimes_R L(\wp)$$

and likewise

$$M_\wp \otimes_{R_\wp} N(\wp) = (M \otimes_R R_\wp) \otimes_{R_\wp} N(\wp) = M \otimes_R N(\wp)$$

we get that the above morphism

$$M \otimes_R L(\wp) = M_\wp \otimes_{R_\wp} L(\wp) \longrightarrow M_\wp \otimes_{R_\wp} N(\wp) = M \otimes_R N(\wp)$$

is actually obtained from $L(\wp) \hookrightarrow N(\wp)$ by tensoring with M over R. But this preserves injectivity since M is R-flat.

Suppose that $M_\wp$ is flat as $R_\wp$-module for all primes $\wp$. If L is a submodule of N, then we need to consider the natural map φ from $M \otimes_R L$ into $M \otimes_R N$. Let $K := \ker \varphi$ and let $0 \neq x \in K$. Let $I := Ann_R(x) = \{r \in R \mid rx = 0\}$. This is an ideal of R different from R, since $x \neq 0$ and hence $1 \in R \setminus I$. Let $\wp$ be a maximal ideal of R containing I. Then $R_\wp \otimes_R K \neq 0$. However,

$$0 \longrightarrow K \longrightarrow M \otimes_R L \longrightarrow M \otimes_R N$$

is exact, and hence, using Lemma 6.5.7, also

$$0 \longrightarrow R_\wp \otimes_R K \longrightarrow R_\wp \otimes_R (M \otimes_R L) \longrightarrow R_\wp \otimes_R (M \otimes_R N)$$

is exact. However,

$$R_\wp \otimes_R (M \otimes_R L) = M_\wp \otimes_{R_\wp} L_\wp$$

and likewise

$$R_\wp \otimes_R (M \otimes_R N) = M_\wp \otimes_{R_\wp} N_\wp$$

and the above map

$$R_\wp \otimes_R (M \otimes_R L) \longrightarrow R_\wp \otimes_R (M \otimes_R N)$$

becomes the natural map φ induced by the inclusion $L \hookrightarrow N$. Since $M_\wp$ is a flat $R_\wp$-module, $R_\wp \otimes_R K = 0$, which is a contradiction. □

Lemma 6.5.9. *Let R be a principal ideal domain. Then a finitely generated R-module M is flat if and only if M is torsion free, i. e. $M_t = 0$*

Proof. If M is torsion free, then by the classification theorem of modules over principal ideal domains the module M is free. Free modules are flat by Example 6.5.3.

Again by the classification theorem of modules over principal ideal domains, $M = M_f \oplus M_t$ for M_f being free. Then for some $m \in M_t \setminus \{0\}$ there is $r \in R \setminus \{0\}$ with $rm = 0$. Consider the injection

$$R \xrightarrow{\cdot r} R$$

given by multiplication by r. Then

$$M \otimes_R R \xrightarrow{\mathrm{id}\otimes \cdot r} M \otimes_R R$$

translates into multiplication by r on M,

$$M \xrightarrow{r\cdot} M$$

which has kernel containing m at least. Hence M is not flat. □

We shall generalise and elaborate greatly the remark from Section 3.2 into the much vaster result Theorem 6.5.10.

A Λ-module is called finitely presented if there are integers n_1 and n_2 and a homomorphism

$$\Lambda^{n_2} \xrightarrow{\varphi} \Lambda^{n_1}$$

such that $M = \mathrm{coker}(\varphi)$.

Theorem 6.5.10. *Let R be an integral domain with field of fractions K, and let Λ be an R-algebra. Then for any two Λ-modules M and N and any prime ideal $\wp$, if M is finitely presented, we get*

$$R_\wp \otimes_R Hom_\Lambda(M, N) \simeq Hom_{R_\wp \otimes_R \Lambda}(R_\wp \otimes_R M, R_\wp \otimes_R N).$$

Proof. Since M is finitely presented, there are integers n_1 and n_2 and an exact sequence

$$\Lambda^{n_2} \xrightarrow{\varphi} \Lambda^{n_1} \longrightarrow M \longrightarrow 0.$$

Since $R_\wp$ is flat over R also

$$\Lambda_\wp^{n_2} \xrightarrow{\varphi_\wp} \Lambda_\wp^{n_1} \longrightarrow M_\wp \longrightarrow 0$$

is an exact sequence where $\Lambda_\wp = R_\wp \otimes_R \Lambda$, $M_\wp = R_\wp \otimes_R M$ and $\varphi_\wp = \mathrm{id} \otimes \varphi$. We shall use the notation and remarks preceding Lemma 6.5.7. Then

$$0 \longrightarrow Hom_\Lambda(M, N) \longrightarrow Hom_\Lambda(\Lambda^{n_1}, N) \xrightarrow{\varphi^*} Hom_\Lambda(\Lambda^{n_2}, N)$$

and

$$0 \longrightarrow Hom_{\Lambda_\wp}(M_\wp, N_\wp) \longrightarrow Hom_{\Lambda_\wp}(\Lambda_\wp^{n_1}, N_\wp) \xrightarrow{\varphi_\wp^*} Hom_{\Lambda_\wp}(\Lambda_\wp^{n_2}, N_\wp)$$

are exact. Since $R_\wp$ is flat as R-module,

$$0 \longrightarrow R_\wp \otimes_R Hom_\Lambda(M, N) \longrightarrow R_\wp \otimes_R Hom_\Lambda(\Lambda^{n_1}, N) \xrightarrow{\mathrm{id}\otimes\varphi^*} R_\wp \otimes_R Hom_\Lambda(\Lambda^{n_2}, N)$$

is exact as well. Now, $Hom_\Lambda(\Lambda^{n_1}, N) \simeq N^{n_1}$ and $Hom_\Lambda(\Lambda^{n_2}, N) \simeq N^{n_2}$. Since φ is given by multiplication by a matrix F with coefficients in Λ, φ^* is given by multiplication by the transposed matrix F^{tr}. Hence, denoting by α_1 and α_2 the natural isomorphisms $N_\wp^{n_1} \simeq R_\wp \otimes_R N^{n_1}$, respectively $N_\wp^{n_2} \simeq R_\wp \otimes_R N^{n_2}$, we get a commutative diagram

$$\begin{array}{ccccccc} 0 & \longrightarrow & Hom_{\Lambda_\wp}(M_\wp, N_\wp) & \longrightarrow & N_\wp^{n_1} & \xrightarrow{\varphi_\wp^*} & N_\wp^{n_2} \\ & & & & \simeq \downarrow \alpha_1 & & \simeq \downarrow \alpha_2 \\ 0 & \longrightarrow & R_\wp \otimes_R Hom_\Lambda(M,N) & \longrightarrow & R_\wp \otimes_R N^{n_1} & \xrightarrow{\mathrm{id}\otimes\varphi^*} & R_\wp \otimes_R N^{n_2}. \end{array}$$

Since kernels are unique up to isomorphism, there is an isomorphism

$$R_\wp \otimes_R Hom_\Lambda(M,N) \simeq Hom_{R_\wp \otimes_R \Lambda}(R_\wp \otimes_R M, R_\wp \otimes_R N)$$

as required. □

Corollary 6.5.11. *Let R be a Noetherian integral domain with field of fractions K, and let Λ be a Noetherian R-algebra. Then for any two Λ-modules M and N, where M is finitely generated, any positive integer i, and any prime ideal $\wp$ of R we get*

$$R_\wp \otimes_R Ext^i_\Lambda(M,N) \simeq Ext^i_{R_\wp \otimes_R \Lambda}(R_\wp \otimes_R M, R_\wp \otimes_R N).$$

Proof. We first observe that for Noetherian R-algebras Λ, any finitely generated Λ-module M is finitely presented, and moreover all its syzygies $\Omega^i(M)$ for all $i \in \mathbb{N}$ are finitely presented. If

$$0 \longrightarrow \Omega_\pi(M) \longrightarrow P_M \longrightarrow M \longrightarrow 0$$

is an exact sequence, then by Lemma 6.5.7

$$0 \longrightarrow R_\wp \otimes_R \Omega_\pi(M) \longrightarrow R_\wp \otimes_R P_M \longrightarrow R_\wp \otimes_R M \longrightarrow 0$$

is still an exact sequence, and if P_M is a projective Λ-module, then $R_\wp \otimes_R P_M$ is a projective $R_\wp \otimes_R \Lambda$-module. Hence, by Schanuel's Lemma 6.2.10 there are projective $R_\wp \otimes_R \Lambda$-modules P_1 and P_2 such that,

$$P_1 \oplus (R_\wp \otimes_R \Omega_\pi(M)) \simeq P_2 \oplus \Omega_{\mathrm{id}\otimes\pi}(R_\wp \otimes_R M)$$

By the definition of $Ext^1_\Lambda(M,N)$, and the Change of Rings Theorem 6.5.10 we get

$$R_\wp \otimes_R Ext^1_\Lambda(M,N) \simeq Ext^1_{R_\wp \otimes_R \Lambda}(R_\wp \otimes_R M, R_\wp \otimes_R N).$$

By induction on the degree i we get

$$R_\wp \otimes_R Ext^i_\Lambda(M,N) \simeq Ext^i_{R_\wp \otimes_R \Lambda}(R_\wp \otimes_R M, R_\wp \otimes_R N).$$

This proves the statement. □

6.6 Exercises

Exercise 6.1 (with solution in Chapter 9). Let A be a ring and let I be a Noetherian ideal of A. Suppose that $f : A \to A$ is a ring homomorphism with $f(I) = I$.

a) Let g be the restriction of f to I and denote $g^1 := g$ as well as $g^n := g \circ g^{n-1}$ for all $n \geq 2$. Show that there is n_0 such that $ker(g^n) = ker(g^{n_0})$ for all $n \geq n_0$.
b) Let $m \in ker(g)$. Show that there is $m_0 \in I$ with $g^{n_0}(m_0) = m$.
c) Show that $m_0 \in ker(g^{n_0+1})$ and deduce that $m = 0$.
d) Show that g is bijective.
e) Deduce that if A is a Noetherian ring and if $f : A \to A$ is a surjective ring homomorphism, then f is bijective.
f) Find a ring B having the following properties:
   - B is not Noetherian
   - B allows a homomorphism $\varphi : B \longrightarrow B$ which is surjective but not injective.

Exercise 6.2. Let A be a commutative ring. In this exercise we show a theorem due to Cohen: If all prime ideals of A are finitely generated, then A is Noetherian.

a) Let Γ be the set of ideals of A which are not finitely generated. Show the $\Gamma \neq \emptyset \Leftrightarrow A$ is not Noetherian.
b) Suppose in the following that A is not Noetherian. Use Zorn's lemma to show that there is a maximal element I of Γ.

   Suppose in the following that there exists $x, y \in A$ such that $x \notin I$ and $y \notin I$ but $xy \in I$.
c) Show that $I + Ay$ is finitely generated, by $u_1, u_2, \dots, u_n, y$, say, where $u_1, \dots, u_n \in I$.
d) Show that $K := \{a \in A|\ ay \in I\}$ is an ideal of A and that $x \in K$.
e) Deduce that K is finitely generated, by $v_1, v_2, \dots, v_m$ say.
f) Show that I is generated by $u_1, u_2, \dots, u_n, v_1y, v_2y, \dots, v_my$.
g) Why is this a contradiction and how does this show Cohen's theorem?

Exercise 6.3. Let R be a commutative ring and let A be an R-algebra.

a) (Nakayama's lemma) If M is a finitely generated A-module and if $I \subseteq \mathrm{rad}(A)$ is an ideal in the Jacobson radical of A (cf. Exercise 1.3), show that $I \cdot M = M \Rightarrow M = 0$. (NB: take a minimal set of generators G of M, and express $g \in G$ as element of $I \cdot M$. Use then Exercise 1.3).
b) Let R be in addition a local ring and let M be a finitely generated projective R-module. Use a) to show that M is free.

Exercise 6.4. Let R be a commutative ring and let A be an R-algebra.

a) Show that if A is artinian, then $\mathrm{rad}(A)$ is nilpotent. We call the smallest integer n with $\mathrm{rad}(A)^n = 0$ the *Loewy length* of A.
b) Show that for all A-modules M we get $\mathrm{rad}(A) \cdot M \subseteq \mathrm{rad}(M)$.
c) Show that for simple A-modules S we have $\mathrm{rad}(A) \cdot S = 0$.

d) Suppose that A is artinian. If the Loewy length of A is n, show that for all $0 < k < n$ the Loewy length of $A/\operatorname{rad}(A)^k$ is k.
e) Show that an artinian algebra A with $\operatorname{rad}(A) = 0$ is semisimple, and has a composition series.
f) (Hopkins) Suppose that A is artinian. Show by induction on the Loewy length that A is Noetherian.

Exercise 6.5. Let A be an artinian semisimple algebra over some artinian commutative ring R. Adapt the proof of Theorem 1.2.30 to verify that A is isomorphic to a finite direct product of matrix algebras over skew fields.

Exercise 6.6. Let G be a finite group and let R be a commutative ring. Denoting by R the trivial RG-module. Then $\epsilon : RG \longrightarrow R$ with $\epsilon(g) = 1$ for all $g \in G$ defines an RG-module epimorphism.
a) Show that $\ker(\epsilon)$ is the ideal of RG generated by $1 - g$ for $g \in G$. Call this the augmentation ideal $I_R(G)$.
b) Use this to prove the isomorphism of *Ext*-groups and degree 1 group cohomology
$$Ext^1_{RG}(R, M) = H^1(G, M)$$
for each RG-module M.
c) Show that $RG \otimes_R RG$, where G acts by left multiplication on the left factor, is a projective RG-module.
d) Show that $\partial_1 : RG \otimes_R RG \longrightarrow I_R(G)$ defined by $\partial_1(g \otimes h) := g(h - 1)$ is an RG-module epimorphism.
e) Use this to show the isomorphism of Ext^2-groups and degree 2 group cohomology
$$Ext^2_{RG}(R, M) = H^2(G, M)$$
for all RG-modules M.

Exercise 6.7. Let $G = C_n$ be the cyclic group of order n, let $c \in C_n$ be a generator, and consider the group ring $\mathbb{Z}C_n$ over the integers. Let $T := \mathbb{Z}$ be the trivial $\mathbb{Z}C_n$-module, i. e. the additive structure is just $\mathbb{Z}$, and every $g \in G$ acts on T as multiplication by 1.
a) Show that there is a $\mathbb{Z}C_n$-module epimorphism $\mathbb{Z}C_n \xrightarrow{\pi} T$ with kernel $\Omega^1_\pi(\mathbb{Z})$ being the principal ideal $\mathbb{Z}C_n \cdot (1 - c)$.
b) Find an epimorphism $\mathbb{Z}C_n \xrightarrow{\tau} \Omega^1_\pi(T)$ such that $\ker(\tau) \simeq T$.
c) Deduce that
$$Ext^{2k}_{\mathbb{Z}C_n}(\mathbb{Z}, M) \simeq Ext^2_{\mathbb{Z}C_n}(\mathbb{Z}, M)$$
and
$$Ext^{2k+1}_{\mathbb{Z}C_n}(\mathbb{Z}, M) \simeq Ext^1_{\mathbb{Z}C_n}(\mathbb{Z}, M)$$
for all positive integers k and all $\mathbb{Z}C_n$-modules M.

d) Use Exercise 6.6 to show that $H^1(C_n, T) = 0$ and $H^2(C_n, T) = \mathbb{Z}/n\mathbb{Z}$. Find an expression for $Ext^2_{\mathbb{Z}C_n}(\mathbb{Z}, M)$ and for $Ext^1_{\mathbb{Z}C_n}(\mathbb{Z}, M)$ for each $\mathbb{Z}C_n$-module M. Find a $\mathbb{Z}C_n$-module M with $H^1(C_n, M) \neq 0$.

NB: Modules N with $\Omega^k(N) \simeq N$ for some integer k are called periodic of period k. There is an extensive literature about such modules.

Exercise 6.8 (The snake lemma). Let A be a ring, let M_1, M_2, M_3 and N_1, N_2, N_3 be A-modules. Let $\alpha_1 : M_1 \longrightarrow M_2$, resp. $\beta_1 : N_1 \longrightarrow N_2$ be a monomorphism, suppose that $M_3 = \operatorname{coker}(\alpha_1)$, resp. $N_3 = \operatorname{coker}(\beta_1)$, and let α_2, rep. β_2 be the canonical epimorphism.

a) If $\gamma_1 : M_1 \longrightarrow N_1$ and $\gamma_2 : M_2 \longrightarrow N_2$ are homomorphisms such that $\gamma_2 \circ \alpha_1 = \beta_1 \circ \gamma_1$, show that there is a unique morphism $\gamma_3 : M_3 \longrightarrow N_3$ such that $\gamma_3 \circ \alpha_2 = \beta_2 \circ \gamma_2$.
b) Show that the restriction of α_1 to $\ker(\gamma_1)$ is a monomorphism $\kappa_1 : \ker(\alpha_1) \longrightarrow \ker(\alpha_2)$ and the restriction of α_2 to $\ker(\gamma_2)$ is a homomorphism $\kappa_2 : \ker(\alpha_2) \longrightarrow \ker(\alpha_3)$.
c) Show that applying β_1 to representatives of the residue classes in $\operatorname{coker}(\gamma_1)$ is a homomorphism $\delta_1 : \operatorname{coker}(\alpha_1) \longrightarrow \operatorname{coker}(\alpha_2)$ and applying β_2 to the representatives of the residue classes in $\operatorname{coker}(\gamma_2)$ is an epimomorphism $\delta_2 : \operatorname{coker}(\alpha_2) \longrightarrow \operatorname{coker}(\alpha_3)$.
d) We hence get the commutative diagram

$$\begin{array}{ccccc}
\ker(\gamma_1) & \xhookrightarrow{\kappa_1} & \ker(\gamma_2) & \xrightarrow{\kappa_2} & \ker(\gamma_3) \\
\downarrow & & \downarrow & & \downarrow \\
M_1 & \xhookrightarrow{\alpha_1} & M_2 & \xtwoheadrightarrow{\alpha_2} & M_3 \\
\downarrow{\scriptstyle \gamma_1} & & \downarrow{\scriptstyle \gamma_2} & & \downarrow{\scriptstyle \gamma_3} \\
N_1 & \xhookrightarrow{\beta_1} & N_2 & \xtwoheadrightarrow{\beta_2} & N_3 \\
\downarrow & & \downarrow & & \downarrow \\
\operatorname{coker}(\gamma_1) & \xrightarrow{\delta_1} & \operatorname{coker}(\gamma_2) & \xtwoheadrightarrow{\delta_2} & \operatorname{coker}(\gamma_3)
\end{array}$$

Take $x \in \ker(\gamma_3)$, its image m_x in M_3, a preimage $m(x) \in M_2$ of m_x, and consider $\gamma_2(m(x))$. Show that there is $n_x \in N_1$ with $\gamma_2(m(x)) = \beta_1(n_x)$.
e) Consider the mapping $x \in \ker(\gamma_3)$ to the image $n(x) \in \operatorname{coker}(\gamma_1)$ of n_x. Show that this mapping can be chosen to be a well-defined A-module homomorphism σ.

Show that the sequence

$$\ker(\gamma_1) \xhookrightarrow{\kappa_1} \ker(\gamma_2) \xrightarrow{\kappa_2} \ker(\gamma_3) \xrightarrow{\sigma} \operatorname{coker}(\gamma_1) \xrightarrow{\delta_1} \operatorname{coker}(\gamma_2) \xtwoheadrightarrow{\delta_2} \operatorname{coker}(\gamma_3)$$

is exact.

Exercise 6.9. Let A be an algebra over some commutative ring R and let M be an A-module. A submodule N of M is called *small* if the only submodule H of M with $H + N = M$ is the module $H = M$. The *radical* of an A-module M is the intersection of the maximal submodules of M. It is denoted by $\mathrm{rad}(M)$.

a) Show that if M is a finitely generated A-module, then any submodule of $\mathrm{rad}(M)$ is small.
b) Let M be an A-module. A *projective cover* of M is a couple (P, ρ) where P is a projective A-module and $\rho : P \longrightarrow M$ is an epimorphism, such that $\ker(\rho)$ is small. Show that if the Krull-Schmidt theorem holds for finitely generated projective A-modules and if M is Noetherian and artinian, then M has a projective cover. Moreover, show that the projective cover of M and of $M/\mathrm{rad}(M)$ coincide.

7 Some algebraic number theory

DAS WIEDERSEHEN
Ein Mann, der Herrn K. lange nicht gesehen hatte, begrüßte ihn mit den Worten: "Sie haben sich gar nicht verändert." "Oh!" sagte Herr K. und erbleichte.

Bertolt Brecht, aus: Bertolt Brecht, Werke. Große kommentierte Berliner und Frankfurter Ausgabe, Band 18: Prosa 3. © Bertolt-Brecht-Erben / Suhrkamp Verlag 1995.

MEETING AGAIN
A man, who hasn't met Mister K. for a long time, saluted him saying: "You haven't changed at all". "Oh my God !" Mister K. answered and turned pale.

Bertold Brecht[1]

Since we aim to introduce notions of integral representation theory we also need to recall some statements and concepts from algebraic number theory. We remind the reader however that algebraic number theory is a very well developed field and the results we state in this section is by no means meant to replace a course in algebraic number theory. Instead, we only prove those parts and those results which we will either need for the sequel, or which illustrate results in the non commutative setting to be developed in Chapter 8 below.

7.1 Some supplements on algebraic integers

We have already seen some properties of algebraic integers in Definition 2.4.5. We shall need a few supplements in the sequel. In particular, if K is a finite extension of $\mathbb{Q}$, then $\mathrm{algint}_{\mathbb{Z}}(K)$ is a ring, even finitely generated. This implies that its additive structure is quite simple.

Lemma 7.1.1. *Let K be a finite field extension of $\mathbb{Q}$. Then $\mathrm{algint}_{\mathbb{Z}}(K)$ is free of finite rank $\dim_{\mathbb{Q}}(K)$ as abelian group.*

Proof. Since K is a finite extension of $\mathbb{Q}$, also $R := \mathrm{algint}_{\mathbb{Z}}(K)$ is finitely generated as a ring. Hence $R = \mathbb{Z}[s_1, \dots, s_k]$ for some integer k and some elements $s_1, \dots, s_k \in K$. By Lemma 2.4.3, $\mathbb{Z}[s_1]$ is a finitely generated abelian group. Since the additive group of K is $\mathbb{Z}$-torsion free, i. e. there is no non zero element in the additive group of K which is annihilated by some integer $\mathbb{Z}$, the principal structure theorem on finitely generated modules over principal ideal domains shows that $\mathbb{Z}[s_1]$ is a free abelian group of finite rank. Then we apply inductively Lemma 2.4.3 to the extension $\mathbb{Z}[s_1, \dots, s_i][s_{i+1}]$ for any $i \in \{1, \dots, k-1\}$. Each of the extension $\mathbb{Z}[s_1, s_2, \dots, s_i][s_{i+1}]$ is a finitely generated $\mathbb{Z}[s_1, \dots, s_i]$-module since s_{i+1} is an integral element over $\mathbb{Z}$, whence integral over

1 in der Übersetzung von Susanne Brennecke, translation by Susanne Brennecke.

https://doi.org/10.1515/9783110702446-007

this bigger ring as well. Since by induction $\mathbb{Z}[s_1, s_2, \dots, s_i]$ is a finitely generated free abelian group, also $\mathbb{Z}[s_1, s_2, \dots, s_i][s_{i+1}] = \mathbb{Z}[s_1, s_2, \dots, s_i, s_{i+1}]$ is a finitely generated abelian group. As in the case $\mathbb{Z}[s_1]$ we see that $\mathbb{Z}[s_1, s_2, \dots, s_i, s_{i+1}]$ is a finitely generated free abelian group. We see that R is a finitely generated free abelian group. Now, since $\mathbb{Q} \otimes_{\mathbb{Z}} R = K$ as $\mathbb{Q}$-vector spaces, we are done. □

The proof used iterated algebraic extensions. Here is a useful remark concerning this method.

Lemma 7.1.2. *If S is an integral extension of R, and if T is an integral extension of S, then T is an integral extension of R. In particular, the ring of algebraic integers is integrally closed.*

Proof. The proof is exactly the same as for towers of algebraic extensions of fields. □

We shall use some explicit examples. Most important will be cyclotomic fields.

Lemma 7.1.3. *For a prime p and a primitive p-th root of unity ζ_p in the complex numbers, the ideal $\wp$ of $\mathbb{Z}[\zeta_p]$ generated by $(1-\zeta_p)$ is a maximal ideal that contains p and $\wp^{p-1} = p\mathbb{Z}[\zeta_p]$. Further, $\frac{(1-\zeta_p^k)}{(1-\zeta_p^j)}$ is an invertible element in $\mathbb{Z}[\zeta_p]$ for all relatively prime integers k, j.*

Proof. We have

$$X^p - 1 = \prod_{i=0}^{p-1} (X - \zeta_p^i)$$

and therefore

$$X^{p-1} + X^{p-2} + \cdots + X + 1 = \prod_{i=1}^{p-1} (X - \zeta_p^i).$$

For $X = 1$ we get

$$(*): \quad p = \prod_{i=1}^{p-1} (1 - \zeta_p^i).$$

Now, for any i, which is not divisible by p, the ideal generated by $(1 - \zeta_p^i)$ equals I. Indeed, let k, j be integers not divisible by p, then there is an integer t with p divides $k - jt$. Hence

$$\frac{(1-\zeta_p^k)}{(1-\zeta_p^j)} = \frac{(1-\zeta_p^{jt})}{(1-\zeta^j)} = 1 + \zeta_p^j + \cdots + \zeta_p^{j(t-1)} \in \mathbb{Z}[\zeta_p].$$

Applying this for $(i, 1)$ and then for $(1, i)$ in the role of (k, j) gives that $(1-\zeta_p^i)$ differs from $(1-\zeta_p)$ by the unit $\frac{(1-\zeta_p^i)}{(1-\zeta_p)}$. Therefore $\wp^{p-1} = p\mathbb{Z}[\zeta_p]$. Applying the norm

$$N : \mathbb{Z}[\zeta_p] \longrightarrow \mathbb{Z}$$

to the equation $(*)$ we see that $N(\wp) = p\mathbb{Z}$, and hence $\wp \neq \mathbb{Z}[\zeta_p]$. Consider $\mathbb{Z}[\zeta_p]/\wp =: F$. Since ζ_p is congruent to 1 modulo $\wp$, we get that F is a quotient of $\mathbb{Z}$. Moreover, $p \in \wp$ shows that F is a quotient of the field $\mathbb{Z}/p\mathbb{Z}$, and hence $F = \mathbb{Z}/p\mathbb{Z}$. □

Definition 7.1.4. Let p be a prime and let k, j be integers which are prime to p. Then the unit $\frac{(1-\zeta_p^k)}{(1-\zeta_p^j)}$ of $\mathbb{Z}[\zeta_p]$ is called a *cyclotomic unit.*

Lemma 7.1.5. *We have* $\mathrm{algint}_{\mathbb{Z}}(\mathbb{Q}(\zeta_n)) = \mathbb{Z}[\zeta_n]$ *and* $\mathrm{algint}_{\mathbb{Z}}(\mathbb{Q}(\zeta_n) \cap \mathbb{R}) = \mathbb{Z}[\zeta_n + \zeta_n^{-1}]$.

Proof (sketch). We prove this fact first for a prime power $n = p^m$ for some prime p and $m \in \mathbb{N}$. We use induction here. The polynomial $1+X+\cdots+X^{p-1} = \Phi_p(X)$ is irreducible, as follows by Eisenstein's criterion after substitution $X = Y + 1$. This shows the case $m = 1$. Suppose the statement true for $m-1$. Then the minimal polynomial of a primitive p^m-th root of unity ζ_{p^m} is $\Phi_p(X^{p-1})$, as ζ_{p^m} is a root, and as the degree of the extension is $p^{m-1} \cdot (p-1)$, the Euler φ-function of p^m. Then the algebraic integers over $\mathbb{Z}[\zeta_{p^{m-1}}]$ in $\mathbb{Q}(\zeta_{p^m})$ is $\mathbb{Z}[\zeta_{p^{m-1}}][\zeta_{p^m}] = \mathbb{Z}[\zeta_{p^m}]$ and hence, by Lemma 7.1.2 we are done in this case. The general case needs some more care. For more details see [Was-97, Theorem 2.6].

Now, $\mathrm{algint}_{\mathbb{Z}}(\mathbb{Q}(\zeta_n)) = \mathbb{Z}[\zeta_n]$, and algebraic integer in $\mathbb{Q}(\zeta_n) \cap \mathbb{R}$ are also algebraic in $\mathbb{Q}(\zeta_n)$. They hence belong to $\mathbb{Z}[\zeta_n]$ and since they should be real, the coefficient of ζ^k of some element equals the coefficient of ζ^{-k}. However, $\zeta^k + \zeta^{-k} \in \mathbb{Q}[\zeta + \zeta^{-1}]$ as is easily seen by induction on k and considering $(\zeta + \zeta^{-1})^k$. □

7.2 Primary decomposition

In order to study this kind of rings we shall need some preliminary steps. A generalisation of prime ideals are given by the following notion.

Definition 7.2.1. An ideal I in a commutative ring R is
- *primary* if all zero divisors of R/I are nilpotent,
- *irreducible* if whenever I is the intersection of two ideals $I = J_1 \cap J_2$, then $I = J_1$ or $I = J_2$.

Irreducibility is a geometric concept. For further details on this link, the reader should consult any text on elementary algebraic geometry. The relation of these concepts is illustrated by the following

Proposition 7.2.2. *In a Noetherian commutative ring R every ideal is a finite intersection of irreducible ideals and every irreducible ideal is primary.*

Proof. Let $\mathfrak{X}$ be the set of ideals of R which are not a finite intersection of irreducible ideals. If $\mathfrak{X}$ is not empty, then $\mathcal{X}$ is partially ordered by inclusion. Since R is Noetherian, $\mathcal{X}$ contains a maximal element J. In particular J is reducible, and therefore $J = S_1 \cap S_2$ with ideals S_1 and S_2 and both S_1 and S_2 are strictly bigger than J. Then, S_1 and S_2 are both not in $\mathfrak{X}$, and hence S_1 and S_2 are each finite intersections of irreducible ideals. As a consequence J is a finite intersections of irreducible ideals as well. This contradicts $J \in \mathcal{X}$ and shows that $\mathcal{X} = \emptyset$.

Let I be an irreducible ideal of R and let $x \in R$ such that $x + I$ is a zero divisor of R/I. Hence there is $y \in R \setminus I$ with $xy \in I$. Then for each $n \in \mathbb{N}$,

$$Ann_R(x^n + I) := \{r \in R \mid rx^n \in I\}$$

is an ideal of R. Clearly,

$$I \subseteq Ann_R(x^n + I) \subseteq Ann_R(x^{n+1} + I)$$

for all $n \in \mathbb{N}$, and since R is Noetherian, there is $n_0 \in \mathbb{N}$ such that

$$Ann_R(x^n + I) = Ann_R(x^{n_0} + I)$$

for each $n \geq n_0$. Let $a \in x^{n_0}R \cap yR$. Then

$$a = x^{n_0}b = yc$$

for some $b, c \in R$. Hence $xa = x^{n_0+1}b = xyc \in I$. But this shows

$$b \in Ann_R(x^{n_0+1} + I) = Ann_R(x^{n_0} + I)$$

which in turn implies $a = bx^{n_0} \in I$. Hence $I = (x^{n_0}R + I) \cap (yR + I)$. Since I is irreducible and since $y \notin I$, we need to have $x^{n_0}R + I = I$, whence $x^{n_0}R \subseteq I$, which shows that $x + I$ is nilpotent in R/I. □

More generally we define primary submodules. Here we follow Bourbaki [Bour-85, Chapitre 4].

Definition 7.2.3. Let R be a commutative ring and let M be an R-module. An R-submodule N of M is called *primary* if whenever $x \notin N$ and for an $a \in A$, then $ax \in N$ implies there is $t \in \mathbb{N}$ with $a^tM \subseteq N$.

A prime ideal $\wp$ of R is called *associated to M* if there is $x \in M \setminus \{0\}$ such that $\wp$ equals $Ann_R(x) := \{r \in R \mid rx = 0\}$. The set of associated primes is denoted $Ass(M)$.

If Q is a submodule of M such that $Ass(M/Q) = \{\wp\}$, then Q is called *$\wp$-primary*.

A primary ideal is hence a primary submodule of the regular R-module.

Definition 7.2.4. Let R be a Noetherian commutative ring and let M be an R-module and N a submodule of M. A *primary decomposition* of N is a finite family of submodules $Q_1, \ldots, Q_n$ of M such that Q_i is primary with respect to M for all $i \in \{1, \ldots, n\}$, and such that $N = Q_1 \cap Q_2 \cap \cdots \cap Q_n$.

This primary decomposition is called *reduced* if none of the primary submodules can be left out to obtain N as intersection, and $Ass(M/Q_i) = Ass(M/Q_j) \Rightarrow i = j$.

Theorem 7.2.5. *Let R be a Noetherian ring, and let M is a finitely generated R-module. Then any submodule N admits a primary decomposition. Moreover, if M/N is in addition artinian, then there is a unique reduced primary decomposition of N.*

The existence in case of $M = R$ the regular module follows from Proposition 7.2.2. For a proof in the general case we refer to Bourbaki [Bour-85, Chapitre 4 § 2 no 3 Theorem 1] for the existence and to Bourbaki [Bour-85, Chapitre 4 § 2 no 5 Proposition 8] for the unicity. Bourbaki's treatment Bourbaki [Bour-85, Chapitre 4] is completely self-contained and a true highlight of a rigorous and pedagogical exposition. Since it appears to be difficult to improve their introduction to the subject, and since we do not really need the theorem in this generality, we just refer to Bourbaki's text.

If M/N does not have a composition series, then the primary decomposition is not unique. However, any primary decomposition can be refined to a reduced one.

Note that if $N = Q_1 \cap \cdots \cap Q_n$ is a primary decomposition in M, then M/N embeds into $M/Q_1 \oplus \cdots \oplus M/Q_n$.

7.3 Discrete valuation rings

We now come to the concept of a discrete valuation ring.

Localisation is one of the most employed tool in algebra. Localising rings of algebraic integers leads naturally to discrete valuation rings.

7.3.1 Ideal structure of discrete valuation rings

Definition 7.3.1. An integral domain R is called a *discrete valuation ring* if R is a principal ideal domain with exactly one maximal ideal.

In particular, discrete valuation rings are by definition integral domains. Let R be a discrete valuation ring, then by definition there is a unique maximal ideal $\mathfrak{m}$, which in addition is principal, whence $\mathfrak{m} = tR$ for some $t \in R$. Moreover, since R is an integral domain, any other element t_1 of R with $tR = t_1R$ differs from R by a unit u of R, i. e. $t_1 = ut$. We call t a *uniformiser* of R. Since tR is maximal, $k := R/tR$ is a field. Moreover, since principal ideal rings are always Noetherian (Example 6.1.2. item 2 can be adapted easily), any discrete valuation ring is Noetherian as well.

Lemma 7.3.2. *Let R be a discrete valuation ring with uniformizer $t \in R$. Then*

$$\bigcap_{n\in\mathbb{N}} t^nR = \{0\}.$$

Proof. Let T be an ideal of R. If $tT = T$, we get $T = t^sT$ for all s, and therefore, since $T \subseteq R$, also $t^sT \subseteq t^sR$ and finally $T \subseteq \bigcap_{n\in\mathbb{N}} t^nR$. Let $I := \bigcap_{n\in\mathbb{N}} t^nR$. Since by hypothesis R

is a principal ideal domain, let $xR = I$ for some $x \in R$. By Proposition 7.2.2 $I = L_1 \cap \cdots \cap L_n$ for irreducible ideals $L_1, \ldots, L_n$, which are also primary. But then R/I is a subring of $R/L_1 \times \cdots \times R/L_n$. Since $x \in t^m R$ for each m, there are elements $x_m \in R$ such that $t^m x_m = x \in I \setminus \{0\}$. Hence, t is a zero divisor in each of the rings R/L_i for $i \in \{1, \ldots, n\}$. Since each L_i is irreducible, hence primary, t is nilpotent in each of the rings R/L_i for $i \in \{1, \ldots, n\}$. Therefore there is some integer s such that $t^s \in L_1 \cap \cdots \cap L_n = I$. But then $t^s R = \bigcap_{n \in \mathbb{N}} t^n R$. In particular $t^s R = t^{s+1} R$ and hence $R = tR$, using that R is an integral domain. This shows that t is a unit, which gives a contradiction to the fact that t is a uniformizer generating the unique maximal ideal. □

Lemma 7.3.3. *Let R be a discrete valuation ring with uniformizer t. Then for any ideal I of R there is an integer n such that $I = t^n R$.*

Proof. By Lemma 7.3.2 there is an integer n such that $t^n R \subseteq I \subsetneq t^{n-1} R$. Now $k = R/tR \simeq t^{n-1}R/t^n R$ is a field, and therefore $t^n R = I$. □

7.3.2 Valuations

The behaviour of discrete valuation rings has some similarities with that of polynomial rings. Just as polynomial rings can be embedded in a very convenient bigger ring, namely the power series ring, discrete valuation rings can be completed into a bigger ring with many very convenient properties.

First, consider a discrete valuation ring R with field of fractions K and uniformizer t. Then for all $x \in K$ with $x \neq 0$ there is $s \in \mathbb{Z}$ such that $t^s x \in R$. Indeed, let $x = \frac{a}{b}$ for $a \in R$ and $b \in R \setminus \{0\}$. Then we only need to show that there is $s \in \mathbb{Z}$ such that $\frac{t^s}{b} \in R$. However, bR is an ideal of R and Lemma 7.3.2 shows that there is $s \in \mathbb{Z}$ such that $t^s R \subseteq bR \subsetneq t^{s-1} R$ and hence $\frac{t^s}{b} \in R$.

Definition 7.3.4. Let R be a discrete valuation ring with uniformizer t and field of fractions K. Then we define for any element $r \in K$ by

$$v(r) := \sup\{n \in \mathbb{Z} \mid r \cdot t^{-n} \in R\}.$$

Of course, $v(0) = \infty$ and if $r \neq 0$, then $v(r) \in \mathbb{Z}$. Moreover, $v(t) = 1$. We call the map $2^{-v} : K \longrightarrow \mathbb{R} \cup \{\infty\}$ the *valuation* of R.

Lemma 7.3.5. *Let R be a discrete valuation ring with uniformizer t and field of fractions K. Let 2^{-v} be the valuation defined by t. Then for all $x, y \in K$*
1. $v(x) = \infty \Leftrightarrow x = 0$
2. $v(x + y) \geq min(v(x), v(y))$
3. $R = \{x \in K \mid v(x) \geq 0\}$
4. $v(xy) = v(x) + v(y)$, *where we define* $n + \infty = \infty + \infty = \infty$ *for all* $n \in \mathbb{Z}$
5. $v(x^{-1}) = -v(x)$ *for all* $x \neq 0$
6. $R^\times = \{x \in K \mid v(x) = 0\}$.

Proof. If $v(x) = \infty$, then $x \in t^n R$ for all $n \in \mathbb{N}$ and hence $x = 0$ by Lemma 7.3.2. Since by definition $v(0) = \infty$, this shows item 1.

Let $v(x) \geq v(y)$. Then, since we get $x \cdot t^{-v(x)} \in R$ and $y \cdot t^{-v(y)} \in R$, $x \cdot t^{-v(y)} \in R$. Hence $(x + y) \cdot t^{-v(y)} \in R$. This shows item 2.

We need to show item 3. If $x \in R = t^0 R$, then $v(x) = \sup\{n \in \mathbb{Z} \mid x \in t^n R\} \geq 0$. Hence $R \subseteq \{x \in K \mid v(x) \geq 0\}$. We need to prove the other inclusion. Let $x \in K \setminus \{0\}$ with $v(x) < 0$. Then $xt^{-v(x)} \in R$, and $xt^{-v(x)-1} \notin R$. Hence $x \in t^{v(x)}R \setminus t^{v(x)+1}R$. Since $v(x) < 0$, we have $x \notin t^s R$ for some $s \leq 0$. Since $t \in R$, we have $R \subseteq t^s R$, and hence $x \notin R$. This shows $R = \{x \in K \mid v(x) \geq 0\}$ and we proved item 3.

Consider $x \neq 0$. Then $x \cdot t^{-v(x)} \in R$ and $x \cdot t^{-v(x)-1} \notin R$ by definition of $v(x)$. If $x \cdot t^{-v(x)} \notin R^\times$, then $x \cdot t^{-v(x)}$ belong to the unique maximal ideal tR, and hence $x \cdot t^{-v(x)} \in tR$. Therefore, $x \cdot t^{-v(x)-1} \in R$, a contradiction. This shows $x \cdot t^{-v(x)} \in R^\times$ for all non zero x in the field of fractions of R.

Since $x \cdot t^{-v(x)} \in R^\times$ and $y \cdot t^{-v(y)} \in R^\times$, we get

$$xy \cdot t^{-(v(x)+v(y))} = x \cdot t^{-v(x)} \cdot y \cdot t^{-v(y)} \in R^\times.$$

Hence $v(xy) = v(x) + v(y)$. This shows item 4.

Now, $v(1) = 0$ by definition and item 3, and therefore

$$0 = v(1) = v(xx^{-1}) = v(x) + v(x^{-1})$$

by item 4. This implies $v(x^{-1}) = -v(x)$ for all non zero x which shows item 5.

Moreover, if $v(x) = 0$, then $x \in R^\times$ by item 3 and item 5. Conversely, if $v(x) > 0$, then $v(x^{-1}) < 0$ and by consequence $x^{-1} \notin R$. This shows item 6.

We proved the lemma. □

We shall define more generally a valuation on some field K following the most elegant approach from [CasFro-67]. Let $\mathbb{R}^{\geq 0}$ be the set of non negative real numbers.

Definition 7.3.6. Let K be a field. A *valuation* on K is a map $|\ | : K \longrightarrow \mathbb{R}^{\geq 0}$ such that

- $|x| = 0 \Leftrightarrow x = 0$
- $|xy| = |x||y|$
- There is a constant C such that $|x + 1| \leq C$ for those $x \in K$ which satisfy $|x| \leq 1$.

The valuation is *non archimedean* if we can choose $C = 1$. A valuation which is not non archimedean is archimedean.

Note that for $\mathbb{R}$ the absolute value is an archimedean valuation. The second axiom implies that $|1| = 1$. Two valuations $|\ |_1$ and $|\ |_2$ are *equivalent* if there is $0 \neq c \in \mathbb{R}^{\geq 0}$ with $|x|_1 = |x|_2^c$ for all $x \in K$. Since $|1| = 1$, we always have $C \geq 1$. Note that for any valuation is equivalent to a valuation for which we can assume that $C = 2$.

By Lemma 7.3.5 for a discrete valuation ring R the valuation 2^{-v} is a non archimedean valuation. Putting $|x| = 1$ for all non zero $x \in K$ is a valuation, the trivial valuation. *We exclude from now on the trivial valuation.*

We have two important results describing all valuations we shall need in the sequel. We shall not prove these results since we actually do not need these statements.

Theorem 7.3.7 (Ostrowski). *Up to equivalence the only non trivial valuations on $\mathbb{Q}$ are*
- *for a prime p of $\mathbb{Z}$ the p-adic valuation $|\ |_p$ coming from the discrete valuation ring $\mathbb{Z}_p$, which are all non archimedean valuations*
- *and the absolute value on $\mathbb{Q}$. which is the only archimedean valuation.*

Theorem 7.3.8 (Gelfand-Tornheim). *Let K be a subfield of $\mathbb{C}$. Then any archimedean valuation on K is equivalent to the restriction of the absolute value on $\mathbb{C}$.*

We fix $\mathcal{P}_{na} := \{|\ |_p \ :\ p\mathbb{Z} \in Spec(\mathbb{Z})\}$ the non archimedean valuations on $\mathbb{Q}$, and $\mathcal{P}_a := \{|\ |\}$ the unique archimedean valuation on $\mathbb{Q}$. We just use these two sets, and do not use that these are all the possible valuations. Here we denote by $Spec(R)$ the set of prime ideals of an integral domain R. Occasionally, by abuse of notation, for principal ideal domains such as $\mathbb{Z}$, we use the notion $Spec(R)$ also for the set of prime elements of R.

Lemma 7.3.9. *Suppose that L is a field with a valuation $|\ |_L$ and suppose that we can choose the constant C to be 2. Then $|x+y|_L \le |x|_L + |y|_L$ for any two elements $x, y \in L$.*

Proof. Assume that $C = 2$. But then for all $x, y \in L$ we have $|x|_L \le |y|_L$ or $|x|_L \ge |y|_L$. Suppose without loss of generality that we are in the first case. Then $|xy^{-1}|_L \le 1$ and hence

$$|x+y|_L = |y|_L \cdot \left|1 + \frac{x}{y}\right|_L \le 2 \cdot |y|_L$$

Hence, for a sequence $x_1, x_2, \ldots$ in L we have

$$\left|\sum_{i=1}^{2^r} x_i\right|_L \le 2^r \cdot \max\{|x_i|_L \mid i \in \{1, \ldots, 2^r\}\}.$$

Let n be any integer and choose r such that $2^{r-1} < n \le 2^r$. Then adding $2^r - n$ terms 0, we get

$$\left|\sum_{i=1}^{n} x_i\right|_L \le 2^r \cdot \max\{|x_i|_L \mid i \in \{1, \ldots, r\}\} \le 2n \cdot \max\{|x_i|_L \mid i \in \{1, \ldots, r\}\}.$$

In particular, when all $x_i = 1$ we have $|x_i|_L = 1$ and we obtain

$$(\dagger) \quad |n|_L \le 2n|1|_L = 2n$$

for any integer $n > 0$. Therefore

$$|x+y|_L^n = \left|\sum_{j=1}^{n} \binom{n}{j} x^j y^{n-j}\right|_L$$

$$\leq 2(n+1)\cdot \max\left\{\left|\binom{n}{j}x^jy^{n-j}\right|_L \mid i \in \{1,\dots,n\}\right\}$$
$$\leq 4(n+1)\left|\binom{n}{j}\right|_L \cdot \max\{|x|_L^j|y|_L^{n-j} \mid i \in \{1,\dots,n\}\} \text{ using } (\dagger)$$
$$\leq 4(n+1)(|x|_L+|y|_L)^n$$

Hence,

$$|x+y|_L = (|x+y|_L^n)^{\frac{1}{n}}$$
$$\leq (4(n+1)(|x|_L+|y|_L)^n)^{\frac{1}{n}}$$
$$\leq (4(n+1))^{\frac{1}{n}}\cdot(|x|_L+|y|_L)$$

Since n is arbitrary, we may consider any very big n. Since $\lim_{n\to\infty}(4(n+1))^{\frac{1}{n}} = 1$ we are done. □

Recall that a *metric* d on a set X is a map $d : X\times X \longrightarrow \mathbb{R}$ with non negative values satisfying

- $d(x,y) = d(y,x)$,
- $d(x,y) = 0 \Leftrightarrow x = y$,
- $d(x,y) \leq d(x,z) + d(z,y)$ for all $x,y,z \in X$.

A metric d is an *ultrametric* if the third condition can be replaced by the stronger $d(x,y) \leq max(d(x,z),d(z,y))$ for all $x,y,z \in X$.

Lemma 7.3.10. *Let K be a field and let $|\ |$ be a valuation on the field. For all $x,y \in K$ let*

$$d(x,y) := |x-y|.$$

Then $d : K\times K \longrightarrow \mathbb{R}$ is a metric, and if $|\ |$ is a non archimedean valuation, then d is actually an ultrametric.

Proof. The values of d are non negative by definition. Lemma 7.3.5 item 2 gives $d(x,y) \leq max(d(x,z),d(z,y))$ for all $x,y,z \in K$. By definition $|-x| = |-1|\cdot|x|$ for all x and $1 = |(-1)\cdot(-1)| = |-1|^2$ gives $|-1| = 1$. Hence d is symmetric. By definition, $d(x,y) = 0 \Leftrightarrow x = y$. □

7.3.3 Completions

Metric spaces (X,d) are topological space, by putting as a basis for the topology the open balls $B_\epsilon(x) = \{y \in X \mid d(x,y) < \epsilon\}$. As is readily verified, equivalent valuations induce the same topology, and actually also the converse is true. Two valuations inducing the same topology are equivalent.

Metric fields are known from elementary analysis. They can be completed, just as real numbers are constructed from rational numbers by considering equivalence classes of Cauchy sequences.

Recall briefly the construction, in the vocabulary of quotient fields of discrete valuation domains. A *Cauchy sequence in* K is a sequence $(x_n)_{n\in\mathbb{N}}$ of elements of K such that for all real $\epsilon > 0$ there is $N(\epsilon)$ such that $d(x_n, x_m) < \epsilon$ for all $n, m > N(\epsilon)$. A sequence $(x_n)_{n\in\mathbb{N}}$ converges to 0 if for all real $\epsilon > 0$ there is $N(\epsilon)$ such that $d(x_n, 0) < \epsilon$ for all $n > N(\epsilon)$.

On the set of Cauchy sequences in K we impose an equivalence relation by saying that $(x_n)_{n\in\mathbb{N}}$ and $(y_n)_{n\in\mathbb{N}}$ are equivalent if the sequence $(x_n - y_n)_{n\in\mathbb{N}}$ converges to 0. Then the set of Cauchy sequences is a field again, called $\widehat{K}_v$. This field has the property that for each Cauchy sequence $(x_n)_{n\in\mathbb{N}}$ in $\widehat{K}_v$ there is an element $x \in \widehat{K}_v$ such that $(x_n - x)_{n\in\mathbb{N}}$ converges to 0. A proof of these facts are standard arguments and can be found in virtually any standard textbook in analysis.

We can extend v to $\widehat{K}_v$, and denote the extended valuation, for simplicity, again by v. Actually, if $(x_n)_{n\in\mathbb{N}}$ is a Cauchy sequence in K, then $(d(x_n, 0))_{n\in\mathbb{N}}$ is a Cauchy sequence with values in $\mathbb{R}$, hence converges to some value $\delta_{(x_n)_{n\in\mathbb{N}}}$. Put

$$v((x_n)_{n\in\mathbb{N}}) := -\log_2(\delta_{(x_n)_{n\in\mathbb{N}}}).$$

Now, put

$$\widehat{R}_v := \{x \in \widehat{K}_v \mid v(x) \geq 0\}.$$

This is a ring, using Lemma 7.3.5 and the fact that v is continuous on K, and the extension of the valuation v to $\widehat{K}$ is a continuous extension of the valuation on K.

Further, there is an embedding ι of K into $\widehat{K}_v$ by mapping $x \in K$ to the constant sequence in $\widehat{K}_v$. The image of K in $\widehat{K}_v$ is dense. Indeed, if $(x_n)_{n\in\mathbb{N}}$ is a Cauchy sequence with values in K, and let $x \in \widehat{K}_v$ be the limit of $(x_n)_{n\in\mathbb{N}}$. Then for all $\epsilon > 0$ there is $N(\epsilon)$ with $2^{-v(x_n - x_m)} < \epsilon$ for all $n, m > N(\epsilon)$. Hence, in the field $\widehat{K}_v$ we have

$$2^{-v(\iota(x_n) - x)} = \lim_{m\to\infty} 2^{-v(x_n - x_{n+m})} \leq \epsilon$$

and so for any $x \in \widehat{K}_v$ and any ϵ-neighbourhood of x contains an element in the image of ι.

Corollary 7.3.11. *Let A be a finite dimensional K-algebra. Then there is a dense embedding $\iota : A \longrightarrow \widehat{K}_v \otimes_K A$.*

Indeed, this follows by taking a K-basis, and using that K is dense in $\widehat{K}_v$. □

Suppose now that the image of v is a subset of $\mathbb{Z}$. Then $\widehat{R}_v$ is again local. Indeed,

$$\widehat{\mathfrak{m}}_v := \{x \in \widehat{K}_v \mid v(x) > 0\}$$

is an ideal, again using Lemma 7.3.5, and $\{x \in \widehat{K}_v \mid v(x) = 0\}$ are the units of $\widehat{R}_v$. This follows from the fact that the valuation is continuous, and that $v(K)$ is the discrete set $\mathbb{Z}$. Hence also $v(\widehat{K}_v) = \mathbb{Z}$. Take $t \in \widehat{K}$ with $v(t) = 1$. Then by Lemma 7.3.5 and the continuity of v again, we see that $\widehat{R}_v$ is a local principal ideal domain and the ideals of $\widehat{R}_v$ are precisely the principal ideals $t^n\widehat{R}_v$ for $n \in \mathbb{N}$.

By definition $\widehat{R}_v$ is a discrete valuation domain again.

Definition 7.3.12. We call $\widehat{R}_v$ the *completion of R at v*. A discrete valuation domain is *complete* if it is isomorphic to its completion.

For simplicity, denote by $\widehat{K}_\wp$ the completion of the field of fractions K of an integral domain R with respect to some prime ideal $\wp$.

Definition 7.3.13. Let K be a field and let $|\ |$ be a valuation. Then the valuation $|\ |$ is *discrete* if the possible values $\{\log_2(|x|) \mid x \in K \setminus \{0\}\}$ form a discrete subset of $\mathbb{R}$. In any case we call $\{\log_2(|x|) \mid x \in K \setminus \{0\}\}$ *the value group* of $|\ |$.

Since $|xy| = |x| \cdot |y|$ for all $x, y \in K$ the value group of a valuation is indeed a subgroup of $(\mathbb{R}, +)$. Moreover, the value group is a discrete subset of $\mathbb{R}$ if and only if the valuation is discrete.

Lemma 7.3.14. *Let K be a finite extension of $\mathbb{Q}$ and let $|\ |$ be a non archimedean valuation on K. Let $R := \{x \in K \mid |x| \le 1\}$ and $\wp := \{x \in K \mid |x| < 1\}$. Then R is a ring, $\wp$ is the unique maximal ideal of R, and $R \setminus \wp = R^\times$. Moreover, $|\ |$ is discrete if and only if $\wp$ is principal. If $\wp$ is principal, then all non zero ideals of R are of the form $\wp^t$ for some t, and hence are principal as well.*

Proof. The fact that R is a ring and that $\wp$ is an ideal follow directly from the definition of a valuation. Note that $|1| = |1|^2$, and hence $|1| = 1$ and $1 \in R$. Further, $R^\times = \{x \in R \mid |x| = 1\}$. Indeed, if $x \in R^\times$, then there is $y \in R$ with $xy = 1$. Hence $|x| \cdot |y| = |xy| = |1| = 1$ and since both $x, y \in R$, hence $|x| \le 1$ and $|y| \le 1$, we get $|x| = |y| = 1$. Further, if $|x| = 1$, then the $y \in K$ with $xy = 1$ also satisfies $|y| = 1$. Hence, also $y \in R$ and $x \in R^\times$. Since $R^\times = R \setminus \wp$, the ideal $\wp$ is the ideal of non units, and hence the unique maximal ideal of R. The ring R is hence local.

Suppose that $|\ |$ is discrete. Then let $s := \sup\{|x| \mid x \in R \setminus R^\times\}$. Since $|\ |$ is discrete, $s < 1$ and there is π with $|\pi| = s$. We claim that $\wp = \pi R$. Clearly, $\pi R \subseteq \wp$. If $x \in R$ with $|x| < 1$. Then there is $t \in \mathbb{N}$ such that $|\pi^{t+1}| < |x| \le |\pi^t|$. Therefore

$$s = |\pi| = \left|\frac{\pi^{t+1}}{\pi^t}\right| < \left|\frac{x}{\pi^t}\right| \le 1$$

and hence $x\pi^{-t} \in R$. If $|x| < |\pi^t|$, we get $|\pi^{-t} \cdot x| < 1$, and hence $s < |\pi^{-t} \cdot x| < 1$ which contradicts the maximality of s. Hence $|x\pi^{-t}| = 1$ and $u = x \cdot \pi^{-t} \in R^\times$. This unit u satisfies $\pi^t \cdot u = x$. If now I is an ideal of R, then by the same argument there is $x \in I$ such that $|x|$ is maximal possible, and by the above, the principal ideal generated by

x is $\wp^t$ for some t. If $y \in I$ is another element, then $|x \cdot y^{-1}| \le 1$ and hence $x \cdot y^{-1} \in R$. Since the valuation is discrete, there is a unit u of R and $s \in \mathbb{N}$ such that $y = x\pi^s u$. This shows that all non zero ideals of R are of the form $\wp^t$ for some integer t.

If $\wp = \pi R$ is principal, then $s := |\pi| < 1$ and $s = \sup\{|x| \mid x \in R \setminus R^\times\}$. Let $t \neq 0$ be an accumulation point in the set $\{|x| \mid x \in K \setminus \{0\}\}$. Then there are $x, y \in K$ with (without loss of generality) $|x| > |y| > 0$ and $\frac{s}{2} > \frac{x}{y}$. But then $|xy^{-1}| < 1$ and hence $xy^{-1} \in R$. However, since all elements in $\wp$ are multiples of π, all elements z of R are either units (and satisfy therefore $|z| = 1$) or are in $\wp$, and satisfy therefore $|z| < s$. However $xy^{-1} \in R$ is not a unit and $|xy^{-1}| > s$. This is a contradiction. Hence the valuation is discrete. □

In case R is a complete discrete valuation ring with maximal ideal $\wp$, and if $R/\wp$ is a finite field, then we have a very nice description of completions.

Lemma 7.3.15. *Let R be a complete discrete valuation ring with maximal ideal $\wp = \pi R$, and suppose that $R/\wp$ is a finite field. Let $R/\wp = \{0 + \wp, 1 + \wp, r_2 + \wp \cdots, r_n + \wp\}$ and define $r_0 := 0$ and $r_1 := 1$. Then for any $x \in R$ there is a uniquely defined sequence $(x_i)_{i \in \mathbb{N}}$ with $x_i \in \{r_0, \ldots, r_n\} =: S$ and*

$$x = \sum_{n=0}^{\infty} x_i \pi^i .$$

The ring structure is given by the usual ring structure laws from power series rings.

Proof. Let $x \in R$. Then $x + \wp \in R/\wp$ and there is $x_0 \in S$ with $x - x_0 \in \wp = \pi R$. Hence $\frac{x - x_0}{\pi} \in R$. We claim that we already have constructed $x_0, x_1, \ldots, x_n \in S$ with

$$x - \sum_{i=0}^{n} x_i \pi^i \in \pi^{n+1} R.$$

Then

$$\frac{x - \sum_{i=0}^{n} x_i \pi^i}{\pi^{n+1}} \in R$$

and

$$\left(\frac{x - \sum_{i=0}^{n} x_i \pi^i}{\pi^{n+1}} \right) + \wp \in R/\wp .$$

Let $x_{n+1} \in S$ with

$$\frac{x - \sum_{i=0}^{n} x_i \pi^i}{\pi^{n+1}} - x_{n+1} \in \wp = \pi R$$

and therefore

$$\frac{x - \sum_{i=0}^{n+1} x_i \pi^i}{\pi^{n+2}} \in R.$$

The fact that the ring structure is given by the same formulas as for power series rings comes from the fact that up to fixed degree we have multiplication as in polynomial rings. This shows the lemma. □

We need to recall a result which is well-known from analysis, but which also holds in our more general setting.

Recall that for a field K, which is equipped with a valuation $|\ |$, and a vector space V a *norm on* V is a map $\|\ \|: V \longrightarrow \mathbb{R}$ such that $\| v \| > 0$ for all $v \in V \setminus \{0\}$, such that $\| xv \| = |x| \cdot \| v \|$ for all $v \in V$ and $x \in K$, and such that $\| v + w \| \leq \| v \| + \| w \|$ for all $v, w \in V$.

Lemma 7.3.16. *Let K be a complete field and let V be a finite dimensional K-vector space. Then any two norms on V are equivalent.*

Proof. The well-known proof from classical analysis, using that any norm is equivalent to the L_∞-norm, also know as the max-norm, carries over. (Cf. [CasFro-67, Chapter II, Section 8] for more details.) □

Proposition 7.3.17. *Let R be a complete discrete valuation ring with maximal ideal $\wp = \pi R$, and suppose that $R/\wp$ is a finite field. Then R is compact with respect to the topology defined by $|\ |$. Moreover, the field of fraction K of R is locally compact.*

Proof. Let $R = \bigcup_{\lambda\in\Lambda} U_\lambda$ with open sets U_λ in R. Let again

$$R/\wp = \{0 + \wp, 1 + \wp, r_2 + \wp \cdots, r_n + \wp\},$$

define $r_0 := 0$ and $r_1 := 1$, and put $\{r_0, \ldots, r_n\} =: S$. If we cannot cover R by opens U_λ for λ in a finite subset of Λ then

$$R = \bigcup_{i=0}^{n} (r_i + \pi R)$$

there is $x_0 \in S$ with $x_0 + \pi R$ is not covered by only finitely many open sets in the family $(U_\lambda)_{\lambda\in\Lambda}$. Similarly, there is $x_1 \in S$ such that $x_0 + x_1\pi + \pi^2 R$ is not covered by only finitely many open sets in the family $(U_\lambda)_{\lambda\in\Lambda}$. By an induction there is a sequence $(x_i)_{i\in\mathbb{N}}$ such that $\sum_{i=0}^{n} x_i\pi^i$ is not covered by only finitely many open sets in the family $(U_\lambda)_{\lambda\in\Lambda}$. Let now by Lemma 7.3.15 be

$$x := \sum_{i=0}^{\infty} x_i\pi^i \in R$$

Then there is $\lambda_0 \in \Lambda$ with $x \in U_{\lambda_0}$. Since U_{λ_0} is open, by the definition of a metric topology there is $m \in \mathbb{N}$ such that $x + \pi^m R \subseteq U_{\lambda_0}$. This is a contradiction to the construction of x_{m+1}.

We need to show that K is locally compact, i. e. we need to show that each point has a compact neighborhood. Since the topological spaces R and $\pi^m R$ are homeomorphic for all $m \in \mathbb{Z}$ (with the homeomorphism given by multiplication by π^m), we see that for each $z \in K$ there is $m \in \mathbb{Z}$ with $z \in \pi^m R$. This proves the result. □

7.3.4 Extension of valuations

For a finite field extension L over K we denote by $N_K^L : L \longrightarrow K$ the (algebraic) norm map. It associates to any $x \in L$ the determinant of the K-linear endomorphism on L given by multiplication with x.

Proposition 7.3.18. *Let K be a finite field extension of $\mathbb{Q}$ and let $|\ |$ be a valuation on K. Suppose that K is locally compact with respect to the topology given by the valuation $|\ |$. Let L be a finite field extension of K. If $\dim_K L = N$, then $|\ |_L := |N_K^L(\ )|^{\frac{1}{N}}$ is the unique valuation on L such that for all $x \in K$ we have $|x|_L = |x|$. We say that $|\ |_L$ is the unique extension of $|\ |$ to L.*

Proof. For the unicity we use an elementary argument from analysis (or normed vector spaces). Given a valuation $\widetilde{|\ |}_L$ on L. Then $(L, \widetilde{|\ |}_L)$ is a normed K-vector space. Since $\dim_K(L) < \infty$, any two norms $\widetilde{|\ |}_{L,1}$ and $\widetilde{|\ |}_{L,2}$ on L are equivalent. Hence they induce the same topology on L. Any two valuations inducing the same topology are equivalent as valuations. Therefore there is $0 \neq c \in \mathbb{R}$ such that

$$\widetilde{|\ |}_{L,1} = (\widetilde{|\ |}_{L,2})^c.$$

If we consider the extension not only up to equivalence, but a true extension restricting to the original valuation, then for all $x \in K$ we get

$$(\widetilde{|x|}_{L,2})^c = \widetilde{|x|}_{L,1} = |x| = \widetilde{|x|}_{L,2}$$

and we obtain $c = 1$.

We show that $|\ |_L$ is indeed by a valuation. The fact that $|x|_L \geq 0$ for all $x \in L$ is trivial, such as $|xy|_L = |x|_L \cdot |y|_L$ for all $x, y \in L$. If $|x|_L = 0$, then $N_K^L(x) = 0$, and this implies $x = 0$ by definition of the norm. Hence, the only remaining property is to show the third property of a norm.

But now, we fix any norm $||\ ||$ on the finite dimensional K-vector space L. Since K is locally compact by hypothesis, the unit sphere $B_1^{||\ ||} := \{x \in L \mid ||x|| = 1\}$ is compact. Since $|\ |_L$ is a continuous function on the compact set $B_1^{||\ ||}$, and since $B_1^{||\ ||}$ does not contain 0, there are real positive constants Δ and δ such that

$$\Delta \geq |x|_L \geq \delta > 0 \quad \forall x \in B_1^{||\ ||}.$$

Hence, since for any non zero $x \in L$ the element $x \cdot ||x||^{-1}$ is in $B_1^{||\ ||}$, we obtain

$$\Delta \geq \frac{|x|_L}{||x||} \geq \delta > 0 \quad \forall x \in L.$$

If now $x \in L$ with $|x|_L \leq 1$, then

$$||x|| \leq \frac{|x|_L}{\delta} \leq \frac{1}{\delta}$$

and hence for all $x \in L$ with $|x|_L \leq 1$ we get

$$
\begin{aligned}
|1+x|_L &\leq \Delta \cdot (||1+x||) \\
&\leq \Delta \cdot (||1|| + ||x||) \quad \text{since } || \; || \text{ is a norm and satisfies triangle equality} \\
&\leq \Delta \cdot \left(1 + \frac{1}{\delta}\right) =: C
\end{aligned}
$$

where $C > 1$. □

Remark 7.3.19. Note that this result holds for archimedean valuations as well as for non archimedean valuations.

Recall that the field of fractions of a complete discrete valuation ring is locally compact by Proposition 7.3.17.

For fields which are not locally compact we have a slightly more complicated situation.

Proposition 7.3.20. *Let L be a separable extension of the field K and suppose that $\dim_K(L) = N < \infty$. Then for any valuation $| \; |_K$ on K there are at most N extensions of $| \; |_K$ to valuations $| \; |_{L,1}, \cdots, | \; |_{L,s}$ of L. If $\widehat{K}$ is the completion of K with respect to $| \; |_K$, then*

$$\widehat{K} \otimes_K L \simeq L_1 \oplus \cdots \oplus L_s$$

for complete fields $L_1, \ldots, L_s$ with respect to the valuations $| \; |_{L,1}, \cdots, | \; |_{L,s}$. This isomorphism is an isomorphism of algebras and a homeomorphism of topological spaces.

Proof. Since the degree of L over K is finite, there is a primitive element $x \in L$, i. e. $L = K(x)$. Let $f(X)$ be the minimal polynomial of x over K. Then $L = K[X]/f(X)K[X]$ and since tensor product with $\widehat{K}$ over K preserves injections ($\widehat{K}$ is flat over K since K is a field and hence $\widehat{K}$ contains a K-basis), and since $\widehat{K} \otimes_K K[X] = \widehat{K}[X]$,

$$\widehat{K} \otimes_K L \simeq \widehat{K}[X]/f(X)\widehat{K}[X].$$

Now, since L is separable over K, there are no multiple roots of f in any algebraic closure. Hence since the degree of f is N the polynomial $f(X)$ decomposes in $\widehat{K}[X]$ into at most N distinct irreducible factors. Therefore

$$\widehat{K}[X]/f(X)\widehat{K}[X] \simeq L_1 \oplus \cdots \oplus L_s$$

for field extensions $L_1, \ldots, L_s$ of $\widehat{K}$. Note that this is actually the Wedderburn decomposition of the semisimple algebra $\widehat{K} \otimes_K L$. This shows the algebraic isomorphism. We shall need to show the topological part. Since for all i the field L_i is a finite extension of the complete field $\widehat{K}$, each of the finite dimensional $\widehat{K}$-vector spaces L_i is again complete. By Proposition 7.3.18 the valuation $| \; |_K$ on $\widehat{K}$ extends uniquely to a valuation $| \; |_{L,i}$ on L_i.

We need to show that these are the only possible extensions of $|\ |_K$. Suppose that $|\ |_K$ is non archimedean. Let $||\ ||$ be a valuation of L extending $|\ |_K$. Then, by continuity $||\ ||$ extends to a real valued function $\widetilde{||\ ||}$ on $\widehat{K} \otimes_K L$. Again by continuity, $\widetilde{||\ ||}$ satisfies

$$\widetilde{||x+y||} \leq \max(\widetilde{||x||}, \widetilde{||y||})$$

and

$$\widetilde{||x \cdot y||} = \widetilde{||x||} \cdot \widetilde{||y||}$$

for all $x, y \in \widehat{K} \otimes_K L$. Since

$$\widehat{K} \otimes_K L \simeq L_1 \oplus \cdots \oplus L_s,$$

for field extensions $L_1, \ldots, L_s$ of $\widehat{K}$ and of L, we may restrict $\widetilde{||\ ||}$ to each of the L_i. If there is $x \in L_i$ such that $\widetilde{||x||} \neq 0$, then for all non zero $y \in L_i$ we have

$$0 \neq \widetilde{||x||} = \widetilde{||y||} \cdot \widetilde{||xy^{-1}||}$$

and hence the restriction of $\widetilde{||\ ||}$ to L_i is either identical 0, or a non zero non archimedean valuation on L_i. If e_i is the primitive idempotent corresponding to L_i in the Wedderburn decomposition of $\widehat{K} \otimes_K L$, then $e_i e_j = 0$ whenever $i \neq j$. Hence

$$0 = \widetilde{||e_i e_j||} = \widetilde{||e_i||} \cdot \widetilde{||e_j||}$$

which implies that the restriction of $\widetilde{||\ ||}$ to either L_i is 0 or to L_j is 0. This shows that there is a unique $i_0 \in \{1, \ldots, s\}$ such that the restriction of with $\widetilde{||\ ||}$ to L_{i_0} is non zero. Such a non zero valuation is hence an extension of the valuation $|\ |_K$ on $\widehat{K}$ to L_{i_0}. Since these extensions are unique, we get that the restriction of $\widetilde{||\ ||}$ to L_{i_0} to L_{i_0} equals $||\ ||_{L,i_0}$.

The case of archimedean valuations is completely analogous.

We still need to show that the algebraic isomorphism is also a homeomorphism. For $x := (x_1, \ldots, x_s) \in L_1 \oplus \cdots \oplus L_s = \widehat{K} \otimes_K L$ we put

$$||x||_0 := \max(\{||x_1||_{L,1}, \ldots, ||x_s||_{L,s}\}$$

which comes from the fact that $x = \sum_{i=1}^{s} x_i$. Then $||\ ||_0$ is a norm on the finite dimensional $\widehat{K}$-vector space $\widehat{K} \otimes_K L$. By Lemma 7.3.16 any two norms are equivalent, since $\widehat{K}$ is complete. Hence the topologies are the same. □

By Proposition 7.3.18 and Proposition 7.3.20 for finite field extension K of $\mathbb{Q}$

- for any prime number p there are finitely many equivalence classes of valuations $|\ |_{p,1}, \ldots, |\ |_{p,n_p}$ extending $|\ |_p$ on $\mathbb{Q}$
- and there are finitely many equivalence classes of archimedean valuations $|\ |_{\infty,1}, \ldots, |\ |_{\infty,n_\infty}$ extending the absolute value $|\ |$ on $\mathbb{Q}$.

We let

$$\mathcal{P}_{na}(K) := \{|\ |_{p,1}, \dots, |\ |_{p,n_p} \mid p \in Spec(\mathbb{Z})\}$$

the non archimedean valuations on K and

$$\mathcal{P}_a(K) := \{|\ |_{\infty,1}, \dots, |\ |_{\infty,n_\infty}\}$$

the archimedean valuations on K. Finally, let

$$\mathcal{P}(K) := \mathcal{P}_{na}(K) \cup \mathcal{P}_a(K)$$

the set of (representatives of equivalence classes) of valuations on K. We also call these the *places* on K (cf. Definition 7.6.4 below).

7.4 Fractional ideals

For integral domains a sort of arithmetic of ideals is possible, and will be most useful for Dedekind domains, to be introduced in Definition 7.5.1 below. In Section 8.5.2 we shall see that fractional ideals and valuations are closely related for a special class of algebras. This link is going to be of high importance for us. We shall now introduce this concept of fractional ideals.

Definition 7.4.1. Let R be an integral domain and let K be its field of fractions. Then K is an R-module. A *fractional ideal* of R is a finitely generated R-submodule of K.

Let F_1 and F_2 be fractional ideals of R, then the module sum $F_1 + F_2$ is a fractional ideal of R, as well as the product $F_1 \cdot F_2$ being defined as the R-submodule of K generated by the elements $f_1 f_2$, where $f_1 \in F_1$ and $f_2 \in F_2$, and where the product is taken inside K. Moreover, if F_1 and F_2 are fractional ideals of R, then let

$$(F_1 : F_2) := \{r \in K \mid r \cdot F_2 \subseteq F_1\}.$$

Lemma 7.4.2. *Let R be an integral domain. For a fractional ideal F of R there is a non zero element $d \in R$ such that $d \cdot F$ is an ideal (in the usual sense) of R.*

Proof. For any $x \in K$ the set $xR = \{xr \mid r \in R\}$ is a fractional ideal of R. Let F be a fractional ideal of R, the ideal F is finitely generated as R-module, and hence there are $x_1, \dots, x_n$ in K such that

$$F = Rx_1 + \dots + Rx_n.$$

Since K is the field of fractions of R, there are non zero elements $d_1, \dots, d_n$ in R such that $d_i x_i \in R$ for all $i \in \{1, \dots, n\}$. Hence with $d := d_1 \cdot d_2 \cdot \dots \cdot d_n$ we get $d \cdot F \subseteq R$. □

Definition 7.4.3. A fractional ideal F of an integral domain R is *invertible* if there is another fractional ideal G of R such that $F \cdot G = R$.

Lemma 7.4.4. *Let R be an integral domain with field of fractions K. Then for all fractional ideals F of R we get*

$$(R : F) \cdot F \subseteq R$$

and if F is an invertible fractional ideal, then

$$(R : F) \cdot F = R$$

Proof. First, by the Lemma 7.4.2 we get that there is a non zero element $d \in R$ such that $dF = I$ is an ideal of R. Moreover,

$$z \in (R : dI) \Leftrightarrow zdI \leq R \Leftrightarrow dz \in (R : I) \Leftrightarrow z \in \frac{1}{d}(R : I)$$

Hence

$$d \cdot (R : dI) = (R : I)$$

This shows that we may assume that $F = I$ is a usual ideal of R. Let $z \in (R : I)$ and $t \in I$, then by definition $zt \in R$. Hence

$$(R : I) \cdot I \subseteq R.$$

If I is an invertible fractional ideal, and $J \cdot I = R$ for another fractional ideal. Then also $I \cdot J = R$ and hence

$$(R : I) = (R : I) \cdot R = (R : I) \cdot (I \cdot J) = ((R : I) \cdot I) \cdot J \subseteq R \cdot J = J.$$

But since $J \cdot I = R$ we also get that $J \subseteq (R : I)$, since $(R : I)$ is precisely the set of elements x of K with $xI \subseteq R$. Hence we get equality. □

Corollary 7.4.5. *In an integral domain the set of invertible fractional ideals form an abelian group with group law being multiplication of ideals.*

Proof. Indeed, multiplication of fractional ideals is clearly associative and commutative, the neutral element is R, and by Lemma 7.4.4 for an invertible fractional ideal F the inverse is $(R : F)$. □

Fractional principal ideals, i. e. fractional ideals generated by a single element, of integral domains R are invertible. Indeed, let x be a non zero element of the field of fractions of R, then $x \cdot (R : Rx) = (R : R) = R$ by the proof of Lemma 7.4.4 and hence

$$xR \cdot (R : Rx) = xR \cdot \frac{1}{x}R = R$$

which shows that xR is invertible. Since for any non zero $x, y \in R$ we get $xR \cdot yR = xyR$, the class of fractional principal ideals form a subgroup of the group of invertible fractional ideals of R.

Proposition 7.4.6. *Let R be an integral domain with field of fractions K. Then a fractional ideal J is invertible if and only if J is projective. In this case J is finitely generated.*

Proof. Let J be an invertible fractional ideal. Then there is another fractional ideal J' such that $J \cdot J' = R$. Hence there are $a_1, \ldots, a_n \in J$ and $b_1, \ldots, b_n \in J'$ with $\sum_{i=1}^n a_i b_i = 1$. define $f_i : J \longrightarrow R$ given by $f_i(a) = ab_i$. This gives an R-module homomorphism

$$J \xrightarrow{\begin{pmatrix} f_1 \\ \vdots \\ f_n \end{pmatrix}} R^n$$

and an R-module morphism

$$R^n \xrightarrow{\rho} J$$

given by

$$\rho\left(\begin{pmatrix} x_1 \\ \vdots \\ x_n \end{pmatrix}\right) = \sum_{i=1}^n x_i a_i.$$

Since for any $a \in J$ we get

$$a = a \cdot 1 = a \cdot \sum_{i=1}^n a_i b_i = \sum_{i=1}^n f_i(a) \cdot a_i,$$

we have

$$\rho \circ \begin{pmatrix} f_1 \\ \vdots \\ f_n \end{pmatrix} = \mathrm{id}_J$$

and J is finitely generated projective.

Suppose that J is a projective fractional ideal. Then there is an R-module homomorphism $f : J \longrightarrow R^I$ for some index set I and an R-module homomorphism $\sigma : R^I \longrightarrow J$ such that $\sigma \circ f = \mathrm{id}_J$. As usual there is a non zero $c \in R$ such that $cJ \subseteq R$ is an integral ideal. For any $x, y \in J$ we have

$$f(cxy) = cxf(y) = cyf(x)$$

and hence $xf(y) = yf(x)$. Therefore for any two non zero $x, y \in J$ we get

$$\frac{1}{x} f(x) = \frac{1}{y} f(y) \in K^I.$$

Let $q_i = (\frac{1}{y}f(y))_i$ be the component of index $i \in I$. Since for a non zero $x \in J$ only a finite number of coefficients $f(x)_i$ is non zero. Therefore there are $q_1, \ldots, q_n \in K$ with

$$x = (\sigma \circ f)(x) = \sigma\left(\begin{pmatrix} f_1(x) \\ \vdots \\ f_n(x) \end{pmatrix}\right).$$

However, with $a_i = \sigma(e_i)$ with e_i being the element of R^n such that $(e_i)_k = 0$ whenever $k \neq i$ but $(e_i)_i = 1$, we get

$$x = \sum_{i=1}^{n} f_i(x)a_i = \sum_{i=1}^{n} xq_ia_i = x\sum_{i=1}^{n} q_ia_i.$$

Therefore,

$$1 = \sum_{i=1}^{n} q_ia_i.$$

Since the image of σ is J we get that J is generated as an ideal by $a_1, \ldots, a_n$. If we define the ideal J' to be the R-submodule of K generated by $q_1, \ldots, q_n$, then $JJ' = R$ and J is invertible. □

Definition 7.4.7. Let R be an integral domain. Then the quotient of the group of invertible fractional ideals of R modulo the subgroup of principal fractional ideals of R is the *ideal class group of* R, denoted by $Cl_I(R)$.

We shall come back to this group in Corollary 8.5.14. In particular the ideal class group of the ring of algebraic integers in finite extensions of $\mathbb{Q}$ is a very classical object in algebraic number theory. In this case we shall prove in Corollary 8.5.14 that $Cl_I(R)$ is a finite group. Its order is called the *class number*. Besides algebraic methods also analytic number theory is used in some cases to determine this group. For an algorithm to compute class numbers of algebraic number fields we refer to Cohen, Diaz y Diaz and Oliver [CoDiOl-97].

7.5 Dedekind domains

What are the integral domains for which
- every fractional ideal is invertible?
- the localisation at every prime ideal is principal?
- each ideal is an essentially unique product of prime ideals?

These properties are all equivalent. The integral domains which have one of these properties are called Dedekind domains. Most useful, the ring of integers in finite extensions of $\mathbb{Q}$ are Dedekind domains. Hence, the integral domains which occur in Wedderburn components of integral group rings of finite groups are of this kind. This section will provide proofs of these statements and some of the properties of Dedekind domains and their modules. As usual, our choice is guided by the needs of the further treatment.

For the definition we follow the monograph of Auslander-Buchsbaum [AuBu-74].

Definition 7.5.1. A commutative ring R is a *Dedekind domain* if it is an integral domain which is in addition hereditary.

Recall from Theorem 6.4.4 that an integral domain R is a Dedekind domain if and only if each non zero ideal of R is a projective module, and this is equivalent to the statement that each submodule of a projective R-module is again projective.

However, the fact that R is an integral domain implies that each projective module M is faithful. Indeed, take a non zero $x \in M$. Then Rx is actually projective and since there is an epimorphism $R \longrightarrow Rx$, given by $r \mapsto rx$, the module Rx is (isomorphic to) a direct factor of R, whence a non zero ideal. Since R does not have zero divisors, Rx is faithful. Since modules with a faithful submodule are also faithful, all non zero submodules of projective R-modules are faithful.

Example 7.5.2. A principal ideal domain is Dedekind. Indeed, if I is an ideal, then $I = Ra$ for some $a \in R$. Then $I \simeq R$ as R-module, since R is an integral domain, and hence multiplication by a is invertible. Hence each ideal is projective.

Lemma 7.5.3. *Any Dedekind domain is Noetherian.*

Proof. Let I be any non zero ideal of R. Then I is projective as R-module, and hence I is a direct factor of some free module $F := \bigoplus_{i\in J} R$, for some index set J. Let $\iota : I \longrightarrow F$ be the injection and let $\rho : F \longrightarrow I$ be the projection with $\rho \circ \iota = \mathrm{id}_I$. We denote $\iota(x) = (x_j)_{j\in J}$ for coefficients $x_j \in R$ such that $|\{j \in J \mid x_j \neq 0\}| < \infty$. Then, $I \ni x \mapsto x_j \in R$ defines an R-module homomorphism $f_j : I \longrightarrow R$. Further let e_k be the element in F with $(e_k)_k = 1$, and $(e_k)_j = 0$ whenever $j \neq k$. Let $m_k := \rho(e_k)$ for all $k \in J$. Then for all $x \in I$ we get

$$x = (\rho \circ \iota)(x) = \rho(x_j)_{j\in J} = \sum_{j\in J} f_j(x)\rho(e_j) = \sum_{j\in J} x_j m_j.$$

Note that

$$(\dagger) \quad y \cdot f_j(x) = f_j(yx) = f_j(xy) = x f_j(y)$$

for all $x, y \in I$. Then $J_1(x) := \{j \in J \mid x_j \neq 0\}$ is a finite set. Hence for a non zero $x \in I$ we get

$$x = \sum_{j\in J_1(x)} f_j(x)m_j = \sum_{j\in J_1(x)} x \cdot f_j(m_j) = x \cdot \Big(\sum_{j\in J_1(x)} f_j(m_j) \Big)$$

using (†). Since R is an integral domain,

$$1 = \sum_{j\in J_1(x)} f_j(m_j).$$

For any $y \in I$ we have

$$y = y \cdot \Big(\sum_{j\in J_1(x)} f_j(m_j) \Big) = \sum_{j\in J_1(x)} y \cdot f_j(m_j) = \sum_{j\in J_1(x)} f_j(y) m_j$$

again by (†) and hence the finite set $\{m_j \mid j \in J_1(x)\}$ generates I, and hence I is finitely generated. As we have shown now that all ideals of R are finitely generated, R is Noetherian by Proposition 6.1.5. □

Lemma 7.5.4. *Let R be a Noetherian integral domain and let S be a multiplicative subset of R. Then for any ideal $I(S)$ of R_S the ideal $I := R \cap I(S)$ of R satisfies $R_S \cdot I = I(S)$.*

If $\wp$ is a prime ideal and $S = R \setminus \wp$, and if J is a prime ideal of R with $J \subseteq \wp$, then $J = (R_S \cdot J) \cap R$.

Proof. First, observe that since R is an integral domain, R is a subring of R_S. Let $I(S)$ be an ideal of R_S. Then $I := I(S) \cap R$ is an ideal of R, and therefore I is finitely generated by $g_1, \ldots, g_n$, say. Then $J(S) := R_S \cdot g_1 + \cdots + R_S \cdot g_n$ is an ideal of R_S and since $g_1, \ldots, g_n \in I = I(S) \cap R \subseteq I(S)$, we get $J(S) \subseteq I(S)$. Let $x \in I(S)$. Then $x = \frac{y}{z}$ for some $y \in R$ and $z \in S$. Hence, $zx \in R \cap I(S) = I$. Therefore $zx = \sum_{j=1}^{n} x_j g_j$ and since z is invertible in R_S, we get $x \in J(S)$. This shows $J(S) \subseteq I(S) \subseteq J(S)$.

Note that $J \subseteq \wp$ is equivalent to $S \cap J = \emptyset$. Let $(R_S \cdot J) \cap R =: L$. This is an ideal of R containing J. Let $x \in L$. Since J is a finitely generated ideal of R, generated by generators $h_1, \ldots, h_m$, say, we have $x = \sum_{i=1}^{m} x_i h_i$ for some $x_i \in R_S$. Hence there is $s \in S$ with $sx_i \in R$ for all i and therefore $sx \in J$. Since J is a prime ideal, $s \in J$ or $x \in J$. Since $J \subseteq \wp$, the case $s \in J$ contradicts the choice of s, and hence we get $x \in J$. Therefore $L = J$. This proves the statement. □

Remark 7.5.5. We encountered the case of two prime ideals $J \subseteq \wp$ of R. The maximal length of inclusions of prime ideals in R gives rise to the important notion of *Krull dimension* of a commutative ring. We do not need this concept here.

Lemma 7.5.6. *Let R be a commutative ring. If R is Noetherian, then the localisation R_S is Noetherian for each multiplicative subset S of R.*

Proof. If

$$I(S)_1 \leq I(S)_2 \leq I(S)_3 \leq \ldots$$

is an ascending chain of ideals of R_S, then

$$I(S)_1 \cap R \leq I(S)_2 \cap R \leq I(S)_3 \cap R \leq \ldots$$

is an ascending chain of ideals of R. Since R is Noetherian, there is n_0 such that $I(S)_{n_0} \cap R = I(S)_k \cap R$ for all $k > n_0$. Hence

$$I(S)_{n_0} = (I(S)_{n_0} \cap R) \cdot R_S = (I(S)_k \cap R) \cdot R_S = I(S)_k$$

for all $k > n_0$ by Lemma 7.5.4. Hence R_S is Noetherian. □

Lemma 7.5.7. *If R is a Dedekind domain, if M is a finitely generated R-module, and if $R_\wp \otimes_R M = 0$ for all prime ideals $\wp$, then $M = 0$.*

Proof. Let $x \in M \setminus \{0\}$ and let $Ann_R(M) := \{r \in R \mid rx = 0\}$. Clearly this is an ideal of R. Since $x \neq 0$, we have $1 \notin Ann_R(x)$, and therefore there is a maximal ideal $\wp$ of R with $Ann_R(x) \subseteq \wp$. Consider $\lambda_\wp : M \longrightarrow R_\wp \otimes_R M$ given by $y \mapsto 1 \otimes y$. If $\lambda_\wp(x) = 0$, then, since $1 \otimes x = 0 \otimes x$ there is $t \in R \setminus \wp$ with $t \cdot x = 0$. Hence $t \in Ann_R(x) \subseteq \wp$, a contradiction. □

Lemma 7.5.8. *Let R be a Dedekind domain and let Λ be an R-algebra which is f. g. projective as an R-module. If L is a finitely generated Λ-module such that $R_\wp \otimes_R L$ is a free $R_\wp \otimes_R \Lambda$-module for all prime ideals $\wp$ of R, then L is projective.*

Proof. Let L be a Λ-module such that $R_\wp \otimes_R L$ is a free $R_\wp \otimes_R \Lambda$-module for all prime ideals $\wp$ of R. Then for all prime ideals $\wp$ and all $R_\wp \otimes_R \Lambda$-modules $X(\wp)$ we get

$$Ext^1_{R_\wp \otimes_R \Lambda}(R_\wp \otimes_R L, X(\wp)) = 0.$$

By Corollary 6.5.11 we get

$$R_\wp \otimes_R Ext^1_\Lambda(L, M) \simeq Ext^1_{R_\wp \otimes_R \Lambda}(R_\wp \otimes_R L, R_\wp \otimes_R M) = 0.$$

By Lemma 7.5.7 this implies $Ext^1_\Lambda(L, M) = 0$ and by Lemma 6.3.4 we then get that L is projective. □

Lemma 7.5.8 motivates the following definition. For an R-algebra Λ we say that a Λ-module L is *locally free* if $R_\wp \otimes_R L$ is a free $R_\wp \otimes_R \Lambda$-module for all prime ideals $\wp$ of R. We should emphasize that locally free modules are in general not free. Examples are non principal ideals in rings of algebraic integers in finite extensions of $\mathbb{Q}$.

Lemma 7.5.9. *If R is a Dedekind domain and if $\wp$ is a prime ideal of R, then the localisation $R_\wp$ of R at $\wp$ is a principal ideal domain, whence Dedekind again.*

If R is an integral domain such that for all prime ideals $\wp$ of R we have $R_\wp$ is a principal ideal domain, then R is a Dedekind domain.

Proof. Let $I(\wp)$ be an ideal of $R_\wp$. By Lemma 7.5.4 we get $I(\wp) = R_\wp \cdot (I(\wp) \cap R)$. Since $(I(\wp) \cap R)$ is an ideal of R, whence projective, hence a direct factor of a free R-module F, also $I(\wp) = R_\wp \cdot (I(\wp) \cap R)$ is a direct factor of the free $R_\wp$-module $R_\wp F$, whence projective. Hence $R_\wp$ is a Dedekind domain. By definition $R_\wp$ is local with unique maximal ideal $\wp R_\wp$. By Exercise 6.3 every finitely generated projective module over a local ring is free.

Hence any ideal $I(\wp)$ of $R_\wp$ is free. Tensoring with the field of fractions implies that any ideal $I(\wp)$ of $R_\wp$ is free of rank 1. Hence $I(\wp)$ is generated by one element $a \in R_\wp$ which shows that $I(\wp)$ is a principal ideal domain.

Suppose that for all prime ideals $\wp$ of R the localisation $R_\wp$ is a principal ideal domain. Let I be an ideal of R. Then $I_\wp := R_\wp \otimes_R I$ is an ideal of $R_\wp$. Since $R_\wp$ is a principal ideal domain, $I_\wp$ is free. Hence I is a locally free ideal. By Lemma 7.5.8 we see that I is projective and hence R is a Dedekind domain. □

Lemma 7.5.10. *If R is a Dedekind domain, then a finitely generated module M is torsion free if and only if it is projective, if and only if it is flat.*

Proof. By Lemma 6.5.8 a module M is flat over R if and only if its localisation is flat over the localised ring at any prime ideal. By Lemma 7.5.9 if R is Dedekind, then the localisation of R at any prime ideal is a principal ideal domain. By Lemma 6.5.9 a finitely generated module M over a principal ideal domain is flat if and only if it is free, and this is equivalent with the module being torsion free. Locally free modules are projective by Lemma 7.5.8. This shows the lemma. □

Definition 7.5.11. Let R be an integral domain with field of fractions K. Then R is called *integrally closed* if all elements in K which are integral over R are actually already in R.

Example 7.5.12. By Lemma 7.1.2 we see that the ring of algebraic integers in a finite field extension of $\mathbb{Q}$ is integrally closed.

Lemma 7.5.13. *Let R be an integrally closed integral domain with field of fractions K. If $f(X)$ and $g(X)$ are monic polynomials in $R[X]$, and if $f(X) = g(X)h(X)$ in $K[X]$, then $h(X) \in R[X]$ as well.*

Proof. Let L be a finite extension of K such that $f(X)$ decomposes into a product of linear factors in $L[X]$. Let $S := \text{algint}_R(L)$ be the algebraic integers over R in L. Then

$$f(X) = g(X)h(X) = \prod_{i=1}^{\text{degree}(f)} (X - s_i)$$

for some $s_i \in L$. Since $f(s_i) = 0$ we get $s_i \in S$. Since S is an integral domain by Lemma 2.4.4, we get $g(X) \in S[X]$ and $h(X) \in S[X]$. Since $g(X) \in K[X]$ and $h(X) \in K[X]$, and since $K \cap S = R$, using that R is integrally closed, we get the result. □

Lemma 7.5.14. *Dedekind domains are integrally closed.*

Proof. We first assume that R is a principal ideal domain with field of fractions K. Let $f(X) = X^n + a_{n-1}X^{n-1} + \cdots + a_1X + a_0 \in R[X]$ be a monic polynomial and $a = \frac{u}{v} \in K$ with $f(a) = 0$. Since R is a principal ideal domain we may assume that u and v do not have any common divisor. Hence u and v are relatively prime. Then

$$\left(\frac{u}{v}\right)^n + a_{n-1}\left(\frac{u}{v}\right)^{n-1} + \cdots + a_1\left(\frac{u}{v}\right) + a_0 = 0$$

which implies

$$u^n + a_{n-1}u^{n-1}v + \cdots + a_1uv^{n-1} + a_0v^n = 0.$$

Hence

$$u^n = v(-a_{n-1}u^{n-1} - \cdots - a_1uv^{n-2} - a_0v^{n-1})$$

Since u and v are relatively prime, and since the above equation implies that v divides u^n, we need to have v is a unit in R. Hence $a \in R$.

Suppose now that R is a Dedekind domain with field of fractions K. Let $f(X) = X^n + a_{n-1}X^{n-1} + \cdots + a_1X + a_0 \in R[X]$ be a monic polynomial and $a = \frac{u}{v} \in K$ with $f(a) = 0$. Since by Lemma 7.5.9 for any prime ideal $\wp$ the ring $R_\wp$ is a principal ideal domain, we get $a \in R_\wp$ for any prime ideal $\wp$. Let $I = \{y \in R \mid yu \in Rv\}$. This is an ideal of R and if we can show that $I = R$, then $1 \in I$, which shows that $u = rv$ for some $r \in R$, which gives $x = \frac{u}{v} = r \in R$. Suppose to the contrary that $I \neq R$, and hence there is a maximal ideal $\wp$ containing I. Since $v \in I$ we have $v \in \wp$. But $a = \frac{u}{v} \in R_\wp$ by the above, and hence $v \notin \wp$. This contradiction shows that $a \in R$. □

Lemma 7.5.15. *Any non zero prime ideal of a Dedekind domain is a maximal ideal.*

Proof. By Lemma 7.5.4 there is an inclusion preserving bijection between the set of ideals $J(\wp)$ of $R_\wp$ and the set of those ideals J of R which are contained in $\wp$. The bijection is given by intersection with the smaller ring, respectively multiplication with the bigger ring. By Lemma 7.5.9 $R_\wp$ is a principal ideal domain. If $\wp$ is a prime ideal which is not maximal. Then let $\mathfrak{m}$ be a maximal ideal containing strictly $\wp$. We see that

$$R \subsetneq R_\mathfrak{m} \subsetneq R_\wp$$

is an inclusion of Dedekind domains. Then $\wp R_\mathfrak{m} = x \cdot R_\mathfrak{m}$ for some non invertible $x \in R_\mathfrak{m}$. Since $\wp R_\mathfrak{m}$ is not maximal, x is reducible in $R_\mathfrak{m}$. Therefore there is $y = \frac{a}{b}$, $z = \frac{c}{d} \in R_\mathfrak{m}$ both non invertible in $R_\mathfrak{m}$, such that $x = yz = \frac{ac}{bd}$. Hence $b, d \notin \mathfrak{m}$ and $a, c \in \mathfrak{m}$. Since then $\wp R_\wp = xR_\wp$ is a maximal ideal in $R_\wp$. Hence x is an irreducible element in $R_\wp$. Since $b, d \notin \mathfrak{m}$ we get that $b, d \notin \wp$, and $y, z \in R_\wp$. Since x is irreducible in $R_\wp$, y or z must be invertible in $R_\wp$, but then it is also invertible in $R_\mathfrak{m}$. This gives a contradiction. □

Example 7.5.16. Lemma 7.5.15 shows that Dedekind domains have Krull dimension 1. Moreover, for any field K the ring $K[X]$ is a principal ideal domain, and hence Dedekind. However, $K[X, Y]$ admits a sequence of prime ideals

$$0 \subsetneq X \cdot K[X, Y] \subsetneq X \cdot K[X, Y] + Y \cdot K[X, Y] \subsetneq K[X, Y]$$

(the quotient is an integral domain in all cases) and hence $K[X, Y]$ is not Dedekind. This corresponds to the fact that polynomials in two variables correspond to two-dimensional planes, whereas polynomial rings in one variable correspond to one-dimensional lines.

A converse to Lemma 7.5.15 is true as well under additional hypotheses. The proof unfortunately is somewhat lengthy.

Proposition 7.5.17. *A Noetherian, integrally closed integral domain such that all non zero prime ideals are maximal is a Dedekind domain.*

Proof. We will first reduce to the case of R being local. Let K be the field of fractions of R and $R_\wp$ for any prime ideal $\wp$ of R.

- If R is Noetherian, then by Lemma 7.5.6 the localisation $R_\wp$ is Noetherian as well.
- If R is integrally closed, then $R_\wp$ is integrally closed as well. Indeed, if $u \in K$ is integral over $R_\wp$, then there is a monic polynomial

$$f(X) = X^n + \frac{a_{n-1}}{s_{n-1}}X^{n-1} + \cdots + \frac{a_1}{s_1}X + \frac{a_0}{s_0}$$

  with $a_i \in R$ and $s_i \notin \wp$ for all $i \in \{0, \ldots, n-1\}$ and with $f(u) = 0$. Let $s = s_{n-1} \cdot \cdots \cdot s_0$. Then multiplying $f(u) = 0$ by s^n we see that su is integral over R. Since R is integrally closed, $su \in R$ and hence $u \in R_S$.
- If all prime ideals $\mathfrak{q}$ of R are maximal, the same is true for $R_\wp$. Indeed, if $\mathfrak{q}_\wp$ is a non zero prime ideal of $R_\wp$, then $\mathfrak{q}_\wp \cap R$ is a prime ideal of R contained in $\wp$. Since $\mathfrak{q}_\wp \cap R$ is prime, hence maximal, $\mathfrak{q}_\wp \cap R = \wp$ and therefore $\mathfrak{q}_\wp = (\mathfrak{q}_\wp \cap R)R_\wp = \wp R_\wp$ by Lemma 7.5.4.

Now, we may suppose that R is a Noetherian, integrally closed integral domain such that there is only one non zero prime ideal $\wp$. By Lemma 7.5.9 we need to see that R is a principal ideal domain.

We first show that $\wp$ is a principal ideal. Let $x \in \wp \setminus \{0\}$. Then $S := \{x^n \mid n \in \mathbb{N}\}$ is a multiplicative subset of R (containing 1 since $0 \in \mathbb{N}$). Then $R \subseteq R_S \subseteq K$. Let $\mathfrak{m}$ be a maximal ideal of R_S. If $\mathfrak{m}$ is not zero, then $R \cap \mathfrak{m}$ is a maximal ideal of R. Since $\wp$ is the only non zero prime ideal of R, we get $\wp = R \cap \mathfrak{m}$. By construction $x \in \wp$ and so $x \in \mathfrak{m}$. But x is a unit of R_S, which is a contradiction. Hence R_S is a field. This proves that $K = R_S$. Therefore, for any $z \in K$ we get $z = \frac{y}{x^n}$ for some $n \in \mathbb{N}$ and some $y \in R$. If $b \in \wp \setminus \{0\}$, then there is $r \in R$ and $n \in \mathbb{N}$ such that $\frac{1}{b} = \frac{r}{x^n}$. Hence $x^n = rb$. Therefore, for any $b \in \wp \setminus \{0\}$ there is $n \in \mathbb{N}$ with $x^n \in bR$. Since R is Noetherian, $\wp$ is finitely generated by $\{x_1, \ldots, x_n\}$, say. Since x from above was arbitrary non zero in $\wp$, we obtain integers $m_1, \ldots, m_n$ such that $x_i^{m_i} \in bR$. If $m > (n-1) \cdot \max(m_1, \ldots, m_n)$, then $\wp^m \subseteq bR$. Let $\widehat{m}$ be minimal with $\wp^{\widehat{m}} \subseteq bR$. Since $\wp^{\widehat{m}-1} \nsubseteq bR$, there is $c \in \wp^{\widehat{m}-1} \setminus bR$. But

$$c\wp \subseteq \wp^{\widehat{m}} \subseteq bR$$

we get $\frac{c}{b}\wp \subseteq R$, whereas

$$(\dagger) \quad z := \frac{c}{b} \notin R$$

since $c \notin bR$. Since $\frac{c}{b}\wp$ is an ideal of R, either $\frac{c}{b}\wp = R$ or $\frac{c}{b}\wp \subseteq \wp$. Then $z\wp \subseteq \wp$ implies that $\wp$ is an $R[z]$-module, clearly faithful and finitely generated. Let $\{u_1, \dots, u_s\}$ be a generating system of $\wp$ as $R[z]$-module. Then multiplication by z gives elements $c_{i,j} \in R$ with

$$zu_i = \sum_{j=1}^{s} c_{i,j}u_j.$$

Hence

$$(*) \quad \sum_{j=1}^{s}(z\delta_{i,j} - c_{i,j})u_j = 0,$$

where $\delta_{i,j}$ is the Kronecker symbol $\delta_{i,i} = 1$ and $\delta_{i,j} = 0$ whenever $i \neq j$. Define an epimorphism

$$\begin{aligned} R^s &\xrightarrow{\mu} \wp \\ (r_1, \dots, r_s)^{tr} &\mapsto \sum_{j=1}^{s} r_j u_j \end{aligned}$$

of R-modules. Let $\gamma : R^s \longrightarrow R^s$ be the endomorphism of R^s given by the matrix

$$D := (z\delta_{i,j} - c_{i,j})_{1 \le i,j \le s}.$$

By the equation $(*)$ there is a unique homomorphism $\text{coker}(\gamma) \xrightarrow{\rho} \wp$ satisfying $\mu = \rho \circ \nu$, where $\nu : R^s \longrightarrow \text{coker}(\gamma)$ is the natural homomorphism. Hence, since μ is an epimorphism, ρ is an epimorphism as well. Since $det(D) \cdot \text{coker}(\gamma) = 0$, we get $det(D) \cdot \wp = 0$ as well. Now, $det(D) \in R[z]$, and since $\wp$ is faithful as $R[z]$-module, $det(D) = 0$. Consider the characteristic polynomial of the matrix $(c_{i,j})_{1 \le i,j \le s}$. This is a monic polynomial with coefficients in R, and z is a root of it. Hence z is integral over R.

Since R is integrally closed, $z \in R$. This is a contradiction to (†). We therefore obtain $\frac{c}{b}\wp = R$. Since R is an integral domain, $\frac{b}{c}R = \wp$, and since $\frac{b}{c} \cdot 1 = \frac{b}{c} \in \wp \subseteq R$, we get that $\wp$ is a principal ideal.

Let I be an ideal of R. Since R is Noetherian, I is finitely generated, and let $\{y_1, \dots, y_n\}$ be a minimal set of generators. If now $\sum_{i=1}^{n} r_i y_i = 0$, and some $r_{i_0} \notin \wp$, then r_{i_0} is not in the unique maximal ideal, whence r_{i_0} is invertible. This contradicts the minimality of the set $\{y_1, \dots, y_n\}$ since we would be able to express y_{i_0} as a linear combination of the other members of the set. Hence $r_1, \dots, r_n \in \wp$. We define an epimorphism of R-modules

$$\begin{aligned} R^n &\xrightarrow{g} I \\ (r_1, \dots, r_n)^{tr} &\mapsto \sum_{i=1}^{n} r_i y_i \end{aligned}$$

and let $N = \ker(g)$.

Let $z = (r_1, \dots, r_n)^{tr} \in N$. Therefore

$$0 = g(z) = \sum_{i=1}^{n} r_i y_i$$

which shows that $r_i \in \wp$ for all $i \in \{1, \dots, n\}$. Since $\wp$ is principal, there is $a \in R$ with $\wp = aR$. Hence there are $s_1, \dots, s_n \in R$ with $r_i = as_i$ and therefore $z = a \cdot (s_1, \dots, s_n)^{tr}$. Still

$$0 = g(z) = a \cdot \sum_{i=1}^{n} s_i y_i = a \cdot g((s_1, \dots, s_n)^{tr})$$

and since R has no zero divisors, either $a = 0$ or

$$g((s_1, \dots, s_n)^{tr}) = \sum_{i=1}^{n} s_i y_i = 0.$$

Since we assumed $\wp \neq 0$ we get $\sum_{i=1}^{n} s_i y_i = 0$ and hence $(s_1, \dots, s_n)^{tr} \in N$. Therefore $z = a \cdot (s_1, \dots, s_n)^{tr} \in \wp N$ which shows that $N \subseteq \wp N$. The other inclusion is trivial, and hence $N = \wp N$. Nakayama's lemma (cf. Exercise 6.3) gives $N = 0$. Since $N = 0$, we get that I is a free submodule of R, and since R is an integral domain, I is free of rank 1, and hence principal. We are done. □

As an important application we prove the following lemma.

Lemma 7.5.18. *Let K be a finite field extension of $\mathbb{Q}$, and let $R = \mathrm{algint}_{\mathbb{Z}}(K)$ be the ring of algebraic integers in K. Then R is a Dedekind domain.*

Proof. We know that R is an integrally closed integral domain. Let $\wp$ be a non zero prime ideal of R. Then $\wp \cap \mathbb{Z}$ is a prime ideal $p\mathbb{Z}$ of $\mathbb{Z}$. If $\wp$ is not maximal, $\wp \subsetneq \mathfrak{m}$ for some maximal ideal $\mathfrak{m}$ of R. Then $\mathbb{Z} \cap \mathfrak{m}$ is again a prime ideal of $\mathbb{Z}$ containing $p\mathbb{Z}$. Hence

$$\mathbb{Z} \cap \mathfrak{m} = p\mathbb{Z} = \wp \cap \mathbb{Z}.$$

Since R is integral over $\mathbb{Z}$, also $R/\wp$ is integral over $\mathbb{Z}/p\mathbb{Z}$. We now consider the integral extension $R/\wp$ of $\mathbb{Z}/p\mathbb{Z}$. Let $a \in \mathfrak{m} \setminus \wp$. Since $R/\wp$ is integral over $\mathbb{Z}/p\mathbb{Z}$, there is a monic polynomial $f(X) = X^n + c_{n-1}X^{n-1} + \cdots + c_1 X + c_0 \in \mathbb{Z}/p\mathbb{Z}[X]$ with $f(a) = 0$. Since $\wp$ is a prime ideal, $R/\wp$ is an integral domain and hence we may suppose that $c_0 \neq 0$. However,

$$c_0 = -a \cdot (a^{n-1} - c_{n-1}a^{n-2} - \cdots - c_1) \in (\mathfrak{m}/\wp) \cap \mathbb{Z}/p\mathbb{Z} = p\mathbb{Z}/p\mathbb{Z} = 0.$$

This is a contradiction and we see that $\wp$ is a maximal ideal. By Proposition 7.5.17 and the fact that by Lemma 7.1.2 algebraic integers are integrally closed we get that R is a Dedekind domain. □

Proposition 7.5.19. *An integral domain is a Dedekind domain if and only if all non zero ideals are invertible.*

Proof. Suppose that I is a fractional ideal of the Dedekind domain R. Then there is a non zero $x \in R$ with xI is a 'honest' ideal of R. Hence xI is projective since R is a Dedekind domain. By Proposition 7.4.6 a fractional ideal is invertible if and only if it is projective. Hence xI is invertible, and therefore also I is invertible. If all fractional ideals are invertible, then all ideals are projective, again by Proposition 7.4.6 and hence R is Dedekind. □

Corollary 7.5.20. *If R is a Dedekind domain, then each localisation R_S at a multiplicatively closed subset S of R is a Dedekind domain.*

Proof. Indeed, the let $I(S)$ be a non zero ideal of R_S. Then $R \cap I(S) =: I$ is an ideal of R, whence invertible. By Lemma 7.5.4 the ideal $I(S) = R_S I$ is invertible as well. Hence by Proposition 7.5.19 R_S is Dedekind again. □

Lemma 7.5.21. *Let R be a Dedekind domain with field of fractions K. If for prime ideals $\wp_1, \dots, \wp_s$ and $\mathfrak{q}_1, \dots, \mathfrak{q}_t$ we have*

$$\wp_1 \cdot \cdots \cdot \wp_t = \mathfrak{q}_1 \cdot \cdots \cdot \mathfrak{q}_s,$$

then $s = t$ and there is $\sigma \in \mathfrak{S}_s : \wp_i \simeq \mathfrak{q}_{\sigma(i)}$ for all $i \in \{1, \dots, s\}$.

Proof. We proceed by induction on t. If $t = 1$, then a prime ideal $\wp_1$ is a product of prime ideals $\wp_1 = \mathfrak{q}_1 \cdot \cdots \cdot \mathfrak{q}_s$, which implies, by the definition of a prime ideal, that $\wp_1 = \mathfrak{q}_i$ for some i.

Suppose the statement is true for $t - 1$. Since

$$\wp_1 \supseteq \wp_1 \cdot \cdots \cdot \wp_t = \mathfrak{q}_1 \cdot \cdots \cdot \mathfrak{q}_s,$$

there is i with $\wp_1 \supseteq \mathfrak{q}_i$. By Lemma 7.5.15 the prime ideal $\mathfrak{q}_i$ is maximal and hence we get equality. We now multiply by $\wp_1^{-1}$ (by Proposition 7.5.19 in Dedekind domains all fractional ideals are invertible) and obtain

$$\wp_2 \cdot \cdots \cdot \wp_t = \mathfrak{q}_1 \cdot \cdots \cdot \mathfrak{q}_{i-1} \cdot \mathfrak{q}_{i+1} \cdot \cdots \cdot \mathfrak{q}_s.$$

By induction hypothesis $s = t$ and there is $\sigma \in \mathfrak{S}_s : \wp_i \simeq \mathfrak{q}_{\sigma(i)}$ for all $i \in \{1, \dots, s\}$. □

Proposition 7.5.22. *If R is a Dedekind domain, then any non zero ideal I of R is the product of non zero prime ideals of R. These prime ideals are unique up to isomorphism and up to permutation of factors.*

Proof. If any ideal of R is a product of prime ideals, then Lemma 7.5.21 shows the unicity.

Let $\mathcal{N}$ be the set of ideals of R which are not a product of prime ideals. If $\mathcal{N} \neq \emptyset$, since R is Noetherian, $\mathcal{N}$ contains a maximal element I (cf. Proposition 6.1.5). The ideal

I cannot be maximal, since maximal ideals are prime ideals. Let $\wp$ be a maximal ideal containing I. Then $I \subseteq \wp$ implies $\wp^{-1} \cdot I \subseteq \wp^{-1} \cdot \wp = R$. Hence $\wp^{-1} \cdot I$ is an ideal of R. By Lemma 7.4.4 we see that

$$\wp^{-1} = \{x \in K \mid x \cdot \wp \subseteq R\}.$$

This implies $R \subseteq \wp^{-1}$. Hence $I \subseteq I \cdot \wp^{-1}$ and by the maximality of I, the ideal $I \cdot \wp^{-1} = \wp_2 \cdot \cdots \cdot \wp_t$ is a product of prime ideals of R. Now, multiplying with $\wp$ we get

$$I = \wp \cdot \wp_2 \cdot \cdots \cdot \wp_t.$$

This shows the statement. □

Corollary 7.5.23. *For a Dedekind domain R, which is not a field, we have the following.*
- *If I and J are ideals of R. Then $I \subseteq J$ if and only if there is an ideal L of R with $J \cdot L = I$.*
- *Any ideal I of R is of the form $I = \wp_1^{e_1} \cdot \cdots \cdot \wp_n^{e_n}$ for pairwise different prime ideals $\wp_1, \ldots, \wp_n$ and integers $e_1, \ldots, e_n$. This expression is unique up to reordering.*

Proof. The last item follows directly from Proposition 7.5.22. For the first item we observe that

$$I = \wp_1^{e_1} \cdot \cdots \cdot \wp_n^{e_n} \subseteq \wp_1^{f_1} \cdot \cdots \cdot \wp_n^{f_n} = J$$

if and only if $e_i \geq f_i$ for all i. Hence

$$L := \wp_1^{e_1 - f_1} \cdot \cdots \cdot \wp_n^{e_n - f_n}$$

is the required ideal. □

Two ideals I and J of the Dedekind domain R. Then I and J are relatively prime if $I + J = R$.

Corollary 7.5.24. *For any Dedekind domain R and pairwise different prime ideals $\wp_1, \ldots, \wp_n$ let $I = \prod_{i=1}^n \wp_i^{e_i}$ and $J = \prod_{i=1}^n \wp_i^{f_i}$. Then*

$$I + J = \prod_{i=1}^n \wp_i^{\min(e_i, f_i)} \quad \text{and} \quad I \cap J = \prod_{i=1}^n \wp_i^{\max(e_i, f_i)}$$

Further, if $I + J = R$, then $R/IJ \simeq R/I \times R/J$.

Proof. This follows immediately from Proposition 7.5.22. □

Lemma 7.5.25. *If R is a Dedekind domain and if I is a non zero ideal of R. Then there are only finitely many prime ideals of R containing I.*

Proof. By Proposition 7.5.22 we get

$$I = \wp_1^{e_1} \cdot \dots \cdot \wp_n^{e_n}$$

for pairwise different prime ideals $\wp_1, \dots, \wp_n$ and positive integers $e_1, \dots, e_n$. This expression is essentially unique by Proposition 7.5.22 and hence the prime ideals containing I are precisely the ideals $\wp_1, \dots, \wp_n$. □

Remark 7.5.26. Recall a general statement on commutative rings, usually called *Chinese Remainder Theorem* for integers. Let I and J be ideals of a commutative ring satisfying $I + J = R$. Then $R/(I \cap J) \simeq R/I \times R/J$. Indeed, $R \ni x \overset{\pi}{\mapsto} (x + I, x + J) \in R/I \times R/J$ is a ring homomorphism with kernel $I \cap J$. Since $I + J = R$, there is $a \in I$ and $b \in J$ with $a + b = 1$. Then $\pi(ay + bx) = (x + I, y + J)$. Indeed, $ay + bx = (1 - b)y + bx \in y + J$ and $ay + bx = ay + (1 - a)x \in x + I$. Hence π is surjective.

Proposition 7.5.27 (Steinitz Theorem). *Let R be a Dedekind domain and let $\mathfrak{a}$ and $\mathfrak{b}$ be two (fractional) ideals of R. Then*

$$\mathfrak{a} \oplus \mathfrak{b} \simeq R \oplus \mathfrak{a} \cdot \mathfrak{b}.$$

Proof. We first multiply by non zero elements of R in order to have proper ideals of R. If R is a Dedekind domain and $\mathfrak{a}$ and $\mathfrak{b}$ are ideals of R, then we consider $\mathfrak{a} \oplus \mathfrak{b}$ and want to prove that this is isomorphic to $R \oplus \mathfrak{a} \cdot \mathfrak{b}$. Then there is a map

$$\begin{aligned} \mathfrak{a} \oplus \mathfrak{b} &\longrightarrow R \\ (x, y) &\mapsto x - y \end{aligned}$$

which has image $\mathfrak{a} + \mathfrak{b}$. The kernel of this epimorphism is $\mathfrak{a} \cap \mathfrak{b}$. Since R is Dedekind, the ideal $\mathfrak{a} + \mathfrak{b}$ is projective, and we hence obtain an isomorphism

$$\mathfrak{a} \oplus \mathfrak{b} \simeq (\mathfrak{a} + \mathfrak{b}) \oplus (\mathfrak{a} \cap \mathfrak{b}).$$

If $\mathfrak{a}$ and $\mathfrak{b}$ are relatively prime, we are done since then $(\mathfrak{a} + \mathfrak{b}) = R$ and $(\mathfrak{a} \cap \mathfrak{b}) = \mathfrak{a} \cdot \mathfrak{b}$, using Corollary 7.5.24.

Else, multiplying by an element of K if necessary we may assume that $\mathfrak{a}$ and $\mathfrak{b}$ are both proper ideals of R. Let $\mathfrak{a} = \prod_{i=1}^n \wp_i^{s_i}$ for prime ideals $\wp_1, \dots, \wp_n$ and integers $s_1, \dots, s_n$, where we can enumerate in such a way that $\wp_1, \dots, \wp_k$ are those prime ideals containing $\mathfrak{b}$ as well. For each $i \le k$, choose $x_i \in \wp_i^{s_i} \setminus \wp_i^{s_i+1}$. Let $\mathcal{P}$ be the set of prime ideals $\mathfrak{r}$ such that $\mathfrak{a} \cdot \mathfrak{b} \subseteq \mathfrak{r}$. This is a finite set by Lemma 7.5.25, containing $\{\wp_1, \dots, \wp_k\}$. Hence, by Corollary 7.5.24 or the Chinese Reminder Theorem there is $a \in R$ such that $a - x_i \in \wp^{s_i+1}$ for all $i \le k$, and $a \notin \mathfrak{r}$ for all maximal ideals $\mathfrak{r}$ with $\mathfrak{c} \in \mathcal{P} \setminus \{\wp_1, \dots, \wp_k\}$. Then define an ideal $\mathfrak{d}$ by $aR = \mathfrak{a} \cdot \mathfrak{d}$. Now $aR \subseteq \wp_i^{s_i}$ for all $i \in \{1, \dots, k\}$, but $aR \not\subseteq \wp_i^{s_i+1}$ for each $i \in \{1, \dots, k\}$. Further, $aR \not\subseteq \wp$ for all $\wp \in \mathcal{P} \setminus \{\wp_1, \dots, \wp_k\}$. Hence, considering the composition of aR into a product of prime ideals, when we define $\mathfrak{d}$ by $\mathfrak{d} \cdot \mathfrak{a} = aR$, then $\mathfrak{d}$ and $\mathfrak{a}\mathfrak{b}$ are relatively prime, and so $\mathfrak{d} + \mathfrak{a}\mathfrak{b} = R$. Decompose $\mathfrak{d} = \mathfrak{q}_1^{v_1} \cdot \dots \cdot \mathfrak{q}_t^{v_t}$ for prime ideals $\mathfrak{q}_1, \dots, \mathfrak{q}_t$. Again by Corollary 7.5.24 or the Chinese Reminder Theorem let $b \in \mathfrak{q}_i^{v_i} \setminus \mathfrak{q}_i^{v_i+1}$ for all $i = 1, \dots, t$, and such that $b - 1 \in \mathfrak{b}$. Then define an ideal $\mathfrak{c}$ by $bR = \mathfrak{d} \cdot \mathfrak{c}$, and see that $\mathfrak{c} + \mathfrak{b} = R$ and $a\mathfrak{c} = \mathfrak{a}\mathfrak{d}\mathfrak{c} = b\mathfrak{a}$, and hence $\mathfrak{c} \simeq \mathfrak{a}$. We may replace $\mathfrak{a}$ by $\mathfrak{c}$. We are then in the situation of relatively prime ideals $\mathfrak{a}$ and $\mathfrak{b}$. □

As a consequence (of a generalisation) of Steinitz theorem we get the following result.

Proposition 7.5.28. *Let E and F be finite extensions of $\mathbb{Q}$ and suppose that $F \subseteq E$. Let $\mathcal{O}_E := \operatorname{algint}_{\mathbb{Z}}(E)$ and $\mathcal{O}_F := \operatorname{algint}_{\mathbb{Z}}(F)$. Then $\mathcal{O}_F \subseteq \mathcal{O}_E$ and therefore $\mathcal{O}_E$ is naturally an $\mathcal{O}_F$-module. Furthermore, there is an ideal $\mathfrak{a}$ of $\mathcal{O}_F$ such that*

$$\mathcal{O}_E \simeq \mathcal{O}_F^{|E:F|-1} \oplus \mathfrak{a}$$

as $\mathcal{O}_F$-modules.

Proof. By Lemma 7.1.1 we get that both $\mathcal{O}_E$ and $\mathcal{O}_F$ are finitely generated free abelian groups. Moreover, $\mathcal{O}_E$ does not have any $\mathcal{O}_F$-torsion elements. By Lemma 7.5.10 the $\mathcal{O}_F$-module $\mathcal{O}_E$ is finitely generated projective. By Auslander-Buchsbaum's Theorem 6.4.4 we get that $\mathcal{O}_E$ is isomorphic, as an $\mathcal{O}_F$-module to a direct sum of ideals of $\mathcal{O}_F$. The result follows from Proposition 7.5.27. □

Lemma 7.5.29. *Let R be a Dedekind domain and let I be a non zero ideal. Then R/I is artinian.*

Proof. By Proposition 7.5.22 we get

$$I = \wp_1^{e_1} \cdot \dots \cdot \wp_n^{e_n}$$

for prime ideals $\wp_1, \dots, \wp_n$ and positive integers $e_1, \dots, e_n$. Hence the ideals of R/I are the images of the ideals $\wp_1^{f_1} \cdot \dots \cdot \wp_n^{f_n}$ for $0 \le f_i \le e_i$ for all $i \in \{1, \dots, n\}$. Hence R/I allows a composition series, and is hence artinian and Noetherian (cf. Remark 6.1.12). □

Example 7.5.30. We continue the discussion from Example 1.2.20 item 2. Let $R = \mathbb{Z}[\sqrt{-5}]$ and we state without proof that $R = \operatorname{algint}_{\mathbb{Z}}(\mathbb{Q}(\sqrt{-5}))$. The proof can be found in any classical text in number theory, such as [Has-50], and is not important for us at this stage. Hence, R is Dedekind by Lemma 7.5.18. We consider again the ideal $6R$ and know by Proposition 7.5.22 that there are prime ideals $\wp_1, \dots, \wp_n$ of R such that

$$6R = \wp_1 \cdot \dots \cdot \wp_n.$$

Since we know that

$$6 = 2 \cdot 3 = (1 - \sqrt{-5})(1 + \sqrt{-5})$$

are two distinct decompositions of 6 into irreducible elements of R, we may consider prime ideals generated by these irreducible elements. Let $\wp_1 := 2R + (1+\sqrt{-5})R$, $\wp_2 = 3R+(1+\sqrt{-5})R$ and $\wp_3 = 3R+(1-\sqrt{-5})R$. Then we first note that $1-\sqrt{-5} = 2-(1+\sqrt{-5}) \in \wp_1$. Hence

$$6 \in \wp_1^2 \cap \wp_1\wp_2 \cap \wp_1\wp_3 \cap \wp_2\wp_3.$$

Then, note that $R/\wp_1$ is a ring of characteristic 2, whereas $R/\wp_2$ and $R/\wp_3$ are of characteristic 3. Further, since in all three cases at least one of the elements $1+\sqrt{-5}$ or $1-\sqrt{-5}$ is in the ideal, $\sqrt{-5}$ is identified with 1 or with -1 in the quotient. Hence

$$R/\wp_1 = \mathbb{F}_2 \quad \text{and} \quad R/\wp_2 = R/\wp_3 = \mathbb{F}_3$$

which is a field in all three cases, and we see that $\wp_1, \wp_2, \wp_3$ are maximal ideals of R. Moreover, this shows as well that $\wp_3 \neq \wp_1 \neq \wp_2$. Since $2 = (1+\sqrt{-5})+(1-\sqrt{-5}) \in \wp_2+\wp_3$, which gives $R = \wp_2+\wp_3$ from $2R+3R = R$ and $3 \in \wp_2$. Hence $\wp_2 = \wp_3$ implies $R = \wp_2 = \wp_3$ which is absurd. By considering the cardinality of the quotients we see that $2R = \wp_2^2$ and $3R = \wp_2\wp_3$. Hence we get by Corollary 7.5.24

$$\wp_1^2 \cap \wp_1\wp_2 \cap \wp_1\wp_3 \cap \wp_2\wp_3 = \wp_1^2\wp_2\wp_3 = 6R.$$

We say that 2 is *totally ramified* in R since $2R$ is a power of a maximal ideal, and 3 is *unramified* in R since $3R$ is a product of pairwise distinct prime ideals of R. For more details on this concept see Section 8.7.1 below.

7.6 The strong approximation theorem

We now generalise the Chinese Remainder Theorem into the important Strong approximation theorem. The result and its proof is relatively involved. We will use a topological approach and follow closely Kneser's proof as it is presented in [CasFro-67, Chapter 2, Sections 13, 14, 15]. We only need the result once, namely for Theorem 8.5.11 which clarifies the structure of class groups of orders.

A somehow weaker result, the weak approximation theorem has short algebraic proofs. We follow the proof of this result given in [ZaSa-60].

Proposition 7.6.1 (Weak Approximation Theorem). *If $\wp_1, \dots, \wp_n$ are pairwise different prime ideals of the Dedekind domain R, and let K be the field of fractions of R, then for all $a_1, \dots, a_n \in K$ and integers $r_1, \dots, r_n \in \mathbb{Z}$ there is $b \in K$ with $b - a_i \in \wp_i^{r_i} R_{\wp_i}$ for all $i \in \{1, \dots, n\}$.*

Proof. We will reformulate the result in terms of valuations. Following Definition 7.3.4 the prime ideals $\wp_1, \dots, \wp_n$ of R yield discrete valuations $2^{-v_1}, \dots, 2^{-v_n}$ on K. Then we need to show that for all $a_1, \dots, a_n \in K$ and integers $r_1, \dots, r_n \in \mathbb{Z}$ there is $b \in K$ with

$v_i(b - a_i) \geq r_i$ for all $i \in \{1,\ldots,n\}$. We shall show that for any integer m there is $x \in K$ with $v_i(x - a_i) \geq m$ for all $i \in \{1,\ldots,n\}$. Suppose this has been shown. Then let $m > r_i$ for all $i \in \{1,\ldots,n\}$ and for each i fix x_i with $v_i(x_i) = r_i$. We supposed to have proven that there is $y \in K$ with $v_i(y - a_i) \geq m$ for all $i \in \{1,\ldots,n\}$. Since $y = (y - x_i) + x_i$ for all $i \in \{1,\ldots,n\}$, and since $v_i(y - x_i) \geq v_i(x_i) = r_i$ we get $v_i(y) = r_i$ for all $i \in \{1,\ldots,n\}$. Let $x \in K$ be such that $v_i(x - a_i) \geq m$ for all $i \in \{1,\ldots,n\}$ and put $b := x + y$. Then

$$b - a_i = (x - a_i) + y$$

and

$$v_i(y) = r_i < m \leq v_i(x - a_i).$$

Hence

$$v_i(b - a_i) = v_i(y) = r_i \quad \forall i \in \{1,\ldots,n\}.$$

This proves that we only need to show that there is $x \in K$ with $v_i(x - a_i) \geq m$.

By Corollary 7.5.24 we see that for pairwise different prime ideals $\wp_1,\ldots,\wp_n$ and positive integers $r_1,\ldots,r_n$ and for $I := \wp_1^{r_1} \cdot \wp_2^{r_2} \cdot \cdots \cdot \wp_n^{r_n}$ we obtain

$$R/I = R/\wp_1^{r_1} \times R/\wp_2^{r_2} \times \cdots \times R/\wp_n^{r_n}.$$

Hence we can find elements $\eta_1,\ldots,\eta_n$ with $v_i(\eta_i) = 0$ and $v_i(\eta_j) > 0$ for all $i,j \in \{1,\ldots,n\}$ with $i \neq j$. Put

$$\zeta_i := \frac{\eta_i}{\eta_1 + \cdots + \eta_n}$$

for all $i \in \{1,\ldots,n\}$. Still $v_i(\zeta_i) = 0$ and $v_i(\zeta_j) > 0$ for all $i,j \in \{1,\ldots,n\}$ with $i \neq j$, but in addition η_i maps to 1 under the canonical morphism $R \longrightarrow R/\wp_i$. Hence $v_i(\zeta_i - 1) > 0$ for all $i \in \{1,\ldots,n\}$. Let k be a positive integer such that

$$\begin{aligned} k \cdot v_i(\zeta_i - 1) + v_i(a_i) &\geq m \quad \forall i \in \{1,\ldots,n\} \\ k \cdot v_j(\zeta_i) + v_j(a_i) &\geq m \quad \forall \in \{1,\ldots,n\} \text{ with } i \neq j. \end{aligned}$$

Let

$$\xi_i := 1 - (1 - \zeta_i^k)^k \quad \text{for all } i \in \{1,\ldots,n\}.$$

Then

$$v_i(a_i(\xi_i - 1)) = v_i(a_i) + k \cdot v_i(1 - \zeta_i^k) \geq v_i(a_i) + k \cdot v_i(1 - \zeta_i) \geq m$$

for each $i \in \{1,\ldots,n\}$. Since by construction $\xi_i = \zeta_i^k \cdot (\sum_{s=1}^{k-1} \binom{k}{s}(-1)^s \zeta_i^{ks})$, we get for $j \neq i$

$$v_j(a_i\xi_i) \geq v_j(a_i) + k \cdot v_j(\zeta_i) \geq m$$

by construction of k. The element $x := \sum_{t=1}^{n} a_t\xi_t$ satisfies $v_i(x - a_i) \geq m$. for all $i \in \{1,\ldots,n\}$. ☐

The strong approximation theorem looks very much like the weak approximation theorem Proposition 7.6.1, but we require that for all other primes $\wp$ the constructed element is in $R_\wp$. This additional condition needs a lot of additional preparation.

Definition 7.6.2. Let Λ be a set and for each $\lambda \in \Lambda$ let Ω_λ be a topological space. Fix for each $\lambda \in \Lambda$ an open subset Θ_λ of Ω_λ, denote $\Theta := \prod_{\lambda\in\Lambda} \Theta_\lambda$, and consider

$$\Omega := \left\{(x_\lambda)_{\lambda\in\Lambda} \in \prod_{\lambda\in\Lambda} \Omega_\lambda \mid |\{\lambda \in \Lambda \mid x_\lambda \notin \Theta_\lambda\}| < \infty\right\}$$

the subset of points x in $\prod_{\lambda\in\Lambda} \Omega_\lambda$ whose components x_λ are all but a finite number of them in Θ_λ. This set Ω is a topological space with a basis of open sets given by $\prod_{\lambda\in\Lambda} \Gamma_\lambda$ where for each $\lambda \in \Lambda$ the set Γ_λ is open in Ω_λ and $|\{\lambda \in \Lambda \mid \Gamma_\lambda \neq \Theta_\lambda\}| < \infty$, i. e. all but a finite number of Γ_λ equal Θ_λ. This is the *restricted topological product of* Ω *with respect to* Θ.

Remark 7.6.3.
1. If S is a finite subset of Λ, then
$$\Omega_S := \prod_{\lambda\in S} \Omega_\lambda \times \prod_{\lambda\in\Lambda\setminus S} \Theta_\lambda$$
is open in Ω and the subset topology of Ω induced on Ω_S equals the product topology.
2. If we have for all $\lambda \in \Lambda$ open sets $\Theta'_\lambda \subseteq \Theta_\lambda$, such that for all but a finite number of $\lambda \in \Lambda$ we get $\Theta'_\lambda = \Theta_\lambda$, then denoting $\Theta' := \prod_{\lambda\in\Lambda} \Theta'_\lambda$, the restricted topological product of Ω with respect to Θ equals restricted topological product of Ω with respect to Θ'.
3. If in the restricted topological product of Ω with respect to Θ all Ω_λ are locally compact and all Θ_λ are compact, then the restricted topological product of Ω with respect to Θ is locally compact. For all finite subsets S of Λ the sets Ω_S are open, and $\Omega = \bigcup_{S\subseteq\Lambda} \Omega_S$. By Tychonoff's theorem we are done.
4. Suppose that μ_λ is a measure on Ω_λ for each $\lambda \in \Lambda$ and $\mu_\lambda(\Theta_\lambda) = 1$. Then define a basis of measurable sets on $\prod_{\lambda\in\Lambda} \Omega_\lambda$ to be formed by the sets $\prod_{\lambda\in\Lambda} M_\lambda$ for $M_\lambda \subseteq \Omega_\lambda$ measurable in Ω_λ and $\mu_\lambda(M_\lambda) < \infty$ for all $\lambda \in \Lambda$ and for all but a finite number of $\lambda \in \Lambda$ we have $M_\lambda = \Theta_\lambda$. Then
$$\mu\left(\prod_{\lambda\in\Lambda} M_\lambda\right) := \prod_{\lambda\in\Lambda} \mu_\lambda(M_\lambda)$$
is a measure on the restricted topological product of Ω with respect to Θ. Note that for each finite subset S of Λ the restriction of μ on Ω_S is the product measure.

Let now K be a finite extension of $\mathbb{Q}$ and let $R = \mathrm{algint}_{\mathbb{Z}}(K)$. For each prime ideal $\wp$ of R we have the localisation $R_\wp$, the completion $\widehat{R}_\wp$ of $R_\wp$ and its field of fractions $\widehat{K}_\wp$, and the corresponding valuation $2^{-v_\wp}$ on K and on $\widehat{K}_\wp$. The completion at an infinite place is either $\mathbb{R}$ or $\mathbb{C}$.

Definition 7.6.4. A *place* on a field K is a representant of an equivalence class of valuations on K. The *adèle ring* V_K of K is the restricted topological product of the topological spaces $\widehat{K}_\wp$ for $\wp \in Spec(R)$ and the completions of K at the infinite places with respect to the subspaces $\widehat{R}_\wp$ of $\widehat{K}_\wp$.

The adèle ring V_K is a topological ring in the following sense. First, it is a ring since if $x = (x_\wp)_{\wp \in Spec(R)}$ and $y = (y_\wp)_{\wp \in Spec(R)}$ are elements of V_k, then also $x + y := (x_\wp + y_\wp)_{\wp \in Spec(R)}$ and $x \cdot y := (x_\wp \cdot y_\wp)_{\wp \in Spec(R)}$ are in V_K. Clearly $1 = (1)_{\wp \in Spec(R)}$ is in V_K and if $x \in V_K$, also $-x \in V_K$. The ring axioms follow from the ring axioms in each component. Since '+' and '·' are both continuous operations in each component, these stay continuous in the restricted product.

Lemma 7.6.5. *Let K be a finite extension of $\mathbb{Q}$. Then V_K is locally compact.*

Proof. Indeed, for each prime ideal $\wp$ the ring $\widehat{R}_\wp$ is compact and $\widehat{K}_\wp$ is locally compact by Proposition 7.3.17. Closed intervals in $\mathbb{R}$ and closed discs in $\mathbb{C}$ are compact. Hence also the completion of K at infinite places is locally compact. The rest follows from Remark 7.6.3. item 3. □

Proposition 7.6.6. *Let K be a finite extension of $\mathbb{Q}$ of degree N. Then $V_\mathbb{Q} \otimes_\mathbb{Q} K \simeq V_K$ as topological spaces and as rings and in particular the additive group $(V_K, +)$ is isomorphic to $(V_\mathbb{Q}, +)^N$.*

Proof. Using Lemma 7.5.25 and Lemma 7.5.18 for any prime p of $\mathbb{Q}$ there are a finite number of prime ideals of $\mathrm{algint}_\mathbb{Z}(K)$ containing p. Each of these correspond to a discrete valuation. Proposition 7.3.20 describes these valuations precisely, and shows that also for the archimedean places there are at most N extensions of these valuations, given by the tensor product. By Proposition 7.3.18 for any of these primes $\wp$ there is a unique extension of the p-adic valuation on $\widehat{\mathbb{Q}}_p$ to the $\wp$-adic valuation on $\widehat{K}_\wp$. This gives a topology on V_K. Given a $\mathbb{Q}$-basis $\{\omega_1, \ldots, \omega_n\}$ of K we see that $\omega_i \cdot V_\mathbb{Q} \subseteq V_K$ is a homeomorphism of $V_\mathbb{Q}$ to its image in V_K. Hence summing up over the basis this gives a homeomorphism with the product topology. □

Let K be a finite extension of $\mathbb{Q}$. Since K is a subfield of $\widehat{K}_\wp$ for each prime ideal $\wp$, and similarly K is a subfield of the completion at any archimedean place, the diagonal map induces a map $\iota : K \longrightarrow V_K$.

Proposition 7.6.7. *Let K be a finite extension of $\mathbb{Q}$. Then $\iota(K)$ is a discrete subset of V_K and considering the quotient as additive groups, $V_K/\iota(K)$ is compact in the quotient topology. Moreover, there is a compact subset W of V_K defined as the set of $\zeta \in V_K$ satisfying inequalities $|\zeta_v|_v \le \delta_v$ for all places v of K and for real numbers δ_v which equal 1 for all but a finite number of places, such that for any $x \in V_K$ there is $z \in K$ with $x - \iota(z) \in W$.*

Proof. By Proposition 7.6.6 it is sufficient to show the result in case $K = \mathbb{Q}$. Let U be the open set given by the conditions

$$\{x \in V_{\mathbb{Q}} \mid |x_p|_p \leq 1\ \forall p \in Spec(\mathbb{Z}) \text{ and } |x|_\infty < 1\}.$$

Here $|\ |_\infty$ is the unique archimedean valuation of $\mathbb{Q}$, namely the absolute value (cf. Theorem 7.3.7). Let $x \in \mathbb{Q}$ with $\iota(x) \in U$. Then $|x|_p \leq 1$ for all p, whence

$$x \in \bigcap_{p \in Spec(\mathbb{Z})} \mathbb{Z}_p = \mathbb{Z}.$$

Since $|x|_\infty < 1$ we have $x = 0$. Therefore for any $x \in \mathbb{Q}$ the open neighborhood of $\iota(x)$ in $V_{\mathbb{Q}}$ is $\iota(\mathbb{Q}) \cap (\iota(x) + U) = \{\iota(x)\}$. Hence $\iota(\mathbb{Q})$ is discrete in $V_{\mathbb{Q}}$.

Let

$$W_0 := \left\{x \in V_{\mathbb{Q}} \;\middle|\; |x_p|_p \leq 1\ \forall p \in Spec(\mathbb{Z}) \text{ and } |x_\infty|_\infty \leq \frac{1}{2}\right\}$$

Since $\widehat{\mathbb{Z}}_p$ is compact by Proposition 7.3.17, and since closed intervals of the real numbers are compact, W_0 is compact by Tychonoff's theorem. We claim that for every adèle $x \in V_{\mathbb{Q}}$ there is an adèle $y \in W_0$ such that $y - x \in \iota(\mathbb{Q})$. Indeed, for each $p \in Spec(\mathbb{Z})$ there is $n_p \in \mathbb{N}$ and $u_p \in \mathbb{Z}$ such that

$$\left|x_p - \frac{1}{p^{n_p}} u_p\right| \leq 1$$

and since $x \in V_{\mathbb{Q}}$ we may choose $u_p = 0$ for almost all p. Then

$$u := \sum_{p \in Spec(\mathbb{Z})} \frac{1}{p^{n_p}} u_p$$

is a well-defined element in $\mathbb{Q}$ and satisfies

$$|x_p - \iota(u)|_p \leq 1 \quad \text{for all } p \in Spec(\mathbb{Z}).$$

Moreover, since any real number is at distance at most $\frac{1}{2}$ from an integer, there is $s \in \mathbb{Z}$ such that

$$|x_\infty - u - s|_\infty \leq \frac{1}{2}.$$

Putting $z := u + s$ the adèle $x - \iota(z) \in W_0$. Then the continuous map $V_{\mathbb{Q}} \twoheadrightarrow V_{\mathbb{Q}}/\iota(\mathbb{Q})$ restricts to a continuous and surjective map

$$W_0 \hookrightarrow V_{\mathbb{Q}} \twoheadrightarrow V_{\mathbb{Q}}/\iota(\mathbb{Q}).$$

Since W_0 is compact, also $V_{\mathbb{Q}}/\iota(\mathbb{Q})$ is compact.

Clearly W_0 is contained in a W as required. □

Proposition 7.6.8. *Let K be a finite extension of $\mathbb{Q}$. Then there is a constant $C > 0$ which only depends on K such that for any $x = (x_v)_{v \text{ place of } K} \in V_K$ with $\prod_{v \text{ place of } K} |x_v|_v > C$ we always get a $y \in K \setminus \{0\}$ such that $|\iota(y)|_v \le |x_v|_v$ for all places v of K.*

In particular for v_0 any fixed valuation of K and for all places v of K different from v_0 let $\delta_v > 0$ with $\delta_v = 1$ for all but a finite number of them. Then there is $x \in K \setminus \{0\}$ with $|x|_v \le \delta_v$ for all places $v \neq v_0$.

Proof. First of all, since for all but a finite number of places v we get $|x_v|_v \le 1$ the inequality $\prod_{v \text{ place of } K} |x_v|_v > C$ implies that $|x_v|_v = 1$ for all but a finite number of places v. By Proposition 7.6.7 the additive group of V_K is locally compact, and the additive group of $V_K/\iota(K)$ is compact. By Theorem 1.2.12 there is an up to scalar unique Haar measure μ on V_K, and a Haar measure on $V_K/\iota(K)$. The Haar measure on V_K is actually the product of the Haar measures on $\widehat{K}_v$, where v is taken over the places v of K, in the sense that the measure of a product space $\prod_{v \text{ place of } K} S_v$ is the product of the measures of the spaces S_v. We normalise the measures on each of the complete fields in the way that the measure of the set of x of $\widehat{K}_v$ with $|x|_v \le 1$ has measure 1. Let

$$W := \left\{ y = (y_v)_{v \text{ place of } K} \in V_K \;\middle|\; \begin{array}{ll} |y_v|_v \le \frac{1}{10} & \text{if } v \text{ is archimedean} \\ |y_v|_v \le 1 & \text{if } v \text{ is non archimedean} \end{array} \right\}$$

and denote $d := \mu((W + \iota(K))/(\iota(K)))$ and $D := \mu(V_K/\iota(K))$. Since by Proposition 7.3.20 there are only finitely many archimedean places, $D \ge d > 0$, and since μ is a Haar measure on a compact space, $d \le D < \infty$. Let $T := x \cdot W$ where multiplication is understood componentwise. If the restriction of the map $V_K \longrightarrow V_K/\iota(K)$ to W is injective, then with $C := \frac{D}{d}$, and since the measure on V_K is the product measure of the measures on the completed fields, we get

$$\mu(T) = d \cdot \prod_{v \text{ place of } K} |x_v|_v > \frac{dD}{d} = D = \mu(V_K/\iota(K)).$$

This contradiction shows that there is a $y \neq z \in W$ which map to the same element in $V_K/\iota(K)$. Let $k \in K$ with $y - z = \iota(k)$. Then $|k|_v = |y_v - z_v|_v \le |x_v|_v$ for all v.

For the second part just choose y_v with $|y_v|_v \le \delta_v$ whenever $\delta_v \neq 1$ and $|y_v|_v = 1$ if $\delta_v = 1$. Then there is $y_{v_0} \in \widehat{K}_{v_0}$ with sufficiently big valuation such that the product over all valuations is bigger than C. This shows the proposition □

Theorem 7.6.9 (Strong Approximation Theorem). *If $\wp_1, \dots, \wp_n$ are pairwise different prime ideals of the Dedekind domain R, and let K be the field of fractions of R, then for all $a_1, \dots, a_n \in K$ and integers $r_1, \dots, r_n \in \mathbb{Z}$ there is $b \in K$ with $b - a_i \in \wp_i^{r_i} R_{\wp_i}$ for all $i \in \{1, \dots, n\}$, and $b \in R_\wp$ for all $\wp \notin \{\wp_1, \dots, \wp_n\}$.*

Proof. Let v_0 be any archimedean valuation and choose a minimal $\varepsilon \ge 2^{-r_i}$ for all $i \in \{1, \dots, n\}$. By Proposition 7.6.7 there is a set W defined as the set of $\zeta \in V_K$ satisfying inequalities $|\zeta_v|_v \le \delta_v$ for all places v of K and for real numbers δ_v which are equal to

1 for all but a finite number of places, such that for any $x \in V_K$ there is $z \in K$ with $x - \iota(z) =: w \in W$. Define $a \in V_K$ by $a_{\wp_i} = a_i$ for all $i \in \{1, \dots, n\}$ and $a_v = 0$ for all other valuations.

By Proposition 7.6.8 there is a non zero $q \in K$ such that

$$|q|_v \le \frac{\varepsilon}{\delta_v} \text{ if } v \in S \quad \text{and} \quad |q|_v \le \frac{1}{\delta_v} \text{ if } v \notin S \text{ and } v \neq v_0.$$

Put $x := \iota(q)^{-1}a$ and then the above equation $x - \iota(z) = w \in W$ gives

$$a = \iota(q)x = \iota(q)w + \iota(q)\iota(z) = \iota(q)w + \iota(qz) \in \iota(q)W + \iota(qz).$$

Then $b = qz$ has the required properties. □

7.7 Exercises

Exercise 7.1. Find an integral domain R which is not Noetherian but has a unique maximal ideal and for which there are subrings R_n, for $n \in \mathbb{N}$, such that $R_n \subseteq R_{n+1}$ for all n and R_n are discrete valuation rings, and such that $R = \bigcup_{n\in\mathbb{N}} R_n$. Why R cannot be a principal ideal domain?

Exercise 7.2. Let R be a unique factorisation domain. Show that if any non zero prime ideal of R is of height one (i. e. any prime ideal $\mathfrak{q}$ strictly included in the prime ideal $\wp$ is 0), then any prime ideal is principal.

Exercise 7.3 (with solution in Chapter 9). Let R be an integral domain with field of fractions K and suppose that for any non zero $x \in K$ one gets $x \in R$ or $x^{-1} \in R$.

a) Let $M := \{x \in R \setminus \{0\} \mid x^{-1} \notin R\} \cup \{0\}$. Show that if $x \in M$, then for all $y \in R$ we have $xy \in M$.
b) Show that $1 \notin M$.
c) Show that M is an ideal of R.
d) Show that if $x \in M$, then $1 + x \in R \setminus M$.
e) Show that M is the unique maximal ideal of R.
f) If R is Noetherian, show that each ideal of R is principal. (Consider a partial order on elements given by $x < y \Leftrightarrow yx^{-1} \in R$).
g) Deduce that R is a discrete valuation ring.

NB: A discrete valuation ring R certainly has the property that any non zero element of the field of fractions is either in R or its inverse is in R. Hence the exercise gives an alternative definition of discrete valuation rings.

Exercise 7.4. Consider $R = \mathbb{Z}[\sqrt{-5}] = \{a + b\sqrt{-5} \in \mathbb{Q}(\sqrt{-5}) \mid a, b \in \mathbb{Z}\}$.

a) Show that in the factorisation $9 = 3 \cdot 3 = (2 + \sqrt{-5}) \cdot (2 - \sqrt{-5})$ all three factors are irreducible.

b) Show that $\wp_1 := 3R + (2 + \sqrt{-5})R$ and $\wp_2 := 3R + (2 - \sqrt{-5})R$ are both prime ideals of R.
c) Show $3R = \wp_1 \cdot \wp_2$, $(2 + \sqrt{-5})R = \wp_1^2$ and $(2 - \sqrt{-5})R = \wp_2^2$.
d) Decompose $9R$ into an (essentially unique) product of prime ideals.

Exercise 7.5 (with solution in Chapter 9). Let R be a Dedekind domain and let $\mathfrak{a}$ and $\mathfrak{b}$ be non zero ideals of R. Let $\mathfrak{b} = \prod_{j=1}^{k} \wp_j^{m_j}$ be its decomposition into prime ideals and let $\mathfrak{a} = \prod_{j=1}^{k} \wp_j^{n_j}$.
a) For each $\alpha_j \in \wp_j^{n_j} \setminus \wp_j^{n_j+1}$ show that there is $\alpha \in R$ such that $\alpha - \alpha_j \in \wp_j^{n_j+1}$ for all $j \in \{1, \ldots, k\}$.
b) Show that $\alpha R \subseteq \mathfrak{a}$.
c) Show that $\alpha \cdot \mathfrak{a}^{-1} = \prod_{i=1}^{t} \mathfrak{q}^{s_i}$ for prime ideals $\mathfrak{q}_1, \ldots, \mathfrak{q}_t$, and each of the ideals $\mathfrak{q}_i$ is different from any of the ideals $\wp_1, \ldots, \wp_k$.
d) Deduce that $\alpha \cdot \mathfrak{a}^{-1} + \mathfrak{b} = R$.
e) Deduce that if $\mathfrak{a}$ is a non zero ideal of R and if β is a non zero element of $\mathfrak{a}$, then there is an element α of $\mathfrak{a}$ such that $\mathfrak{a} = \alpha R + \beta R$.

Exercise 7.6. Let R be a complete discrete valuation ring with radical $\pi R = \wp$. Let A be an R-algebra, finitely generated as R-module, and denote by $\nu_n : A \longrightarrow A/\pi^n A$, as well as $\mu_{r,r-1} : A/\pi^r A \longrightarrow A/\pi^{r-1} A$ the natural projections. Let $\overline{e}^2 = \overline{e} \in A/\pi^n A$. Let $f \in A/\pi^{n+1} A$ be such that $\mu_{n+1,n}(f) = \overline{e}$.
a) Show that

$$\mu_{n+1,n}(3f^2 - 2f^3) = \overline{e}$$

and

$$(3f^2 - 2f^3)^2 - (3f^2 - 2f^3) = -(3 - 2f)(1 + 2f)(f^2 - f)^2.$$

Deduce

$$(3f^2 - 2f^3)^2 - (3f^2 - 2f^3) \in \ker(\mu_{n+1,n})^2 = 0.$$

b) Construct by induction a Cauchy sequence $(f_r)_{r \geq n}$ defining an element $e \in A$ with $e^2 = e$ and $\nu_n(e) = \overline{e}$.
c) Let A be an R-algebra, finitely generated as R-module, and let M be a finitely generated A-module. Show that the A-module M is indecomposable if and only if the $A/\pi^n A$-module $M/\pi^n M$ is indecomposable.

NB: We call this property the *lifting of idempotents*.

Exercise 7.7. Consider the field $\mathbb{C}(t)$ of rational functions with coefficients in $\mathbb{C}$.

a) For any $x \in \mathbb{C}(t)$ there are polynomials $f(t), g(t) \in \mathbb{C}[t]$ such that $x = \frac{f(t)}{g(t)}$. Put $v_\infty(x) := \text{degree}(g) - \text{degree}(f)$ and $v_\infty(0) = \infty$. Show that $|\ |_\infty := 2^{-v_\infty}$ is a non archimedean, discrete valuation. Here, we define $|0|_\infty := 0$.

b) Fix $a \in \mathbb{C}$. Then for any $x \in \mathbb{C}(t)$ there is a unique $n_x(a) \in \mathbb{Z}$ such that $x = \frac{f(t)}{g(t)}$ as in a), and such that $x = (t-a)^{n_x(a)} \frac{f_0(t)}{g_0(t)}$ and such that $f_0(t), g_0(t) \in \mathbb{C}[t]$ and such that $f_0(a) \neq 0 \neq g_0(a)$. Put $v_a(x) := n_x(a)$, and show that $|\ |_a := 2^{-v_a}$ is a non archimedean, discrete valuation on $\mathbb{C}(t)$.

c) Embed $\mathbb{C}$ into its projective complex line $P^1(\mathbb{C})$ by joining a point at infinity. Show that $-v_a(x)$ is the order of the pole of x at a (or minus the order of the zero) and $-v_\infty(x)$ is the order of the pole (or minus the order of the zero) at infinity.

d) Determine the valuation rings for v_a and for v_∞, and determine the maximal ideal in each of the cases.

8 Some notions of integral representations

SILBERDISTEL

Sich zurückhalten
auf der erde

Keinen schatten werfen
auf andere

Im schatten der anderen
leuchten.

Reiner Kunze, Gedichte @ 2001, S. Fischer Verlag Frankfurt am Main

SILVER THISTLE

Refraining
on earth

Not casting a cloud
over others,

In the shadow of others
gleaming.

Reiner Kunze, poetry;[1]

After having assimilated representation theory of finite groups over fields whose characteristic is not dividing the group order, this semisimple theory can be deepened into various directions. Most frequently one starts with the modular representation theory, that is studying group rings over fields whose characteristic does divide the group order. Many textbooks deal with this theory, and one might like to consult e. g. [Ser-78], [Mül-80], [Kow-14] for a pedagogical account, e. g. [Fei-82], [NaTsu-88] for a high level treatment, or simply [Zim-14, Chapter 2].

Since the wealth of choice of literature on modular representation theory textbooks is already very satisfactory we decided to concentrate on a less frequently found subject, which nevertheless is a logical continuation of the elementary theory, namely integral representations, that is representations of finite groups over some Dedekind domain. We recall that we already studied the case $\mathbb{Z}\mathfrak{S}_3$ in detail in Example 1.2.43.(4).

1 in der Übersetzung von Susanne Brennecke, translation by Susanne Brennecke.

https://doi.org/10.1515/9783110702446-008

8.1 Classical orders and their lattices

8.1.1 Basic definitions and examples

We start with an example.

Example 8.1.1. Let G be a finite group and let K be a field of characteristic 0. Then by Maschke's theorem and Wedderburn's theorem there is some integer s and an algebra isomorphism

$$KG \simeq \prod_{i=1}^{s} Mat_{n_i}(D_i)$$

for skew fields D_i with $K \subseteq D_i$ for all $i \in \{1, \ldots, s\}$. Let R be an integral domain with field of fractions K. Then RG is a subring of KG, hence isomorphic to a subring Λ of the above product. Moreover, by definition RG is free as R-module, and since

$$K \otimes_R RG \simeq KG \simeq \prod_{i=1}^{s} Mat_{n_i}(D_i)$$

we get that Λ contains a K-basis of the semisimple artinian algebra $KG \simeq \prod_{i=1}^{s} Mat_{n_i}(D_i)$.

Example 8.1.1 gives motivation of the following definition.

Definition 8.1.2. Let R be an integral domain with field of fractions K. An R-algebra Λ is called an *R-order* in a semisimple K-algebra D if
- $K \otimes_R \Lambda \simeq D$ as K-algebra
- Λ is finitely generated projective as an R-module.

Corollary 8.1.3. *Let R be an integral domain of characteristic 0 with field of fractions K. Then, for any finite group G the algebra RG is an R-order in the semisimple K-algebra KG.*

Indeed, free R-modules are projective by Proposition 6.2.5.

Example 8.1.4.
1. Let R be an integral domain with field of fractions K. Then R is an R-order in K. More generally $Mat_{n\times n}(R)$ is an R-order in $Mat_{n\times n}(K)$.
2. More tricky, let again R be an integral domain with field of fractions K, and let T be an invertible element in $Mat_{n\times n}(K)$. Then $T \cdot Mat_{n\times n}(R) \cdot T^{-1}$ is an R-order in $Mat_{n\times n}(K)$.
3. Let R be an integral domain with field of fractions K. Then for each ideal I of R the subring

$$\begin{pmatrix} R & R & \dots & \dots & R \\ I & R & \ddots & & \vdots \\ \vdots & \ddots & \ddots & \ddots & \vdots \\ \vdots & & \ddots & \ddots & R \\ I & \dots & \dots & I & R \end{pmatrix}$$

of $d \times d$ matrices over K is an R-order in the semisimple K-algebra of $d \times d$ matrices over K.

4. Let R be an integral domain with field of fractions K and a non zero ideal I. Then define the subalgebra Λ of the ring of three copies of 2×2 matrix rings over K as

$$\left\{ \begin{pmatrix} a_1 & b_1 \\ c_1 & d_1 \end{pmatrix} \times \begin{pmatrix} a_2 & b_2 \\ c_2 & d_2 \end{pmatrix} \times \begin{pmatrix} a_3 & b_3 \\ c_3 & d_3 \end{pmatrix} \;\middle|\; \forall_{i \in \{1,2,3\}}\ a_i - d_{i-1} \in I; c_i \in I \right\}$$

where, of course, we consider indices modulo 3. This again is an R-order in the semisimple K-algebra $\prod_{i=1}^{3} Mat_{2\times 2}(K)$.

5. Recall from Example 1.2.43.(4) the structure of $\mathbb{Z}\mathfrak{S}_3$.

$$\mathbb{Z}\mathfrak{S}_3 \simeq \left\{ \left(d_0, \begin{pmatrix} a_1 & b_1 \\ c_1 & d_1 \end{pmatrix}, a_2 \right) \in \mathbb{Z} \times Mat_{2\times 2}(\mathbb{Z}) \times \mathbb{Z} \;\middle|\; a_2 - d_0 \in 2\mathbb{Z}; a_1 - d_0 \in 3\mathbb{Z}; a_2 - d_1 \in 3\mathbb{Z}; b_1 \in 3\mathbb{Z} \right\}$$

This is a $\mathbb{Z}$-order in the semisimple $\mathbb{Q}$-algebra $\mathbb{Q}\mathfrak{S}_3$.

6. Note that we have an epimorphism of groups

$$\mathfrak{S}_3 \twoheadrightarrow C_2,$$

which induces an epimorphism of the corresponding group rings

$$\mathbb{Z}\mathfrak{S}_3 \twoheadrightarrow \mathbb{Z}C_2$$

and this can be seen on the level of the above description of the group ring $\mathbb{Z}\mathfrak{S}_3$ by the projection onto the first and the third component. In particular,

$$\mathbb{Z}C_2 \simeq \{(a,b) \in \mathbb{Z} \times \mathbb{Z} \mid b - a \in 2\mathbb{Z}\}.$$

7. Let K be an algebraic number field, that is an extension of $\mathbb{Q}$ of finite degree. Then $R := \mathrm{algint}_{\mathbb{Z}}(K)$, the ring of algebraic integers in $\mathbb{Q}$, is a $\mathbb{Z}$-order in K, using Lemma 7.1.1.

In order to study algebras we used modules over the algebra.

Remark 8.1.5. If R is an integral domain, and $\mathfrak{m}$ is a maximal ideal of R, then $k := R/\mathfrak{m}$ is a field. If Λ is an R-order, then $k\otimes_R \Lambda =: A$ is a finite dimensional k-algebra. Indeed, Λ is finitely generated projective as an R-module, and hence a direct factor of R^n for some integer n, as R-module. Therefore A is a direct factor of k^n for some n, as a k-module. So, there is a surjective ring homomorphism

$$\Lambda \xrightarrow{\pi} A.$$

As a consequence, any A-module is naturally also a Λ-module.

In order to study special kind of algebras we use a special kind of modules, namely those reflecting the specific property. We are mostly interested in those modules which reflect the defining property of Λ, which is being projective as R-module. Note that since there is a homomorphism $R \longrightarrow \Lambda$ given by the R-algebra structure of Λ, any Λ-module is also an R-module.

Definition 8.1.6. Let R be an integral domain with field of fractions K, and let Λ be an R-order in the semisimple K-algebra A. Then a Λ-module L is said to be a Λ-*lattice* if L is finitely generated projective as an R-module.

Example 8.1.7. Recall Example 8.1.4.(7). Let K be a finite extension over $\mathbb{Q}$ and let $R := \mathrm{algint}_{\mathbb{Z}}(K)$ be the ring of algebraic integers over $\mathbb{Q}$. Then we have seen that R is an order in $\mathbb{Q}$. But there is much more structure behind. If K is moreover Galois over $\mathbb{Q}$, with Galois group $G = \mathrm{Gal}(K : \mathbb{Q})$, then $R = \mathrm{algint}_{\mathbb{Z}}(K)$ is stable under the action of G. Indeed, if $\alpha \in R$, then there is $p(X)$ a polynomial with integer coefficients and leading coefficient 1, such that $p(\alpha) = 0$. Let $g \in G$. Then

$$0 = g(0) = g(p(\alpha)) = p(g(\alpha))$$

since all coefficients are in $\mathbb{Z} \subseteq \mathbb{Q}$ and $\mathbb{Q}$ is fixed under the action of G. Hence $g(\alpha)$ is again root of $p(X)$, and therefore $g(\alpha) \in R$. It follows that R is a $\mathbb{Z}G$-module. Moreover, since $R \subseteq K$, and since K is of characteristic 0, we get that R is $\mathbb{Z}$-free. By Lemma 7.1.1 we see that R is finitely generated as abelian group. Hence R is a $\mathbb{Z}G$-lattice. We call this the *Galois module*. We shall give more details on this type of module in Section 8.7.

Let R be an integral domain with field of fractions K and let Λ be an R-order in a semisimple K-algebra. Let $\mathfrak{m}$ be a maximal ideal in R, and let $k := R/\mathfrak{m}$. Then, there is a surjective homomorphism $R \xrightarrow{\pi} k$. For any Λ-lattice L this induces a surjective homomorphism of Λ-modules

$$L \longrightarrow k \otimes_R L =: M.$$

Then, $k\otimes_R \Lambda =: A$ is a finite dimensional k-algebra using Remark 8.1.5, and M is a finite dimensional A-module by construction. We say that M is a reduction modulo $\mathfrak{m}$ of a Λ-lattice.

However, not all finite dimensional A-modules are reductions, as defined above, of Λ-lattices. We give a criterion from representation theory of groups.

Theorem 8.1.8 (Green). *Let R be a complete discrete valuation ring of characteristic 0 and let k be its residue field of characteristic $p > 0$. Let G be a finite group and let M be a finitely generated kG-module with $\mathrm{Ext}^2_{kG}(M, M) = 0$. Then there is an RG-lattice L with $k \otimes_R L \simeq M$ as kG-modules.*

We shall not need this result in the sequel, and omit the proof. The proof can be found in e. g. [Bens-91, Theorem 3.7.7], together with a further result in this direction. See also Example 8.1.9 below for modules which cannot be lifted.

Example 8.1.9. Let R be a discrete valuation ring with maximal ideal $\mathfrak{m}$. Denote $k := R/\mathfrak{m}$ and K the field of fractions of R. Let

$$\Lambda := \begin{pmatrix} R & R \\ \mathfrak{m} & R \end{pmatrix}$$

which is an R-order in the 2×2 matrix ring over the field of fractions of R. Then

$$\mathrm{rad}(\Lambda) := \begin{pmatrix} \mathfrak{m} & R \\ \mathfrak{m} & \mathfrak{m} \end{pmatrix} \quad \text{and} \quad I := \begin{pmatrix} \mathfrak{m} & \mathfrak{m} \\ \mathfrak{m} & \mathfrak{m} \end{pmatrix}$$

are 2-sided ideals of Λ and $A := \Lambda/I$ is a finite dimensional k-algebra. There are precisely two indecomposable Λ-lattices, namely

$$L_1 := \begin{pmatrix} R \\ \mathfrak{m} \end{pmatrix} \quad \text{and} \quad L_2 := \begin{pmatrix} R \\ R \end{pmatrix}.$$

However,

$$\Lambda/I = A = \begin{pmatrix} k & k \\ 0 & k \end{pmatrix}$$

admits precisely three indecomposable A-modules, namely the two projective modules P_1 and P_2, where $P_1 \subseteq P_2$ and P_1 is simple and projective, and a third module P_2/P_1, which is simple. This second simple module cannot be lifted to a Λ-lattice.

Note however that A is not isomorphic to $\Lambda/\mathfrak{m}\Lambda$. Indeed, $\mathfrak{m}\Lambda \subsetneq I$. Note that we have inclusions of Λ-lattices

$$\cdots \subsetneq \mathfrak{m}^3 L_2 \subsetneq \mathfrak{m}^2 L_1 \subsetneq \mathfrak{m}^2 L_2 \subsetneq \mathfrak{m} L_1 \subsetneq \mathfrak{m} L_2 \subsetneq L_1 \subsetneq L_2$$

with successive quotients $L_2/L_1 =: S_2$ respectively $L_1/\mathfrak{m}L_2 =: S_1$ in alternating order. These quotients are simple Λ-modules. $L_1/\mathfrak{m}L_1$, resp. $L_2/\mathfrak{m}L_2$ both have a simple submodule S_2, resp. S_1 with quotient being a simple module of the respectively other type. By definition

$$\Lambda/\mathfrak{m}\Lambda = L_1/\mathfrak{m}L_1 \oplus L_2/\mathfrak{m}L_2$$

whereas

$$A = L_1/\mathfrak{m}L_2 \oplus L_2/\mathfrak{m}L_2$$

as left Λ-modules.

Lemma 8.1.10. *Let R be a Dedekind domain with field of fractions K and let Λ be an R-order in the finite dimensional semisimple K-algebra A. If V is a finite dimensional A-module. Then there is a Λ-lattice L such that $KL = V$.*

Proof. Let $v_1, \ldots, v_s$ be a generating set of V as A-module. Then $L := \Lambda v_1 + \cdots + \Lambda v_s$ is a Λ-lattice. Indeed, $L \subseteq V$, and since K is R-torsion free, L is R-torsion free as well. Moreover, L is a Λ-module and $\varphi : \Lambda^s \longrightarrow L$ defined by $\varphi(\lambda_1, \ldots, \lambda_s) = \lambda_1 v_1 + \cdots + \lambda_s v_s$ is an epimorphism. Since Λ is finitely generated as R-module, since R is Noetherian by Lemma 7.5.3, we have that L is Noetherian and finitely generated as R-module. Torsion free finitely generated R-modules are projective by Lemma 7.5.10. Since $K\Lambda = A$, we have

$$KL = K\Lambda \cdot L = A \cdot L \supseteq Av_1 + \cdots Av_s = V.$$

This shows the statement. □

8.1.2 Orders as pullbacks

We may ask how general the Examples 8.1.4 are. Again let R be an integral domain and let K be its field of fractions. Let moreover Λ be an R-order in the semisimple K-algebra D. We shall use the fact that Λ is a subring of the semisimple K-algebra D.

Recall the construction of pullbacks from Definition 6.2.7. This describes actually Example 8.1.4.(4), Example 8.1.4.(5) and Example 8.1.4.(6).

We now come to the promised explanation why pullbacks naturally arise when we deal with lattices over orders.

Proposition 8.1.11. *Let R be a Dedekind domain with field of fractions K, and let Λ be an R-order in the semisimple K-algebra $D = K \otimes_R \Lambda$. Let $e^2 = e \in Z(D)$ be a central idempotent of D, and let L be a Λ-lattice. Consider L as an R-submodule of $K \otimes_R L = D \otimes_\Lambda L$. Then $e \otimes L =: eL$ and $(1-e) \otimes L =: (1-e)L$ are Λ-lattices as well, we get that the map*

$$\varphi : eL/(L \cap eL) \longrightarrow (1-e)L/(L \cap (1-e)L)$$

given for each $\ell \in L$ by $\varphi(e\ell) := (1-e)\ell$, is an isomorphism of Λ-modules, and the diagram

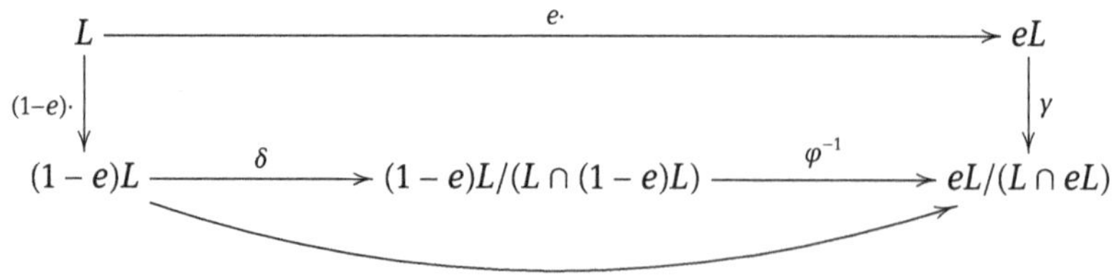

with the canonical projection on the right and on the bottom (followed by the above isomorphism), is a pullback diagram.

Proof. eL is an R-submodule of the D-module $e \cdot (K \otimes_R L)$. The latter is a finite dimensional K-module. Since K is projective as R-module, and since R is a Dedekind domain, eL is a submodule of a projective R-module, hence projective as well. Moreover, eL is an epimorphic image of L, whence eL is finitely generated. Therefore, eL is a Λ-lattice. An analogous argument shows that $(1-e)L$ is a Λ-lattice.

We now show that φ is a well-defined isomorphism. For each $\ell \in L$ we get

$$\ell = 1 \cdot \ell = e\ell + (1-e)\ell$$

and if $e\ell \in L$, then also $(1-e)\ell \in L$, and vice versa. Hence, suppose

$$e\ell + (L \cap eL) = e\ell' + (L \cap eL).$$

Then $e(\ell - \ell') \in L$, which implies $(1-e)(\ell - \ell') \in L$, and therefore

$$\begin{aligned} eL/(L \cap eL) &\xrightarrow{\varphi} (1-e)L/(L \cap (1-e)L) \\ e\ell + (L \cap eL) &\mapsto (1-e)\ell + (L \cap (1-e)L) \end{aligned}$$

is well-defined and clearly a morphism of Λ-modules. By the same argument the inverse of the mapping is a well-defined morphism of Λ-modules.

We now prove that the above diagram is a pullback diagram. Suppose we have a Λ-module X and morphisms of Λ-modules $\xi : X \longrightarrow eL$ as well as $\eta : X \longrightarrow (1-e)L$ such that

$$\varphi \circ \gamma \circ \xi = \delta \circ \eta.$$

Then for each $x \in X$ we have

$$e\xi(x) + (L \cap eL) = e\eta(x) + (L \cap eL)$$

and hence first $e(\xi(x) - \eta(x)) \in L$, and second $(1-e)(\xi(x) - \eta(x)) \in L$. Therefore $\xi(x) + \eta(x) \in L$, and the morphism

$$X \ni x \mapsto \xi(x) + \eta(x) \in L$$

makes the diagram commutative. By Lemma 6.2.8 we proved the statement. □

Corollary 8.1.12. *Let Λ be an R-order in the semisimple algebra D, and let $e^2 = e \in Z(D)$. Then the pullback of the regular Λ-module ${}_\Lambda\Lambda$ with respect to the idempotent e from Proposition 8.1.11 is a pullback of orders, in the sense that $e\Lambda$ and $(1-e)\Lambda$ are orders and the morphisms in the square are morphisms of rings.*

Proof. This follows by direct inspection. □

Example 8.1.13.

1. We study Proposition 8.1.11 for the very easy case $\mathbb{Z}C_2$. We know that $\mathbb{Q}C_2 \simeq \mathbb{Q} \times \mathbb{Q}$ and hence $\mathbb{Z}C_2$ is a subring of $\mathbb{Q} \times \mathbb{Q}$. Denote by c the non neutral element of C_2. We know precisely the central idempotents by Proposition 2.2.8. Actually

$$e_+ = \frac{1}{2}(1+c) \quad \text{and} \quad e_- = \frac{1}{2}(1-c).$$

Then, the regular $\mathbb{Z}C_2$-module $L := \mathbb{Z}C_2$ contains $2\mathbb{Z} \times 2\mathbb{Z}$ since $2e_+ \in \mathbb{Z}C_2$ and $2e_- \in \mathbb{Z}C_2 = L$ and therefore $(2,0) \in \mathbb{Z}C_2$ and $(0,2) \in \mathbb{Z}C_2 = L$. Since 2 is prime, and since $e_+ \cdot \mathbb{Z}C_2 \simeq \mathbb{Z}$, we get $e_+L \cap L = 2\mathbb{Z}$ and therefore

$$\begin{array}{ccc} \mathbb{Z}C_2 & \longrightarrow & \mathbb{Z} \\ \downarrow & & \downarrow \\ \mathbb{Z} & \longrightarrow & \mathbb{Z}/2\mathbb{Z} \end{array}$$

is a pullback. This gives the description of $\mathbb{Z}C_2$ from Example 8.1.4.(6).

2. Let p be some odd prime. Then $\mathbb{Q}C_p \simeq \mathbb{Q} \times \mathbb{Q}(\zeta_p)$ for a primitive p^{th} root of unity ζ_p. The central idempotent corresponding to the trivial module is

$$e_+ = \frac{1}{p} \sum_{c \in C_p} c.$$

Then, as in item (1) we get $\mathbb{Z}C_p e_+ = \mathbb{Z}$ since all group elements are identified after multiplication by e_+. The group element c^j acts as ζ_p^j on $\mathbb{Q}(\zeta_p)$, and hence

$$\mathbb{Z}C_p(1 - e_+) = \mathbb{Z}[\zeta_p]$$

is the smallest subring of $\mathbb{Q}(\zeta_p)$ containing all ζ_p^j. Again

$$\mathbb{Z}C_p \cap \mathbb{Z}C_p e_+ = p\mathbb{Z}C_p e_+ = p\mathbb{Z}$$

and we get a pullback diagram

$$\begin{array}{ccc} \mathbb{Z}C_p & \longrightarrow & \mathbb{Z}[\zeta_p] \\ \downarrow & & \downarrow \pi_p \\ \mathbb{Z} & \longrightarrow & \mathbb{Z}/p\mathbb{Z} \end{array}.$$

What is the morphism π_p? This is the reduction modulo the ideal $\wp$ generated by $1-\zeta_p$. By Lemma 7.1.3 this is a maximal ideal in $\mathbb{Z}[\zeta_p]$ containing p, and $p\mathbb{Z}[\zeta_p] = \wp^{p-1}$.

8.1.3 Orders, lattices and localisation

For a Dedekind domain R and an R-order Λ we will start to give some elementary properties linking for each prime $\wp$ of R properties of Λ with properties of $R_\wp \otimes_R \Lambda$ and likewise for Λ-lattices L with respect to the $R_\wp \otimes_R \Lambda$-lattice $R_\wp \otimes_R L$. Recall that if R is a Dedekind domain, then by Lemma 7.5.9 the localisation $R_\wp$ is a Dedekind domain again.

Proposition 8.1.14. *Let R be a Dedekind domain with field of fractions K, and let Λ be an R-order in a semisimple K-algebra A. Let* $\mathrm{Spec}(R)$ *be the set of all prime ideals of R. For each $\wp \in \mathrm{Spec}(R)$ denote $R_\wp \otimes_R \Lambda =: \Lambda_\wp$ and for each Λ-lattice L let $R_\wp \otimes_R L =: L_\wp$. Then*

1. *If Λ is an R-order in A, then $\Lambda_\wp$ is an $R_\wp$-order in A for each prime $\wp$ of R.*
2. *If L is a Λ-lattice, then $L_\wp$ is a $\Lambda_\wp$-lattice for each prime $\wp$ of R.*
3. *For any Λ-lattice L we have*

$$L = \bigcap_{\wp \in Spec(R)} L_\wp,$$

   where we consider $R_\wp \subseteq K$ and hence $L_\wp = R_\wp \otimes_R L \subseteq K \otimes_R L$. The intersection is taken inside $K \otimes_R L$.
4. *For each $\wp \in \mathrm{Spec}(R)$ let $M(\wp)$ be a $\Lambda_\wp$-lattice, and suppose that $KM(\wp) = V$ is a fixed K-vector space. If there is a Λ-lattice N such that*

$$N_\wp = M(\wp)$$

   for all but a finite number of $\wp \in \mathrm{Spec}(R)$, then there is a Λ-lattice M such that $M_\wp = M(\wp)$ for all primes $\wp$ of R.
5. *Items 1, 2, 3, 4 hold analogously when we replace localisation by completion.*

Proof. If Λ is an R-order in A, then Λ contains a K-basis B of A, and since B is still contained in $\Lambda_\wp$, we have $K \otimes_{R_\wp} \Lambda_\wp = A$. Moreover, since we only need to show that $\Lambda_\wp$ is finitely generated projective as $R_\wp$-module, and since for item 2 we need to prove the same for a Λ-lattice L, we may, and will, pass directly to the proof of item 2. Since L is finitely generated projective as R-module, L is a direct factor of R^m for some m, in other words, $L \oplus M = R^m$ as R-module, for some R-module M. Then, using Lemma 3.1.10 and Lemma 3.1.11,

$$R_\wp^m = R_\wp \otimes_R R^m = R_\wp \otimes_R (L \oplus M) = (R_\wp \otimes_R L) \oplus (R_\wp \otimes_R M).$$

Hence $L_\wp$ is finitely generated projective as $R_\wp$-module. This shows at once item 1 and item 2.

Let us show now item 3. We consider the map

$$L \xrightarrow{\lambda} K \otimes_R L$$
$$x \mapsto 1 \otimes x$$

Since we only use the R-module structure, we can identify L with a direct factor of R^m for some integer m, and observing $K \otimes_R R^m \simeq K^m$, we need to consider the natural map $R^m \hookrightarrow K^m$ instead of λ. Therefore, the statement in the item 3 is equivalent with the statement that

$$\bigcap_{\wp \in \mathrm{Spec}(R)} R_\wp = R$$

and where the left hand side is taken as subsets $R_\wp$ of K. Clearly $R \subseteq \bigcap_{\wp \in \mathrm{Spec}(R)} R_\wp$. Let

$$x = \frac{y}{z} \in K$$

for some $y, z \in R$, and suppose that $x \in R_\wp$ for all primes $\wp$. Hence, for each prime $\wp$ there is $y_\wp \in R$, $z_\wp \in R \setminus \wp$ with

$$x = \frac{y}{z} = \frac{y_\wp}{z_\wp} \in R_\wp.$$

Hence $yz_\wp = zy_\wp$, and if $z \in \wp$, also $y \in \wp$ since $\wp$ is a prime ideal. By Lemma 7.5.25 there are only finitely many prime ideals $\wp_1, \dots, \wp_n$ of R containing zR. Let I be the ideal of R generated by $z, z_{\wp_1}, \dots, z_{\wp_n}$. We claim that $I = R$. Else I is contained in a maximal ideal $\wp$. Since $z \in I$, there is an $i \in \{1, \dots, n\}$ such that $I \subseteq \wp_i$. But also $z_{\wp_i} \in I \subseteq \wp_i$, which is a contradiction to the choice of $z_\wp$. Hence there are $r, r_1, \dots, r_n \in R$ with

$$1 = rz + r_1 z_{\wp_1} + \cdots + r_n z_{\wp_n}$$

and

$$\begin{aligned} y &= yrz + yr_1 z_{\wp_1} + \cdots + yr_n z_{\wp_n} \\ &= yrz + y_{\wp_1} r_1 z + \cdots + y_{\wp_n} r_n z \\ &= z(yr + y_{\wp_1} r_1 + \cdots + y_{\wp_n} r_n). \end{aligned}$$

Hence

$$x = \frac{y}{z} = yr + y_{\wp_1} r_1 + \cdots + y_{\wp_n} r_n \in R.$$

This shows item 3.

We need to show item 4. Since $N_\wp = M(\wp)$ for all but a finite number of primes $\wp_1, \dots, \wp_n$ of R, there is an $0 \neq r \in R$ is such that $r \cdot M(\wp) \subseteq N_\wp$ for all primes $\wp$. We now replace N by the isomorphic copy $\frac{1}{r}N$ and obtain therefore $M(\wp) \subseteq N_\wp$ for all primes $\wp$. Let

$$M := N \cap M(\wp_1) \cap M(\wp_2) \cap \cdots \cap M(\wp_n).$$

Then M is a lattice since N is a Λ-lattice and $M(\wp_i)$ are $\Lambda_{\wp_i}$-sub lattices of $N_{\wp_i}$ for all $i \in \{1, \ldots, n\}$. Moreover, if $i \neq j$, then $M(\wp_i)_{\wp_j} = V$ since when we localise at two different primes one after the other, we get the entire vector space. Hence, using $M(\wp) \subseteq N_\wp$ for all primes $\wp$, we get

$$M_\wp = N_\wp \cap M(\wp) = M(\wp)$$

for all primes $\wp \in \{\wp_1, \ldots, \wp_n\}$ of R. For the other primes $\wp \notin \{\wp_1, \ldots, \wp_n\}$ of R, we get

$$M_\wp = N_\wp = M(\wp)$$

and obtain this way item 4.

In order to show item 5 we need to show the following

Lemma 8.1.15. *Let V be a finite dimensional K-vector space and let $\widehat{V} := \widehat{K}_\wp \otimes_K V$. Then for any $\widehat{R}_\wp$-lattice $\widehat{T}$ we have that $\widehat{T} \cap V$ is an $R_\wp$-lattice in V. We get $\widehat{R}_\wp \otimes_{R_\wp} (\widehat{T}_\wp \cap V) = \widehat{T}$ for any $\widehat{R}_\wp$-lattice $\widehat{T}$ and $V \cap (\widehat{R}_\wp \otimes_R T) = T$ for any $R_\wp$-lattice T. The construction gives a bijection between $R_\wp$-lattices containing a basis of V and $\widehat{R}_\wp$-lattices containing a basis of $\widehat{V}$.*

Proof. Note that K is embedded into $\widehat{K}$ as constant (Cauchy) sequences with values in K. Hence $\widehat{R}_\wp \cap K = R$. Replacing V by $K \otimes_R T$ we may and will assume that all lattices are full, i. e. contain a K-basis of V. Since finitely generated projective $R_\wp$-modules are free, let $T := \bigoplus_{i=1}^n R_\wp x_i$ for a basis $\{x_1, \ldots, n\}$. Then $V = \bigoplus_{i=1}^n Kx_i$, $\widehat{T} = \bigoplus_{i=1}^n \widehat{R}_\wp x_i$ gives

$$\widehat{T} \cap V = \bigoplus_{i=1}^n (K \cap \widehat{R}_\wp)x_i = \bigoplus_{i=1}^n R_\wp x_i = T.$$

Furthermore if $\widehat{V} := \widehat{K}_\wp \otimes_K V$ and $\widehat{T} = \bigoplus_{i=1}^n \widehat{R}_\wp y_i$, $\widehat{V} = \bigoplus_{i=1}^n \widehat{K}_\wp y_i$, then $V = \bigoplus_{i=1}^n Ky_i$. Now, applying a base change from the basis $\{y_1, \ldots, y_n\}$ to $\{x_1, \ldots, x_n\}$ we also obtain the second statement of the lemma. □

Using Lemma 8.1.15 we obtain the same results for completions as for localisations. This shows Proposition 8.1.14. □

8.2 Reduced norms, traces, characteristic polynomials

Let K be a field and let A be a semisimple K-algebra. We want to define the trace and the norm of an element $a \in A$. The most natural attempt would be to consider the

K-linear map

$$\begin{aligned} A &\xrightarrow{\mu_a} A \\ b &\mapsto ab \end{aligned}$$

Choosing a K-basis B of A, then μ_a is given by a matrix $M_{a,B}$ and we could define the trace of a as $trace(M_{a,B})$ and the norm as $det(M_{a,B})$. Elementary linear algebra shows that these concepts do not depend on the choice of B. However, the concept is not the concept which is most appropriate. Consider for the moment $A = Mat_n(K)$. Then $A = S^n$ as left module, for a simple A-module S. We may as well consider

$$\begin{aligned} S &\xrightarrow{\mu_a^S} S \\ x &\mapsto ax \end{aligned}$$

and the associated matrix or a choice of a basis C is $N_{a,C}$. But the fact that $A \simeq S^n$ we get $det(N_{a,C})^n = det(M_{a,B})$ and $n \cdot trace(N_{a,C}) = trace(M_{a,B})$.

We start with an important result which is interesting in its own right.

Theorem 8.2.1 (Skolem Noether theorem). *Let K be a field and let A be a central simple K-algebra. Then any K-linear algebra automorphism of A is inner.*

Proof. Let $\varphi \in Aut_K(A)$ be an algebra automorphism of A. Let S be a simple A-module and let $D = End_A(S)$. By Schur's Lemma 1.2.21 we know that D is a skew-field. Then S is an $A-D$-bimodule. We consider the $A-D$-bimodule ${}_{\varphi}S_1$ where we twist the action on the left by the automorphism φ. More precisely, the K-vector spaces S and ${}_{\varphi}S_1$ coincide. However, let $d \in D$, $a \in A$ and $x \in S$. Then $a \cdot x \cdot d := \varphi(a)xd$ where we use the notion $\cdot$ to denote the action on ${}_{\varphi}S_1$. Recall that A is a matrix algebra over a skew-field D'. Since $D \otimes_K D'$ is simple by Proposition 5.4.1, we obtain that $A \otimes_K D$ is a matrix algebra over a finite dimensional simple algebra hence is a simple algebra as well (cf. Remark 1.2.42). Therefore the simple $A \otimes_K D$-modules S and ${}_{\varphi}S_1$ are isomorphic. Let $\alpha : S \longrightarrow {}_{\varphi}S_1$ be an isomorphism. Then

$$\alpha(axd) = a \cdot \alpha(x) \cdot d = \varphi(a)\alpha(x)d$$

for all $a \in A$, $d \in D$ and $x \in S$. But α is in particular a D-module homomorphism, and S has the double centralizer property by Lemma 5.4.2 and Corollary 5.4.3. Therefore, $End_D(S) \simeq A$ and α is given by left multiplication by some $t \in A$. Since α is an isomorphism, t is invertible. Therefore the above equation gives $taxd = \varphi(a)txd$. Choosing $d = 1$ and the fact that S is a faithful A-module we get $ta = \varphi(a)t$ as required. □

Suppose now that A is a simple K-algebra. Then there is a skew-field D containing K such that $A = Mat_m(D)$ for some integer m. We now use Theorem 5.4.6. There is hence a splitting field L for A, i. e. an extension field L of K such that there is an isomorphism

$L \otimes_K A \xrightarrow{\varphi} Mat_n(L)$ for some integer n, and such that $n = \dim_K L$. Then we may define a trace of an element in a as the trace of $\mu_a(S_L)$, for S_L being a simple $A \otimes_K L$-module. Let $L \otimes_K A \xrightarrow{\psi} Mat_n(L)$ be another isomorphism, then $\psi \circ \varphi^{-1}$ is an L-linear automorphism of $Mat_n(L)$. By the Skolem Noether Theorem 8.2.1 this automorphism is inner. Hence, there is an element $T \in Gl_n(L)$ such that $\varphi(a) = T \cdot \psi(a) \cdot T^{-1}$ for any $a \in L \otimes_K A$. Therefore the characteristic polynomial of $\varphi(a)$ equals the characteristic polynomial of $\psi(a)$.

Definition 8.2.2. Let A be a simple finite dimensional K-algebra and let L be a splitting field of A. Let $L \otimes_K A \xrightarrow{\varphi} Mat_n(L)$ be an isomorphism of L-algebras. Then for any $a \in A$ the *reduced characteristic polynomial* $chaRed_a$ is the characteristic polynomial of $\varphi(1 \otimes a)$. If

$$chaRed_a(X) = X^n - t_{n-1}X^{n-1} + \cdots + (-1)^{n-1}t_1X + (-1)^n t_0$$

then we say that $t_{n-1} = tRed(a)$ is the *reduced trace* of A, and $t_0 = noRed(a)$ is the *reduced norm* of a.

Proposition 8.2.3. *Let K be a field of characteristic 0, and let A be a central simple finite dimensional K-algebra and let L be a splitting field of A. Then $chaRed_a$ does not depend on the choice of L, nor does it depend on the choice of the isomorphism $L \otimes_K A \xrightarrow{\varphi} Mat_n(L)$. Moreover, $chaRed_a(X) \in K[X]$.*

Proof. If E is another splitting field for A, then there is a field F containing both E and L. Indeed, let $C := E \otimes_K L$. This is a commutative K-algebra. Let $\mathfrak{m}$ be a maximal ideal of C and consider $F := C/\mathfrak{m}$. The map $E \longrightarrow C$ given by $e \mapsto e \otimes 1$ and $L \longrightarrow C$ given by $\ell \mapsto 1 \otimes \ell$ compose with the natural projection $C \longrightarrow F$ to ring homomorphisms $E \longrightarrow F$ and $L \longrightarrow F$. Since E and L are fields, these are injective. Now, if L is a splitting field, so is F. Hence in order to prove the independence of the reduced characteristic polynomial it is sufficient to show the statement in case $L \subseteq F$. However, then the isomorphism $F \otimes_K A \xrightarrow{\psi} Mat_n(F)$ is a composition

$$F \otimes_K A = F \otimes_L L \otimes_K A \xrightarrow{\simeq} F \otimes_L Mat_n(L) \xrightarrow{\simeq} Mat_n(F)$$

and for any $a \in A$ we have that $\varphi(1 \otimes a)$ and $\psi(1 \otimes a)$ are the same matrices when we identify $Mat_n(L)$ inside $Mat_n(F)$ by the identification of L inside F. Hence the reduced characteristic polynomial is independent of the choice of the splitting field.

We need to show that the coefficients of the reduced characteristic polynomial are in K. Since K is of characteristic 0 all field extensions are separable. Extending L if necessary we may assume that L is a Galois extension over K. Let $Gal(L : K)$ be the Galois group. Hence, for any $g \in Gal(L : K)$ the isomorphism $L \otimes_K A \xrightarrow{\varphi} Mat_n(L)$ induces another isomorphism

$$L \otimes_K A \xrightarrow{g \otimes id_A} L \otimes_K A \xrightarrow{\varphi} Mat_n(L)$$

and since the reduced characteristic polynomial is independent of the isomorphism by the remarks preceding Definition 8.2.3 the group $Gal(L : K)$ acts trivially on the reduced characteristic polynomial of a. Hence all coefficients are in the fixed field, which is K. □

Corollary 8.2.4. *Let A be a central simple K-algebra for a field K of characteristic 0. Then for all $a, b \in A$ we have $noRed(a \cdot b) = noRed(a) \cdot noRed(b)$ and $tRed(a + b) = tRed(a) + tRed(b)$.*

Proof. Since, up to a sign depending only on the degree, $tRed$ is the second highest coefficient and since $noRed$ is the constant coefficient of the reduced characteristic polynomial the result follows from general properties of the characteristic polynomial. Indeed these coefficients are the traces and determinants of the corresponding matrices. Traces are additive and determinants are multiplicative. □

Corollary 8.2.5. *Let A be a central simple K-algebra for a field K of characteristic 0. Then*

$$\begin{aligned} A \times A &\xrightarrow{\tau} K \\ (a, b) &\mapsto tRed(a \cdot b) \end{aligned}$$

is a symmetric non degenerate bilinear form on A. Moreover, $\tau(a, bc) = \tau(ab, c)$ for all $a, b, c \in A$.

Proof. Let E be a splitting field for A and let $1 \otimes a, 1 \otimes b \in Mat_n(E)$ be the images of a, b in $Mat_n(E)$. Then

$$\begin{aligned} tRed(ab) &= trace(1 \otimes ab) \\ &= trace((1 \otimes a) \cdot (1 \otimes b)) \\ &= trace((1 \otimes b) \cdot (1 \otimes a)) \\ &= trace(1 \otimes ba) \\ &= tRed(ba) \end{aligned}$$

and τ is symmetric. Since by Corollary 8.2.4 $tRed$ is additive as well, τ is bilinear, and moreover $\{a \in A \mid \tau(a, b) = 0 \ \forall b \in A\}$ is a twosided ideal of A. Since A is simple this is 0, and hence τ is non degenerated. The equation $\tau(a, bc) = \tau(ab, c)$ for all $a, b, c \in A$ is clear. □

Consider now the central simple L-algebra A as a simple K-algebra for some subfield K of L such that $\dim_K L < \infty$. Then, since L is a K-vector space, we choose a K-basis of L and for each $\ell \in L$ consider the matrix M_ℓ defining the multiplication by ℓ on the K-vector space L with respect to the chosen basis. Let $N_K^L : L \longrightarrow K$ be defined as $N_K^L(\ell) := det(M_\ell)$ and $tr_K^L(\ell) := trace(M_\ell)$ for all $\ell \in L$. If $f(X) = \sum_{i=0}^m a_i X^i \in L[X]$, then each coefficient a_i can be replaced by its matrix M_{a_i} with coefficients in K, giving hence a matrix $M_{f(X)}$ with coefficients in $K[X]$. Then we define $N_K^L(f(X)) = det(M_{f(X)})$. Observe that we first have to form the matrix $M_{f(X)}$ and then compute the determinant. The polynomial we obtain is *not the same* as the polynomial formed by the norms of the coefficients!

Definition 8.2.6. Let A be a simple K-algebra over some field of characteristic 0 and suppose that $L = Z(A)$ is a finite dimensional K-vector space. Then, the *reduced characteristic polynomial relative to* K is

$$chaRedpol_K(a) := N_K^L(chaRedpol(a))$$

is the reduced characteristic polynomial of the K-algebra A. The *reduced trace relative to* K is the opposite of the coefficient of $chaRedpol_K(a)$ of second highest degree and the *reduced norm relative to* K is $(-1)^n$ of the constant term of $chaRedpol_K(a)$.

Note that by definition and a little elementary linear algebra we get

$$NoRed_K(a) = N_K^{Z(A)}(noRed(a))$$

and

$$tRed_K(a) = tr_K^{Z(A)}(tRed(a)).$$

Corollary 8.2.7. *Let K be a field of characteristic 0 and let A be a finite dimensional simple K-algebra. Then*

$$\begin{aligned} A \times A &\xrightarrow{\tau} K \\ (a,b) &\mapsto tRed_K(a \cdot b) \end{aligned}$$

is a symmetric non degenerate bilinear form on A.

Proof. By Corollary 8.2.5 the result holds in case $A = Z(A)$. Since $tRed_K(a) = tr_K^{Z(A)}(tRed(a))$, the result then follows from the fact that $(a,b) \mapsto tr_K^L(ab)$ is a non degenerate bilinear form for separable field extensions L over K. □

We come to the semisimple case.

Let A be a finite dimensional semisimple K-algebra for a field K of characteristic 0. Then by Wedderburn's Theorem 1.2.30 we have

$$A \simeq A_1 \times \cdots \times A_s$$

for simple algebras $A_1, \ldots, A_s$. Then each $a \in A$ defines $a = (a_1, \ldots, a_s) \in A_1 \times \cdots \times A_s$. We define

$$chaRedpol_K(a) := \prod_{i=1}^{s} chaRedpol_K(a_i)$$

and $tRed_K(a)$ to be the opposite of the coefficient of second highest degree and $noRed_K(a)$ to be $(-1)^n$ times the constant coefficient of $chaRedpol_K(a)$, where $n = \dim_K A$.

Corollary 8.2.8. *Let K be a field of characteristic 0 and let A be a finite dimensional semisimple simple K-algebra. Then*

$$\begin{aligned} A \times A &\xrightarrow{T} K \\ (a, b) &\mapsto tRed_K(a \cdot b) \end{aligned}$$

is a symmetric non degenerate bilinear form on A.

Proof. By Corollary 8.2.7 this holds for every simple component of A. Then the result also follows for the direct product. □

8.3 Maximal orders

8.3.1 Definition, existence and examples

Let R be a Dedekind domain with field of fractions K. Let A be a finite dimensional semisimple K-algebra. Then the set of R-orders $\mathfrak{O}_A$ in A is partially ordered by inclusion.

Definition 8.3.1. Let R be a Dedekind domain with field of fractions K and let Λ be an R-order in the semisimple K-algebra A. Then a maximal element in $\mathfrak{O}_A$ is called a *maximal R-order*.

Lemma 8.3.2. *Let R be a Dedekind domain with field of fractions K. If Δ is a maximal order in the semisimple K-algebra A, then $Mat_n(\Delta)$ is a maximal order in $Mat_n(A)$.*

Proof. Suppose Γ is an R-order in $Mat_n(D)$ containing $\Lambda := Mat_n(\Delta)$. Then there is an element $\gamma = (\gamma_{i,j})_{1\le i,j\le n} \in \Gamma \setminus \Lambda$. Hence there must be some index (s, t) with $\gamma_{s,t} \notin \Delta$. Let e_k be the matrix whose only one non zero coefficient 1 is in position (k, k). Then consider the set $e_i\Lambda e_j =: \Sigma_{i,j}$. We claim that Σ is an order in D. For this we first note that permutation matrices are in Λ. Then, by multiplication with a permutation matrix exchanging the first and the i-th column, and from the left with a permutation matrix exchanging the j-th row with the first row, we see that actually we have $\Sigma_{i,j} = \Sigma_{1,1} =: \Sigma$, and therefore we may assume $i = j = 1$. But then it is clear that the set Σ is a subring of D and also contains Δ. We need to show that Σ is an order in Δ. Since Δ is a maximal order in D, and since $\Delta \subseteq \Sigma$, we get $\Delta = \Sigma$ once we know that Σ is an order. We hence need to show that Σ is an order in D. But since $\Delta \subseteq \Sigma$, and since $K \otimes_R \Delta = D$, the order Δ contains a K-basis of D. Hence also Σ contains a K-basis of D, and $K \otimes_R \Sigma = D$. Now,

Γ is an R-order, hence finitely generated R-projective. Since

$$\Gamma = \bigoplus_{1\le i,j\le n} e_i\Gamma e_j$$

as R-modules, also the direct factor $e_1\Gamma e_1 = \Sigma$ is finitely generated R-projective as well. □

By Wedderburn's theorem 1.2.30 we see that

$$A \simeq \prod_{i=1}^{\ell} Mat_{n_i}(D_i)$$

for finite dimensional skew fields D_i whose centre contain K. If Δ_i is a maximal order in D_i for each i, then

$$\Lambda := \prod_{i=1}^{\ell} Mat_{n_i}(\Delta_i)$$

is a maximal order in A. We hence need to find maximal orders in skew fields.

Proposition 8.3.3. *Let R be a Dedekind domain and let Λ be an R-order. Then any element in Λ is integral over R. Moreover, its minimal polynomial and also its reduced characteristic polynomial are both in $R[X]$.*

Proof. Let $a \in \Lambda$. Then the subring $R[a]$ of Λ generated by a and R is again finitely generated since Λ is a finitely generated R-module, and since R is Noetherian by Lemma 7.5.3. Therefore a is integral over R by Lemma 2.4.3. Let $g(X) \in K[X]$ be the minimal polynomial of a. Since a is integral, there is a monic polynomial $f(X) \in R[X]$ with $f(a) = 0$. Since $g(X)$ is the minimal polynomial of a, we get $f(X) = g(X)h(X)$ for some monic polynomial $h(X) \in K[X]$. By Lemma 7.5.14 R is integrally closed. By Lemma 7.5.13 we get the result for the reduced minimal polynomial. The statement for the reduced characteristic polynomial is analogous. □

Remark 8.3.4. Suppose now that no skew field occurs in the Wedderburn decomposition of a finite dimensional semisimple K-algebra A. Then

$$A \simeq \prod_{i=1}^{\ell} Mat_{n_i}(K_i)$$

for finite field extensions K_i of K. Proposition 8.3.3 then implies that

$$\Lambda := \prod_{i=1}^{\ell} Mat_{n_i}(\mathrm{algint}_R(K_i))$$

is a maximal order in A.

Remark 8.3.5. Note that not every maximal order is of the type described in Remark 8.3.4. Indeed, if

$$\Lambda := \prod_{i=1}^{\ell} Mat_{n_i}(\text{algint}_R(K_i))$$

and if $u \in A^\times$ is any invertible element in A, then $u\Lambda u^{-1}$ is a maximal order as well. An element u is just the direct product of invertible matrices in each of the algebras $Mat_{n_i}(K_i)$. However, it can be shown that

Remark 8.3.6. Let R be a Dedekind domain and let Λ be an R-order in a semisimple K-algebra A. Let $e^2 = e \in Z(A)$ be a central idempotent in A. Then again $e\Lambda + (1-e)\Lambda$ is an R-order in A. Clearly, $\Lambda \subseteq e\Lambda + (1-e)\Lambda$, but in general the inclusion is strict as it is seen by Example 8.1.4.(4) for example. Hence a maximal order in Λ is a direct product of maximal orders in matrix algebras over skew-fields.

The next result is a very nice characterisation of orders in the spirit of Proposition 8.3.3.

Theorem 8.3.7. *Let R be a Dedekind domain of characteristic 0 with field of fractions K, and let Λ be an R-subalgebra of the semisimple K-algebra A containing a K-basis of A. Then Λ is an R-order if and only if any element of Λ is integral over R.*

Proof. If Λ is an R-order, then Proposition 8.3.3 shows that all elements of Λ are integral over R. The converse also holds. Indeed, we only need to show that Λ is an R-lattice. By hypothesis, there is $B := \{u_1, \dots, u_n\} \subseteq \Lambda$, a K-basis of A. Consider the matrix

$$U := (\text{tRed}_K(u_i u_j))_{1 \le i,j \le n}$$

and let

$$\alpha := \det(U)$$

be the determinant of this matrix. By Proposition 8.3.3 the reduced characteristic polynomial of each element $u_i u_j$ is in $R[X]$. Hence, the reduced trace of the element $u_i u_j$, which is, up to sign, the coefficient of the second highest degree in the polynomial, is in R as well. Hence, the matrix U has coefficients in R, and so is its determinant α. Using that K is of characteristic 0, by Corollary 8.2.8 the bilinear form

$$\begin{aligned} A \times A &\longrightarrow K \\ (x, y) &\mapsto \text{tRed}_K(xy) =: \langle x, y \rangle \end{aligned}$$

is non degenerate. Since U is the Gram matrix of this bilinear form with respect to the basis B, and hence the determinant α of U is non zero. We shall prove that

$$(*) \quad \Lambda \subseteq \frac{1}{\alpha}(Ru_1 + \cdots + Ru_n).$$

This will show our result since by Definition 7.5.1 and Theorem 6.4.4 the ring R is hereditary, and therefore any R-submodule of the free (and therefore projective) R-module

$$\frac{1}{\alpha}(Ru_1 + \cdots + Ru_n) \simeq (Ru_1 + \cdots + Ru_n)$$

is projective. We shall now show the inclusion $(*)$. Let $x = \sum_{i=1}^{n} x_i u_i$ for coefficients $x_i \in K$. Then for each $j \in \{1, \ldots, n\}$ we have

$$\langle x, u_j \rangle = \mathrm{tRed}_K(xu_j) = \sum_{i=1}^{n} x_i \cdot \mathrm{tRed}_K(u_j u_i) = \sum_{i=1}^{n} x_i \cdot \langle u_i, u_j \rangle.$$

Since x and u_j are in Λ, also

$$\langle x, u_j \rangle = \mathrm{tRed}_K(xu_j) \in R.$$

Hence, we have a linear equation

$$\begin{pmatrix} \langle x, u_1 \rangle \\ \vdots \\ \langle x, u_n \rangle \end{pmatrix} = U \cdot \begin{pmatrix} x_1 \\ \vdots \\ x_n \end{pmatrix}$$

where the coefficients of the vector on the left are in R, and where the coefficients of U are in R. The determinant of U is α, and hence U^{-1} has coefficients in $\frac{1}{\alpha}R$ by Cramer's rule. This shows that

$$x_i \in \frac{1}{\alpha}R$$

which shows the result. □

In the proof of Theorem 8.3.7 we used in most prominent way the element $\alpha = \det(U)$ for U being the matrix $(\mathrm{tRed}_K(u_i u_j))_{1 \le i,j \le n}$ for a K-basis of A in Λ.

Definition 8.3.8. If Λ is an R-order in the semisimple K-algebra A and if $\dim_K(A) = n$ we call

$$d(\Lambda) := R \cdot \{\det(\mathrm{tRed}_K(u_i u_j))_{1 \le i,j \le n} \mid (u_1, \ldots, u_n) \in \Lambda^n\}$$

the ideal of R generated by the set $\{\det(\mathrm{tRed}_K(u_i u_j))_{1 \le i,j \le n} \mid (u_1, \ldots, u_n) \in \Lambda^n\}$, the *discriminant* of the order Λ.

Lemma 8.3.9. *Let R be a Dedekind domain with field of fractions K, and let Λ and Γ be R-orders in the semisimple K-algebra A. Then*

$$\Lambda \subseteq \Gamma \Rightarrow d(\Lambda) \subseteq d(\Gamma)$$

and

$$\Lambda \subsetneq \Gamma \Rightarrow d(\Lambda) \subsetneq d(\Gamma)$$

Proof. If $\Lambda \subseteq \Gamma$, then

$$d(\Lambda) = R \cdot \{\det(\mathrm{tRed}_K(u_i u_j))_{1\le i,j\le n} \mid (u_1,\dots,u_n) \in \Lambda^n\}$$
$$\subseteq R \cdot \{\det(\mathrm{tRed}_K(u_i u_j))_{1\le i,j\le n} \mid (u_1,\dots,u_n) \in \Gamma^n\} = d(\Gamma).$$

We first suppose that R is a discrete valuation domain. Then R is a principal ideal domain, and by the main structure theorem on principal ideal domains we can find an R-basis $\{\lambda_1,\dots,\lambda_n\}$ of Γ such that there are non zero elements $r_1,\dots,r_n$ of R (the elementary divisors) such that $\{r_1\lambda_1,\dots,r_n\lambda_n\}$ is an R-basis of Λ. Then let

$$\alpha_\Gamma := \det(\mathrm{tRed}_K(\lambda_i\lambda_j))_{1\le i,j\le n}$$

and observe that

$$\alpha_\Lambda := \det(\mathrm{tRed}_K(r_i\lambda_i \cdot r_j\lambda_j))_{1\le i,j\le n} = \prod_{i=1}^n r_i^2 \alpha_\Gamma.$$

Now, since $\{\lambda_1,\dots,\lambda_n\}$ is an R-basis of Γ, each other n-tuple $\{\delta_1,\dots,\delta_n\}$ in Γ^n is image of $\{\lambda_1,\dots,\lambda_n\}$ under multiplication by an n by n square matrix T with coefficients in R. Therefore

$$\det(\mathrm{tRed}_K(\delta_i\delta_j))_{1\le i,j\le n} \in R\alpha_\Gamma.$$

Hence $d(\Gamma) = R\alpha_\Gamma$ and likewise $d(\Lambda) = R\alpha_\Lambda$. This shows that $d(\Lambda) = d(\Gamma)$ implies that $r_1,\dots,r_n$ are all invertible in R, and therefore $\Lambda = \Gamma$.

For a prime ideal $\wp$ of R and an R-lattice L we denote $L_\wp := R_\wp \otimes_R L$. We first note that for all prime ideals $\wp$ of R we get by definition

$$d(\Lambda)_\wp = d(\Lambda_\wp).$$

Using Proposition 8.1.14 the case of local R gives the statement here as well. □

Theorem 8.3.10. *Let R be a Dedekind domain and let Λ be an R-order in a semisimple algebra A. Then there is a maximal R-order Γ in A containing Λ.*

Proof. We consider the set $\mathfrak{O}_A(\Lambda)$ of R-orders in A containing Λ. Clearly $\Lambda \in \mathfrak{O}_A(\Lambda)$. Let $(\Lambda_i)_{i\in I}$ be a totally ordered subset in $\mathfrak{O}_A(\Lambda)$. Then let

$$\widehat{\Lambda} := \bigcup_{i\in I} \Lambda_i.$$

We first observe that $\widehat{\Lambda}$ is a subring of A. Indeed, let $a,b \in \widehat{\Lambda}$. Then $a \in \Lambda_{i_a}$ and $b \in \Lambda_{i_b}$ for some $i_a, i_b \in I$. Without loss of generality we may suppose that $\Lambda_{i_a} > \Lambda_{i_b}$ and then $a - b \in \Lambda_{i_a} \subseteq \widehat{\Lambda}$. Likewise $ab \in \Lambda_{i_a} \subseteq \widehat{\Lambda}$, and also $1 \in \widehat{\Lambda}$ since it is contained in any of the Λ_i. Let $a \in \widehat{\Lambda}$. Then again $a \in \Lambda_{i_a}$ for some $i_a \in I$, and hence a is integral over R by Proposition 8.3.3. Now, Theorem 8.3.7 then shows that $\widehat{\Lambda}$ is an order and this then implies by Zorn's lemma that $\mathfrak{O}_A(\Lambda)$ has maximal elements. Any such maximal element satisfies the properties needed for the statement. □

Remark 8.3.11. If R is a Dedekind domain with field of fractions K, and if A is a commutative semisimple K-algebra. Then there is a unique maximal R-order in A. Indeed, the set of algebraic integers over R in A is a ring (cf. Definition 2.4.5). Hence by Theorem 8.3.7 we get that the ring of algebraic integers over R is the unique maximal R-order in A.

Corollary 8.3.12. *Let R be a Dedekind domain and let Λ be an R-order in a semisimple algebra. If its discriminant ideal $d(\Lambda)$ is all of R, then Λ is maximal.*

Indeed, let Γ be a maximal order containing Λ. By $d(\Lambda) \subseteq d(\Gamma)$ with equality if and only if $\Lambda = \Gamma$. If now $d(\Lambda) = R$, then $d(\Gamma) = R$ as well, and hence $\Lambda = \Gamma$ is maximal. □

How different is an order compared to a maximal one which it contains? A first rough approximation is the following statement.

Proposition 8.3.13. *Let R be a Dedekind domain with field of fractions K and let Λ and Γ be R-orders in a semisimple K-algebra A. Then there is $t \in R\setminus\{0\}$ such that $t\Gamma \subseteq \Lambda \subseteq \Gamma$. Moreover, if $\Lambda \subseteq \Gamma$, then Γ/Λ is a finitely generated R/tR-module. Similarly, if L and M are Λ-lattices with $KL = KM$, then there is a non zero $r \in R$ with $rM \subseteq L$.*

Proof. Since Γ is a finitely generated projective R-module, let $\{u_1, \dots, u_m\}$ be a generating set of Γ with $\Gamma = Ru_1 + \dots + Ru_m$. Since $K\Lambda = A = K\Gamma$, there are non zero elements $k_1, \dots, k_m \in K$ with $k_s u_s \in \Lambda$ for all $s \in \{1, \dots, m\}$. Since $k_s = \frac{r_s}{t_s}$ for all s, with $t := t_1 \cdot \dots \cdot t_m$ we get

$$t\Gamma = t(Ru_1 + \dots + Ru_m) = \sum_{s=1}^{m} Rtu_s \subseteq \sum_{s=1}^{m} R \cdot t \cdot r_s \cdot u_s = \sum_{s=1}^{m} R \cdot \frac{t}{t_s} \cdot k_s \cdot u_s \subseteq \Lambda$$

since $\frac{t}{t_s} \in R$ for all s. Hence Γ/Λ is a quotient of $\Gamma/t\Gamma$. Clearly $\Gamma/t\Gamma$ is a finitely generated R/tR-module. The case of lattices is proved completely analogously. This shows the statement. □

Proposition 8.3.13 enables us to progress further and to introduce the following classical and important construction.

Lemma 8.3.14. *Let R be a Dedekind domain with field of fractions K and let A be a finite dimensional semisimple K-algebra. Let L be an R-lattice such that $KL = A$. Then*

$$\mathcal{O}_\ell(L) := \{\lambda \in A \mid \lambda L \subseteq L\}$$

and

$$\mathcal{O}_r(L) := \{\lambda \in A \mid L\lambda \subseteq L\}$$

are R-orders in A.

Proof. We show the statement for $\mathcal{O}_\ell(L)$. The statement for $\mathcal{O}_r(L)$ is analogous. Since L is an R-lattice, for each $\lambda \in A$ we get that λL is an R-lattice again. By Proposition 8.3.13 there is $r \in R$ with $r\lambda L \subseteq L$. Hence, $r\lambda \in \mathcal{O}_\ell(L)$. Therefore $K\mathcal{O}_\ell(L) = A$. Moreover, there is $s \in R$ such that $s \cdot 1 \in L$. Hence $\mathcal{O}_\ell(L) \cdot s \subseteq L$ which shows $\mathcal{O}_\ell(L) \subseteq \frac{1}{s}\Lambda$. Since Λ is an R-lattice, also $\frac{1}{s}\Lambda$ is Noetherian, and $\mathcal{O}_\ell(L)$ is a Noetherian R-module, and hence also a finitely generated R-module. Since R is a Dedekind domain, R is hereditary, and therefore also $\mathcal{O}_\ell(L)$ is projective as R-module. Hence $\mathcal{O}_\ell(L)$ is a lattice and a subalgebra of A containing a K-basis of A, which shows that it is an R-order. □

Definition 8.3.15. Let R be a Dedekind domain with field of fractions K and let A be a finite dimensional semisimple K-algebra. Let L be an R-lattice such that $KL = A$. Then $\mathcal{O}_\ell(L)$ is called the *left order of L* and $\mathcal{O}_r(L)$ is called the *right order of L.*

Proposition 8.3.16. *Let R be a Dedekind domain with field of fractions K, and let Λ be an R-order in a semisimple K-algebra A. For each prime $\wp$ of R let $R_\wp$ be its localisation and $\Lambda_\wp := R_\wp \otimes_R \Lambda$. Then the following are equivalent:*
- *Λ is maximal.*
- *$\Lambda_\wp$ is a maximal order in A for all prime ideals $\wp$ of R.*

Moreover, the following are equivalent
- *Λ is hereditary.*
- *$\Lambda_\wp$ is a hereditary order in A for all prime ideals $\wp$ of R.*

Proof. We start with the maximal order case. Let Λ be a maximal order over a Dedekind domain R. Then by Lemma 7.5.9 (or Corollary 7.5.20) $R_\wp$ is a Dedekind domain again. Moreover, by Proposition 8.1.14 $R_\wp \otimes_R \Lambda$ is an $R_\wp$-order. Suppose that $\Gamma(\wp)$ is a maximal $R_\wp$-order strictly larger than $\Lambda_\wp$. Then let $x \in R$ such that $x\Gamma(\wp) \subseteq \Lambda_\wp$. Then put $\Delta := x\Gamma(\wp) \cap \Lambda$, where the intersection is taken in A. Then $R_\wp \cdot \Delta = x\Gamma(\wp)$, and Δ is a $\Lambda - \Lambda$-bimodule. Since Δ is an R-lattice, for each $\lambda \in A$ we get that $\Delta\lambda$ is an R-lattice again. By Proposition 8.3.13 there is $r \in R$ with $\Delta\lambda r \subseteq \Delta$. Hence, $r\lambda \in \mathcal{O}_r(\Delta)$. Therefore $K\mathcal{O}_r(\Delta) = A$. By Lemma 8.3.14 the right order of Δ

$$\mathcal{O}_r(\Delta) = \{\lambda \in A \mid \Delta\lambda \subseteq \Delta\}$$

is an R-order. Since $\Lambda \subseteq \mathcal{O}_r(\Delta)$, and since Λ is a maximal order, $\Lambda = \mathcal{O}_r(\Delta)$. We claim that

$$R_\wp\mathcal{O}_r(\Delta) = \mathcal{O}_r(R_\wp\Delta).$$

Indeed, the inclusion "$\subseteq$" is trivial. Let $\lambda \in A$ be such that $R_\wp\Delta\lambda \subseteq R_\wp\Delta$. Then, since $R_\wp\Delta$ is finitely generated projective as $R_\wp$-module, there is a non zero $r \in R \setminus \wp$ such that $\Delta \cdot r\lambda \subseteq \Delta$. Hence $\lambda \in R_\wp\mathcal{O}_r(\Delta)$, proving our claim. This then gives the statement by the following equation.

$$R_\wp\Lambda = R_\wp\mathcal{O}_r(\Delta) = \mathcal{O}_r(R_\wp\Delta) = \mathcal{O}_r(x\Gamma(\wp)) = \Gamma(\wp)$$

where the last equation holds since $\Gamma(\wp)$ is maximal and is clearly contained in $\mathcal{O}_r(x\Gamma(\wp))$.

Suppose now that $\Lambda_\wp$ is a maximal order for all prime ideals $\wp$ of R. If there is an R-order Γ with $\Lambda \subseteq \Gamma$, then $\Lambda_\wp \subseteq \Gamma_\wp$ is an inclusion of $R_\wp$-orders for all primes $\wp$. Since $\Lambda_\wp$ is a maximal $R_\wp$-order, $\Lambda_\wp = \Gamma_\wp$. Hence

$$R_\wp \otimes_R (\Gamma/\Lambda) = \Gamma_\wp/\Lambda_\wp = 0$$

for all primes $\wp$. By Lemma 7.5.7 we proved the statement.

Suppose that $\Lambda_\wp$ is a hereditary $R_\wp$-order for all $\wp \in \mathrm{Spec}(R)$. Let L be an ideal of Λ and let M be any Λ-module. Then

$$R_\wp \otimes_R Ext^1_\Lambda(L, M) \simeq Ext^1_{\Lambda_\wp}(R_\wp \otimes_R L, R_\wp \otimes_R M) = 0$$

by Corollary 6.5.11. Since $Ext^1_\Lambda(L, M)$ is a finitely generated R-module for any ideal L and any finitely generated Λ-module M, we get by Lemma 7.5.7 that $Ext^1_\Lambda(L, M) = 0$. By Lemma 6.3.4 we see that L is projective, and hence Λ is hereditary.

If Λ is hereditary let S and T be simple Λ-module. Then there is some prime ideal $\wp$ of R such that S (resp. T) is an $R/\wp$-module. Indeed, $\wp S$ is a Λ-submodule of S. Hence $\wp S = S$ or $\wp S = 0$. If $\wp S = S$ for a prime ideal $\wp$ of R, then $\wp S_\wp = S_\wp$ for $\wp$, where as usual we denote $S_\wp = R_\wp \otimes_R S$, and by Nakayama's lemma (cf. Exercise 6.3) we get $S_\wp = 0$ for this prime ideal $\wp$. If this happens to be true for all prime ideals $\wp$, then we get $S = 0$ by Lemma 7.5.7. This yields a contradiction. Hence there is a prime ideal $\wp$ of R with $\wp S = 0$ and we get that S is an $R/\wp$-module. If there are two prime ideals $\wp_1$ and $\wp_2$ of R, which are hence also maximal ideals of R, with $\wp_1 S = 0 = \wp_2 S$, since $\wp_1 + \wp_2 = R$, we get a $p_1 \in \wp_1$ and $p_2 \in \wp_2$ with $1 = p_1 + p_2$ and hence $S = 1 \cdot S = p_1 S + p_2 S = 0$.

This shows that S is a $\Lambda_\wp$-module for some unique prime ideal $\wp$.

Since Λ is hereditary, by Theorem 6.4.6 we get that $Ext^2_\Lambda(S, M) = 0$ for all Λ-modules M, and in particular for $M = T$. If also $\wp T = 0$, then again and therefore, using that $R_\wp \otimes_R S = S$ and $R_\wp \otimes_R T = T$, by Theorem 6.4.6 we see that

$$Ext^2_{\Lambda_\wp}(S, T) = R_\wp \otimes_R Ext^2_\Lambda(S, T) = 0.$$

If $\wp S = 0$ and $\wp T = T$, then $R_\wp T = T_\wp = 0$ and hence

$$R_\wp \otimes_R Ext^2_\Lambda(S, T) = Ext^2_{\Lambda_\wp}(S, T_\wp) = Ext^2_{\Lambda_\wp}(S, 0) = 0.$$

For any prime ideal $\wp' \neq \wp$ we get

$$R_{\wp'} \otimes_R Ext^2_\Lambda(S, T) = Ext^2_{\Lambda_{\wp'}}(0, R_{\wp'} \otimes_R T) = 0.$$

Therefore

$$Ext^2_{\Lambda_\wp}(S, T) = 0.$$

By Theorem 6.4.6 $\Lambda_\wp$ is hereditary for all primes $\wp$. □

If Λ is not a maximal order, then by Proposition 8.3.16 there is some prime ideal $\wp$ such that $\Lambda_\wp$ is not a maximal order. However, there are only finitely many such prime ideals.

Proposition 8.3.17. *Let R be a Dedekind domain with field of fractions K and let Λ be an R-order in some semisimple K-algebra A. Then*

$$S(\Lambda) := \{\wp \in Spec(R) \mid \Lambda_\wp \text{ is not a maximal } R_\wp\text{-order}\}$$

is a finite set. Moreover, $S(\Lambda) = \emptyset$ if and only if Λ is a maximal R-order.

Proof. By Proposition 8.3.16 we get that $S(\Lambda) = \emptyset$ if and only if Λ is maximal.

If Λ is not a maximal R-order, then using Theorem 8.3.10 let Γ be a maximal R-order containing Λ. By Proposition 8.3.13 there is some non zero $r \in R$ with

$$r\Gamma \subseteq \Lambda \subseteq \Gamma.$$

Since by Lemma 7.5.25 the ideal rR is contained in a finite number of maximal ideals only,

$$R_\wp r\Gamma = R_\wp\Gamma$$

whenever $rR \nsubseteq \wp$. Hence $\Lambda_\wp = \Gamma_\wp$ for all prime ideals $\wp$ with $r \notin \wp$. Since $\Gamma_\wp$ is maximal for all prime ideals by Proposition 8.3.16 we showed that $S(\Lambda)$ is finite. □

Proposition 8.3.18. *Let R be a Dedekind domain with field of fractions K and let G be a finite group. Then for any R order Γ in KG we have*

$$RG \subseteq \Gamma \Rightarrow \Gamma \subseteq \frac{1}{|G|}RG.$$

Proof. Consider the regular representation KG of G, and its the character $\chi : KG \longrightarrow K$. Then for all $g \in G$ we have

$$\chi(g) = \begin{cases} |G| & \text{if } g = 1 \\ 0 & \text{if } g \neq 1 \end{cases}$$

Let $y \in \Gamma$. Since $K\Gamma = KG$, there are coefficients $y_g \in K$ such that $y = \sum_{g\in G} y_g g \in \Gamma$. By Theorem 8.3.7 all elements of Γ are integral. We may adapt Lemma 2.4.7 to our situation. Indeed, by Proposition 8.3.3 the characteristic polynomial of each element y of Γ has coefficients in R, and hence the character of y, which is the opposite of the coefficient of second highest degree, is in R. Since $y \cdot g^{-1} \in \Gamma$ for all $g \in G$ and since all elements in Γ are integral, also $\chi(yg^{-1})$ are elements of R. Therefore, by the above

$$\chi(yg^{-1}) = y_g \cdot |G| \in R.$$

This shows $y \in \frac{1}{|G|}RG$ and hence $\Gamma \subseteq \frac{1}{|G|}RG$. □

Remark 8.3.19. Maximal R-orders over complete discrete valuation rings R are explicitly known (cf. [Rei-75, Theorem 17.3]). First one reduces to the case of orders in a finite dimensional simple algebra, hence a matrix algebra over a skew field D. The first step then is to show that there is a unique maximal order Δ in the skew field D, namely the integral closure of R in D. The matrix ring over Δ and all its conjugates are then the only possible maximal orders. If we drop the hypothesis that R is complete, and assume only that R is a discrete valuation domain, then still two maximal orders are conjugate (cf. [Rei-75, Theorem 18.7]).

8.3.2 Maximal orders are hereditary: the Auslander Buchsbaum theorem

We continue with a very important structure theorem for maximal orders, and in particular with respect to their lattices in view of Theorem 6.4.4.

Theorem 8.3.20 (Auslander-Buchsbaum [AuBu-74]). *Let R be a Dedekind domain with field of fractions K, and let Λ be a maximal R-order in a semisimple finite dimensional K-algebra A. Then Λ is (left and right) hereditary.*

The proof of Theorem 8.3.20 needs a lot of preparation. We will give the proof at the end of Section 8.3.2.

Let R be a Dedekind domain with field of fractions K. Let Λ be an R-order in the central simple K-algebra A. In the proof of Lemma 8.3.14 we encountered a special type of algebraic structure, namely a $\Lambda - \Lambda$-bimodule Δ in A containing a K-base.

Definition 8.3.21. Let R be a Dedekind domain with field of fractions K and let Λ be an R-order in the finite dimensional simple K-algebra A. Then a *fractional left (resp. right) ideal* of A is a finitely generated Λ-module (resp. Λ^{op}-module) M in A with $KM = A$. A fractional $\Lambda - \Lambda$-bimodule is a fractional left and right ideal, i. e. a $\Lambda - \Lambda$-bimodule M with $M \subseteq A$ and $KM = A$. A fractional ideal contained in Λ is called *integral.*

An integral $\Lambda - \Lambda$-bimodule P with $KP = A$ is called *prime* if for any two integral $\Lambda - \Lambda$-bimodules S and T with $KS = KT = A$ we get

$$ST \subseteq P \Rightarrow (S \subseteq P \quad \text{or} \quad T \subseteq P)$$

Note that a Λ^{op}-module is a Λ-right module.

Let R be a Dedekind domain with field of fractions K and let Λ be an R-order in the finite dimensional simple K-algebra A. For a fractional ideal M of Λ define

$$M^{-1} := \{x \in A \mid MxM \subseteq M\}.$$

Note that by Lemma 8.3.14 we have that $\mathcal{O}_\ell(M)$ is an order containing Λ. If Λ is a maximal order, then $\mathcal{O}_\ell(M) = \Lambda$. Hence, in this case $Mx \subseteq \Lambda \Rightarrow MxM \subseteq M$ and therefore

$$M^{-1} = \{x \in A \mid Mx \subseteq \Lambda\}.$$

This shows that M^{-1} is a Λ-right module.

Proposition 8.3.22. *Let R be a Dedekind domain with field of fractions K and let Λ be a maximal R-order in the finite dimensional simple K-algebra A. Then for a fractional left ideal M we have*

$$MM^{-1} = \Lambda.$$

Proof. By the above, M is a left module and M^{-1} is a right module. Hence MM^{-1} is a $\Lambda - \Lambda$-bimodule.

We will proceed in several steps.

Step 1: An ideal P of Λ is prime if and only if P is a maximal twosided ideal of Λ, i. e. Λ/P is a simple algebra.

Let P be a prime ideal. Since $KP = K\Lambda = A$ we have Λ/P is a finitely generated R/tR-module for some non zero $t \in R$. Hence Λ/P is artinian. For two-sided ideals I/P and J/P of Λ/P we get

$$(\dagger) \quad (I/P) \cdot (J/P) = (IJ)/P = P/P \Leftrightarrow IJ = P \Leftrightarrow (I = P \text{ or } J = P)$$

by the definition of a prime ideal. Hence, any nilpotent two-sided ideal N/P of Λ/P is 0, choosing in the above $I = N$ and $J = N^m$, when $(N/P)^{m+1} = 0$, and induction on m. Note that Exercise 1.3 holds under the weaker assumption of being artinian. Then $\operatorname{rad}(N/P) = 0$. Since this implies that the intersection of all maximal left ideals of Λ/P is 0, and hence Λ/P is the direct sum of its simple quotient modules. Indeed, the kernel of an epimorphism $\Lambda/P \longrightarrow S$ for a simple submodule is a maximal left ideal, and hence the maximal left ideals correspond to simple quotients. Since the intersection of all those maximal left ideals is 0, Λ/P is a direct sum of simple modules, is hence semisimple, and artinian by the above. By Exercise 6.5 Wedderburn's theorem applies in this case as well, and hence

$$\Lambda/P = M_1 \times \cdots \times M_\ell$$

for simple algebras $M_1, \ldots, M_\ell$. Since in this case, $M_1M_2 = 0$ which contradicts the first observation $(\dagger)$, we are done.

Step 2: Any twosided ideal M of Λ contains a product of prime ideals.

Since Λ is a finitely generated R-module, since R is a Dedekind domain, and hence Noetherian by Lemma 7.5.3, Λ is Noetherian and hence

$$\{J \text{ twosided ideal of } \Lambda \mid J \text{ does not contain a product of prime ideals}\}$$

is empty or contains a maximal element M. We suppose the set is not empty. Any prime ideal is not in the above set. Therefore there are twosided ideals S and T with $ST \subseteq M$ and $S \not\subseteq M$ and $T \not\subseteq M$. Since $ST \subseteq M$ implies

$$(S + M)(T + M) \subseteq M$$

and hence we can replace S by $S + M$, and T by $T + M$. Therefore we can assume that $M \subsetneq S$ and $M \subsetneq T$. Since M was maximal in the above set, S and T contain a product of primes each, and hence ST does. Since $ST \subseteq M$ we obtain a contradiction. Hence we are done.

Step 3: If S and T are fractional twosided ideals of Λ with $ST \subseteq \Lambda$, then also $TS \subseteq \Lambda$. Indeed,

$$(TS)T = T(ST) \subseteq T\Lambda = T$$

Hence $TS \subseteq \mathcal{O}_\ell(T)$. Since $\Lambda \subseteq \mathcal{O}_\ell(T)$, and since by Lemma 8.3.14 we have that $\mathcal{O}_\ell(M)$ is an order containing the maximal Λ, $\Lambda = \mathcal{O}_\ell(T)$. Hence $TS \subseteq \Lambda$.

Step 4: Let $M \subsetneq \Lambda$ be a proper twosided ideal of Λ with $KM = A$. Then $M^{-1} \neq \Lambda$.

Indeed, since $KM = A$ we get $R \cap M \neq \{0\}$, where we identify R with $R \cdot 1_A$. Moreover, localising $M \cap R$ at the multiplicative system $S := R \setminus \{0\}$ gives

$$(M \cap R)_S = M_S \cap R_S = A \cap K = K.$$

The ideal M is contained in a twosided maximal ideal P of Λ, and by the above $P \cap R \neq \{0\}$. Let $u \in P \cap R$ be a non zero element, and using Step 2, we choose prime ideals $P_1, \ldots, P_r$ of Λ such that

$$P_1 \cdot \cdots \cdot P_r \subseteq u\Lambda \subseteq P,$$

and choose r minimal with this property. Since P is a prime ideal, there is $i \in \{1, \ldots, r\}$ with $P_i \subseteq P$. Since P_i is prime, hence maximal by Step 1, we have $P = P_i$. Let $B := P_1 \cdot \cdots \cdot P_{i-1}$ and $C := P_{i+1} \cdot \cdots \cdot P_r$ such that $BPC \subseteq u\Lambda$. Hence

$$(u^{-1}B)(PC) \subseteq \Lambda$$

and therefore, recalling that u is central and hence PC as well as $u^{-1}B$ being twosided ideals, $(PC)(u^{-1}B) \subseteq \Lambda$ by Step 3. By the remark preceding the statement of Proposition 8.3.22 we have $P^{-1} = \{x \in A \mid Px \subseteq \Lambda\}$ and hence

$$Cu^{-1}B \subseteq P^{-1}.$$

However, since $M \subseteq P$ we get

$$P^{-1} = \{x \in A \mid Px \subseteq \Lambda\} \subseteq \{x \in A \mid Mx \subseteq \Lambda\} = M^{-1}$$

and hence, if $M^{-1} = \Lambda$, we get $P^{-1} \subseteq \Lambda$, which in turn implies

$$Cu^{-1}B \subseteq P^{-1} \subseteq \Lambda.$$

Since u is central, $CB \subseteq u\Lambda$, contradicting the minimality of r.

Final Step 5: We finish the proof of Proposition 8.3.22.
As seen before, $X := MM^{-1}$ is a twosided ideal inside Λ. Since

$$MM^{-1}X^{-1} = XX^{-1} \subseteq \Lambda,$$

we get $M^{-1}X^{-1} \subseteq M^{-1}$ by the remark preceding the statement of Proposition 8.3.22. Hence $X^{-1} \subseteq \mathcal{O}_r(M^{-1})$. However, Λ is maximal and since $\Lambda \subseteq \mathcal{O}_r(M^{-1})$ trivially, we get $\Lambda = \mathcal{O}_r(M^{-1})$. Since $\Lambda \subseteq M^{-1}$ we get therefore $X = \Lambda$ by Step 4. □

Lemma 8.3.23. *Let R be a Dedekind domain with field of fractions K and let Λ be a maximal R-order in the simple finite dimensional K-algebra A. Then a twosided Λ-ideal is projective as a Λ-left and as a Λ-right module.*

Proof. Let M be a twosided ideal of Λ. By Proposition 8.3.22 we get $MM^{-1} = \Lambda$. Likewise applying Proposition 8.3.22 to the opposite algebra instead, we see that $M^{-1}M = \Lambda$. Then there are $m_1, \dots, m_s \in M$ and $n_1, \dots, n_s \in M^{-1}$ with $\sum_{i=1}^{s} n_i m_i = 1$. Consider the map

$$\begin{aligned} M &\xrightarrow{\alpha} \underbrace{\Lambda \oplus \cdots \oplus \Lambda}_{s \text{ copies}} \\ m &\mapsto (mn_1, \dots, mn_s) \end{aligned}$$

which is a well-defined Λ-left module homomorphism since $n_i \in M^{-1}$ and since $MM^{-1} = \Lambda$. Consider further the map

$$\begin{aligned} \underbrace{\Lambda \oplus \cdots \oplus \Lambda}_{s \text{ copies}} &\xrightarrow{\beta} M \\ (\lambda_1, \dots, \lambda_s) &\mapsto \lambda_1 m_1 + \cdots + \lambda_s m_s \end{aligned}$$

Then by construction,

$$\beta \circ \alpha(x) = \sum_{j=1}^{s} x n_i m_i = x \sum_{j=1}^{s} n_i m_i = x$$

and hence, using that $\beta \circ \alpha = \mathrm{id}$ implies that $\alpha \circ \beta$ is an idempotent endomorphism of Λ^s,

$$\Lambda^s = \ker(\beta) \oplus \operatorname{im}(\beta) \simeq \ker(\beta) \oplus M$$

by Lemma 1.2.8. This shows that M is a direct factor of a free module, and using Proposition 6.2.5 is hence projective as a Λ-module. The case of the right module structure is dual. □

Proposition 8.3.24. *Let R be a Dedekind domain with field of fractions K and let Λ be a maximal order in a simple finite dimensional K-algebra A. Then the set of fractional twosided Λ-ideals form a free abelian group with generators the prime ideals, and multiplication law the product of ideals. Furthermore, integral twosided ideals are a product of prime ideals.*

Proof. By Proposition 8.3.22 the set of 2-sided fractional ideals is a group, where the inverse of an ideal M of Λ is M^{-1}.

The set of prime ideals is a generating set in the sense that any fractional twosided Λ-ideal is a product of prime ideals and inverses of prime ideals. Indeed, let M be a twosided integral ideal, and suppose that M is not a product of prime ideals. We suppose that M is maximal with this property. Then there is some prime ideal P with $M \subseteq P$. Now $M = P^{-1}M$ implies $\Lambda = MM^{-1} = P^{-1}MM^{-1} = P^{-1}$, which gives $P = P\Lambda = PP^{-1} = \Lambda$, which in turn is absurd. Therefore we get

$$M \subsetneq P^{-1}M \subseteq P^{-1}P = \Lambda.$$

Hence, by maximality of M we see that $P^{-1}M$ is a product of prime ideals $P_1 \cdot \dots \cdot P_s$, and therefore

$$M = P(P^{-1}M) = P \cdot P_1 \cdot \dots \cdot P_s.$$

If M is a fractional twosided ideal of Λ, then, since M is finitely generated as R-module, there is a non zero $u \in R$ such that $uM =: L$ is an integral twosided ideal. By the above there are prime ideals $P_1, \dots, P_s$ with $L = P_1 \cdot \dots \cdot P_s$ and there are prime ideals $Q_1, \dots, Q_t$ such that $u\Lambda = Q_1 \cdot \dots \cdot Q_t$. Hence

$$M = Q_t^{-1} \cdot \dots \cdot Q_1^{-1} \cdot P_1 \cdot \dots \cdot P_s.$$

If P and Q are prime ideals of Λ, then

$$Q \supseteq QP = P(P^{-1}QP)$$

which implies $P \subseteq Q$ or $P^{-1}QP \subseteq Q$, using that Q is a prime ideal. In the first case $P = Q$, and in the second case $PQ = QP$. In any case any two prime ideals commute. Hence the group of fractional ideals is commutative.

The group of fractional ideals is free. Indeed, let $P_1, \dots, P_s$ and $Q_1, \dots, Q_t$ be prime ideals such that

$$\Lambda = Q_t^{-1} \cdot \dots \cdot Q_1^{-1} \cdot P_1 \cdot \dots \cdot P_s.$$

Then

$$P_1 \cdot \dots \cdot P_s = Q_1 \cdot \dots \cdot Q_t \subseteq Q_1$$

Since Q_1 is prime there is an i with $P_i \subseteq Q_1$, and since P_i is prime, whence maximal, $P_i = Q_1$. By induction on $s + t$ we get the result. □

Proof of Theorem 8.3.20. Since Λ is a maximal order, we may assume that A is actually simple. Let M be an ideal of Λ. Hence KM is an ideal of $K\Lambda = A$. Since A is semisimple by Proposition 1.2.39, there is an ideal W of A such that $A = KM \oplus W$. By Lemma 8.1.10

there is a Λ-lattice N in W such that $M \oplus N$ is an ideal of Λ with $K(M \oplus N) = A$. We hence may and will suppose that $KM = A$. This implies that there is a non zero $r \in R$ with $r\Lambda \subseteq M$. Therefore Λ/M is a $\Lambda/r\Lambda$-module.

Since R/rR is artinian, and since Λ is a finitely generated R-module, $\Lambda/r\Lambda$ is artinian. Hence Λ/M has a composition series. We shall show that $\Lambda/M \simeq X/Y$ for a projective Λ-module X and a projective Λ-submodule Y. In order to do so, it is sufficient to show this fact for maximal ideals M. Indeed, if this is shown, we may proceed by induction on the composition length of Λ/M. If the result is true for ideals I of Λ, with composition length at most $\ell - 1$, then M is contained in a maximal ideal $\widehat{M}$ and $\Lambda/\widehat{M}$ is of composition length 1. Moreover, Λ/M has a submodule $\widehat{M}/M$ and

$$(\Lambda/M)/(\widehat{M}/M) \simeq \Lambda/\widehat{M}.$$

The composition length of $\widehat{M}/M$ is by 1 smaller than the composition length of M, and by assumption

$$\widehat{M}/M \simeq \widehat{X}/\widehat{Y}$$

for two projective Λ-modules $\widehat{X}$ and $\widehat{Y}$. Since the composition length of $\Lambda/\widehat{M}$ is 1, again

$$\Lambda/\widehat{M} \simeq X_s/Y_s$$

for two projective Λ-modules X_s and Y_s. By the Horseshoe Lemma 6.3.11 we get that

$$\Lambda/M \simeq \operatorname{coker}((\widehat{Y} \oplus Y_s) \hookrightarrow (\widehat{X} \oplus X_s))$$

for some embedding $(\widehat{Y} \oplus Y_s) \hookrightarrow (\widehat{X} \oplus X_s)$.

Suppose now that M is a maximal left ideal of Λ and hence $S := \Lambda/M$ is a simple Λ-module. By Proposition 8.3.24 there are prime ideals $P_1, \dots, P_\ell$ such that

$$r\Lambda = P_1 \cdot \dots \cdot P_\ell.$$

Since S is simple, for any twosided Λ-ideal J we have $JS = S$ or $JS = 0$. Since $rT = 0$ there must be a $j_0 \in \{1, \dots, \ell\}$ with $P_{j_0}S = 0$. Hence S is a simple Λ/P_{j_0}-module. But P_{j_0} is a maximal twosided Λ-ideal, and Λ/P_{j_0} is a simple artinian algebra. By Proposition 1.2.39 simple algebras are semisimple, and hence S is a direct factor of the rank one free Λ/P_{j_0}-module. Moreover, there is an integer t such that $\Lambda/P_{j_0} \simeq S^t$ as Λ-modules. By Theorem 6.4.6 we get that Λ is hereditary if and only if $Ext^2_\Lambda(S, N) = 0$ for all simples S and all Λ-modules N. But $Ext^2_\Lambda(S, N) = 0$ is equivalent with $Ext^2_\Lambda(S^t, N) = 0$. It is therefore sufficient to show the statement for Λ/P_{j_0} instead of S. But by Lemma 8.3.23 we get that P_{j_0} is a projective Λ-left module and Λ as well as P_{j_0} are projective. Hence S^t has a projective resolution of length 1, namely the embedding of P_{j_0} into Λ, and the second syzygy of S^t is 0. Therefore $Ext^2_\Lambda(S^t, N) = 0$. This shows the theorem. □

8.4 The Jordan Zassenhaus theorem

The main result of this section is the statement that, simplifying a bit, there are only finitely many lattices over an R-order with fixed R-rank.

We start with a technical lemma which allows to pass to maximal orders.

Lemma 8.4.1. *Let R be the ring of algebraic integers in a finite extension K of $\mathbb{Q}$ and let Λ and $\widehat{\Lambda}$ be R-orders in the same semisimple K-algebra A with $\dim_K A < \infty$. Let $d \in \mathbb{N}$ and suppose $\Lambda \subseteq \widehat{\Lambda}$.*

Then there only exists a finite number of isomorphism classes of Λ-lattices L satisfying $\dim_K(K \otimes_R L) < d$ if and only if there only exists a finite number of isomorphism classes of $\widehat{\Lambda}$-lattices $\widehat{L}$ satisfying $\dim_K(K \otimes_R \widehat{L}) < d$

Proof. By Proposition 8.3.13 there is a non zero $t \in R$ with

$$t\widehat{\Lambda} \subseteq \Lambda \subseteq \widehat{\Lambda}.$$

We claim that if L is a Λ-lattice, then, forming $\widehat{\Lambda} \cdot L =: \widehat{L}$ inside KL, the $\widehat{\Lambda}$-module $\widehat{L}$ is a $\widehat{\Lambda}$-lattice. Indeed, first, $\widehat{\Lambda}$ is a finitely generated R-module, generated by $\{\lambda_1, \dots, \lambda_k\}$, say, and L is a finitely generated R-module, by $\{l_1, \dots, l_s\}$, say, then $\widehat{L}$ is finitely generated by $\{\lambda_j l_i \mid 1 \le i \le s; 1 \le j \le k\}$. Since $\widehat{L}$ is a submodule of KL, we get that $\widehat{L}$ is R-torsion free. By Lemma 7.5.10 we see that $\widehat{L}$ is finitely generated projective as R-module.

Then

$$t\widehat{\Lambda} \subseteq \Lambda \subseteq \widehat{\Lambda}$$

implies

$$t\widehat{L} = t\widehat{\Lambda}L \subseteq \Lambda L = L \subseteq \widehat{\Lambda}L = \widehat{L}.$$

Since $\widehat{L}$ is a lattice, and since R is the ring of algebraic integers in a finite extension K of $\mathbb{Q}$, by Lemma 7.1.1 the quotient R/tR is a finite abelian group, and therefore also $\widehat{L}/t\widehat{L}$ is a finite abelian group. This then shows that for any fixed $\widehat{\Lambda}$-lattice M there are only finitely many Λ-lattices L with $\widehat{\Lambda}L = M$.

Conversely, suppose that there are only finitely many Λ-lattices $L_1, \dots, L_s$ with $\dim_K(K \otimes_R L_i) = d$ for all i. Let M be a $\widehat{\Lambda}$-lattice with $\dim_K(K \otimes_R M) = d$. Then the restriction of the action to Λ again makes M into a Λ-lattice. Hence, there is a Λ-module isomorphism $\varphi : M \longrightarrow L_i$ for some i. Then,

$$\varphi_K := id_K \otimes_R \varphi : K \otimes_R M \longrightarrow K \otimes_R L_i$$

is an isomorphism of A-modules. Hence, since $M = \widehat{\Lambda}M$, the restriction to $\widehat{\Lambda}M$ gives an isomorphism

$$\varphi_K|_{\widehat{\Lambda}M} : \widehat{\Lambda}M \longrightarrow \widehat{\Lambda}L_i,$$

whence M is one of the lattices $\widehat{\Lambda}L_1, \dots, \widehat{\Lambda}L_s$. □

Remark 8.4.2. Note that the last part of the statement does not use that R is the ring of algebraic integers in a finite extension of $\mathbb{Q}$. A Dedekind domain R is sufficient.

Hence, under this weaker hypothesis, if there are only finitely many isomorphism classes of Λ-lattices of R-rank at most d, then there only finitely many isomorphism classes of $\hat{\Lambda}$-lattices of R-rank at most d.

Lemma 8.4.1 indicates that it is reasonable to study first the case of maximal orders in skew fields. The proof follows the treatment given in Roggenkamp [RoVol2-70, Part VI].

Proposition 8.4.3. *Suppose that K is a finite extension of $\mathbb{Q}$, let R be the ring of algebraic integers in K, and let Δ be a maximal R-order in a skew field D. Suppose that the centre of D contains K, and that D is finite dimensional as K-vector space. Then, for each integer $d \in \mathbb{N}$ there is only a finite number of isomorphism classes of Δ-lattices L with $\dim_K(K \otimes_R L) < d$.*

Proof. Let L be a Δ-lattice, then KL is a finitely generated D-module, and hence $KL = D^n$ for some $n \in \mathbb{N}$. Hence, using that L is finitely generated as R-module, multiplying L with some $r \in R \setminus \{0\}$ if necessary, L is isomorphic to a submodule of Δ^n. Since Δ is a maximal order, by Auslander-Buchsbaum's Theorem 8.3.20, Δ is hereditary. Hence, L is projective as Δ-module, and by Theorem 6.4.4 the Δ-lattice L is a direct sum of ideals of Δ.

We may, and will therefore assume that L is an ideal of Δ.

Since K is an extension field of $\mathbb{Q}$ of finite degree, and since the skew field D is finite dimensional over K, the skew field D is also finite dimensional over $\mathbb{Q}$. By Lemma 7.1.1 the ring R is a free abelian group of finite rank. Since Δ is projective as R-module, Δ is free of finite rank as $\mathbb{Z}$-module. This shows that Δ is a $\mathbb{Z}$-order.

We claim that $\mathbb{Q}L = D$. Indeed, Δ contains R as multiples of 1, just as D contains K. Now, L is an ideal of Δ. Since $\mathbb{Q}R \subseteq K$ is an integral domain containing $\mathbb{Q}$ and R, and since K is a finite dimensional $\mathbb{Q}$-vector space, also $\mathbb{Q}R$ is a finite dimensional vector space and we shall see that $\mathbb{Q}R$ is a field. Indeed, for any $x \in \mathbb{Q}R \setminus \{0\}$, multiplication by x on $\mathbb{Q}R$ is injective, and hence the ideal $x\mathbb{Q}R$ is a $\mathbb{Q}$-sub vector space of $\mathbb{Q}R$ of the same dimension as $\mathbb{Q}R$. Therefore $x\mathbb{Q}R = \mathbb{Q}R$, and hence x is invertible. Since any non zero element in the commutative ring $\mathbb{Q}R$ is invertible, $\mathbb{Q}R$ is a field, and therefore $\mathbb{Q}R = K$ since K is the field of fractions of R. Now, since Δ is an order in D, we get $K\Delta = D$. Therefore $\mathbb{Q}L$ is a D-ideal in the skew field D. This shows $\mathbb{Q}L = D$. Hence, by the structure theorem on finitely generated abelian groups, Δ/L is a direct product of finitely many finite cyclic groups. Consider $N(L) := \sharp(\Delta/L)$ the number of elements of the abelian group Δ/L. For a fixed $\mathbb{Z}$-basis $B = \{b_1, \ldots, b_n\}$ of Δ and any fixed $\mathbb{Z}$-basis $\{\ell_1, \ldots, \ell_n\}$ of L, by the structure theorem on finitely generated abelian groups the morphism of abelian groups $\varphi : \Delta \longrightarrow L$ defined by $\varphi(b_i) = \ell_i$ for all $i \in \{1, \ldots, n\}$ has a determinant with absolute value

$$|\det(\varphi)| = N(L).$$

We shall prove now that there is a $c \in \mathbb{R}$, $c > 0$, which only depends on L and on D, such that there is $x \in L \setminus \{0\}$ with

$$N(\Delta x) \leq c \cdot N(L).$$

For each $b \in B$ let M_b be the square matrix with integer coefficients of the $\mathbb{Z}$-linear endomorphism of Δ given by multiplication of b, with respect to the basis B. Then, for variables $X_1, \ldots, X_n$ let

$$F(X_1, \ldots, X_n) := \det(X_1 M_{b_1} + \cdots + X_n M_{b_n}) \in \mathbb{Z}[X_1, \ldots, X_n].$$

The polynomial F is homogeneous. Hence there is $c \in \mathbb{R}$, that only depends on $M_{b_1}, \ldots, M_{b_n}$, such that as polynomials of with real coefficients over real variables,

$$|F(X_1, \ldots, X_n)| \leq c \cdot a^n$$

whenever $|X_i| \leq a$ for all $i \in \{1, \ldots, n\}$. Consider the set

$$V := \left\{ \sum_{i=1}^{n} k_i b_i \;\middle|\; k_i \in \mathbb{Z} \text{ and } 0 \leq k_i \leq N(L)^{\frac{1}{n}} \right\}.$$

We get that V contains more than $N(L)$ elements and hence there must exist two elements of V with difference in L. Therefore there is some $x \in L$ with $0 \neq x = \sum_{i=1}^{n} m_i b_i$, with $m_i \in \mathbb{Z}$ and with $|m_i| \leq N(L)^{\frac{1}{n}}$. This shows then

$$N(\Delta x) \leq c \cdot N(L),$$

using the interpretation of $N(L)$ as determinant of the linear endomorphism φ defined above.

Since D is a skew field, x is invertible in D, and we consider the Δ-lattice $L' := L \cdot x^{-1}$. Clearly, $L' \simeq L$ as Δ-lattices, since right multiplication with x gives an isomorphism $L' \longrightarrow L$. Since $\Delta x \subseteq L$, we get $\Delta \subseteq L'$. But then, again denoting by $\sharp(A)$ the order of the abelian group A,

$$\sharp(L'/\Delta) = \sharp(L/\Delta x) = \frac{\sharp(\Delta/\Delta x)}{\sharp(\Delta/L)} = \frac{N(\Delta x)}{N(L)} \leq c.$$

Hence, any Δ-ideal L admits an isomorphic Δ-lattice L' containing Δ and $\sharp(L'/\Delta) \leq c$. But for any positive integer $m \leq c$ we get $\sharp(L'/\Delta) = m$ implies $mL' \subseteq \Delta$. Since $\Delta \subseteq L'$ we have

$$m\Delta \subseteq mL' \subseteq \Delta$$

and since $\Delta/m\Delta$ is a finite abelian group, there are only finitely many such Δ-lattices L'. We hence parameterised the isomorphism classes of Δ-ideals by elements of the finite abelian group $\Delta/m\Delta$. This shows the result. □

The following statement is one of the main reduction steps of the general Jordan Zassenhaus theorem. It reduces the statement for orders in finite dimensional semisimple K-algebras to the statement for maximal orders in skew fields.

Proposition 8.4.4. *Let R be a Dedekind domain with field of fractions K, suppose that for all skew-fields D with centre containing K, and such that $\dim_K(D) < \infty$ we have that for all maximal orders Δ in D and for each $d \in \mathbb{N}$ there are only finitely many isomorphism classes of Δ-lattices L with $dim_K(K \otimes_R L) < d$.*

Then for all maximal R-orders in a semisimple K-algebra A with $\dim_K A < \infty$ there are only finitely many isomorphism classes of Λ-lattices L with $dim_K(K \otimes_R L) < d$.

Proof. **First step:** We may assume that $\Lambda = Mat_n(\Delta)$ for some integer n and some maximal order Δ in a skew field D.

Indeed, By Theorem 8.3.10 there is a maximal order $\hat{\Lambda}$ containing Λ. Moreover, Lemma 8.4.1 showed that if the theorem is proved for $\hat{\Lambda}$, then it is true for Λ as well.

We know by Wedderburn's Theorem 1.2.30 that a finite dimensional semisimple K-algebra A is a direct product

$$A = A_1 \times \cdots \times A_\ell$$

for finite dimensional simple K-algebras $A_1, \ldots, A_\ell$. By Remark 8.3.6 a maximal order in A is the direct product of maximal orders Λ_i in the simple algebras A_i, for all $i \in \{1, \ldots, \ell\}$. If $e^2 = e \in Z(\Lambda)$ is a central idempotent of Λ, and if M is a Λ-lattice, using Proposition 8.1.11 we get eM and $(1-e)M$ are Λ-lattices again and $M = eM \times (1-e)M$. Hence, if $\Lambda = \Lambda_1 \times \Lambda_2$ in order to count the number n_d of isomorphism classes of Λ-lattices L such that $K \otimes_R L$ is of dimension d, it is sufficient to count the number k_{d_1} of isomorphism classes of Λ_1-lattices L_1 with $K \otimes_R L_1$ being of dimension d_1, and the number m_{d_2} of isomorphism classes Λ_2-lattices L_2 with $K \otimes_R L_2$ being of dimension d_2. Then

$$n_d = \sum_{d_1+d_2=d} m_{d_2} + k_{d_1}.$$

We may hence suppose that Λ is a maximal R-order in the simple K-algebra A.

Applying Wedderburn's Theorem 1.2.30 again, we see that there is an integer n and a skew field D such that $A = Mat_n(D)$. If Δ is a maximal order in D, then by Lemma 8.3.2 we see that $Mat_n(\Delta)$ is a maximal order in $Mat_n(D)$. Moreover, $\Lambda \cap Mat_n(\Delta)$ is an order in A. Since $\Lambda \cap Mat_n(\Delta)$ is included in Λ as well as in $Mat_n(\Delta)$, by Lemma 8.4.1 the result is true for Λ if and only if the result is true for $\Lambda \cap Mat_n(\Delta)$, and using Lemma 8.4.1 this is again equivalent to the fact that the theorem is true for $Mat_n(\Delta)$.

Second step: We may assume that $n = 1$, i. e. Λ is a maximal order in a skew field D.

We assume now that $\Lambda = Mat_n(\Delta)$. Let L be the natural Λ-lattice, denoted by Δ^n, i. e. the Λ-lattice given by

$$\left\{\begin{pmatrix} d_1 \\ \vdots \\ d_n \end{pmatrix} \in D^n \;\middle|\; d_1,\dots,d_n \in \Delta\right\}.$$

Then $End_\Lambda(L) = \Delta^{op}$. Therefore L is a Λ-Δ-bimodule and $Hom_\Delta(L,\Delta)$ is a Δ-Λ-bimodule, which can be written explicitly as row vectors

$$Hom_\Delta(L,\Delta) = \{(d_1,\dots,d_n) \mid d_i \in \Delta\}.$$

For any Λ-lattice M let

$$F(M) := Hom_\Delta(L,\Delta) \otimes_\Lambda M.$$

Since $Hom_\Delta(L,\Delta)$ is a Δ-left module, $F(M)$ is a Δ-left module.

If N is a Δ-module, then

$$G(N) := L \otimes_\Delta N \simeq \left\{\begin{pmatrix} d_1 \\ \vdots \\ d_n \end{pmatrix} \in D^n \;\middle|\; d_1,\dots,d_n \in N\right\}$$

is a Λ-module, where the last isomorphism is just the natural multiplication $\Delta \otimes_\Delta N \simeq N$ on the components. Trivially, since $G(N) \simeq N^n$ as R-modules, N is a Δ-lattice if and only if $G(N)$ is a Λ-lattice. (†)

Further, elementary linear algebra gives that

$$Hom_\Delta(L,\Delta) \otimes_\Lambda L \simeq \Delta$$

by the map

$$(\delta_1,\dots,\delta_n) \otimes_\Lambda \begin{pmatrix} \delta'_1 \\ \vdots \\ \delta'_n \end{pmatrix} \mapsto \sum_{j=1}^{n} \delta_j \delta'_j \in \Delta.$$

Moreover,

$$L \otimes_\Delta Hom_\Delta(L,\Delta) \simeq \Lambda$$

by the map

$$\begin{pmatrix} \delta'_1 \\ \vdots \\ \delta'_n \end{pmatrix} \otimes_\Delta (\delta_1,\dots,\delta_n) \mapsto \begin{pmatrix} \delta'_1\delta_1 & \delta'_1\delta_2 & \dots & \delta'_1\delta_n \\ \delta'_2\delta_1 & \delta'_2\delta_2 & \dots & \delta'_2\delta_n \\ \vdots & \vdots & & \vdots \\ \delta'_n\delta_1 & \delta'_n\delta_2 & \dots & \delta'_n\delta_n \end{pmatrix}.$$

Hence

$$FG(N) = Hom_\Delta(L,\Delta) \otimes_\Lambda L \otimes_\Delta N \simeq \Delta \otimes_\Delta N \simeq N$$

and

$$GF(M) = L \otimes_\Delta Hom_\Delta(L,\Delta) \otimes_\Lambda M \simeq \Lambda \otimes_\Lambda M \simeq M.$$

Further, if there is an isomorphism of Λ-lattices $M_1 \simeq M_2$ then $F(M_1) \simeq F(M_2)$ as Δ-lattices. If $N_1 \simeq N_2$ as Δ-lattices, then $G(N_1) \simeq G(N_2)$ as Λ-lattices. Hence there is an isomorphism of Λ-lattices $M_1 \simeq M_2$ if and only if $F(M_1) \simeq F(M_2)$ as Δ-lattices, and likewise for G.

Finally, if M is a Λ-lattice, then $F(M)$ is a Δ-lattice. Indeed, since M is a Λ-lattice, and since $GF(M) \simeq M$, also $G(F(M))$ is a Λ-lattice. By the observation (†) we get that $F(M)$ is a Δ-lattice.

This shows that the isomorphism classes of Λ-lattices are in bijection with the isomorphism classes of Δ-lattices. □

Remark 8.4.5. The proof of the second step is a special case of what is known to be a Morita equivalence. This classical, but somehow abstract theory is not needed here in full generality, though it explains more clearly the above proof. The interested reader may like to consult [Zim-14, Chapter 4] for a presentation of the theory, and [Zim-14, Chapter 5 and Chapter 6] for vast generalisations. In the literature, to my best knowledge only marginal special attention of these generalisations to the case of orders is paid so far.

We come to the main result of this section.

Theorem 8.4.6 (Jordan-Zassenhaus). *Let R be a Dedekind domain, and let Λ be an R-order in a semisimple K-algebra A such that* $\dim_K(A) < \infty$. *Suppose that for each $d \in \mathbb{N}$ and for all maximal R-orders in A there are only finitely many isomorphism classes of Λ-lattices L with $\dim_K(K \otimes_R L) < d$. Then*
- *for each $d \in \mathbb{N}$ there is a finite number of isomorphism classes of Λ-lattices L with $\dim_K(K \otimes_R L) < d$.*
- *The hypothesis holds true if R is the ring of algebraic integers in a finite extension K of $\mathbb{Q}$.*

Proof of Theorem 8.4.6. We prove the first item. Lemma 8.4.1 shows that we may assume that Λ is a maximal order in A. Proposition 8.4.4 then shows the statement.

We now prove the second item. By Lemma 7.5.18 the ring of algebraic integers R in a finite extension K of $\mathbb{Q}$ is a Dedekind domain. Proposition 8.4.3 and the proof of Proposition 8.4.4 shows that for each $d \in \mathbb{N}$ and for all maximal R-orders in A there are only finitely many isomorphism classes of Λ-lattices L with $\dim_K(K \otimes_R L) < d$ in case R is the ring of algebraic integers in a finite extension K of $\mathbb{Q}$. □

Remark 8.4.7. We should remark that the Jordan Zassenhaus theorem is valid under less restrictive hypotheses. Generally one assumes that R is a Dedekind domain. Mostly one asks that K is a finite extension of $\mathbb{Q}$. Two lattices L and M *are in the same genus* if and only if $R_{\wp} \otimes_R L \simeq R_{\wp} \otimes_R M$ as $R_{\wp} \otimes_R \Lambda$-modules, for all primes $\wp$ of R. A recent result of L. Fuchs and Vamos [FuVa-00] shows that for a Dedekind domain R of characteristic 0 the following statements are equivalent:

- every torsion free R-module of finite rank has, up to equivalence, but a finite number of direct decompositions
- for every R-order Λ the genus classes of Λ-lattices are finite.

The idea to prove this result is to first reduce to the case of a maximal order. By Theorem 8.3.10 there is a maximal order Γ containing Λ. By Proposition 8.3.13 there is a non zero $r \in R$ such that $r\Gamma \subseteq \Lambda$ and Γ/Λ is a finitely generated R/rR-module.

Remark 8.4.8. One should carefully note that the Jordan Zassenhaus theorem *does not* tell that there are only finitely many isomorphism types of indecomposable lattices. The Jordan Zassenhaus theorem just give information of the finiteness of isomorphism types of lattices *when one fixes the rank over the base ring*. There are orders with infinitely many isomorphism classes of indecomposable lattices (cf. e. g. [Rog-71]). However, using the Jordan Zassenhaus theorem, any infinite family of pairwise non isomorphic lattices has to have unbounded R-rank.

8.5 Class groups for orders

In Definition 7.4.7 we defined the class group of integral domains. It parameterizes invertible ideals up to principal ideals. Now, we try to transpose this to the non commutative world of orders and lattices. This section is also an application of the Jordan Zassenhaus Theorem, which is used to associate a most interesting finite group to an order over the ring of algebraic integers of a finite extension of $\mathbb{Q}$.

8.5.1 Definitions and elementary properties

In Definition 5.1.2 we defined the Grothendieck group of modules over an algebra A. Let R be a Dedekind domain with field of fractions K, and let Λ be an R-order in the semisimple K-algebra A. Consider the subgroup $K_0(\Lambda)$ of $G_0(\Lambda)$ generated by projective Λ-modules. Note that if P_2 is a projective Λ-module, then $P_1/P_0 \simeq P_2$ implies that $P_1 \simeq P_2 \oplus P_0$. Hence $K_0(\Lambda)$ is the free abelian group generated by isomorphism classes of indecomposable projective Λ-lattices modulo the subgroup generated by the expressions

$$[P_0 \oplus P_2] - [P_0] - [P_2]$$

for projective Λ-lattices P_0, P_2.

Then, for all primes $\wp$ of R

$$\begin{array}{rcl} K_0(\Lambda) & \xrightarrow{\lambda_\wp} & G_0(R_\wp \otimes_R \Lambda) \\ [L] & \mapsto & [R_\wp \otimes_R L] \end{array}$$

is a group homomorphism. First we need to show that the map is well-defined. If L_1, L_2, L_3 are Λ-lattices such that

$$L_3 \simeq L_2/L_1,$$

then

$$(R_\wp \otimes_R L_3) \simeq (R_\wp \otimes_R L_2)/(R_\wp \otimes_R L_1),$$

since by Lemma 6.5.7 localisation is flat. Hence, $\lambda_\wp$ is well-defined. The fact that $\lambda_\wp$ is additive follows from the trivial computation

$$\lambda_\wp([L] + [L']) = [R_\wp \otimes_R (L \oplus L')] = [R_\wp \otimes_R L] + [R_\wp \otimes_R L'] = \lambda_\wp([L]) + \lambda_\wp([L']).$$

Definition 8.5.1. Two Λ-lattices L and M are in the same *genus* if $R_\wp \otimes_R L \simeq R_\wp \otimes_R M$ as $R_\wp \otimes_R \Lambda$-modules for all prime ideals $\wp$ of R. We write $L \vee M$ in this case.

A Λ-lattice L is called *locally free* of rank one if $L \vee \Lambda$.

Note that we mentioned the concept of being locally free already in the context of Lemma 7.5.8, where we showed that locally free modules are projective. Then consider the subset $LF(\Lambda)$ which is generated by the classes in the Grothendieck group of projective Λ-lattices which are locally free of rank one.

Now, we consider the map

$$LF(\Lambda) \xrightarrow{\kappa} K_0(A)$$

which maps a locally free Λ-lattice L to $A \otimes_\Lambda L$. Then define

$$Cl(\Lambda) := \ker(\kappa)$$

the locally free class group. We observe that defined this way, $Cl(\Lambda)$ is automatically an abelian group.

Definition 8.5.2. We call $Cl(\Lambda)$ the *(locally free) class group* of the R-order Λ.

We have the following very interesting result Theorem 8.5.4 due to Jacobinski. One essential tool is the following technical lemma, essentially due to Chevalley.

Lemma 8.5.3 (Jacobinski [Jac-68, Proposition 1.1]). *Let K be a finite extension of $\mathbb{Q}$ and let R be a Dedekind ring with field of fractions K. Let Γ be a maximal R-order in a semisimple finite dimensional K-algebra A. Then for any Γ-lattices N and M we have that $End_R(M)$ is a maximal R-order in $End_K(KM)$. Further, if $KM = KN$, then there is a unique (fractional) left $End_R(M)$-ideal $\mathfrak{a}$ such that $N = M \cdot \mathfrak{a}$. The ideal $\mathfrak{a}$ is unique with this property.*

We omit the proof.

In the situation of Lemma 8.5.3 let Λ be an R-order in A and suppose that $\Lambda \subseteq \Gamma$. Since Λ and Γ are finitely generated R-modules, there is a twosided Γ-ideal $\mathcal{F}$ of Γ such that $\mathcal{F} \subseteq \Lambda$. For any Λ-lattice L and a fractional $End_R(L)$-ideal $\mathfrak{f}$ we say that $\mathfrak{f}$ is prime to $\mathcal{F}$ if for any $\wp \in Spec(R)$ we have

$$\mathcal{F}_\wp \neq \Gamma_\wp \Rightarrow \mathfrak{f}_\wp = (End_R(L))_\wp .$$

If $\mathfrak{f}$ is prime to $\mathcal{F}$, then there is a non zero scalar α such that $\mathfrak{f}' := \alpha \cdot \mathfrak{f}$ is an integral ideal of $End_R(L)$. Define

$$L_\mathfrak{f} := \frac{1}{\alpha} \cdot (L \cap (\Gamma \cdot L \cdot \mathfrak{f}')).$$

This is a Λ-lattice and by another theorem of Jacobinski [Jac-68, Proposition 2.1] we have $L_1 \vee L_2$ if and only if there is fractional $End_R(L_1)$-ideal $\mathfrak{f}$ such that $L_2 = (L_1)_\mathfrak{f}$.

If $\mathfrak{f} = End_R(L) \cdot \varphi$ is a principal integral ideal of $End_R(L)$, and in addition $\mathfrak{f}$ prime to $\mathcal{F}$, then $L \simeq L_\mathfrak{f}$. Indeed, $(L \cdot \varphi)_\wp = (L_\mathfrak{f})_\wp$ for every $\wp \in SpecR$ and hence $L \simeq L \cdot \varphi = L_\mathfrak{f}$.

Theorem 8.5.4 (Jacobinski [Jac-68, Theorem 3.3]). *Let K be a finite extension of $\mathbb{Q}$ and let R be a Dedekind ring with field of fractions K. Let Λ be an R-order in a semisimple finite dimensional K-algebra. Let X, Y and M be Λ-lattices and suppose that $(X \oplus Y) \vee M$. Then there are Λ-lattices X' and Y' with $X' \vee X$ and $Y' \vee Y$ with $M \simeq X' \oplus Y'$.*

Proof (sketch). Let Γ be a maximal order containing Λ and choose an ideal $\mathcal{F}$ as above. By multiplying with a suitable element $r \in R$ we may suppose that $M \subseteq X \oplus Y$. Then by the above there is an integral ideal $\mathfrak{a}$ of $End_R(\Gamma \otimes_\Lambda (X \oplus Y))$ such that $M = (X \oplus Y)_\mathfrak{a}$. We can suppose $\mathfrak{a}$ to be prime to $\mathcal{F}$. Further,

$$\Gamma \otimes_\Lambda M = ((\Gamma \otimes_\Lambda X) \oplus (\Gamma \otimes_\Lambda Y)) \cdot \mathfrak{a} \subseteq \Gamma \otimes_\Lambda (X \oplus Y).$$

Let

$$X_\Gamma := (\Gamma \otimes_\Lambda M) \cap (\Gamma \otimes_\Lambda X).$$

Then there is an ideal $\mathfrak{b}$ of $End_R(\Gamma \otimes_\Lambda X)$ prime to $\mathcal{F}$ such that

$$X_\Gamma = (\Gamma \otimes_\Lambda X) \cdot \mathfrak{b}.$$

Then $(\Gamma \otimes_\Lambda M)/((\Gamma \otimes_\Lambda X) \cdot \mathfrak{b})$ is a Γ-lattice again. Now, Γ is a maximal order, and since by Auslander-Buchsbaum's Theorem 8.3.20 maximal orders are hereditary, and since therefore all Γ-lattices are projective, we get

$$\Gamma \otimes_\Lambda M \simeq ((\Gamma \otimes_\Lambda X) \cdot \mathfrak{b}) \oplus T$$

for some Γ-lattice T. Since $M \subseteq X \oplus Y$, consider the projection $\pi : X \oplus Y \longrightarrow X$ and the image of $\Gamma \otimes_\Lambda M$ under $\mathrm{id}_\Gamma \otimes \pi$ in $\Gamma \otimes_\Lambda X$. Again there is an ideal $\mathfrak{b}'$ of $End_R(\Gamma \otimes_\Lambda X)$ such that

$$(\mathrm{id}_\Gamma \otimes \pi)(\Gamma \otimes_\Lambda M) = (\Gamma \otimes_\Lambda X) \cdot \mathfrak{b}'.$$

Since

$$\Gamma \otimes_\Lambda X \supseteq (\Gamma \otimes_\Lambda X) \cdot \mathfrak{b}' \supseteq (\Gamma \otimes_\Lambda X) \cdot \mathfrak{b}$$

and since $\mathfrak{b}$ is prime to $\mathcal{F}$, also $\mathfrak{b}'$ is prime to $\mathcal{F}$. Hence there is $\delta \in R$ with $\delta - 1 \in \mathcal{F}$ such that $\mathfrak{b}' \cdot \delta \subseteq \mathfrak{b}$. Then there is a unique $End_R(\Gamma \otimes_\Lambda X)$-ideal $\mathfrak{d}$ prime to $\mathcal{F}$ such that

$$(\Gamma \otimes_\Lambda M) \cdot \mathfrak{d} = (\Gamma \otimes_\Lambda X) \cdot \mathfrak{b} \oplus T\delta.$$

Again there is a $End_R(\Gamma \otimes_\Lambda Y)$-ideal $\mathfrak{c}$ prime to $\mathcal{F}$ such that

$$(\Gamma \otimes_\Lambda M) \cdot \mathfrak{d} = (\Gamma \otimes_\Lambda X) \cdot \mathfrak{b} \oplus (\Gamma \otimes_\Lambda Y) \cdot \mathfrak{c}.$$

This then implies

$$M_\mathfrak{d} = X_\mathfrak{b} \oplus Y_\mathfrak{c}.$$

By the remarks preceding the theorem (actually Jacobinski's theorem [Jac-68, Proposition 2.1]) we get $X_\mathfrak{b} \vee X$ and $Y_\mathfrak{c} \vee Y$. But $M \simeq M_\mathfrak{d}$ since $\mathfrak{d}$ is generated by $d = \mathrm{id}_{(\Gamma\otimes_\Lambda X)\mathfrak{b}} \oplus \delta$. Since $\delta - 1 \in \mathcal{F}$ we get $d \in End_R(M)$. □

Lemma 8.5.5. *Let K be a finite extension of $\mathbb{Q}$ and let R be a Dedekind domain with field of fractions K. Let Λ be an R-order in a semisimple K-algebra. Then the locally free class group $Cl(\Lambda)$ is generated by $[L] - [\Lambda]$ for L a locally free ideal of Λ.*

Proof. Let L be locally free indecomposable. Then, by Theorem 8.5.4 we get that $L \vee \Lambda$. The elements $[L] - [\Lambda]$ for rank one locally free Λ-lattices L are elements in $Cl(\Lambda)$, and actually theses form a generating set. This shows the statement. □

Corollary 8.5.6. *Let K be a finite extension of $\mathbb{Q}$ and let R be the ring of algebraic integers in K and let Λ be an R-order in a semisimple K-algebra. Then $Cl(\Lambda)$ is a finitely generated abelian group.*

Proof. Indeed, this comes from the fact that $Cl(\Lambda)$ is generated by the expressions $[L] - [\Lambda]$ for L being a locally free ideal of Λ. The Jordan Zassenhaus Theorem 8.4.6 implies that there are only finitely many isomorphism classes of rank one locally free Λ-lattices. This shows the statement. □

Note that since we use the Grothendieck group to define $Cl(\Lambda)$ we obtain for two locally free Λ-ideals L and M, then for any locally free Λ-ideal X we get

$$L \oplus X \simeq M \oplus X \Rightarrow [L] = [M] \in K_0(\Lambda).$$

Definition 8.5.7. Let L and M be two Λ-modules. Then we say that L is *stably isomorphic* to M if there is some integer $n \in \mathbb{N}$ with

$$L \oplus \Lambda^n \simeq M \oplus \Lambda^n.$$

Remark 8.5.8. Since we did not really prove Jacobinski's results, the reader may feel some unease with the approach we chose for the definition of the class group. However, our method explains how the class group fits in a general framework. Alternatively one may well define right away the locally free class group of an order by idèles, as is done in the next Section 8.5.2.

8.5.2 Idèle class groups

We now pass to a different description of locally free class groups. This description is at once well-suited for computations of examples and also gives a very interesting further insight in the structure of the class group.

Let R be a Dedekind domain with field of fractions K and let Λ be an R-order in a semisimple K-algebra. Let L be a rank one locally free Λ-lattice. Then, multiplying with a non zero $r \in R$ we may suppose that $L \subseteq \Lambda$ is a locally free ideal, and denote by $\Lambda_\wp := R_\wp\Lambda$, as well as $L_\wp := R_\wp L$ for any prime ideal $\wp$ of R. Further, for any prime ideal $\wp$ of R let $\widehat{R}_\wp$ be the completion of $R_\wp$ with field of fractions $\widehat{K}_\wp$ and $\widehat{\Lambda}_\wp := \widehat{R}_\wp \otimes_R \Lambda$ as well as $\widehat{A}_\wp := \widehat{K}_\wp \otimes_K A$.

By Proposition 8.3.13 there is a non zero $r \in R$ with

$$r\Lambda \subseteq L \subseteq \Lambda.$$

By Lemma 7.5.25 there are only a finite number of prime ideals containing r. Since for each prime ideal $\wp$ with $r \notin \wp$ we have $rR_\wp = R_\wp$ and hence if $r \notin \wp$, then $L_\wp = \Lambda_\wp$.

Since $L_\wp \simeq \Lambda_\wp$ for all prime ideals $\wp$ of R, also $\widehat{L}_\wp \simeq \widehat{\Lambda}_\wp$ for all prime ideals $\wp$ of R, and we fix an isomorphism

$$\alpha_\wp : L_\wp \longrightarrow \Lambda_\wp$$

of $\Lambda_\wp$-modules. Since

$$End_{\Lambda_\wp}(\Lambda_\wp) = \Lambda_\wp^{op}$$

given by right multiplication with elements in $\Lambda_\wp$ for any prime ideal $\wp$ of R there is an element $u_\wp \in \Lambda_\wp$ such that

$$L_\wp = \Lambda_\wp \cdot u_\wp.$$

Moreover, since $\alpha_\wp$ is an isomorphism, $u_\wp$ is invertible in A. Finally, for all but a finite number of prime ideals $\wp$ we get $u_\wp$ is an invertible element of $\Lambda_\wp$, respectively in $\widehat{\Lambda}_\wp$, and actually can be supposed to be 1. The same holds for $\widehat{L}_\wp$ and $\widehat{\Lambda}_\wp$ and we can consider the image of $u_\wp$ in $\widehat{\Lambda}_\wp$.

Fix now

$$u := (u_\wp)_{\wp\in Spec(R)} \in \prod_{\wp\in Spec(R)} A^\times \subseteq \prod_{\wp\in Spec(R)} \widehat{A}_\wp^\times$$

and denote

$$J(A) := \left\{(u_\wp)_{\wp\in Spec(R)} \in \prod_{\wp\in Spec(R)} A^\times \;\middle|\; \sharp(\{\wp \in Spec(R) \mid u_\wp \notin \Lambda_\wp^\times\}) < \infty\right\}$$

as well as the completed version

$$\widehat{J}(A) := \left\{(u_\wp)_{\wp\in Spec(R)} \in \prod_{\wp\in Spec(R)} \widehat{A}_\wp^\times \;\middle|\; \sharp(\{\wp \in Spec(R) \mid u_\wp \notin \widehat{\Lambda}_\wp^\times\}) < \infty\right\}$$

the set of all sequences $(u_\wp)_{\wp\in Spec(R)}$ of elements in $A^\times$, respectively $\widehat{A}_\wp$, indexed by the set of prime ideals of R, such that all but a finite number of the $u_\wp$ are invertible elements of $\Lambda_\wp$. Put

$$U(A) := \prod_{\wp\in Spec(R)} \Lambda_\wp^\times \quad\text{and}\quad \widehat{U}(A) := \prod_{\wp\in Spec(R)} \widehat{\Lambda}_\wp^\times$$

and observe that $U(A)$, respectively $\widehat{U}(A)$, is a group under multiplication, which acts on the left on $J(A)$, respectively on $\widehat{J}(A)$ by multiplication componentwise.

Two rank one locally free Λ-lattice L and M are isomorphic if and only if there is an isomorphism

$$L \xrightarrow{\alpha} M .$$

Now, α induces

$$A = KL \xrightarrow{\mathrm{id}_K \otimes_R \alpha} KM = A$$

However, $End_A(A) = A^{op}$ given by multiplication from the right with an element $x \in A^\times$. Hence $A^\times$ is a group by multiplication, and acts on $J(A)$ on the right by diagonal action on each of the components:

$$\begin{array}{ccc} \prod_{\wp \in Spec(R)} A^\times \supseteq J(A) & \xrightarrow{\cdot x} & J(A) \subseteq \prod_{\wp \in Spec(R)} A^\times \\ (u_\wp)_{\wp \in Spec(R)} & \mapsto & (u_\wp x)_{\wp \in Spec(R)} \end{array}$$

The analogous formula for completions holds. Still, a fixed $x \in A^\times$ is in $\Lambda_\wp^\times$ for almost all prime ideals $\wp$. Indeed, $x \in A^\times \subseteq K\Lambda$ and expressing x as well as x^{-1} with respect to any fixed K-basis $B \subseteq \Lambda$ of A involves only finitely many coefficients in K. The denominators of these coefficients belong to only finitely many prime ideals, and hence x and x^{-1} belong to $\Lambda_\wp$ for almost all $\wp$, namely those to which the denominators of all the coefficients do not belong.

Definition 8.5.9. The group $\widehat{J}(A)$ is called the idèle group of A. An idèle is called *integral* if it belongs to $\prod_{\wp \in Spec(R)} \widehat{\Lambda}_\wp$. The group $\widehat{U}(A)$ is called the group of *unit idèles.*

Note that $\widehat{J}(A)$ does not depend on the order Λ, though it was used for the definition. Indeed, if Λ_1 and Λ_2 are two R-orders in A, then by Proposition 8.3.13 there is a non zero $r \in R$ with $r\Lambda_1 \subseteq \Lambda_2$. Since by Lemma 7.5.25 r is contained in only finitely many prime ideals, $(\Lambda_1)_\wp = (\Lambda_2)_\wp$ for almost all prime ideals $\wp$.

Theorem 8.5.10. *There is a bijection between the set $LF_1(\Lambda)$ of isomorphism classes of locally free ideals of Λ and elements of the double class*

$$U(A)\backslash J(A)/A^\times$$

and a bijection with elements in the double class

$$\widehat{U}(A)\backslash\widehat{J}(A)/A^\times$$

This bijection is given by

$$\begin{array}{ccc} U(A)\backslash J(A)/A^\times & \xrightarrow{\quad\Phi\quad} & LF_1(\Lambda) \\ U(A)\cdot(u_\wp)_{\wp \in Spec(R)} \cdot A^\times & \longmapsto & \bigcap_{\wp \in Spec(A)} \Lambda_\wp \cdot u_\wp \end{array}$$

and likewise for the completed version.

Proof. Let $u \in J(A)$. Then by Proposition 8.1.14 item 3 and item 4 the space

$$\Phi(u) = \bigcap_{\wp \in Spec(A)} \Lambda_\wp u_\wp$$

is a Λ-lattice with $\Phi(u)_\wp = \Lambda_\wp u_\wp$. By Proposition 8.1.14 item 5 we also get that $\bigcap_{\wp\in Spec(A)} \widehat{\Lambda}_\wp u_\wp$ is a Λ-lattice with

$$\widehat{\left(\bigcap_{\wp\in Spec(A)} \widehat{\Lambda}_\wp u_\wp\right)}_\wp = \widehat{\Lambda}_\wp u_\wp.$$

Hence $\Phi(u)$ is locally free of rank 1. Clearly two elements $u, v \in J(A)$ with $U(A)\cdot u = U(A)\cdot v$ have the very same image under Φ, i. e. $\Phi(u) = \Phi(v)$. Further,

$$\Phi(u) \simeq \Phi(v) \Leftrightarrow U(A)\cdot u\cdot A^\times = U(A)\cdot v\cdot A^\times$$

by the discussion preceding the statement of the theorem.

If L is a rank one locally free Λ-ideal, then by the development preceding the statement of the theorem, there is a $u \in J(A)$ such that $\Phi(u) \simeq L$. □

Seemingly the description of locally free Λ-ideals by double classes of elements in $J(A)$ is a lot more complicated than the original definition. However, we shall see that the map Φ in Theorem 8.5.10 induces actually an isomorphism

$$Cl(\Lambda) \simeq U(A)\backslash J(A)/A^\times.$$

Denote

$$\Lambda\alpha := \bigcap_{\wp\in Spec(R)} \widehat{\Lambda}\alpha_\wp$$

for an idèle $\alpha = (\alpha_\wp)_{\wp\in Spec(R)} \in \widehat{J}(A)$. With this notation we get the most important

Theorem 8.5.11. *Let R be a Dedekind domain with field of fractions K and let Λ be an R-order in the semisimple K-algebra A. Then for all idèles $\alpha, \beta \in \widehat{J}(A)$ we get*

$$\Lambda\alpha \oplus \Lambda\beta \simeq \Lambda \oplus \Lambda\alpha\beta.$$

Proof. Since we only need to consider $\Lambda\alpha$ and $\Lambda\beta$ and their isomorphism class, we may replace α by some $ru\alpha$ for some non zero $r \in R$ and $u \in \widehat{U}(A)$ such that $\Lambda ru\alpha$ is an ideal of Λ and such that $ru_\wp\alpha_\wp = 1$ for all but a finite number of prime ideals $\wp$. Suppose hence that

$$S_0 := \{\wp \in Spec(R) \mid \alpha_\wp \neq 1\}$$

is a finite set.

If $S_0 = \emptyset$, then $\alpha = 1$ and the statement is trivial.

For each prime ideal $\wp$ the algebra A is dense in $A_\wp$ by Corollary 7.3.11.

Since S_0 is finite there is a positive integer k such that $\alpha_\wp^{-1} \in \wp^{-k}\widehat{\Lambda}_\wp$ for all $\wp \in S_0$. Hence $\alpha_\wp^{-1} = \pi_\wp^{-k}\cdot\lambda_\wp^{(0)}$ for some $\lambda_\wp^{(0)} \in \widehat{\Lambda}_\wp$, and where $\pi_\wp$ a uniformizer of $\widehat{R}_\wp$.

Hence, by Theorem 7.6.9 for actually any choice of a positive integer t there is $x \in A^\times$ such that for any $\wp \in S_0$ we have

$$\beta_\wp x - 1 \in \wp^t \widehat{\Lambda}_\wp$$

and such that βx is an integral idèle. We may and will hence assume that $\beta_\wp - 1 \in \wp^t \widehat{\Lambda}_\wp$ for all $\wp \in S_0$. We choose $t = k + 1$ and therefore

$$\beta_\wp = 1 + \lambda \pi_\wp^t$$

for some $\lambda \in \widehat{\Lambda}_\wp$, and where we denote by $\pi_\wp$ a uniformizer of $\widehat{R}_\wp$.

We now use that $\widehat{R}_\wp$ is complete! Then, using that $\widehat{R}_\wp$ is complete, and hence power series in powers of $\pi_\wp$ converge in $\widehat{\Lambda}_\wp$

$$\begin{aligned}\beta_\wp^{-1} &= (1 + \lambda \pi_\wp^t)^{-1} \\ &= 1 - \lambda \pi_\wp^t + \lambda^2 \pi_\wp^{2t} - \lambda^3 \pi_\wp^{3t} + \cdots \\ &= 1 + \lambda_1 \pi_\wp^t\end{aligned}$$

for some $\lambda_1 \in \widehat{\Lambda}_\wp$. We compute

$$\begin{aligned}\beta_\wp \alpha_\wp \beta_\wp^{-1} \alpha_\wp^{-1} &= (1 + \lambda \pi_\wp^t) \alpha_\wp (1 + \lambda_1 \pi_\wp^t) \alpha_\wp^{-1} \\ &= (\alpha_\wp + \lambda \alpha_\wp \pi_\wp^t) \cdot (\alpha_\wp^{-1} + \lambda_1 \lambda_\wp^{(0)} \pi_\wp^{t-k}) \\ &= (\alpha_\wp + \lambda \alpha_\wp \pi_\wp^t) \cdot (\alpha_\wp^{-1} + \lambda_2 \pi_\wp^{t-k})\end{aligned}$$

for $\lambda_2 := \lambda_1 \lambda_\wp^{(0)}$. As a whole

$$\beta_\wp \alpha_\wp \beta_\wp^{-1} \alpha_\wp^{-1} = 1 + \lambda_3 \pi_\wp^{t-k} = 1 + \lambda_3 \pi_\wp$$

for some $\lambda_3 \in \widehat{\Lambda}_\wp$, and for our choice of t as $t = k + 1$. By Exercise 8.1, or actually simpler by observing that power series in $\pi_\wp$ and coefficients in $\widehat{\Lambda}_\wp$ converge since $\widehat{R}_\wp$ is complete, in the $\widehat{R}_\wp$-algebra $\widehat{\Lambda}_\wp$, we see that

$$\beta_\wp \alpha_\wp \beta_\wp^{-1} \alpha_\wp^{-1} \in \widehat{\Lambda}_\wp^\times$$

for each $\wp \in S_0$.

Hence

$$\widehat{\Lambda}_\wp \alpha_\wp \beta_\wp = \widehat{\Lambda}_\wp \beta_\wp \alpha_\wp$$

for all $\wp \in S_0$.

Since α and β are both integral ideals, $\Lambda\alpha + \Lambda\beta \subseteq \Lambda$ is an ideal of Λ.

If $\wp \notin S_0$, then $\alpha_\wp = 1$ and hence

$$\widehat{\Lambda}_\wp \alpha_\wp + \widehat{\Lambda}_\wp \beta_\wp = \widehat{\Lambda}_\wp.$$

If $\wp \in S_0$, then $\beta_\wp - 1 \in \wp^t \widehat{\Lambda}_\wp$, and therefore again by Exercise 8.1, or actually simpler by observing that power series in $\pi_\wp$ and coefficients in $\widehat{\Lambda}_\wp$ converge since $\widehat{R}_\wp$ is complete, in the $\widehat{R}_\wp$-algebra $\widehat{\Lambda}_\wp$, we see that $\beta_\wp \in \widehat{\Lambda}_\wp^\times$. Hence also there

$$\widehat{\Lambda}_\wp \alpha_\wp + \widehat{\Lambda}_\wp \beta_\wp = \widehat{\Lambda}_\wp.$$

By Proposition 8.1.14 we see that this shows that

$$\Lambda\alpha + \Lambda\beta = \Lambda.$$

Hence the map

$$\begin{array}{rcl} \Lambda\alpha \oplus \Lambda\beta & \xrightarrow{f} & \Lambda \\ (x,y) & \mapsto & x+y \end{array}$$

is an epimorphism with $\ker(f) =: L$, say.

Then

$$\ker(f) = \Lambda\alpha \cap \Lambda\beta = \bigcap_{\wp \in Spec(R)} \widehat{\Lambda}_\wp \alpha_\wp \cap \widehat{\Lambda}_\wp \beta_\wp.$$

Since

$$\widehat{\Lambda}_\wp \alpha_\wp \beta_\wp = \widehat{\Lambda}_\wp \beta_\wp \alpha_\wp$$

for all $\wp \in S_0$ we have

$$\widehat{\Lambda}_\wp \alpha_\wp \cap \widehat{\Lambda}_\wp \beta_\wp \supseteq \widehat{\Lambda} \alpha_\wp \beta_\wp$$

for all $\wp \in S_0$. But if $\wp \in S_0$, then $\beta_\wp \in \widehat{\Lambda}_\wp^\times$, and therefore

$$\widehat{\Lambda}_\wp \beta_\wp = \widehat{\Lambda}_\wp.$$

This give for $\wp \in S_0$ that

$$\begin{aligned} \widehat{\Lambda}_\wp \alpha_\wp \cap \widehat{\Lambda}_\wp \beta_\wp &= \widehat{\Lambda}_\wp \alpha_\wp \cap \widehat{\Lambda}_\wp \\ &= \widehat{\Lambda}_\wp \alpha_\wp \\ &= \widehat{\Lambda}_\wp \beta_\wp \alpha_\wp \\ &= \widehat{\Lambda}_\wp \alpha_\wp \beta_\wp \end{aligned}$$

For $\wp \notin S_0$ we have $\alpha_\wp = 1$ and hence

$$\begin{aligned}\widehat{\Lambda}_\wp \alpha_\wp \cap \widehat{\Lambda}_\wp \beta_\wp &= \widehat{\Lambda}_\wp \cap \widehat{\Lambda}_\wp \beta_\wp \\ &= \widehat{\Lambda}_\wp \beta_\wp \\ &= \widehat{\Lambda}_\wp \alpha_\wp \beta_\wp\end{aligned}$$

Therefore as a whole we have $L = \Lambda\alpha\beta$. But f is an epimorphism and Λ is a projective Λ-module, which shows by Proposition 6.2.5 that f splits and we get the statement. □

Corollary 8.5.12. *Let R be a Dedekind domain with field of fractions K and let Λ be an R-order in the semisimple K-algebra A. Then*

$$Cl(\Lambda) \simeq \widehat{U}(A)\backslash\widehat{J}(A)/A^\times$$

sending $\alpha \in \widehat{J}(A)$ to $[\Lambda\alpha] - [\Lambda] \in Cl(\Lambda)$.

Proof. Indeed, by Theorem 8.5.10 the locally free ideals of Λ are given by $\Lambda\alpha$ for some idèle α. By Lemma 8.5.5 the class group $Cl(\Lambda)$ is generated by expressions $[L]-[\Lambda]$ in the Grothendieck group of locally free Λ-modules. Hence $Cl(\Lambda)$ is generated by expressions $[\Lambda\alpha] - [\Lambda]$ for idèles α in the Grothendieck group of locally free Λ-modules. What is the sum of two of them?

$$\begin{aligned}([\Lambda\alpha] - [\Lambda]) + ([\Lambda\beta] - [\Lambda]) &= [\Lambda\alpha \oplus \Lambda\beta] - 2[\Lambda] \\ &= [\Lambda \oplus \Lambda\alpha\beta] - 2[\Lambda] \\ &= [\Lambda] + [\Lambda\alpha\beta] - 2[\Lambda] \\ &= [\Lambda\alpha\beta] - [\Lambda]\end{aligned}$$

where the second equation uses Theorem 8.5.11. Therefore the map

$$\begin{aligned}\widehat{U}(A)\backslash\widehat{J}(A)/A^\times &\longrightarrow Cl(\Lambda) \\ \alpha &\mapsto [\Lambda\alpha] - [\Lambda]\end{aligned}$$

is a group isomorphism. □

Corollary 8.5.13. *Let R be a Dedekind domain, and let Λ be an R-order in a semisimple K-algebra A such that $\dim_K(A) < \infty$. Suppose that for each $d \in \mathbb{N}$ and for all maximal R-orders in A there are only finitely many isomorphism classes of Λ-lattices L with $\dim_K(K \otimes_R L) < d$. Then $Cl(\Lambda)$ is a finite group.*

The hypothesis holds true if R is the ring of algebraic integers in a finite extension K of $\mathbb{Q}$.

Proof. The Jordan Zassenhaus theorem 8.4.6 shows that there are only finitely many isomorphism classes of rank one locally free Λ-lattices. Corollary 8.5.12 shows that $Cl(\Lambda)$ is parameterised by rank one locally free Λ-lattices. This shows the statement. □

We rediscover (by our much more involved method) the classical Steinitz theorem Proposition 7.5.27.

Corollary 8.5.14 (Steinitz theorem). *Let R be the ring of integers in a finite extension K of $\mathbb{Q}$. Then $Cl_I(R)$ is a finite group with group law given by $[I]\cdot[J] := [IJ]$ for any two isomorphism classes $[I]$ and $[J]$ of fractional ideals of R. Moreover,*

$$I \oplus J \simeq R \oplus IJ.$$

Proof. This follows from Corollary 8.5.13 and the fact that the class group of fractional ideals is isomorphic to the class group of the R-order R. The isomorphism from the group of fractional ideals to the class group of the order is given by

$$Cl_I(R) \ni [I] \mapsto [I] - [R] \in Cl(R).$$

This shows the Corollary. □

Remark 8.5.15. The theory of adèles and idèles is a highly sophisticated branch of algebraic number theory. In particular, these objects carry a complicated topology, such that the ring of adèles is a topological ring, and the group of idèles is a topological group. These additional structures play an important role in the theory. We do not use it here.

Remark 8.5.16. Usually it is very hard to compute class groups of orders, and even class groups of algebraic integers. However, there are methods to obtain at least some computations, or to reduce the computation of class groups of orders to the computation of class groups of certain rings of algebraic integers.

For class groups of algebraic integers methods from analytic number theory are frequently used. Moreover, there is a link to unit groups of rings as we shall indicate now. We note without proof one very useful result in this direction.

Let R be a Dedekind domain of characteristic 0 with field of fractions K, and let Λ be an R-order. Suppose that $e^2 = e$ is a central idempotent of $K\Lambda$ and put $f = 1 - e$. By Corollary 8.1.12 we obtain the following pullback diagram.

$$\begin{array}{ccc} \Lambda & \xrightarrow{e} & \Lambda e \\ \downarrow{\scriptstyle f} & & \downarrow{\scriptstyle \pi_e} \\ \Lambda f & \xrightarrow{\pi_f} & \Lambda f/\Lambda \cap \Lambda f \end{array}$$

If A is separable over K, i. e. A is a projective $A \otimes_K A^{op}$-module, then Reiner-Ullom show [ReUl-74] that there is an exact sequence

$$\Lambda^\times \longrightarrow (\Lambda e)^\times \times (\Lambda f)^\times \longrightarrow (\Lambda f/\Lambda \cap \Lambda f)^\times \xrightarrow{\delta} Cl(\Lambda) \longrightarrow Cl(\Lambda e) \times Cl(\Lambda f) \longrightarrow 0$$

All the maps are explicit, and are mostly induced from the canonical maps in the pullback diagram. Only $\delta(u) := \{(\lambda_1, \lambda_2) \in \Lambda e \times \Lambda f \mid \pi_e(\lambda_1) = u\pi_f(\lambda_2)\}$ is not an obvious map. This exact sequence parallels Mayer-Vietoris sequences from algebraic topology. The sequence comes from Milnor's approach to low-dimensional algebraic K-theory. This result allows to reduce the computation of the class groups of an order to the computation of the class groups of its components in the simple Wedderburn factors.

Example 8.5.17. Consider the integral group ring $\mathbb{Z}C_p$ for prime numbers p. Recall from Example 8.1.13 that we have the Wedderburn decomposition $\mathbb{Q}C_p \simeq \mathbb{Q} \times \mathbb{Q}(\zeta_p)$ for $\zeta_p = e^{\frac{2\pi i}{p}}$ being a primitive p-th root of unity. Let $e := \frac{1}{p}\sum_{g \in C_p} g$ be the primitive idempotent of $\mathbb{Q}C_p$ corresponding to the trivial module. Then $\mathbb{Z}C_p$ is a subring of $\mathbb{Z} \times \mathbb{Z}[\zeta_p]$ since $\mathbb{Z}[\zeta_p]$ is the ring of algebraic integers in $\mathbb{Q}(\zeta_p)$ (cf. Lemma 7.1.5) and hence $\mathbb{Z} \times \mathbb{Z}[\zeta_p]$ is the unique maximal order in $\mathbb{Q}C_p$ (cf. Remark 8.3.6 and Remark 8.3.11). Moreover, as shown in Example 8.1.13

$$\begin{array}{ccc} \mathbb{Z}C_p & \xrightarrow{\ e\ } & \mathbb{Z}[\zeta_p] \\ \downarrow{\scriptstyle f} & & \downarrow{\scriptstyle \pi_{1-e}} \\ \mathbb{Z} & \xrightarrow{\ \pi_e\ } & \mathbb{Z}/p\mathbb{Z} \end{array}$$

is a pullback. Hence in order to obtain the class group we may use Reiner-Ullom's exact sequence from Remark 8.5.16. Of course $(\mathbb{Z}/p\mathbb{Z})^\times = \mathbb{Z}/p\mathbb{Z} \setminus \{0\}$ and $\mathbb{Z}^\times = \{\pm 1\}$. Further, by Lemma 7.1.3 we see that $\frac{1-\zeta_p^i}{1-\zeta_p}$ is a unit in $\mathbb{Z}[\zeta_p]$. Moreover, $\pi_{(1-e)}(\zeta_p) = 1$. Since for all positive integers $i < p$ we get $\frac{1-\zeta_p^i}{1-\zeta_p} = 1 + \zeta_p + \cdots + \zeta_p^{i-1}$ we observe that $\pi_{(1-e)}(\frac{1-\zeta_p^i}{1-\zeta_p}) = i + p\mathbb{Z}$. Therefore $\delta = 0$ in Reiner-Ullom's exact sequence from Remark 8.5.16. Since $\mathbb{Z}$ is a principal ideal domain, $Cl(\mathbb{Z}) = 0$ and hence

$$Cl(\mathbb{Z}C_p) \simeq Cl(\mathbb{Z}[\zeta_p]).$$

Determining $Cl(\mathbb{Z}[\zeta_p])$ is a highly sophisticated problem in algebraic number theory. Its order h_p is a product $h_p^+ \cdot h_p^-$ of two integers. The number h_p^+ is the order of the class group of the maximal real subfield. There is a formula, due to Iwasawa giving the second number h_p^-, called relative class number. Iwasawa uses analytic number theory for a formula linking h_p^- with expressions involving Dirichlet L-functions. It can be shown that $h_p = 1 \Leftrightarrow p \leq 19$. This is a result due to Montgomery and Uchida. Concerning non prime indices m, there is an explicit (and small) list of integers m with $h_m = 1$ (cf. [Was-97, Chapter 11])

8.6 Swan's example

In this section we follow [Swa-62] and we will give Swan's famous example of a finite group G and a non free ideal I of $\mathbb{Z}G$ such that $\mathbb{Z}G \oplus I \simeq \mathbb{Z}G \oplus \mathbb{Z}G$.

There are deeper theoretical reasons, coming from a theorem by Eichler, showing that we need to stay close to quaternion groups.

Theorem 8.6.1. *Let*

$$Q_{32} := \langle x, y \mid yxy^{-1}x, x^8y^{-2}, y^4\rangle$$

be the group generated by two elements, x and y, subject to the relations

$$yxy^{-1} = x^{-1}, \quad x^8 = y^2, \quad y^2 \text{ generates the centre}, \quad y^4 = 1$$

Then there is an ideal I of $\mathbb{Z}Q_{32}$ which is not a free $\mathbb{Z}Q_{32}$-module, and there is an isomorphism of $\mathbb{Z}Q_{32}$-modules

$$I \oplus \mathbb{Z}Q_{32} \simeq \mathbb{Z}Q_{32} \oplus \mathbb{Z}Q_{32}$$

The proof of this result is going to cover the entire section. Note that

$$D_8 = Q_{32}/Z(Q_{32})$$

is the dihedral group of order 16. Be aware of the somehow inconsistent notation on the indices. However, we used this notation everywhere else, and so we think it is most appropriate to keep it here. Then Q_{32} is called the *generalised quaternion group of order* 32 and $Q_{16} := \langle x, y \mid yxy^{-1}x, x^4y^{-2}, y^4\rangle$ is the *generalised quaternion group of order* 16. Denote as usual by Q_8 the quaternion group of order 8.

Let $\mathbb{H}$ be the quaternion algebra over the real numbers with $\mathbb{R}$-basis $\{1, i, j, k\}$ subject to the relations with $i^2 = j^2 = k^2 = -1$ being central and $ij = k$.

We start with a preliminary lemma.

Lemma 8.6.2 (Coxeter 1940). *The finite subgroups of the unit group of the real quaternion algebra $\mathbb{H}$ are either cyclic, generalised quaternion, or one of three exceptional groups of order 24, 48, or 120.*

Proof. Recall that $\mathbb{H}$ is equipped with a norm $N_{\mathbb{R}}^{\mathbb{H}}$ defined as

$$N_{\mathbb{R}}^{\mathbb{H}}(x_1 + x_i i + x_j j + x_{ij} ij) = x_1^2 + x_i^2 + x_j^2 + x_{ij}^2.$$

This quadratic form induces a scalar product on the 4-dimensional real vector space $\mathbb{H}$.

Let ζ_n be a primitive n^{th} root of unity in $\mathbb{C} = \mathbb{R} + \mathbb{R}i \subseteq \mathbb{H}$. Then ζ_n generates a cyclic group of order n in the multiplicative group of $\mathbb{H}$.

The elements ζ_{4m} and j generated a subgroup of the multiplicative group of $\mathbb{H}$. Now $\zeta_{4m}^{2m} = -1 = j^2$ and $\zeta_{4m}^m = i$. The *generalised quaternion group* of order $8m$ has a presentation

$$\langle x, y \mid x^{4m} = y^4 = 1, x^{2m} = y^2, y^{-1}xy = x^{-1}\rangle.$$

If we identify x with ζ_{4m} and y with j we get at least a homomorphism. It is clear that the homomorphism is surjective, and it is not hard to see that the homomorphism is an isomorphism.

The binary tetrahedral group of order 24 is the group of units of the Hurwitz integers (cf. Exercise 8.2)

$$\Big\{1,-1,i,-i,j,-j,ij,-ij,\pm\frac{1}{2}(1+i+j+ij),\pm\frac{1}{2}(1-i-j-ij),$$
$$\pm\frac{1}{2}(1-i+j+ij),\pm\frac{1}{2}(1+i-j+ij),\pm\frac{1}{2}(1+i+j-ij),$$
$$\pm\frac{1}{2}(1-i-j+ij),\pm\frac{1}{2}(1-i+j-ij),\pm\frac{1}{2}(1+i-j-ij)\Big\}$$

In order to study the situation more systematically, let $\mathbb{S} := \{q \in \mathbb{H} \mid N(q) = 1\}$ and let $P := \mathbb{R}i + \mathbb{R}j + \mathbb{R}ij$. The space P equipped with $N^{\mathbb{H}}_{\mathbb{R}}$ is actually an euclidien 3-dimensional space. Then it is a classical theorem (see e. g. [CoSm-03]) that the map

$$\begin{aligned} \mathbb{H}\setminus\{0\}\times P &\xrightarrow{\alpha} P \\ (q,x) &\mapsto q^{-1}xq \end{aligned}$$

is an action of the unit group of the quaternions on the 3-dimensional euclidien space P. This action is an action as isometries. Indeed $N^{\mathbb{H}}_{\mathbb{R}}(q_1q_2) = N^{\mathbb{H}}_{\mathbb{R}}(q_2q_1)$ for any $q_1, q_2 \in \mathbb{H}$, and hence conjugation leaves invariant the value of $N^{\mathbb{H}}_{\mathbb{R}}$. Of course, for any $\lambda \in \mathbb{R}\setminus\{0\}$ we get $\alpha(\lambda\cdot q,x) = \alpha(q,x)$. Hence, the action α factorizes through the restricted action

$$\begin{aligned} \mathbb{S}\times P &\xrightarrow{\overline{\alpha}} P \\ (q,x) &\mapsto q^{-1}xq \end{aligned}$$

and the action is 2 by 1, i. e. $\overline{\alpha}(q,x) = \overline{\alpha}(q',x)$ for all $x \in P$ if and only if $q' \in \{q,-q\}$. Hence, the finite subgroups of $\mathbb{H}$ are central extensions by a group of order 2 of the finite subgroups of the orthogonal group $SO_3(\mathbb{R})$. More precisely, for a finite subgroup G of $\mathbb{H}\setminus\{0\}$ there is a subgroup $C_2 \le Z(G)$, where C_2 is of order 2, with $G/C_2 \le SO_3(\mathbb{R})$. Finite subgroups of $SO_3(\mathbb{R})$ are classified (cf. Exercise 8.6) as the cyclic groups, the dihedral groups, the alternating group of degree 12 (symmetry group of the cube=symmetry group of the octahedron), the alternating group of degree 5 (symmetry group of the dodecahedron=symmetry group of the icosahedron), or the symmetric group of degree 4 (the symmetry group of the tetrahedron).

Hence the finite subgroups of $\mathbb{H}$ are as claimed. □

We continue with the study of the integral group ring of Q_{32}. By the results from Section 1.2.3

$$\mathbb{Q}Q_{32} \simeq \mathbb{Q}D_8 \oplus \mathbb{Q}Q_{32}/((y^2+1)\mathbb{Q}G)$$

since y^2 is of order 2 and $\frac{1}{2}(y^2+1)$ is an idempotent of $\mathbb{Q}Q_{32}$. Put

$$A := \mathbb{Q}Q_{32}/((y^2+1)\mathbb{Q}G).$$

Then we have an embedding of $\mathbb{Q}$-algebras

$$\begin{aligned} A &\longrightarrow \mathbb{H} \\ x &\mapsto e^{\frac{2\pi i}{16}} =: \zeta \\ y &\mapsto j \end{aligned}$$

Hence A can be considered as a $\mathbb{Q}$-subalgebra of $\mathbb{H}$. The centre of A is

$$Z(A) = \mathbb{R} \cap \mathbb{Q}(\zeta) = \mathbb{Q}(\tau) =: K$$

where $\tau = \zeta + \zeta^{-1}$. Indeed, we work inside $\mathbb{H}$. Since $yxy^{-1} = j \cdot \zeta \cdot j^{-1} = \zeta^{-1}$, we have $\zeta + \zeta^{-1} = \tau \in Z(A)$ and τ generates $K = Z(A)$ as $\mathbb{Q}$-algebra inside $\mathbb{H}$. Note that

$$\tau = 2\cos\frac{\pi}{8} = \sqrt{2+\sqrt{2}},$$

which follows from $\tau^2 = \zeta_8 + \zeta_8^{-1} + 2$ and therefore $(\tau^2-2)^2 = \zeta_4 + \zeta_4^{-1} + 2 = i - i + 2 = 2$. Here we denote $\zeta_m := e^{\frac{2\pi i}{m}}$ for any positive integer m. Put

$$\tau' := \sqrt{2-\sqrt{2}} = \frac{\sqrt{2}}{\tau}.$$

Further, A is the quaternion algebra with base field K:

$$A = K\langle i,j\rangle/\langle (i^2+1, j^2+1; (ij)^2+1)\rangle \subseteq \mathbb{H}$$

Let $R = \mathrm{algint}_{\mathbb{Z}}(K)$. By Lemma 7.1.5 we see that $R = \mathbb{Z}[\tau]$, and $\tau' \in R$ since $\tau' \in \mathbb{Q}(\zeta) \cap \mathbb{R}$.

Consider the elements

$$\alpha := \frac{\sqrt{2}+1+i}{\sqrt{2}\tau}, \quad \beta := \frac{(1+j)}{\tau'}, \quad \gamma = \alpha\beta = \frac{(\sqrt{2}+1+i)(1+j)}{2}$$

and consider the R-module Λ generated by $1, \alpha, \beta, \gamma$. Note that Λ is a submodule of $\mathbb{H}$, and not of $R \cdot 1 + R \cdot i + R \cdot j + R \cdot ij$.

Lemma 8.6.3. *Λ is a maximal R-order in A. Mapping x to α and y to j gives a monomorphism of groups $Q_{32} \longrightarrow \Lambda^\times$, and hence Λ contains a subring isomorphic to $\mathbb{Z}Q_{32}$.*

Proof.

$$\beta^2 = \frac{(1+j)^2}{2-\sqrt{2}} = \frac{2j}{2-\sqrt{2}} = \frac{2j(2+\sqrt{2})}{(2-\sqrt{2})(2+\sqrt{2})} = \frac{2j(2+\sqrt{2})}{4-2} = \tau^2 j.$$

Since we consider Λ as an R-order, $\tau \in R$ and also $\tau' \in R$. Hence $\beta \in \Lambda$ implies $j \in \Lambda$, by the definition of β. Moreover,

$$\alpha^2 = \frac{(\sqrt{2}+1)^2 + 2i(\sqrt{2}+1) - 1}{2(2+\sqrt{2})} = \frac{2+2\sqrt{2}+2i(\sqrt{2}+1)}{2(2+\sqrt{2})} = \frac{(1+\sqrt{2})(1+i)}{(2+\sqrt{2})}$$

$$\alpha^4 = \frac{(1+\sqrt{2})^2(1+i)^2}{(2+\sqrt{2})^2} = \frac{(3+2\sqrt{2})\cdot 2i}{(6+4\sqrt{2})} = i$$

Hence α is of order 16, and $\alpha^8 = -1$. Moreover,

$$j\alpha j^{-1}\alpha = -j\alpha j\alpha = \frac{\sqrt{2}+1+i}{\sqrt{2}\tau}\cdot\frac{\sqrt{2}+1-i}{\sqrt{2}\tau} = \frac{(\sqrt{2}+1)^2+1}{2(2+\sqrt{2})} = \frac{4+2\sqrt{2}}{4+2\sqrt{2}} = 1$$

Hence, using that y^2 is the unique non trivial minimal subgroup in Q_{32}, therefore contained in any possible non trivial kernel, the map $y \mapsto \alpha$ and $x \mapsto j$ is a group monomorphism of Q_{32} into the algebra $\widehat{\Lambda}$ generated by $1, \alpha, \beta$.

In order to see that $\Lambda := R + R\alpha + R\beta + R\gamma$ is an order, and as a consequence $\widehat{\Lambda} = \Lambda$, we need to show that $\alpha^2 \in \Lambda$, $\beta\alpha \in \Lambda$, $\beta^2 \in \Lambda$. Since $\beta \in \Lambda$, and since $\tau' \in R$, also $1 + j \in \tau'\Lambda \subseteq \Lambda$. But then also $j \in \Lambda$. But this implies that

$$(\sqrt{2}+1+i) \in \sqrt{2}\tau\Lambda = \tau'\tau^2\Lambda \subseteq \Lambda.$$

Since $\sqrt{2}+2 = \tau^2 \in \tau^2\Lambda$, this shows that $i-1 \in \tau^2\Lambda$. Since $j\cdot(1-i)\cdot(-j) = 1+i$, the fact that $j \in \Lambda$ implies that also $i+1 \in \tau^2\Lambda$, which shows that $\frac{i+1}{2+\sqrt{2}} \in \Lambda$ and also $i \in \Lambda$. Therefore,

$$\alpha^2 = (1+i) - \frac{1+i}{2+\sqrt{2}} \in \Lambda.$$

Since $\beta \in \Lambda$, also $\tau'\beta = (1+j) \in \Lambda$ and therefore $\beta^2 = \tau^2 j \in \Lambda$. By definition $\gamma = \alpha\beta \in \Lambda$. We still need to show that $\beta\alpha \in \Lambda$. Since $i \in \Lambda$,

$$\begin{aligned}\beta\alpha &= \frac{(1+j)(\sqrt{2}+1+i)}{2}\\ &= \frac{(\sqrt{2}+1+i) + j(\sqrt{2}+1+i)}{2}\\ &= \frac{(\sqrt{2}+1+i) + (\sqrt{2}+1-i)j}{2}\\ &= \gamma + ij \in \Lambda\end{aligned}$$

Since Λ is an R-order we may compute the discriminant of Λ. We compute the determinant of the Gram matrix given by the reduced trace bilinear form of the generating set $1, \alpha, \beta, \gamma$. The matrix formed by the products of the basis elements is

$$\begin{pmatrix} 1 & \alpha & \beta & \gamma \\ \alpha & \alpha^2 & \alpha\beta & \alpha\gamma \\ \beta & \beta\alpha & \beta^2 & \beta\gamma \\ \gamma & \gamma\alpha & \gamma\beta & \gamma^2 \end{pmatrix}$$

and we computed these products above. We need to form reduced traces of these. For this purpose we realise the quaternion algebra as 2×2 matrix ring over a splitting field, cay $\mathbb{C}$, via Pauli matrices as in Exercise 1.12. The traces of the elements σ_i mentioned in the exercise are 0, whereas the trace of the identity matrix is 2. Since traces are R-linear we observe that the determinant of the resulting matrix is -1, whence the discriminant ideal is R. By Corollary 8.3.12 we see that Λ is a maximal order.

Now, Q_{32} is isomorphic to the group generated by α and j, identifying x with α and y with β. Hence $\mathbb{Z}Q_{32}$ is a sub-order of Λ. □

Let Λ' be the R-submodule of $\mathbb{H}$ generated by

$$\delta_0 := 1; \quad \delta_1 := \frac{1+i}{\sqrt{2}}; \quad \delta_2 := \frac{1+j}{\sqrt{2}}; \quad \delta_3 := \frac{1+i+j+ij}{2}.$$

Lemma 8.6.4. *Λ' is a maximal R-order in A.*

Proof. We first note that $\delta_1^2 = i \in \Lambda'$, $\delta_2^2 = j \in \Lambda'$ and $\delta_3 = \delta_1\delta_2$. Further,

$$\delta_3^3 = \frac{1}{4}(-2 + 2i + 2j + 2ij) \cdot \frac{1}{2}(1 + i + j + ij) = -1.$$

In particular $\delta_3^2 = \delta_3 - 1 \in \Lambda'$, δ_1, δ_2 are of order 8, δ_3 is of order 6, and all three are algebraic integers over $\mathbb{Z}$. Since $\delta_3 - \delta_2\delta_1 = ij \in \Lambda'$, we have $\delta_2\delta_1 \in \Lambda'$. Hence Λ' is a ring.

We can compute the discriminant. The matrix $(\delta_i\delta_j)_{0\le i,j\le 3}$ is

$$\begin{pmatrix} 1 & \frac{1}{\sqrt{2}}(1+i) & \frac{1}{\sqrt{2}}(1+j) & \frac{1}{2}(1+i+j+ij) \\ \frac{1}{\sqrt{2}}(1+i) & i & \frac{1}{2}(1+i+j+ij) & \frac{1}{\sqrt{2}}(i+ij) \\ \frac{1}{\sqrt{2}} & \frac{1}{2}(1+i+j-ij) & j & \frac{1}{\sqrt{2}}j \\ \frac{1}{2}(1+i+j+ij) & \frac{1}{\sqrt{2}}i & \frac{1}{\sqrt{2}}(j-ij) & \frac{1}{\sqrt{2}}(j+ij) \end{pmatrix}$$

Then, we need to tensor with a splitting field and compute the traces of the representing matrices. The splitting field $\mathbb{C}$ is appropriate, and for computing the traces we realize the quaternion skew field by Pauli matrices as in Exercise 1.12. Therefore the matrix formed by the reduced traces of the coefficients in the matrix above is

$$\begin{pmatrix} 2 & \sqrt{2} & \sqrt{2} & 1 \\ \sqrt{2} & 0 & 1 & 0 \\ \sqrt{2} & 1 & 0 & 0 \\ 1 & 0 & 0 & 0 \end{pmatrix}$$

whose determinant is 1. The discriminant ideal is R, and applying Corollary 8.3.12 shows that Λ' is a maximal order. □

Lemma 8.6.5. *Consider the left ideal $\mathfrak{p} := \Lambda\tau + \Lambda\beta$ of Λ. Then*

$$\Lambda' = \mathcal{O}_r(\mathfrak{p}) = \{\lambda \in A \mid \mathfrak{p}\lambda \subseteq \mathfrak{p}\}.$$

Proof. Since $\delta_3 = \delta_1\delta_2$, in order to see $\Lambda' \subseteq \mathcal{O}_r(\mathfrak{p})$ we need to show that

$$\tau\delta_1 \in \mathfrak{p}, \tau\delta_2 \in \mathfrak{p}, \beta\delta_1 \in \mathfrak{p}, \beta\delta_2 \in \mathfrak{p}.$$

Note that

$$\frac{\tau}{\tau'} = \frac{\tau^2}{\sqrt{2}} = \frac{2+\sqrt{2}}{\sqrt{2}} = \sqrt{2}+1 \in \operatorname{algint}_{\mathbb{Z}}(K) = R$$

and similarly $\frac{\tau'}{\tau} \in R$, and therefore $\frac{\tau}{\tau'} \in R^{\times}$.

Further,

$$\alpha\tau^2 = \frac{1+i}{\sqrt{2}}\tau + \tau$$

and hence

$$\tau\delta_1 = \frac{1+i}{\sqrt{2}}\tau = \alpha\tau^2 - \tau \in \mathfrak{p}.$$

Further

$$\tau\delta_2 = \frac{1+j}{\sqrt{2}}\tau = \frac{1+j}{\tau\tau'}\tau = \beta \in \mathfrak{p}.$$

Further,

$$\alpha\tau - 1 = \frac{1+i}{\sqrt{2}} = \delta_1.$$

Hence

$$\beta\delta_1 = \beta(\alpha\tau - 1) = \beta\alpha\tau - \beta \in \mathfrak{p}.$$

Now,

$$\beta\delta_2 = \frac{1+j}{\tau'}\frac{1+j}{\tau\tau'} = \frac{2j}{\tau(\tau')^2} = j\tau \in \mathfrak{p}.$$

It follows that $\Lambda' \subseteq \mathcal{O}_r(\mathfrak{p})$.

Since $\mathcal{O}_r(\mathfrak{p})$ is an order and since Lemma 8.6.4 implies that Λ' is a maximal order, we have $\Lambda' = \mathcal{O}_r(\mathfrak{p})$. □

Lemma 8.6.6. *$\mathfrak{p}$ is not principal.*

Proof. Since Λ is an order in the quaternion algebra A over K, we can consider the restriction of the norm $N_{\mathbb{R}}^{\mathbb{H}} : \mathbb{H} \longrightarrow \mathbb{R}$ on elements in the quaternion algebra (denoting $k = ij$):

$$A \xrightarrow{N_K^A} K$$
$$a_0 + a_1 i + a_2 j + a_3 k \mapsto a_0^2 + a_1^2 + a_2^2 + a_3^2$$

Recall that K is a subfield of $\mathbb{R}$, so that $N_K^A(x) \geq 0$ for all $x \in A$, with equality only for $x = 0$. Let $\Lambda^\times$ the group of units in Λ, and likewise $R^\times$ the group of units in R. Let

$$\Lambda_0^\times := \{x \in \Lambda^\times \mid N_K^A(x) = 1\}.$$

We shall prove that $\Lambda_0^\times$ is finite.

By Galois theory there are 4 embeddings of K into $\mathbb{R}$, namely corresponding to the four real roots

$$\sqrt{2+\sqrt{2}},\ \sqrt{2-\sqrt{2}},\ -\sqrt{2+\sqrt{2}},\ -\sqrt{2-\sqrt{2}}$$

of the minimal polynomial $(X^2-2)^2-2$ of τ. Denote by $\sigma_1,\ldots,\sigma_4$ these embeddings. Adjoining i and j we hence obtain 4 embeddings of A into $\mathbb{H}$. Note that as in the proof of Theorem 8.3.7 we see that the discriminant of Λ is non zero. Hence, again as in the proof of Theorem 8.3.7

$$A \xrightarrow{\begin{pmatrix}\sigma_1\\ \sigma_2\\ \sigma_3\\ \sigma_4\end{pmatrix}} \mathbb{H}\times\mathbb{H}\times\mathbb{H}\times\mathbb{H}$$

restricts to an embedding of Λ into some lattice in $\mathbb{H}^4$. Now, $x \in \Lambda_0^\times$ if and only if $N_K^A(x) = 1$. But this implies that

$$N_{\mathbb{R}}^{\mathbb{H}}(\sigma_1(x)) = N_{\mathbb{R}}^{\mathbb{H}}(\sigma_2(x)) = N_{\mathbb{R}}^{\mathbb{H}}(\sigma_3(x)) = N_{\mathbb{R}}^{\mathbb{H}}(\sigma_4(x)) = 1.$$

We claim that there are only finitely many lattice points verifying these conditions. Let $x = x_1 + x_i i + x_j j + x_k k \in \Lambda_0^\times$ with $x_1, x_i, x_j, x_k \in K$. Then

$$N_K^A(x) = x_1^2 + x_i^2 + x_j^2 + x_k^2 = 1$$

Since $K \subseteq \mathbb{R}$, all the terms $x_1^2, x_2^2, x_j^2, x_k^2$ are non negative real numbers. But

$$N_K^A(\sigma_\ell(x)) = \sigma_\ell(x_1)^2 + \sigma_\ell(x_i)^2 + \sigma_\ell(x_j)^2 + \sigma_\ell(x_k)^2 = 1$$

for all $\ell \in \{1,2,3\}$ and also $\sigma_\ell(y)^2 \geq 0$ for all y and ℓ. Hence

$$\mathrm{trace}_{\mathbb{Q}}^K(x_1^2) + \mathrm{trace}_{\mathbb{Q}}^K(x_i^2) + \mathrm{trace}_{\mathbb{Q}}^K(x_j^2) + \mathrm{trace}_{\mathbb{Q}}^K(x_k^2) = 4$$

where $\mathrm{trace}_{\mathbb{Q}}^K = \mathrm{id}_K + \sigma_1 + \sigma_2 + \sigma_3$ has values in $\mathbb{Q}$. Moreover, $\mathrm{trace}_{\mathbb{Q}}^K(y^2)$ is a non negative rational number. Now, by Lemma 7.1.1 the additive group of R, whence of any free R-module of finite rank, and in turn of any direct factor, that is of an R-lattice is a free abelian group of finite rank. Hence, we may multiply with an appropriate integer in order to find all solutions $(y_1, y_2, y_3, y_4) \in \mathbb{N}^4$ of an equation $y_1 + y_2 + y_3 + y_4 = t$ for some positive integer t. This number is finite. Now, $y_i = \mathrm{trace}_{\mathbb{Q}}^K(x_i^2) \cdot \frac{t}{4} = x_i^2 \cdot \frac{t}{4} +$

$\sigma_1(x_i^2) \cdot \frac{t}{4} + \sigma_2(x_i^2) \cdot \frac{t}{4} + \sigma_3(x_i^2) \cdot \frac{t}{4}$. Each of these summands is a positive element of R, and hence we need to prove that there are only a finite number of elements z in R satisfying $0 < z < t$ and $0 < \sigma_\ell(z) < t$ for each $\ell \in \{1,2,3\}$. Else there is a converging sequence $(z_n)_{n\in\mathbb{N}}$ of pairwise distinct elements of R satisfying $0 < z_n < t$. Therefore, there is a converging subsequence $(z_n^{(1)})_{n\in\mathbb{N}}$ such that $z_n^{(1)}$ and $\sigma_1(z_n^{(1)})$ satisfy $0 < z_n^{(1)} < t$ and $0 < \sigma_1(z_n^{(1)}) < t$. Taking again subsequences of this one we get a converging sequence $(z_n^{(2)})_{n\in\mathbb{N}}$ and finally $(z_n^{(3)})_{n\in\mathbb{N}}$ of pairwise distinct elements of R such that $0 < z_n^{(3)} < t$ and $0 < \sigma_\ell(z_n^{(3)}) < t$ for all $\ell \in \{1,2,3\}$. Hence, for large enough n we have that $0 < |z_n^{(3)} - z_{n+1}^{(3)}| < 1$ and $0 < |\sigma_\ell(z_n^{(3)}) - \sigma_\ell(z_{n+1}^{(3)})| < 1$ for all $\ell \in \{1,2,3\}$. Hence $\mathrm{norm}_{\mathbb{Q}}^K(z_n^{(3)} - z_{n+1}^{(3)}) < 1$ is a non zero integer. Contradiction. Therefore $\Lambda_0^\times$ is finite.

Remark 8.6.7. Though Swan suggests within a few lines the above argument, Swan's original proof proceeds as follows. We get

$$\Lambda_0^\times \cap R = \{1,-1\}$$

since when $x \in R$, we get $N_K^A(x) = x^2 = 1$ and hence the statement. A deep result due to Eichler [Eich-38] shows that $\Lambda^\times/R^\times$ is finite. The precise statement is quite technical, and expresses the index in terms of values of the Riemann zeta-function, the class number of R, its discriminant and further arithmetic invariants. The proof uses analytic number theory, and it would lead too far to give the proof in this book. Using this fact however, we have

$$\Lambda_0^\times/\{1,-1\} = \Lambda_0^\times/(R^\times \cap \Lambda_0) \subseteq \Lambda^\times/R^\times$$

and therefore $\Lambda_0^\times$ is finite.

Now, by Lemma 8.6.3 α and j generate a subgroup of $\Lambda_0^\times$ isomorphic to Q_{32}. Hence 32 divides the order of $\Lambda_0^\times$.

If $\mathfrak{p} = \Lambda x$, for some $x \in \Lambda$, we get by Lemma 8.6.5 and by Lemma 8.6.3 that

$$\Lambda' = \mathcal{O}_r(\mathfrak{p}) = x^{-1}\Lambda x.$$

But $\delta_3 \in \Lambda'$ and $\delta_3^3 = -1$. This shows that $x\delta_3 x^{-1}$ is an element of $\Lambda_0^\times$ of order 6. Therefore 96, which is the least common multiple of 32 and 6, divides the order of $\Lambda_0^\times$.

By Lemma 8.6.2 $\Lambda_0^\times$ is hence either cyclic or generalised quaternion. Since $\Lambda_0^\times$ is non abelian, $\Lambda_0^\times$ is generalised quaternion. Since 96 divides the order of $\Lambda_0^\times$, the group has to contain an element v of order $96/2 = 48$. Hence $\mathbb{Q}(v) \simeq \mathbb{Q}(\zeta_{48})$ is a subfield of A. Since

$$\dim_{\mathbb{Q}}(\zeta_{48}) = 16 = \dim_{\mathbb{Q}}(A)$$

we would get $A = \mathbb{Q}(e^{\frac{2\pi i}{48}})$. However A is non commutative, which gives a contradiction. □

We now come to the construction of the non free ideal I.

Consider the left ideal $\mathbb{Z}Q_{32}(y+4)$. Since

$$(y^2+16)(y-4)(y+4) = y^4 - 246 = -255 = -3\cdot 5\cdot 17$$

the ideal $\mathbb{Z}Q_{32}(y+4)$ contains $3\cdot 5\cdot 17\cdot \mathbb{Z}Q_{32}$. We get that

$$\mathbb{Z}Q_{32}/\mathbb{Z}Q_{32}(y+4) \simeq M = M_3 \oplus M_5 \oplus M_{17}$$

for M_3 an $\mathbb{F}_3Q_{32}$-module, for M_5 an $\mathbb{F}_5Q_{32}$-module and M_{17} an $\mathbb{F}_{17}Q_{32}$-module. Localising at primes different from 3, 5 and 17, the element $y+4$ is invertible. Hence the ideal $\mathbb{Z}Q_{32}(y+4)$ has $\mathbb{Z}$-rank 32.

Since Q_{32} is a 2-group, and since 3, 5 and 17 are relatively prime to 2, Maschke's Theorem 1.2.9 implies that each of the modules M_3, M_5, M_{17} is semisimple.

Since M is a quotient of the rank one free $\mathbb{Z}Q_{32}$-module, which is generated by $1\in \mathbb{Z}Q_{32}$, the image of 1 in the quotient M is (e_3, e_5, e_{17}), where e_p are elements in M_p, the image of 1 in the module M_p for all $p\in\{3,5,17\}$.

Recall that A is embedded into $\mathbb{H}$ by mapping y to j and x to $\zeta = e^{\frac{2\pi i}{16}}$. Since Λ is a (maximal) R-order in A, y^2 acts as $j^2=-1$ on Λ. Since

$$(y-4)(y+4) = y^2 - 16$$

and since $16 \equiv 1 \mod 3$ and $16\equiv 1 \mod 5$, the element y^2 acts as 1 on M_3 and as 1 on M_5. Moreover, since y^2 is in the centre of Q_{32}, we get that

$$\Lambda \otimes_{\mathbb{Z}Q_{32}} M_3 = 0 = \Lambda\otimes_{\mathbb{Z}Q_{32}} M_5$$

Indeed, for an element $\lambda\in\Lambda$ and an element $m\in M_3$ or $m\in M_5$ we get

$$\lambda\otimes m = \lambda\otimes 1m = \lambda\otimes y^2\cdot m = \lambda\cdot y^2\otimes m = y^2\lambda\otimes m = -\lambda\otimes m.$$

Similarly since $16\equiv -1 \mod 17$, the element y^2 acts as -1 on M_{17}. Hence y^2+1 acts as 0 on M_{17} and M_{17} is a $\mathbb{Z}Q_{32}/((y^2+1)\mathbb{Z}Q_{32})$-module. However, y is mapped to j in A and $j^2+1=0$ in A. Hence

$$\mathbb{Z}Q_{32}/((y^2+1)\mathbb{Z}Q_{32}) \subseteq \Lambda.$$

By Proposition 8.3.18 we see that

$$32\Lambda \subseteq \mathbb{Z}Q_{32} \subseteq \Lambda.$$

Therefore the abelian group $\Lambda/\mathbb{Z}Q_{32} =: T$ is an abelian 2-group.

We hence obtain

$$\Lambda\otimes_{\mathbb{Z}Q_{32}} M_{17} \simeq (\Lambda/\mathbb{Z}Q_{32})\otimes_{\mathbb{Z}} M_{17} = T\otimes_{\mathbb{Z}} M_{17} \simeq M_{17}$$

since 2 is invertible in $\mathbb{F}_{17}$, and hence multiplication by 2 is an automorphism on M_{17}.

This shows M_{17} is a Λ-module via the isomorphism

$$\Lambda \otimes_{\mathbb{Z}Q_{32}} M_{17} \simeq M_{17}.$$

Since y acts as j on Λ, we get

$$M_{17} \simeq \Lambda/\Lambda(j+4).$$

Recall the ideal $\mathfrak{p} := \Lambda\tau + \Lambda\beta$ of Λ introduced in Lemma 8.6.5. We showed in Lemma 8.6.6 that this ideal is not principal.

Lemma 8.6.8. *There is a short exact sequence*

$$0 \longrightarrow \mathfrak{p} \longrightarrow \Lambda \longrightarrow M_{17} \longrightarrow 0$$

Proof. We start with some arithmetics in R. Recall that $\tau = \sqrt{2+\sqrt{2}}$ and $\tau' = \sqrt{2-\sqrt{2}}$. Then

$$\begin{aligned}(1+2\tau)(1-2\tau)(1+2\tau')(1-2\tau') &= (1-4\tau^2)(1-4(\tau')^2)\\ &= (1-8-4\sqrt{2})(1-8+4\sqrt{2})\\ &= -(4\sqrt{2}-7)(4\sqrt{2}+7)\\ &= -16\cdot 2+49\\ &= 17\end{aligned}$$

is a prime element in $\mathbb{Z}$. Since the product above represents the product of all Galois conjugates of $(1+2\tau)$ (resp. $(1-2\tau)$, resp. $(1+2\tau')$, resp. $(1-2\tau')$), we get that 17 is the norm of each of these factors. Hence, the elements $(1+2\tau)$, $(1-2\tau)$, resp. $(1+2\tau')$, resp. $(1-2\tau')$, are all prime elements in R and the above product is a decomposition is a decomposition of 17 into a product of prime elements in R.

Therefore by the Chinese Reminder Theorem

$$R/17R \simeq \mathbb{Z}/17\mathbb{Z} \times \mathbb{Z}/17\mathbb{Z} \times \mathbb{Z}/17\mathbb{Z} \times \mathbb{Z}/17\mathbb{Z}$$

as rings. Hence,

$$\Lambda/17\Lambda \simeq (\mathbb{Z}/17\mathbb{Z})[i,j] \times (\mathbb{Z}/17\mathbb{Z})[i,j] \times (\mathbb{Z}/17\mathbb{Z})[i,j] \times (\mathbb{Z}/17\mathbb{Z})[i,j].$$

However, we can compute the Legendre quadratic residue symbol $(\frac{-1}{17}) = 1$ and actually $4^2 \equiv -1 \bmod 17$. Hence the quaternion algebra over $\mathbb{Z}/17\mathbb{Z}$ splits as a 2×2 matrix algebra. (cf. Exercise 1.12). More explicitly, using the classical realisation of the quaternions via $-i$ times the Pauli matrices we get an isomorphism

$$\begin{aligned}\mathbb{Z}/17\mathbb{Z}[i,j] &\longrightarrow Mat_{2\times Z}(\mathbb{Z}/17\mathbb{Z})\\ 1 &\mapsto \begin{pmatrix} 1 & 0\\ 0 & 1\end{pmatrix}\end{aligned}$$

$$i \mapsto \begin{pmatrix} 4 & 0 \\ 0 & -4 \end{pmatrix}$$

$$j \mapsto \begin{pmatrix} 0 & 1 \\ -1 & 0 \end{pmatrix}$$

$$ij \mapsto \begin{pmatrix} 0 & 4 \\ 4 & 0 \end{pmatrix}$$

Hence $(\mathbb{Z}/17\mathbb{Z})[i,j] \simeq S \oplus S$ is semisimple with one simple module S. Now consider $0 \neq j+4$ in $(\mathbb{Z}/17\mathbb{Z})[i,j]$. This element is not invertible since

$$(j+4)\cdot(j-4) = -17 = 0 \quad \text{in } (\mathbb{Z}/17\mathbb{Z})[i,j].$$

Hence $(\mathbb{Z}/17\mathbb{Z})[i,j]/(\mathbb{Z}/17\mathbb{Z})(j+4)[i,j]$ is isomorphic to S.

We have seen above that $\Lambda/17\Lambda$ is a direct product of four two by two matrix algebras over $\mathbb{Z}/17\mathbb{Z}$. Hence there are four isomorphism types of simple $\Lambda/17\Lambda$-modules. Let S_1, S_2, S_3, S_4 be simple $\Lambda/17\Lambda$-modules which are two by two non isomorphic. Then

$$\Lambda/17\Lambda \simeq S_1 \oplus S_1 \oplus S_2 \oplus S_2 \oplus S_3 \oplus S_3 \oplus S_4 \oplus S_4.$$

Now, since $17 \in \Lambda(j+4)$, we get $\Lambda/(\Lambda(j+4))$ is therefore isomorphic to the direct sum of one copy of each of S_1, S_2, S_3, S_4:

$$M_{17} = \Lambda/(\Lambda(j+4)) \simeq S_1 \oplus S_2 \oplus S_3 \oplus S_4.$$

Consider

$$\begin{aligned} a_1 = 1 + \frac{i-1}{\tau} &= \frac{((i-1) + \sqrt{2+\sqrt{2}})\cdot\sqrt{2-\sqrt{2}}}{\sqrt{2}} \\ &= ((i-1) + \sqrt{2+\sqrt{2}})\cdot\sqrt{\sqrt{2}-1} \in \Lambda. \end{aligned}$$

Then

$$\left(1 - \frac{i+1}{\tau}\right)a_1 = 1 - \frac{2}{\tau} + \frac{2}{\tau^2} = \frac{\tau'}{2+\tau'}(1+2\tau').$$

Now,

$$\begin{aligned} \frac{\tau'}{2+\tau'} &= \frac{\tau'(2-\tau')}{4-(\tau')^2} \\ &= \frac{\tau'(2-\tau')}{4-(2-\sqrt{2})} \\ &= \frac{\tau'(2-\tau')}{2+\sqrt{2}} \\ &= \frac{\tau'}{\tau}\cdot\frac{(2-\tau')}{\tau} \end{aligned}$$

We have seen already in the proof of Lemma 8.6.5 that $\frac{\tau'}{\tau} \in R^\times$. Hence we compute

$$\begin{aligned}\frac{2-\tau'}{\tau} &= \frac{2}{\tau} - \frac{\tau'}{\tau}\\ &= \frac{(\tau\tau')^2}{\tau} - \frac{\tau'}{\tau}\\ &= \tau(\tau')^2 - \frac{\tau'}{\tau} \in R\end{aligned}$$

but also

$$\begin{aligned}\frac{2+\tau'}{\tau'} &= \frac{2}{\tau'} + 1\\ &= \frac{(\tau\tau')^2}{\tau'} + 1\\ &= \tau^2\tau' + 1 \in R\end{aligned}$$

Therefore

$$\frac{\tau'}{2+\tau'} \in R^\times.$$

Since $17 \in (1+2\tau')R$ by the first step of the proof, we see that $17\Lambda \subseteq \Lambda \cdot a_1$. Moreover, a_1 is not zero in $(\mathbb{Z}/17\mathbb{Z})[i,j]$, and hence not 0 in $\Lambda/17\Lambda$ neither. Further, a_1 is not a unit in $(\mathbb{Z}/17\mathbb{Z})[i,j]$. Hence, $\Lambda/\Lambda a_1$ is isomorphic to a direct sum of simple modules, not all at once. Since a_1 has four conjugates under the action of the Galois group of K over $\mathbb{Q}$,

$$a_1 = 1 + \frac{i-1}{\tau}, \quad a_2 = 1 - \frac{i-1}{\tau}, \quad a_3 = 1 + \frac{i-1}{\tau'}, \quad a_4 = 1 - \frac{i-1}{\tau'}$$

and since $\Lambda/17\Lambda$ has precisely four simple modules, $\Lambda/a_i\Lambda$ is simple and up to renaming the indices

$$\Lambda/a_i\Lambda \simeq S_i$$

for each $i \in \{1,2,3,4\}$. Denoting

$$\overline{\beta} := \frac{1-j}{\tau'} \quad \text{and} \quad \overline{\alpha} := \frac{\sqrt{2}+1-i}{\sqrt{2}\tau},$$

consider the left ideal

$$\mathfrak{b} := \Lambda(\tau\alpha + \overline{\beta}) + \Lambda(\beta\alpha + \tau)$$

of Λ. We see

$$\beta \cdot \overline{\beta} = \frac{(1+j)(1-j)}{(\tau')^2} = \frac{2}{(\tau')^2} = \tau^2$$

and hence

$$\beta \cdot \left(\alpha + \frac{\overline{\beta}}{\tau}\right) = \beta\alpha + \tau \quad \text{and} \quad \tau \cdot \left(\alpha + \frac{\overline{\beta}}{\tau}\right) = \tau\alpha + \overline{\beta}$$

Therefore

$$\mathfrak{b} = \mathfrak{p}\left(\alpha + \frac{\overline{\beta}}{\tau}\right).$$

Note that $\alpha + \frac{\overline{\beta}}{\tau}$ is a unit in A. Indeed, A is a subring of $\mathbb{H}$, and actually a quaternion algebra over K itself. Hence all non zero elements are invertible. More explicitly the inverse of $\alpha + \frac{\overline{\beta}}{\tau}$ in $\mathbb{H}$ is its (quaternion algebra) conjugate in $\mathbb{H}$ multiplied by the inverse of ist norm. The conjugate is $\overline{\alpha} + \frac{\beta}{\tau} \in A$, and its norm is

$$N_A^K\left(\alpha + \frac{\overline{\beta}}{\tau}\right) = \left(\frac{\sqrt{2} + 1 + \tau}{\sqrt{2}\tau}\right)^2 + \frac{1}{\tau^2} + \frac{1}{2} \in K.$$

Hence $\alpha + \frac{\overline{\beta}}{\tau}$ is invertible in A. We therefore get $\mathfrak{b} \simeq \mathfrak{p}$. We remark that $\overline{\alpha} \in \Lambda$ and $\overline{\beta} \in \Lambda$ and

$$(\tau\alpha + \overline{\beta})(\tau\overline{\alpha} + \beta) = \tau \underbrace{(\sqrt{2} + 1)(1 + 2\tau')}_{=:p}.$$

However,

$$(1 + \sqrt{2})(1 - \sqrt{2}) = -1$$

and therefore $(1 + \sqrt{2}) \in R^\times$. Hence

$$\tau(p - \overline{\alpha}(\tau\alpha + \overline{\beta})) = \beta(\tau\alpha + \overline{\beta}) = \tau(\beta\alpha + \tau)$$

which implies

$$(\sqrt{2} + 1)(1 + 2\tau') = p = (\beta\alpha + \tau) + \overline{\alpha}(\tau\alpha + \overline{\beta}) \in \mathfrak{b}.$$

Since $\mathfrak{b} \neq \Lambda p$ and $\mathfrak{b} \neq \Lambda$, $\Lambda/\mathfrak{b}$ must be the simple $\Lambda/17\Lambda$-module on which p acts trivially.

$$\mathfrak{a} := \mathfrak{b}\mathfrak{a}_2\mathfrak{a}_3\mathfrak{a}_4$$

is a left ideal of Λ. The sequence of Λ-ideals

$$\mathfrak{a} \leq \Lambda\mathfrak{a}_2\mathfrak{a}_3\mathfrak{a}_4 < \Lambda\mathfrak{a}_3\mathfrak{a}_4 < \Lambda\mathfrak{a}_4 < \Lambda$$

induces a composition series of $\Lambda/\mathfrak{a}$ whose factors are the simple $\Lambda/17\Lambda$-modules S_1, S_2, S_3, S_4 in this order. They belong to different primary components of R (cf. Definition 7.2.3), and hence, using Lemma 6.3.5, Corollary 6.5.11, and Lemma 7.5.7,

$$\Lambda/\mathfrak{a} = S_1 \oplus S_2 \oplus S_3 \oplus S_4 \simeq M_{17}.$$

Now, a_1, a_2, a_3, a_4 are non zero elements in $\mathbb{C}$. They are hence regular in their action on Λ, i. e. invertible in their action on A, and therefore $\mathfrak{a} \simeq \mathfrak{b}$ as Λ-modules. Since $\mathfrak{a} \simeq \mathfrak{b}$, and since $\mathfrak{b} \simeq \mathfrak{p}$, we are done. □

Consider the epimorphism $\rho : \Lambda \longrightarrow M_{17}$ from Lemma 8.6.8. Then $e'_{17} := \rho(1)$ generates M_{17} as Λ-module. Since the index of $\mathbb{Z}Q_{32}/(y^2+1)\mathbb{Z}Q_{32}$ in Λ is a 2-power, which is invertible modulo 17, e'_{17} generates M_{17} also as $\mathbb{Z}Q_{32}$-module. Hence $M = M_3 \oplus M_5 \oplus M_{17}$ is generated by (e_3, e_5, e'_{17}) as $\mathbb{Z}Q_{32}$-module and $f : \mathbb{Z}Q_{32} \longrightarrow M$ which is defined by $f(1) = (e_3, e_5, e'_{17})$ induces an exact sequence

$$0 \longrightarrow I \longrightarrow \mathbb{Z}Q_{32} \xrightarrow{f} M \longrightarrow 0$$

for some ideal I. This then implies, when we consider the sequence obtained after having tensored by Λ over $\mathbb{Z}Q_{32}$, the sequence

$$0 \longrightarrow \Lambda \otimes_{\mathbb{Z}Q_{32}} I \longrightarrow \Lambda \xrightarrow{\rho} M_{17} \longrightarrow 0$$

which is exactly the exact sequence from Lemma 8.6.8. Hence

$$\Lambda \otimes_{\mathbb{Z}Q_{32}} I \simeq \mathfrak{p}$$

which is not a principal ideal by Lemma 8.6.6. Hence the module $\Lambda \otimes_{\mathbb{Z}Q_{32}} I$ is not free as Λ-module. However, we get another exact sequence by the construction of M

$$0 \longrightarrow \mathbb{Z}Q_{32} \xrightarrow{\cdot(y+4)} \mathbb{Z}Q_{32} \xrightarrow{\cdot(e_3,e_5,e_{17})} M \longrightarrow 0$$

Therefore I must be a projective $\mathbb{Z}Q_{32}$-module as well and by Schanuel's Lemma 6.2.10 we get

$$I \oplus \mathbb{Z}Q_{32} \simeq \mathbb{Z}Q_{32} \oplus \mathbb{Z}Q_{32}.$$

This shows the theorem. □

Remark 8.6.9. The example works by so many coincidences. One gets an idea why the method is really specific to this type of groups. As we already mentioned, there is a theoretical reason for this fact as well, Eichler's theory of totally definite quaternion algebras.

Also, we note that despite its complexity and striking ingeniosity, the proof basically uses some elementary representation theory from Chapter 1, some number theory, a little arithmetics, and the rest is just an elementary computation.

8.7 Galois module structure

We come to a very nice, natural and very deep branch in integral representation theory. The question is very easy to formulate.

Let L be a Galois extension of the field K, and suppose that K is of characteristic 0. Then put $R := \text{algint}_{\mathbb{Z}}(K)$ and $S := \text{algint}_{\mathbb{Z}}(L)$. Let $G = Gal(L : K)$ be the Galois group. Then G acts on L, and as is immediately verified, also on S. Further, since K is fixed under the action of G, also R is fixed under the action of G. Hence S is an RG-module. Galois module structure theory asks to characterise abstractly this module.

Since the methods used in this theory are very involved and are clearly far beyond the scope of this book, we only give some of the main results. The interested reader is advised to consult Fröhlich's book [Fro-83] on the subject, or for a more advanced perspective Snaith's monograph [Sna-94b]. The purpose of this Section 8.7 is not really to explain the details of the theory, but to give an idea how the methods developed in this book in previous chapters are used in more advanced theories. We need results and methods from basically the entire book. We shall use results from Chapter 1 of course. Further, we use Chapter 2 not only in the way that we formulate the main result in terms of characters, but we also use essential methods and results of this chapter. The main result from Chapter 4 gives us one of the main ingredients in the formulation of our main result, but also provides a nice criterion to solve once and for all the case of odd Galois groups. Chapter 5 simplifies the computation of one of the main tools in the main result, the Artin L-function, but also was necessary to define the main object we deal with, the class group, properly introduced in Section 8.5. Chapter 7 is of course necessary since we are dealing with (quite hard) number theory here.

8.7.1 Ramifications

We first recall a classical result from Galois theory.

Theorem 8.7.1. *Let K be a field and let L be a finite separable normal extension of K with Galois group G. Then there is an element $x \in L$ such that $\{gx \in L \mid g \in G\}$ is a K-basis of L. We say that L has a* normal basis.

Note that the normal basis theorem is equivalent to saying that L is a rank one free KG-module.

We restrict our situation even further. Let R be the ring of algebraic integers in a finite Galois extension K over $\mathbb{Q}$. Let $G = Gal(K : \mathbb{Q})$ be the Galois group of this extension. If $\sigma \in G$, then $x \in K$ is integral over $\mathbb{Z}$ if and only if $\sigma(x)$ is integral over $\mathbb{Z}$. This follows directly from the fact that x is integral if and only if x is a root of a monic polynomial with coefficients in $\mathbb{Z}$. Applying σ to the polynomial equation then gives that $\sigma(x)$ is root of the same polynomial, and hence $\sigma(x)$ is integral over $\mathbb{Z}$ as well. Therefore R is a $\mathbb{Z}G$-module, which we call the *Galois module*. By the Normal Basis

Theorem 8.7.1 we see that S is an R-order in KG. Recall that then all elements of S are algebraic integers (cf. Theorem 8.3.7).

We need a new notion for the next statement. Let K be a finite extension of $\mathbb{Q}$ and let $R = \mathrm{algint}_{\mathbb{Z}}(K)$. By Lemma 7.5.18 the ring R is a Dedekind domain. Then for every non zero prime ideal $\wp$ of R we get $\wp \cap \mathbb{Z} = p\mathbb{Z}$ is a prime ideal of $\mathbb{Z}$. Further, the field $R/\wp$ is a finite extension of the field $\mathbb{Z}/p\mathbb{Z}$ and denote by $f_\wp := \dim_{\mathbb{Z}/p\mathbb{Z}}(R/\wp)$ be the degree of this field extension. Now, pR is an ideal of R, and by Proposition 7.5.22 there are pairwise distinct prime ideals $\wp_1, \dots, \wp_n$ of R and positive integers $e_{\wp_1}, \dots, e_{\wp_n}$ such that

$$pR = \wp_1^{e_{\wp_1}} \cdot \dots \cdot \wp_n^{e_{\wp_n}}.$$

If K is a Galois extension of $\mathbb{Q}$, then the Galois group acts transitively on the prime ideals containing a fixed prime element of $\mathbb{Z}$. Then

$$e_{\wp_1} = \dots = e_{\wp_n} \quad \text{and} \quad f_{\wp_1} = \dots = f_{\wp_n}.$$

Lemma 8.7.2. *For any prime ideal $\wp$ of R with $\wp \cap \mathbb{Z} = p\mathbb{Z}$ we get*

$$\dim_{\widehat{\mathbb{Q}}_p} \widehat{K}_\wp = e_\wp \cdot f_\wp.$$

For the global field extension we get

$$\dim_{\mathbb{Q}} K = \sum_{i=1}^{n} e_{\wp_i} \cdot f_{\wp_i}.$$

Definition 8.7.3. We say that $\widehat{K}_\wp$ is
- *unramified* if $e_\wp = 1$ and $R/\wp$ is a separable extension of $\mathbb{Z}/p\mathbb{Z}$,
- *tamely ramified* if $R/\wp$ is a separable extension of $\mathbb{Z}/p\mathbb{Z}$ and p does not divide $e_\wp$,
- *totally ramified* if $f_\wp = 1$.

Let K be a finite extension of $\mathbb{Q}$. Then K is unramified at p if $\widehat{K}_\wp$ is unramified for all primes $\wp$ containing p. Similarly, K is tamely ramified if for each prime $p \in \mathbb{Z}$ there is at least one prime ideal $\wp$ of R such that $p \in \wp$ and $R_\wp$ is tamely ramified over $\mathbb{Z}/p\mathbb{Z}$.

It is not hard to show, using the existence of separable closures, that there is a unique maximal subfield W of K such that the extension $\mathbb{Q} \subseteq W$ is unramified. If K is a Galois extension of $\mathbb{Q}$ with Galois group G, then there is a subgroup G_u of G with $W = K^{G_u}$. The unicity of W implies that G_u is actually normal in G.

Let E be Galois extension of K with Galois group G, and suppose that E is a finite extension of $\mathbb{Q}$. Let $S = \mathrm{algint}_{\mathbb{Z}}(E)$ and let $R = \mathrm{algint}_{\mathbb{Z}}(K)$. Let $\wp$ be a prime ideal of R and let $P_\wp$ be a prime ideal of S such that $\wp \subseteq P_\wp$. Then $D_{P_\wp} := \{g \in G \mid g \cdot P_\wp = P_\wp\}$ is a subgroup of G, the so-called *decomposition group* of $P_\wp$. Further, each element of $D_{P_\wp}$ induces an element in the Galois group of $S/P_\wp$ over $R/\wp$. Since both fields $S/P_\wp$

and $R/\wp$ are finite fields, the Galois group is cyclic, generated by a suitable power of the Frobenius automorphism,

$$\phi_{P_\wp} : S/P_\wp \ni x \mapsto x^{N(\wp)} \in S/P_\wp.$$

Let $I_{P_\wp}$ be the kernel of the natural group homomorphism

$$D_{P_\wp} \longrightarrow Gal(S/P_\wp : R/\wp).$$

It can be shown that this group homomorphism is surjective. A preimage $\sigma_{P_\wp}$ of $\phi_{P_\wp}$ in $D_{P_\wp}$ is called a *Frobenius lift*.

Suppose now that E is unramified over K. Then $I_{P_\wp} = 1$ and there is a unique Frobenius lift for each prime ideal $P_\wp$ containing $\wp$. If $P_\wp$ and $Q_\wp$ are two prime ideals of S containing a fixed prime ideal $\wp$ of R, then these two are conjugate under the Galois action of G. Hence two Frobenius lifts $\sigma_{P_\wp}$ and $\sigma_{Q_\wp}$ have the same image for any character χ of G.

Theorem 8.7.4 (Noether [Noe-32]). *Let K be a Galois extension of $\mathbb{Q}$ and let R be the ring of algebraic integers over $\mathbb{Z}$ in K. Let G be a finite group. Then the Galois module R is projective if and only if the extension K is tamely ramified.*

8.7.2 Taylor's results on Galois module structure

The simplest case is the case of abelian Galois groups.

Theorem 8.7.5 (Taylor [Tay-78]). *Let E be an abelian tame extension K with Galois group G and let E be a finite extension of $\mathbb{Q}$. Then* $\text{algint}_{\mathbb{Z}}(E)$ *is a free $\mathbb{Z}G$-module.*

For a deeper analysis of the Galois module structure we shall need the following deep result due to Swan.

Theorem 8.7.6 ([Swa-60]). *Let R be a Dedekind domain of characteristic* 0 *and let G be a finite group. Then an RG-module M is projective if and only if M is locally free.*

The original proof is very involved, though an alternative, however still complicated proof of this theorem is given in [ReiRo-79].

We conclude that in view of Noether's Theorem 8.7.4 and Swan's Theorem 8.7.6, the Galois module $R = \text{algint}_{\mathbb{Z}}(K)$ of a tamely ramified Galois extension K of $\mathbb{Q}$ is locally free, and hence gives rise to an element $[R]$ in the locally free class group $Cl(\mathbb{Z}G)$ of the integral group ring. Fröhlich conjectured in 1977 that under the hypotheses above $[R]$ is of order at most 2 in $Cl(\mathbb{Z}G)$.

The answer makes use of the (Artin) L-functions $\Lambda(s,\chi)$. We already encountered Dirichlet L-functions in Section 2.5, and Artin L-functions are specific to Galois extensions. Artin introduced this function 1923 in the context of class field theory. For a quite detailed discussion we refer to Cassels-Fröhlich [CasFro-67, Chapter VIII].

Recall from Definition 2.5.6 that for a Dirichlet character χ the Dirichlet L-function is defined to be

$$L(s,\chi) := \sum_{n=1}^{\infty} \frac{\chi(n)}{n^s}.$$

This is a holomorphic function $\{s \in \mathbb{C} \mid Re(s) > 1\} \longrightarrow \mathbb{C}$. By Lemma 2.5.7 and the proof of Proposition 2.5.11 we get the Euler product

$$L(s,\chi) = \prod_{p \text{ prime}} \frac{1}{1 - \frac{\chi(p)}{p^s}}$$

Artin L-functions are defined by such an Euler product.

Let K be a finite extension of $\mathbb{Q}$, let E be a tamely ramified finite Galois extension of K of degree n, and let $G = Gal(E : K)$ be the Galois group. Put $R := \text{algint}_{\mathbb{Z}}(E)$. Let χ be a complex character of G and suppose that χ is afforded by the representation $\rho : G \longrightarrow Gl_m(\mathbb{C})$ for some vector space S.

Then for each prime $\wp$ of R we shall define a function $\Lambda_\wp(s,\chi)$. Since $R/\wp$ is a finite field of characteristic $p \in \mathbb{N}$, the Galois group of $R/\wp$ over $\mathbb{Z}/p\mathbb{Z}$ is cyclic and generated by the Frobenius automorphism $\phi_\wp$

$$R/\wp \ni x \mapsto x^p \in R/\wp.$$

If p is unramified in R, then let $\sigma_\wp$ be the Frobenius lift on R (i. e. such that $\sigma_\wp$ fixes $\wp$ and induces $\phi_\wp$ on $R/\wp$). In this case

$$\Lambda_\wp(s,\chi) = \frac{1}{det(I_m - \frac{\rho(\sigma_\wp)}{N(\wp)^s})}$$

where I_m is the $m \times m$ identity matrix. Since two Frobenius lifts are Galois conjugate the characteristic polynomial in the definition of $\Lambda_\wp(s,\chi)$ does not depend on the lift. If $\wp$ is ramified, then $\chi(\sigma_\wp)$ only depends on the restriction of χ to $D_{P_\wp}$. Since $I_{P_\wp}$ is normal in $D_{P_\wp}$, we can replace $Res^G_{D_{P_\wp}} \chi$ by $Res^G_{D_{P_\wp}} Inf^{D_{P_\wp}}_{D_{P_\wp}/I_{P_\wp}} \chi^{I_{P_\wp}}$, where $\chi^{I_{P_\wp}}$ is the character on the fixed point space, and where $Inf^{D_{P_\wp}}_{D_{P_\wp}/I_{P_\wp}}$ is the representation on $D_{P_\wp}$ which is defined by a representation of $D_{P_\wp}/I_{P_\wp}$ via the natural map $D_{P_\wp} \longrightarrow D_{P_\wp}/I_{P_\wp}$. For notational purpose we denote by V the module affording χ. Hence we put for ramified primes $\wp$

$$\Lambda_\wp(s,\chi) = \frac{1}{det(I_m - \frac{\rho|_{V^{I_{P_\wp}}}(\sigma_\wp)}{N(\wp)^s})}.$$

For more details we refer to [CasFro-67, Heilbronn's chapter]. Finally

$$\Lambda(s,\chi) := \prod_{\wp \in Spec(R)} \Lambda_\wp(s,\chi)$$

is the Artin L-function. Artin proves that $\Lambda(s,\chi)$ converges for all $s \in \mathbb{C}$ with $Re(s) > 1$.

Remark 8.7.7. It can be shown that Artin L-functions $\Lambda(s,\chi)$ behave well under induction. Hence, Brauer's Induction Theorem 5.2.3 applies. This is somehow technical, but it reduces the question to elementary subgroups of the Galois group. We do not need this here.

Recall from Theorem 4.2.2, Lemma 4.1.6 and Corollary 4.1.10 that the Frobenius Schur indicator

$$FS(\chi) =: \frac{1}{|G|}\sum_{g\in G}\chi(g^2)$$

of an irreducible character χ has value -1 if the representation affording χ carries a symplectic non-degenerate G-invariant bilinear form, has value 1 if the representation affording χ carries a symmetric non-degenerated G-invariant bilinear form, and has the value 0 if the representation affording χ does not carry any non-degenerate G-invariant bilinear form.

Then it can be shown that there is a constant $W(\chi)$ such that

$$\Lambda(s,\chi) = W(\chi)\Lambda(1-s,\overline{\chi}).$$

We call $W(\chi)$ the Artin root number. Applying the above equation twice gives $W(\chi)\cdot W(\overline{\chi}) = 1$. Suppose that χ is either orthogonal or symplectic, so that the corresponding simple module carries a non degenerate invariant bilinear form, is hence self-dual, and therefore $\overline{\chi} = \chi$. Then, in this case $W(\chi)^2 = 1$ and hence $W(\chi) \in \{\pm 1\}$. For any orthogonal character, $W(\chi) = 1$, a fact that was conjectured by J.-P. Serre and proved by Fröhlich and Queyrut.

In order to form an element of the class group out of $W(\chi)$ an alternative description of the class group is used, namely Fröhlich's *Hom*-description of class groups. This is again quite technical, and we refer to Fröhlich [Fro-83, I § 6] or for a summary to Taylor [RoTa-92, Part 2, § 2] for more details. Let F be a finite splitting field for KG, and let Ω is the Galois group of F over K. Basically, the class group $Cl(\mathbb{Z}G)$ is identified with a quotient of $Hom_{\mathbb{Z}\Omega}(K_0(\mathbb{C}G), J_F)$, where J_F is the group of unit adèles over F, i. e. the subgroup of the unit group of the adèles where all elements are actually units at each prime. We observe that both $K_0(\mathbb{C}G)$, in its version as abelian group of virtual characters, as well as J_F have a natural action of $\mathbb{Z}\Omega$.

Let S_G be the subgroup of $K_0(\mathbb{C}G)$ generated by the symplectic characters of G. Then, simplifying slightly, S_G is a direct factor of $K_0(\mathbb{C}G)$, and we define a homomorphism $t'_G : Hom_{\mathbb{Z}\Omega}(S_G, \pm 1) \longrightarrow Hom_{\mathbb{Z}\Omega}(K_0(\mathbb{C}G), J_F)$, where J is the group of unit adèles, induced by the natural projection $K_0(\mathbb{C}G) \longrightarrow S_G$. We put for any $f \in Hom_\Omega(S_G, \pm 1)$

$$(t'_G f)(\chi)_\wp := \begin{cases} f(\chi) & \text{if } \chi \text{ is irreducible and symplectic} \\ & \text{and } \wp \text{ is finite above } p \\ 1 & \text{in all other cases} \end{cases}$$

Then by Fröhlich's Hom-description of class groups, the map $f(\chi) = W(\chi)$ for all $\chi \in S_G$ defines an element $\mathbf{t}$ in $Cl(\mathbb{Z}G)$. We call $\mathbf{t}$ the Cassou-Noguès root class. We need the fact that if $t'_G W(\chi) = 1$ for all irreducible characters χ, then also $\mathbf{t} = 1$ and that the square of the numbers $t'_G W(\chi)^2$ give the square of the Cassou-Noguès root class $\mathbf{t}^2$.

Theorem 8.7.8 (Taylor [Tay-81]). *Let K be a finite extension of $\mathbb{Q}$, let E be a tamely ramified finite Galois extension of K of degree n, and let $G = Gal(E : K)$ be the Galois group. Then*

$$[\mathrm{algint}_{\mathbb{Z}}(E)] = [\mathbf{t}]$$

in $Cl(\mathbb{Z}G)$.

This beautiful result confirms the above conjecture of Fröhlich, but is a lot more precise.

Corollary 8.7.9. *Let K be a finite Galois extension of $\mathbb{Q}$ and suppose that the Galois group of K over $\mathbb{Q}$ is of odd order. Then*

$$[\mathrm{algint}_{\mathbb{Z}}(K)] = [\mathbb{Z}G].$$

Proof. Indeed, Exercise 4.2 shows that there is no invariant bilinear form on any non trivial simple $\mathbb{C}G$-module. The trivial module does not allow a symplectic invariant form. Hence, $t_\chi = 1$ for all irreducible characters, which implies that also the Cassou-Noguès root class $\mathbf{t}$ is 1. This then gives the result. □

Remark 8.7.10. We note that Corollary 8.7.9 could be proved using Theorem 2.4.9. Indeed, the character degree for any irreducible character is necessarily odd by Theorem 2.4.9. A symplectic bilinear form exists only on even dimensional vector spaces, since symplectic spaces are a direct sum of hyperbolic planes.

Remark 8.7.11. Corollary 8.7.9 can be refined. Indeed, a deep result by Jacobinski shows that groups satisfy free cancellation, i. e. $X \oplus F \simeq Y \oplus F \Rightarrow X \simeq Y$ for any free module F in case they satisfy the so-called Eichler condition. Odd order groups do satisfy the Eichler condition. For a recent generalisation we refer to Nicholson's paper [Nich-18].

8.8 Exercises

Exercise 8.1. Let R be a discrete valuation domain and let π be a uniformizer. Let Λ be an R-order in a semisimple algebra. Show that $\pi\Lambda \subseteq \mathrm{rad}(\Lambda)$ (cf. Exercise 1.3). Deduce that for all $\lambda \in \Lambda$ elements of the form $1 + \pi\lambda$ are invertible in Λ. If R is in addition complete, construct the inverse to $1 + \pi\lambda$.

Exercise 8.2. Let $\mathbb{H} = \mathbb{Q}\cdot 1+\mathbb{Q}\cdot i+\mathbb{Q}\cdot j+\mathbb{Q}\cdot k$ be the skew field of quaternions with rational coefficients and with $ij = k$ et $i^2 = j^2 = k^2 = -1$. Denoting by $\overline{(x_1 + x_i i + x_j j + x_k k)} = x_1 - x_i i - x_j j - x_k k$ for all $x_1, x_i, x_j, x_k \in \mathbb{Q}$, let $N(z) := z\cdot\overline{z}$ be the usual norm on $\mathbb{H}$. Denote

$$\mathbb{L} := \mathbb{Z}\cdot 1 + \mathbb{Z}\cdot i + \mathbb{Z}\cdot j + \mathbb{Z}\cdot k = \{a + bi + cj + dk \mid a, b, c, d \in \mathbb{Z}\}$$

the *Lipschitz quaternion integers*, and with $\mathbb{Z} + \frac{1}{2} = \{a + \frac{1}{2} \in \mathbb{Q} \mid a \in \mathbb{Z}\}$ denote

$$\mathbb{M} := \{a + bi + cj + dk \mid a, b, c, d \in \mathbb{Z}\} \cup \left\{a + bi + cj + dk \,\middle|\, a, b, c, d \in \left(\mathbb{Z} + \frac{1}{2}\right)\right\}$$

the *Hurwitz quaternion integers*.

a) Let $Z, z \in \mathbb{L}$ be with $z \neq 0$ and $q := Z\cdot z^{-1} = a + bi + cj + dk$ for rational numbers a, b, c, d. Denote by A, B, C, D integers satisfying $|A - a| \le \frac{1}{2}$, $|B - b| \le \frac{1}{2}$, $|C - c| \le \frac{1}{2}$ as well as $|D - d| \le \frac{1}{2}$, and $Q := A + Bi + Cj + Dk$ as well as $R := Z - Qz$. Show that $N(R) \le N(z)$ and $N(R) = N(z)$ if and only if $a, b, c, d \in \mathbb{Z} + \frac{1}{2}$.
b) Let $Z, z \in \mathbb{M}$ with $z \neq 0$ and $q := Z\cdot z^{-1} = a + bi + cj + dk$ for rational numbers a, b, c, d. Denote by A, B, C, D integers with $|A - a| \le \frac{1}{2}$, $|B - b| \le \frac{1}{2}$, $|C - c| \le \frac{1}{2}$ as well as $|D - d| \le \frac{1}{2}$ and $Q := A + Bi + Cj + Dk$ as well as $R := Z - Qz$. Show that, either $Z = Qz + R$ with $N(R) < N(z)$, or $N(R) = N(z)$. Show that if $N(R) = N(z)$, then $q \in \mathbb{M}$ and $Z = qz + 0$ with $N(0) < N(z)$.
c) Show that $u \in \mathbb{L}$ has norm 1 if and only if $u \in \{\pm 1, \pm i, \pm j, \pm k\}$. Show that $u \in \mathbb{M}$ has norm 1 if and only if u is invertible in $\mathbb{L}$ or $2u \in \{\pm 1 \pm i \pm j \pm k\}$.
d) An element $P \in \mathbb{M}$ is a *Hurwitz prime* if $N(P)$ is a prime in $\mathbb{Z}$. If P is a Hurwitz prime, and if $P = P_1 \cdot P_2$ with $P_1, P_2 \in \mathbb{M}$, show that $N(P_1) = 1$ or $N(P_2) = 1$. Deduce that the only factorisations $P = P_1 \cdot P_2$ are $P = PU^{-1}\cdot U$ or $P = V\cdot V^{-1}P$ with U, V one of the 24 elements with norm 1 in $\mathbb{M}$.
e) Let $x = \frac{1}{2}(a + bi + cj + dk)$ be an element of $\mathbb{H}$ and show that its reduced trace is a. If Λ is a $\mathbb{Z}$-order in $\mathbb{H}$ containing $\mathbb{M}$ and if $x = \frac{1}{2}(a + bi + cj + dk) \in \Lambda$, show that $a, b, c, d \in \mathbb{Z}$. Show that if one of the integers a, b, c, d is odd, then all the integers a, b, c, d are odd. Deduce that $\mathbb{M}$ is a maximal $\mathbb{Z}$-order in $\mathbb{H}$.

Exercise 8.3. Let R be a complete discrete valuation ring with $\mathrm{rad}(R) = \pi R$ and field of fractions K, and let Λ be an R-order.

a) Show that for any simple Λ-module S we have $\pi S = 0$. Deduce that for any Λ-lattice L we have $\pi L \subseteq \mathrm{rad}(L)$.
b) Let L and M be two Λ-lattices and let $\alpha : L \longrightarrow M$ be a Λ-module homomorphism. Show that α induces a Λ-module homomorphism $\overline{\alpha} : L/\pi L \longrightarrow M/\pi M$.
c) Suppose that $\overline{\alpha}$ is an isomorphism. Using Nakayama's lemma (Exercise 6.3) show that α is an isomorphism.
d) If M is a Λ-lattice and suppose that

$$M = L_1 \oplus \cdots \oplus L_s = N_1 \oplus \cdots \oplus N_t$$

are two decompositions of M into indecomposable Λ-modules. Show that $L_1, \dots, L_s, N_1, \dots, N_t$ are Λ-lattices again and use Exercise 7.6 to show that $s = t$.

e) Use c) to show that there is an element $\sigma \in \mathfrak{S}_s$ such that $L_i \simeq N_{\sigma(i)}$ for all $i \in \{1, \dots, s\}$.

f) Formulate this as a Krull-Schmidt theorem for lattices over orders over complete discrete valuation domains.

Exercise 8.4 (Roggenkamp). Let R be a Dedekind domain and let K be its field of fractions. For a finite dimensional K-algebra B we call an R-subalgebra Λ of B an R-order if Λ is finitely generated projective as R-module, and if Λ contains a K-basis of B. Note that we did not assume here that B is semisimple. Suppose that $\mathrm{rad}(B) \neq 0$.

a) Recall that $\mathrm{rad}(B)$ is nilpotent, i. e. there is $s \in \mathbb{N}$ such that $\mathrm{rad}(B)^{s+1} = 0 \neq \mathrm{rad}(B)^s$.

b) If $N_i := \mathrm{rad}(B)^i \cap \Lambda$, show that for all non zero $k \in \mathbb{N}$ and all non zero $r \in R \setminus R^\times$ the R-module

$$\Lambda_k := \Lambda + \frac{1}{r^k} N_1 + \frac{1}{r^{2k}} N_2 + \cdots + \frac{1}{r^{sk}} N_s$$

is an R-order in B.

c) Show that for all $k \in \mathbb{N}$ we have $\Lambda_k \subseteq \Lambda_{k+1}$.

d) Show that Λ / N_s is an R-lattice.

e) If $\Lambda_k = \Lambda_{k+1}$, then show that $N_s \subseteq r^s \Lambda$ and $r^s N_s = N_s$. Use localisation and Nakayama's lemma to show that then $N_s = 0$.

f) Deduce that Λ is not contained in a maximal order.

Exercise 8.5. Let R be a discrete valuation ring with field of fractions K and let $\wp = \pi R$ be its non zero prime ideal. Fix an integer $n > 0$ and consider the set of matrices in $Mat_n(R)$

$$\Lambda := \begin{pmatrix} R & R & \dots & \dots & R \\ \wp & R & & & \vdots \\ \vdots & \ddots & \ddots & & \vdots \\ \vdots & & \ddots & \ddots & \vdots \\ \wp & \dots & \dots & \wp & R \end{pmatrix}$$

be formed by those square matrices $(a_{i,j})_{1 \le i,j \le n} \in Mat_n(R)$ of size n with coefficients in $\wp$ whenever $i < j$ and in R for all the others.

a) Show that Λ is an R-order in a simple K-algebra A.

b) Show that each of the n columns $P_1, \dots, P_n$ is a projective Λ-module such that $P_i \subseteq P_{i+1}$ for all $i \in \{1, \dots, n-1\}$ and $\pi P_n \subseteq P_1$.

c) Compute $\mathrm{rad}(\Lambda)$ and show that each simple Λ-module S is isomorphic to one of the modules $P_i / \mathrm{rad}(P_i) =: S_i$ for $i \in \{1, \dots, n\}$.

d) Show that $S_i \not\simeq S_j$ for $i \neq j$.
e) Fix $i \in \{1,\dots,n\}$. Let L be a sublattice of P_i. Show that L is isomorphic to some P_j for some $j \in \{1,\dots,n\}$. Deduce that Λ is hereditary.
f) Show that conjugation by the matrix

$$\begin{pmatrix} 0 & 1 & 0 & \dots & 0 \\ \vdots & \ddots & \ddots & \ddots & \vdots \\ \vdots & & \ddots & \ddots & 0 \\ 0 & \dots & \dots & 0 & 1 \\ \pi & \dots & \dots & 0 & 0 \end{pmatrix}$$

is a non inner automorphism of Λ.
g) Determine all composition series of $P_i/\pi^s P_i$ for any $i \in \{1,\dots,n\}$ and $s \in \mathbb{N}$. NB: Compare with Exercise 1.4.
h) Show that there may be Λ-modules M and N and a Λ-module homomorphism $\alpha : M \longrightarrow N$ inducing an isomorphism $\overline{\alpha} : M/\pi M \longrightarrow N/\pi N$ but α is not an isomorphism.
Compare with Exercise 8.3.c)

Exercise 8.6. We consider the three dimensional euclidien space E^3, and let $SO_3(\mathbb{R})$ its group of isometries with determinant 1 with respect to an orthogonal basis. Let G be a subgroup of $SO_3(\mathbb{R})$ of finite order n. The group $SO_3(\mathbb{R})$ acts on the euclidien 3-space by matrix multiplication.

a) Show that each $g \in G \setminus \{1\}$ fixes two points of $\mathbb{R}^3$. Denote them by $p_1(g)$ and $p_2(g)$ depending on g and call them poles of g. Let

$$\begin{aligned}\Omega(G) &:= \{x \in \mathbb{R}^3 | x = p_1(g) \text{ or } x = p_2(g) \text{ for a } g \in G \setminus \{1\}\} \\ &= \bigcup_{g \in G\setminus\{1\}} \{p_1(g), p_2(g)\}.\end{aligned}$$

b) Show that the action of G on E^3 induces by restriction an action of G on $\Omega(G)$.
c) What is the cardinal of the orbit $G \cdot p_1(g)$ if g is of order m in G, and if there is no element $h \in G$ with $h^k = g$ for some $k \geq 2$? Put then $m = m(p)$ for such an element g.
d) Count the number of elements $g \in G \setminus \{1\}$ having a particular pole. Sum over all poles, and order them in orbits to show

$$n - 1 = \frac{1}{2} \cdot n \cdot \left(\sum_{Gp \in \Omega(G)/G} \frac{(m(p) - 1)}{m(p)} \right)$$

(Note that here $m(p)$ is the maximal order of an element g having pole p.)
e) Deduce $|\Omega(G)/G| \leq 3$.

f) If $|\Omega(G)/G| = 2$, show that in any orbit there is a unique pole and deduce that G is cyclic.
g) If $|\Omega(G)/G| = 3$, and if $n/m_1, n/m_2$ and n/m_3 are the orders of the orbits of the action of G on $\Omega(G)$, show that there is $i \in \{1, 2, 3\}$ with $m_i = 2$. Put $i = 3$ in this case and hence $m_3 = 2$.

If there is $j \in \{1, 2\}$ with $m_j = 2$, show $n = 2m_{3-j}$.

If there is no $j \in \{1, 2\}$ with $m_j = 2$ show there is $j \in \{1, 2\}$ with $m_j = 3$ and in this case $m_{3-j} \in \{3, 4, 5\}$.

Find the order of G in each case.

Exercise 8.7. Let p be a prime number and let ζ_p be a primitive p-th root of unity in $\mathbb{C}$. Denote by $\mathbb{Z}[\zeta_p]$ the smallest subring of $\mathbb{C}$ containing ζ_p. We know that the principal ideal of $\mathbb{Z}[\zeta_p]$ generated by 2 is unramified. Let

$$2 \cdot \mathbb{Z}[\zeta_p] = \wp_1 \cdot \wp_2 \cdot \dots \cdot \wp_{f_p}$$

for prime ideals $\wp_i$ and $\wp_i = \wp_j \Rightarrow i = j$.

a) Show that there are fields $K_1(p), \dots, K_f(p)$ such that

$$\mathbb{Z}[\zeta_p]/(2 \cdot \mathbb{Z}[\zeta_p]) \simeq K_1(p) \times \dots \times K_{f_p}(p).$$

Denote for each $i \in \{1, \dots, f_p\}$ by $\varphi_i^{(p)}$ the composition of the canonical maps

$$\mathbb{Z}[\zeta_p] \longrightarrow \mathbb{Z}[\zeta_p]/(2 \cdot \mathbb{Z}[\zeta_p]) \longrightarrow K_1(p) \times \dots \times K_{f_p}(p) \longrightarrow K_i(p).$$

b) Denote $u_i(p) := \varphi_i^{(p)}(\zeta_p)$. Show that $K_i(p) = \mathbb{F}_2[u_i(p)]$.
c) Use Frobenius morphisms to find f_{17} and f_{13}. Determine the fields $K_i(p)$ for each $p \in \{13, 17\}$.

Exercise 8.8. Let p be a prime number and let ζ_p be a primitive p-th root of unity in $\mathbb{C}$. Show that p is totally ramified in $\mathbb{Z}[\zeta_p]$.

9 Solution to selected exercises

Höre! Höre! Höre!
Alles, was ist, endet.

Richard Wagner, Rheingold

Listen! Listen! Listen!
All being ceases.

Richard Wagner, Rheingold;[1]

Solution to Exercise 1.10

a) We shall show that V is actually a $K\mathfrak{S}_4$-module with this action. For this it is sufficient to show that for all $\sigma_1, \sigma_2 \in \mathfrak{S}_4$ we have $(\sigma_1\sigma_2) \cdot b_i = \sigma_1 \cdot (\sigma_2 \cdot b_i)$ and $\mathrm{id} \cdot b_i = b_i$ for all $i \in \{1, 2, 3, 4\}$. Once this is shown we have $(\sigma_1\sigma_2) \cdot v = \sigma_1 \cdot (\sigma_2 \cdot v)$ and $\mathrm{id} \cdot v = v$ for all $v \in V$ by the fact that $\{b_1, \ldots, b_4\}$ is a basis and linear maps are uniquely defined by the image of a basis. Further, this then shows that $(ab) \cdot v = a \cdot (b \cdot v)$ for all $a, b \in K\mathfrak{S}_4$ and all $v \in V$ by Lemma 1.1.19. However,

$$\mathrm{id} \cdot b_i = b_{\mathrm{id}(i)} = b_i$$

and

$$(\sigma_1\sigma_2) \cdot b_i = b_{(\sigma_1\sigma_2)^{-1}(i)} = b_{\sigma_2^{-1}(\sigma_1^{-1}(i))} = \sigma_2 \cdot b_{\sigma_1^{-1}(i)} = \sigma_1 \cdot (\sigma_2 \cdot b_i).$$

Since the group homomorphism $\mathfrak{A}_4 \hookrightarrow \mathfrak{S}_4$ induces a ring homomorphism $K\mathfrak{A}_4 \hookrightarrow K\mathfrak{S}_4$ the result is proved.

b) Define $T := K \cdot (b_1 + b_2 + b_3 + b_4)$. Then by definition

$$\sigma \cdot (b_1 + b_2 + b_3 + b_4) = (b_1 + b_2 + b_3 + b_4) \ \forall \sigma \in \mathfrak{A}_4$$

and hence T is a trivial submodule of V.

c) The element (1 2)(3 4) acts via the matrix $M_{(1\,2)(3\,4)} := \begin{pmatrix} 0 & 1 & 0 & 0 \\ 1 & 0 & 0 & 0 \\ 0 & 0 & 0 & 1 \\ 0 & 0 & 1 & 0 \end{pmatrix}$ on V and the element (1 2 3) maps 1 to 2, 2 to 3 and 3 to 1. Hence it acts via the matrix

$$M_{(1\,2\,3)} := \begin{pmatrix} 0 & 0 & 1 & 0 \\ 1 & 0 & 0 & 0 \\ 0 & 1 & 0 & 0 \\ 0 & 0 & 0 & 1 \end{pmatrix}^{-1} = \begin{pmatrix} 0 & 1 & 0 & 0 \\ 0 & 0 & 1 & 0 \\ 1 & 0 & 0 & 0 \\ 0 & 0 & 0 & 1 \end{pmatrix}$$

1 in der Übersetzung von Susanne Brennecke, translation by Susanne Brennecke.

https://doi.org/10.1515/9783110702446-009

on V. The characteristic polynomial of $M_{(1\,2)(3\,4)}$ is $(X^2-1)^2$. The eigenspace with respect to the eigen value 1 is $K\cdot(b_1+b_2)+K\cdot(b_3+b_4)$ and the eigenspace with respect to the eigen value -1 is $K\cdot(b_1-b_2)+K\cdot(b_3-b_4)$. The characteristic polynomial of $M_{(1\,2\,3)}$ is $(X^3-1)(X-1)$. The eigenspace for the eigenvalue 1 is $K\cdot(b_1+b_2+b_3)+K\cdot b_4$. Suppose that X^2+X+1 has a root in K. Let hence j be a root of X^2+X+1. Then the eigenspace for the eigenvalue j^2 is $K\cdot(b_1+jb_2+j^2b_3)$ and the eigenspace for the eigenvalue j is $K\cdot(b_1+j^2b_2+jb_3)$. If X^2+X+1 does not have a root in K, then there is no further eigenspace.

d) A $K\mathfrak{A}_4$-submodule W of dimension 1 is generated by some vector w, i. e. $W = Kw$. Hence $M_{(1\,2)(3\,4)}\cdot w=\lambda_1 w$ for some $\lambda_1\in K$ and $M_{(1\,2\,3)}\cdot w=\lambda_2 w$ for some $\lambda_2\in K$. Hence W has to be an eigenspace for both of these matrices, for possible different values. Therefore, examining the eigenspaces we determined in c), $W=T$.

e) We have seen that T is a submodule, and since $K\mathfrak{A}_4$ is semisimple, $V=T\oplus X$ for some $K\mathfrak{A}_4$-module X. Either X is simple or not. If X is not simple, then it has a non zero proper submodule S_1. Hence $X=S_1\oplus S_2$, again since $K\mathfrak{A}_4$ is semisimple. Since X is of dimension 3, then either S_1 or S_2 is of dimension 1, whence a common eigenspace of both matrices. Therefore either $S_1=T$ or $S_2=T$ by d). This is a contradiction to $V=T\oplus X$ (which has as consequence $X\cap T=0$). Hence $V=X\oplus T$.

Solution to Exercise 2.7

a) We have $\tau_x(v)=x+v$ and $\delta_a(v)=av$ for all $a\in\mathbb{F}_7\setminus\{0\}$, and all $x,v\in\mathbb{F}_7$. Then

$$(\delta_a\tau_x)(v)=(\delta_a(\tau_x(v))=a\cdot(x+v)=ax+av=(\tau_{ax}\delta_x)(v)$$

Since $\delta_a^{-1}=\delta_{a^{-1}}$ and since $\tau_x^{-1}=\tau_{-x}$, each element of G is of the form $\prod_{i=1}^n(\delta_{a_i}\tau_{x_i})$ for some $n\in\mathbb{N}$, $a_1,\dots,a_n\in\mathbb{F}_7\setminus\{0\}$, and $x_1,\dots,x_n\in\mathbb{F}_7$. Hence, using $(\delta_a\tau_x)=(\tau_{ax}\delta_x)$ for any a,x, we have

$$\prod_{i=1}^n(\delta_{a_i}\tau_{x_i})=\prod_{i=1}^n\delta_{a_i}\prod_{i=1}^n\tau_{x_i'}=\delta_b\tau_y$$

for some $x_1',\dots,x_n'\in\mathbb{F}_7$, and $b:=\prod_{i=1}^n a_i$, as well as $y:=\sum_{i=1}^n x_i'$. This also shows $G/T=\Delta$ by mapping an element $\delta_a\tau_x$ to δ_a.

b) We have $|T|=7$ and $|\Delta|=6$. By a) we get $|G|=7\cdot 6=42$. Since $(\mathbb{F}_7\setminus\{0\},\cdot)\simeq\Delta\le G$ is a cyclic group of order 6, being the multiplicative group of a finite field, we proved the statement.

c) We compute

$$\delta_a\tau_x\delta_{a^{-1}}=\tau_{xa}\delta_a\delta_{a^{-1}}=\tau_{xa}$$

and hence $\{\delta_x \mid x \in \mathbb{F}_7 \setminus \{0\}\} =: C_2$ is a conjugacy class. Of course, $\{\mathrm{id}\} = \{\tau_0\} =: C_1$ is another conjugacy class and $T = C_1 \cup C_2$.

d) Since $\delta_b(\delta_a\tau_x)\delta_{b^{-1}} = \delta_a\tau_{bx}$ and since $\tau_y\delta_a\tau_x\tau_{-x} = \delta_a\tau_{(a-1)y+x}$ we have, choosing $y = -x/(a-1)$ whenever $a \neq 1$, that the conjugacy class of $\delta_a\tau_x$ for $x \in \mathbb{F}_7$ is of size 7 at least, whenever $a \neq 1$. Since Δ is abelian, and since G maps onto Δ by a), mapping $\delta_a\tau_x$ to δ_a, the class represented by δ_a is different from the class represented by δ_b whenever $a \neq b$. these conjugacy classes are distinct Hence we get 6 conjugacy classes of size 7 represented by δ_a for $a \in \mathbb{F}_7 \setminus \{0\}$, one conjugacy class of size 1, and the class C_2 of size 6.

e) Since $G/T \simeq \Delta$ is cyclic of order 6, there are 6 pairwise distinct one dimensional complex representations of G. There are 7 conjugacy classes, and hence 7 irreducible characters of G. Hence, denoting by n the degree of the unique non linear, seventh, irreducible character

$$|G| = 42 = 1^2 + 1^2 + 1^2 + 1^2 + 1^2 + 1^2 + n^2$$

we get $n^2 = 42 - 6 = 36$ and hence there are 6 irreducible characters of degree 1 and one irreducible character of degree 6.

f) $3^2 = 9 = 2 \bmod 7$. $3^3 = 3 \cdot 2 = 6 = -1 \bmod 7$. Hence the order of 3 in $(\mathbb{F}_7 \setminus \{0\}, \cdot)$ is 6.

g) Since $C_1 \cup C_2 = T$ and since $G/T \simeq \Delta$ is cyclic of order 6, the character values of the classes different from C_2 on the linear characters are those of a cyclic group of order 6. The value of the irreducible characters of degree 1 on C_2 equals the value of the characters on C_1, since these characters come from the epimorphism $G \longrightarrow \Delta$ having kernel $C_1 \cup C_2$. Let j be a primitive third root of unity. Then $-j$ is a primitive 6-th root of unity. Hence, denoting by C_3 the class containing δ_3, C_4 the class containing δ_9, C_5 the class containing δ_{3^3}, C_6 the class containing δ_{3^4}, C_6 the class containing δ_{3^5}, C_7 the class containing δ_{3^6}, we obtain the following part of a table

	C_1	C_2	C_3	C_4	C_5	C_6	C_7
χ_1	1	1	1	1	1	1	1
χ_2	1	1	$-j$	$(-j)^2$	$(-j)^3$	$(-j)^4$	$(-j)^5$
χ_3	1	1	$(-j)^2$	$(-j)^4$	$(-j)^6$	$(-j)^8$	$(-j)^{10}$
χ_4	1	1	$(-j)^3$	$(-j)^6$	$(-j)^9$	$(-j)^{12}$	$(-j)^{15}$
χ_5	1	1	$(-j)^4$	$(-j)^8$	$(-j)^{12}$	$(-j)^{16}$	$(-j)^{20}$
χ_6	1	1	$(-j)^5$	$(-j)^{10}$	$(-j)^{15}$	$(-j)^{20}$	$(-j)^{25}$
χ_7	6						

We evaluate the powers of $-j$ and obtain

	C_1	C_2	C_3	C_4	C_5	C_6	C_7
χ_1	1	1	1	1	1	1	1
χ_2	1	1	$-j$	j^2	-1	j	$-j^2$
χ_3	1	1	j^2	j	1	j^2	j
χ_4	1	1	-1	1	-1	1	-1
χ_5	1	1	j	j^2	1	j	j^2
χ_6	1	1	$-j^2$	j	-1	j^2	$-j$
χ_7	6						

We finish, using the orthogonality relations, that the first and the other columns need to be orthogonal. Hence the character table is

	C_1	C_2	C_3	C_4	C_5	C_6	C_7
χ_1	1	1	1	1	1	1	1
χ_2	1	1	$-j$	j^2	-1	j	$-j^2$
χ_3	1	1	j^2	j	1	j^2	j
χ_4	1	1	-1	1	-1	1	-1
χ_5	1	1	j	j^2	1	j	j^2
χ_6	1	1	$-j^2$	j	-1	j^2	$-j$
χ_7	6	-1	0	0	0	0	0

Solution to Exercise 3.6

a) Indecomposable $\mathbb{C}C_n$-modules M are of dimension 1. Denoting by ζ_n a primitive n-th root of unity, the element a acts as ζ_n^k for some k, and on the module twisted by b the element a acts as ζ_n^{-k}. If b is in the inertia group, these two modules are isomorphic as $\mathbb{C}C_n$-modules, and hence $\zeta_n^k = \zeta_n^{-k}$. This implies $\zeta_n^{2k} = 1$ and therefore, using that n is odd, $k = 0$. This then characterises the trivial module M.

b) By the same argument and notations as above, $\zeta_n^{2k} = 1$ and therefore $k \in \{0, \frac{n}{2}\}$. These two cases do have b in the inertia group.

c) Denote by S_k the indecomposable $\mathbb{C}C_n$-module on which a acts as ζ_n^k. Then by definition the subgroup C_n is always in the inertia group of S_k. The group C_n is of index 2 in D_n, and hence the inertia group is either equal to C_n or to D_n. If k is odd, then the inertia group is equal to C_n for all $k \neq 0$, and is equal to D_n if $k = 0$. If k is even, then the inertia group of S_k is D_n if $k \in \{0, n/2\}$ and the inertia of S_k is C_n if $k \in \{1, \dots, n/2 - 1\} \cup \{n/2 + 1, \dots, n - 1\}$.

d) We keep the notation of c). If $I_{D_n}(S_k) = C_n$, then $S_k \uparrow_{I_{D_n}(S_k)}^{D_n}$ is by Clifford's theorem two-dimensional and indecomposable. Moreover, multiplication by b gives an isomorphism

$$S_k \uparrow_{I_{D_n}(S_k)}^{D_n} \simeq S_{-k} \uparrow_{I_{D_n}(S_{-k})}^{D_n} .$$

If $k = 0$, then $S_0 \uparrow_{I_{D_n}(S_0)}^{D_n}$ is again the trivial module. If n is even and $k = n/2$, then $S_k \uparrow_{I_{D_n}(S_k)}^{D_n}$ is again a one-dimensional module on which a acts as -1 and b acts as 1. Further, if $k = 0$, then $S_0 \uparrow_{C_n}^{D_n}$ is a permutation module, being a direct sum of a trivial submodule and another indecomposable one-dimensional module on which a acts as 1 and b acts as -1. If n is even and $k = n/2$, then

$$\begin{aligned} End_{\mathbb{C}D_n}(S_{n/2} \uparrow_{C_n}^{D_n}) &\simeq Hom_{\mathbb{C}C_n}(S_{n/2}, S_{n/2} \oplus {}^{b}S_{n/2}) \\ &\simeq Hom_{\mathbb{C}C_n}(S_{n/2}, S_{n/2}) \oplus Hom_{\mathbb{C}C_n}(S_{n/2}, {}^{b}S_{n/2}) \end{aligned}$$

by Mackey's theorem, and since b is in the inertia group both factors are 1-dimensional. Hence $S_{n/2} \uparrow_{C_n}^{D_n}$ is a direct sum of two indecomposable, pairwise non isomorphic one-dimensional $\mathbb{C}D_n$-modules. For any group G denote by G' the derived group of G, i. e. the normal subgroup generated by the commutators, the one-dimensional indecomposable $\mathbb{C}D_n$-modules correspond to the modules over $D_n/D_n' = D_n, \langle a^2 \rangle \simeq C_2 \times C_2$ via the natural projection. We get these two factors by multiplying with the idempotents $(1 + b)/2$ and $(1 - b)/2$. On one direct factor a acts as -1 and b acts as 1, on the other direct factor a acts as -1 and b acts as -1 as well.

Solution to Exercise 4.3

a) The order of G is obtained as sum of the squares of the irreducible character degrees, which gives

$$1 + 6^2 + 10^2 + 10^2 + 14^2 + 14^2 + 15^2 + 21^2 + 35^2 = \frac{7!}{2} = 2520.$$

b) A normal subgroup N of G is a union of conjugacy classes, and each irreducible character of G/N is an irreducible character of G as well. Moreover, the value of this character on the conjugacy classes forming N equals the value on the trivial conjugacy class (the one with all values positive integers). There is no such character in the table, except the trivial character, which corresponds to the normal subgroup G of G. Hence G is simple.

c) The size of the conjugacy classes are obtained by the orthogonality relations. The columns have to have length 1, which shows that

$$|C_1| = 1, \qquad |C_2| = \frac{|G|}{(4+5\cdot 4)} = \frac{|G|}{24}$$

$$|C_3| = \frac{|G|}{(5+3\cdot 9+4)} = \frac{|G|}{36}, \qquad |C_4| = \frac{|G|}{(8+4)} = \frac{|G|}{12}$$

$$|C_5| = \frac{|G|}{(5+4)} = \frac{|G|}{9}, \qquad |C_6| = \frac{|G|}{4}$$

$$|C_7| = \frac{|G|}{5}$$

$$|C_8| = \frac{|G|}{(2+2+1+1+1)} = \frac{|G|}{7}, \quad |C_8| = |C_9| = \frac{|G|}{7}.$$

We add these orders to the character table:

	C_1	C_2	C_3	C_4	C_5	C_6	C_7	C_8	C_9
	1	105	70	210	280	630	504	360	360
χ_1	1	1	1	1	1	1	1	1	1
χ_2	6	2	3	−1	0	0	1	−1	−1
χ_3	10	−2	1	1	1	0	0	$\zeta_7^3+\zeta_7^5+\zeta_7^6$	$\zeta_7+\zeta_7^2+\zeta_7^4$
χ_4	10	−2	1	1	1	0	0	$\zeta_7+\zeta_7^2+\zeta_7^4$	$\zeta_7^3+\zeta_7^5+\zeta_7^6$
χ_5	14	2	2	2	−1	0	−1	0	0
χ_6	14	2	−1	−1	2	0	−1	0	0
χ_7	15	−1	3	−1	0	−1	0	1	1
χ_8	21	1	−3	1	0	−1	1	0	0
χ_9	35	−1	−1	−1	−1	1	0	0	0

d) $\chi := (\ 27 \quad -1 \quad 6 \quad 2 \quad 3 \quad 1 \quad 2 \quad -1 \quad -1\ )$ gives

$$(\chi,\chi_1) = \frac{1}{2520}\cdot(27-105+420+420+840+630+1008-360-360) = 1$$

$$(\chi,\chi_2) = \frac{1}{2520}\cdot(162-210+1260-420+1008+360+360) = 1$$

$$(\chi,\chi_3) = (\chi,\chi_4) = \frac{1}{2520}\cdot(270+210+420+420+3\cdot 280+360) = 1.$$

Adding up the degrees we see that $\chi = \chi_1+\chi_2+\chi_3+\chi_4$.

e) The M_i with non degenerate invariant bilinear form are those with only real character values. Hence $M_1, M_2, M_5, M_6, M_7, M_8, M_9$ are those which allow a non degenerate bilinear form.

f) Clearly M_1 is the trivial module, which allows a quadratic bilinear form. We compute the Frobenius Schur indicator. We use the hypothesis that $G \ni x \mapsto x^2 \in G$ fixes C_1, C_3, C_5 et C_7, sends C_2 to C_1, C_4 to C_3, C_6 to C_2, C_8 to C_9, and C_9 to C_8. Denote by

$C_i^{(2)}$ the image of C_i under the map $x \mapsto x^2$.

$$\begin{aligned}
FS(\chi_2) &= \frac{1}{2520}(\chi_2(C_1^{(2)}) + 105 \cdot \chi_2(C_2^{(2)}) + 70 \cdot \chi_2(C_3^{(2)}) + 210 \cdot \chi_2(C_4^{(2)}) \\
&\quad + 280 \cdot \chi_2(C_5^{(2)}) + 630 \cdot \chi_2(C_6^{(2)}) + 504 \cdot \chi_2(C_7^{(2)}) \\
&\quad + 360 \cdot \chi_2(C_8^{(2)}) + 360 \cdot \chi_2(C_9^{(2)})) \\
&= \frac{1}{2520}(\chi_2(C_1) + 105 \cdot \chi_2(C_1) + 70 \cdot \chi_2(C_3) + 210 \cdot \chi_2(C_3) \\
&\quad + 280 \cdot \chi_2(C_5) + 630 \cdot \chi_2(C_2) + 504 \cdot \chi_2(C_7) \\
&\quad + 360 \cdot \chi_2(C_9) + 360 \cdot \chi_2(C_8)) \\
&= \frac{6 + 105 \cdot 6 + 70 \cdot 3 + 210 \cdot 3 + 280 \cdot 0 + 630 \cdot 2 + 504 \cdot 1 - 2 \cdot 360}{2520} \\
&= 1
\end{aligned}$$

Similarly,

$$\begin{aligned}
FS(\chi_5) &= \frac{1}{2520}(\chi_5(C_1^{(2)}) + 105 \cdot \chi_5(C_2^{(2)}) + 70 \cdot \chi_5(C_3^{(2)}) + 210 \cdot \chi_5(C_4^{(2)}) \\
&\quad + 280 \cdot \chi_5(C_5^{(2)}) + 630 \cdot \chi_5(C_6^{(2)}) + 504 \cdot \chi_5(C_7^{(2)}) \\
&\quad + 360 \cdot \chi_5(C_8^{(2)}) + 360 \cdot \chi_5(C_9^{(2)})) \\
&= \frac{1}{2520}(106 \cdot \chi_5(C_1) + 630 \cdot \chi_5(C_2) + 280 \cdot \chi_5(C_3) \\
&\quad + 280 \cdot \chi_5(C_5) + 504 \cdot \chi_5(C_7) + 360 \cdot \chi_5(C_8) + \chi_5(C_9)) \\
&= \frac{1}{2520}(106 \cdot 14 + 630 \cdot 2 + 280 \cdot 2 - 280 - 504) \\
&= 1 \\
FS(\chi_6) &= \frac{1}{2520}(106 \cdot \chi_6(C_1) + 630 \cdot \chi_6(C_2) + 280 \cdot \chi_6(C_3) \\
&\quad + 280 \cdot \chi_6(C_5) + 504 \cdot \chi_6(C_7) + 360 \cdot \chi_6(C_8) + \chi_6(C_9)) \\
&= \frac{1}{2520}(106 \cdot 14 + 630 \cdot 2 - 280 + 280 \cdot 2 - 504) \\
&= 1 \\
FS(\chi_7) &= \frac{1}{2520}(106 \cdot \chi_7(C_1) + 630 \cdot \chi_7(C_2) + 280 \cdot \chi_7(C_3) \\
&\quad + 280 \cdot \chi_7(C_5) + 504 \cdot \chi_7(C_7) + 360 \cdot \chi_7(C_8) + \chi_7(C_9)) \\
&= \frac{1}{2520}(106 \cdot 15 - 630 + 280 \cdot 3 + 720) \\
&= 1 \\
FS(\chi_8) &= \frac{1}{2520}(106 \cdot \chi_8(C_1) + 630 \cdot \chi_8(C_2) + 280 \cdot \chi_8(C_3) \\
&\quad + 280 \cdot \chi_8(C_5) + 504 \cdot \chi_8(C_7) + 360 \cdot \chi_8(C_8) + \chi_8(C_9)) \\
&= \frac{1}{2520}(106 \cdot 21 + 630 - 280 \cdot 3 + 504) \\
&= 1
\end{aligned}$$

$$\begin{aligned}FS(\chi_9) &= \frac{1}{2520}(106\cdot\chi_9(C_1)+630\cdot\chi_9(C_2)+280\cdot\chi_9(C_3)\\&\quad+280\cdot\chi_9(C_5)+504\cdot\chi_9(C_7)+360\cdot\chi_9(C_8)+\chi_9(C_9))\\&=\frac{1}{2520}(106\cdot 35-630-280-280)\\&=1\end{aligned}$$

Hence all self-dual modules are quadratic.

g) We know that the module M_3 is not self-dual since its character values are not real. Hence the dual M_3^* is another simple 10-dimensional module. Since in total there are only two simple 10-dimensional module, $M_3^* \simeq M_4$ and $M_4^* \simeq M_3$. Therefore the module $M_3 \oplus M_4$ is selfdual and carries a non degenerate invariant bilinear form. Since $End_{\mathbb{C}G}(M_3 \oplus M_4) \simeq \mathbb{C} \oplus \mathbb{C}$, the set of non degenerate bilinear forms, taken up to scalar multiples, is parameterised by a projective line over $\mathbb{C}$, where we remove two points, the $[0:1]$ and the $[1:0]$. This is of course in natural bijection with the affine line over $\mathbb{C}$, where we remove the origin. We therefore have many more different cases as only a symplectic or a quadratic form.

NB: $G = \mathfrak{A}_7$, the alternating group of degree 7, and the conjugacy classes can easily identified. C_2 is the class of double transpositions, C_3 is the class of a single 3-cycle, C_5 is the class of a product of two 3-cycles, C_7 is the class of a 5-cycle, C_4 is the class of a the product of a double transposition and a 3-cycle, C_6 is the class of the product of a transposition and a 4-cycle, and C_8 as well as C_9 are classes of 7-cycles.

Solution to Exercise 5.3

a) Since $\dim_K(P_1) = \dim_K(S_2) = 1$ and $\dim_K(P_2) = 2$, any A-module M is of the form

$$M_{(\lambda_1,\lambda_2,\lambda_3)} := P_1^{\lambda_1} \oplus S_2^{\lambda_2} \oplus P_2^{\lambda_3}$$

and $\dim_K(M_{(\lambda_1,\lambda_2,\lambda_3)}) = \lambda_1 + \lambda_2 + 2\lambda_3$. Hence we need to consider the modules $M_{(\lambda_1,\lambda_2,\lambda_3)}$ with $1 \le \lambda_1 + \lambda_2 + 2\lambda_3 \le 5$, where $\lambda_1, \lambda_2, \lambda_3$ are non negative integers.

b) and c) Note that $Hom_A(S_1, S_2) = 0 = Hom_A(S_2, S_1)$ by Schur's lemma. Hence if $\lambda_3 = 0$, then

$$End_A(M_{(\lambda_1,\lambda_2,0)}) \simeq Mat_{\lambda_1\times\lambda_1}(K) \times Mat_{\lambda_2\times\lambda_2}(K).$$

Therefore,

$$End_{End_A(M_{(\lambda_1,\lambda_2,0)})}(M_{(\lambda_1,\lambda_2,0)}) \simeq K \times K$$

since scalar multiplication on a vector space is the only endomorphism which commutes with all other endomorphisms of the vector space. Hence these modules $M_{(\lambda_1,\lambda_2,0)}$ do not have the double centraliser property.

If $\lambda_1 \cdot \lambda_3 \neq 0$, since $A = P_1 \oplus P_2$, then A is a direct factor of $M_{(\lambda_1,\lambda_2,\lambda_3)}$, and therefore Lemma 5.4.2 shows that $M_{(\lambda_1,\lambda_2,\lambda_3)}$ has the double centralizer property.

Suppose now $\lambda_1 = 0$. Further, $Hom_A(S_2, P_2) = 0$ and $Hom_A(P_2, S_2) = K = Hom_A(P_2, P_2)$. If also $\lambda_2 = 0$, then $End_A(M_{(0,0,\lambda_3)}) = Mat_{\lambda_3 \times \lambda_3}(K)$ and hence

$$End_{End_A(M_{(0,0,\lambda_3)})}(M_{(0,0,\lambda_3)} = K.$$

If $\lambda_2 \neq 0$, then

$$End_A(S_2^{\lambda_2} \oplus P_2^{\lambda_3}) = \begin{pmatrix} Mat_{\lambda_2 \times \lambda_2}(K) & 0 \\ Mat_{\lambda_2 \times \lambda_3}(K) & Mat_{\lambda_3 \times \lambda_3}(K) \end{pmatrix}$$

and therefore again only scalar multiplication commutes with all these matrices. Hence $End_{End_A(S_2^{\lambda_2} \oplus P_2^{\lambda_3})}(S_2^{\lambda_2} \oplus P_2^{\lambda_3}) = K$. The module $M_{(0,\lambda_2,\lambda_3)}$ again does not have the double centraliser property.

Solution to Exercise 6.1

a) If $x \in \ker(g^n)$ then $x \in \ker(g^{n+1})$ and hence $\ker(g^n)$ is an increasing sequence of ideals of I. Since I is Noetherien, this sequence is finite.

b) Since $g(I) = I$, also $g^{n_0}(I) = g^{n_0-1}(I) = \cdots = I$. The fact $ker(g) \subseteq I$ shows the result.

c) We have $m = g^{n_0}(m_0)$ and hence $0 = g(m) = g^{n_0+1}(m_0)$. Therefore $m_0 \in \ker(g^{n_0+1}) = \ker(g^{n_0})$, which shows $0 = g^{n_0}(m_0) = m$.

d) g est surjective by definition and injective by c).

e) Apply the preceding parts to the case $A = I$.

f) $B = K[X_i | \ i \in \mathbb{N}\}$ and $\varphi(X_{2i}) = X_i = \varphi(X_{2i+1})$.

Solution to Exercise 7.3

a) If $x = 0$, then $yx = 0$ for all $y \in R$. Else, since $x \in R$, also $yx \in R$. If also $(yx)^{-1} \in R$, then $y \cdot (yx)^{-1} = x^{-1} \in R$, a contradiction. Hence $yx \in M$.

b) Since $1^{-1} = 1$ and since $1 \in R$, we have $1 \notin M$.

c) If $x, y \in M$, then surely $x + y \in R$. But we have that $xy^{-1} \in R$ or $yx^{-1} \in R$. Suppose w. l. o. g. that $xy^{-1} \in R$. Then $1 + xy^{-1} \in R$. Hence $y \cdot (1 + xy^{-1}) = x + y \in M$ by a).

d) If $x \in M$ and $1 + x \in M$, then $(1 + x) - x \in M$ by c), which is a contradiction by b).

e) M is the set of non units in R, and M is an ideal by a) and c). Hence M is the unique maximal ideal.

f) If R is Noetherian, then any ideal I is finitely generated. Hence $I = Rx_1 + \cdots + Rx_n$. The relation $<$ is a total order on $R \setminus \{0\}$ by the hypothesis. Then the minimal element in the set of $\{x_1, \ldots, x_n\}$ generates I.

g) This follows from the definition of a discrete valuation ring.

Solution to Exercise 7.5

a) The existence of α is a consequence of the Chinese reminder theorem for Dedekind domains. Indeed, if I and J are ideals of R such that $I + J = R$, then $R/(I \cdot J) = R/(I \cap J) = R/I \times R/J$, where the first equality can be deduced from the unique decomposition into product of prime ideals. Hence we proceed by induction on the number of prime ideals involved and this gives the existence of α.

b) Since $\alpha \in \wp_i^{n_i} \setminus \wp_i^{n_i+1}$ for all $i \in \{1, \dots, k\}$, we have $\alpha R \subseteq \prod_{i=1}^{k} \wp_i^{n_i} = \mathfrak{a}$ and $\alpha R \nsubseteq \prod_{i=1}^{k} \wp_i^{\ell_i}$ whenever $\ell_i > n_i$ for some $i \in \{1, \dots, k\}$.

c) By b) we have $\alpha\mathfrak{a}^{-1} \subseteq R$, is actually an ideal and hence can be decomposed into a product of prime ideals. Since $\alpha R \nsubseteq \prod_{i=1}^{k} \wp_i^{\ell_i}$ whenever $\ell_i > n_i$ for some $i \in \{1, \dots, k\}$, we get that any prime ideal containing $\alpha\mathfrak{a}^{-1}$ is different from any of the ideals $\wp_1, \dots, \wp_k$.

d) Since any prime ideal containing $\mathfrak{b}$ is one of the ideals $\wp_1, \dots, \wp_k$, by c) $\alpha\mathfrak{a}^{-1}$ and $\mathfrak{b}$ are relatively prime. Hence $\alpha\mathfrak{a}^{-1} + \mathfrak{b} = R$.

e) This is a direct consequence of d), taking $\mathfrak{b} := \beta\mathfrak{a}^{-1}$. Indeed, by d) there is α such that $\alpha\mathfrak{a}^{-1} + \beta\mathfrak{a}^{-1} = R$. Multiplication by $\mathfrak{a}$ gives the result.

Bibliography

[Zim-14] Alexander Zimmermann, REPRESENTATION THEORY: A HOMOLOGICAL ALGEBRA POINT OF VIEW, Springer Verlag, Cham 2014.

[CuRe-82-86] Charles W. Curtis and Irving Reiner, METHODS OF REPRESENTATION THEORY I AND II, J. Wiley and Sons 1982 and 1986.

[ReiRo-79] Irving Reiner and Klaus W. Roggenkamp, INTEGRAL REPRESENTATIONS, Lecture Notes in Mathematics **744**, Springer Berlin 1979.

[Rei-75] Irving Reiner, MAXIMAL ORDERS, Academic Press, London 1975.

[RoHDVol1-70] Klaus W. Roggenkamp, Verena Huber Dyson, LATTICES OVER ORDERS I, Lecture Notes in Mathematics **115**, Springer Berlin 1970.

[RoVol2-70] Klaus W. Roggenkamp, LATTICES OVER ORDERS II, Lecture Notes in Mathematics **142**, Springer Berlin 1970.

[Kuku-07] Aderemi Kuku, REPRESENTATION THEORY AND HIGHER ALGEBRAIC K-THEORY, Chapman and Hall, New York, 2007.

[Bens-91] David Benson, REPRESENTATIONS AND COHOMOLOGY, VOL I AND II, Cambridge University Press, Cambridge 1991.

[Isa-76] I. Martin Isaacs, CHARACTER THEORY OF FINITE GROUPS, Academic Press, New York 1976.

[AlBe-95] Jonathan L. Alperin and Rowen B. Bell, GROUPS AND REPRSENTATIONS, Springer Verlag, New York 1995.

[Ste-12] Benjamin Steinberg, REPRESENTATION THEORY OF FINITE GROUPS, Springer Verlag.

[Kow-14] Emmanuel Kowalski, AN INTRODUCTION TO THE REPRESENTATION THEORY OF GROUPS, Graduate Texts in Mathematics **155**, American Mathematical Society, Providence, Rhode Island, 2014.

[Ser-78] Jean-Pierre Serre, REPRÉSENTATIONS LINÉAIRES DES GROUPES FINIS, Hermann, Éd. des Sciences et des Arts, Paris 1978.

[Seg-41] Irving E. Segal, *The group ring of a locally compact group I*, Proceedings of the National Academy of Sciences **47** (1941) 348–352.

[Has-50] Helmut Hasse, VORLESUNGEN ÜBER ZAHLENTHEORIE, Die Grundlehren der mathematischen Wissenschaften **59**, Springer Berlin 1950.

[Swa-62] Richard G. Swan, *Projective modules over group rings and maximal orders*, Annals of Mathematics **76** (1962) 55–62.

[Nich-18] John K. Nicholson, *A cancellation theorem for modules over integral group rings*, Mathematical Proceedings of the Cambridge Philosophical Society **171** (2021) 317–327.

[Fos-71] Timothy V. Fossum, *Characters and orthogonality in Frobenius algebras*, Pacific Journal of Mathematics **36** (1971) 123–131.

[Land-1903] Edmund Landau *Klassenzahl binärer quadratischer Formen von negativer Diskriminante*, Mathematische Annalen **56** (1903) 674–678.

[Pas-10] Donald Passman, *Character theory and group rings*, Mark L. Lewis et al. (eds.), Character theory of finite groups. Conference in honor of I. Martin Isaacs, Valencia, Spain, June 3–5 2009. American Mathematical Society, Providence, RI. Contemporary Mathematics **524** (2010) 139–148.

[Lang-84] Serge Lang, ALGEBRA, second edition, Addison-Wesley, Menlo Park California 1984.

[Art-72] Emil Artin, GALOISSCHE THEORIE, Verlag Harri Deutsch, Frankfurt am Main 1972.

[Gol-70] David M. Goldschmidt, *A group theoretic proof of the $p^a q^b$-theorem for odd primes*. Mathematische Zeitschrift **113** (1970) 373–375.

[Mat-73] Hiroshi Matsuyama, *Solvability of groups of order $2^a p^b$*. Osaka Journal of Mathematics **10** (1973) 375–378.

https://doi.org/10.1515/9783110702446-010

[Bend-72] Helmut Bender, *A group theoretic proof of Burnside's $p^a q^b$-theorem*. Mathematische Zeitschrift **126** (1972) 327–338.

[Bur-11] William Burnside, THEORY OF GROUPS OF FINITE ORDER, Cambridge University Press, Cambridge 1911.

[Dix-67] John D. Dixon, *High speed computation of group characters*, Numerische Mathematik **10** (1967) 446–450.

[Sch-90] Gerhard J. A. Schneider, *Dixon's character table algorithm revisited*, Journal of Symbolic Computation **9** (1990) 601–606.

[Was-97] Lawrence C. Washington, INTRODUCTION TO CYCLOTOMIC FIELDS, Springer New-York 1997.

[Mon-93] Susan Montgomery, HOPF ALGEBRAS AND THEIR ACTIONS ON RINGS, CBMS Regional Conference Series in Mathematics **82**, American Mathematical Society 1993.

[Bouc-97] Serge Bouc, GREEN FUNCTORS AND *G*-SETS Lecture Notes in Mathematics **1671**, Springer Berlin 1997.

[Ito-51] Noboru Ito, *On the degrees of irreducible representations of a finite group*, Nagoya Mathematical Journal **3** (1951) 5–6.

[CarRog-88] Jon F. Carlson, *Ito's theorem and character degrees, revisited*, Archiv der Mathematik **50** (1988) 214–217.

[Marc-99] Andrei Marcus, REPRESENTATION THEORY OF GROUP GRADED ALGEBRAS, Nova Science publishers, Commack NY 1999.

[AdMi-04] Alejandro Adem and R. James Milgram, COHOMOLOGY OF FINITE GROUPS, second edition, Springer Verlag Berlin 2004.

[SinWi-91] Peter Sin and Wolfgang Willems, *G-invariant quadratic forms*, Journal für die Reine und Angewandte Mathematik **420** (1991) 45–59.

[ClRiWe-92] Gerald Cliff, Jürgen Ritter and Alfred Weiss, *Group representations and integrality*, Journal für die Reine und Angewandte Mathematik, **426** (1992) 193–202.

[CrNe-96] G. Martin Cram and Olaf Neiße, *On integral representations over cyclotomic fields*, Journal of Number Theory **61** (1996) 44–51.

[Bra-47] Richard Brauer, *On Artin L-series with general group characters*, Annals of Mathematics **48** (1947) 502–514.

[Sna-88] Victor Snaith, *Explicit Brauer induction*, Inventiones Mathematicae **94** (1988) 455–478.

[Bol-88] Robert Boltje, *A canonical Brauer induction formula*, in "Représentations linéaires des groupes finis; Luminy 16–21 May 1988", Cabanes (ed.); Astérisque **181–182** (1990) 31–59.

[Sym-91] Peter Symonds, *A splitting principle for group representations*, Commentarii Helvetici **66** (1991) 169–184.

[Sna-94a] Victor Snaith, EXPLICIT BRAUER INDUCTION, WITH AN APPLICATION TO ALGEBRA AND NUMBER THEORY, Cambridge University Press 1994.

[CuRe-62] Charles W. Curtis and Irving Reiner, REPRESENTATION THEORY OF FINITE GROUPS AND ASSOCIATIVE ALGEBRAS, J. Wiley and sons 1962.

[Mül-80] Wolfgang Müller, DARSTELLUNGSTHEORIE VON ENDLICHEN GRUPPEN, Teubner Stuttgart 1980.

[Jam-78] Gordon James, THE REPRESENTATION THEORY OF THE SYMMETRIC GROUP, Lecture Notes in Mathematics, Springer Verlag Berlin 1978.

[Gre-80] J. Alexander Green, POLYNOMIAL REPRESENTATIONS OF GL_n, Lecture Notes in Mathematics **830**, Springer Verlag, Berlin 1980.

[Kles-05] Alexander Kleshchev, LINEAR AND PROJECTIVE REPRESENTATIONS OF SYMMETRIC GROUPS, Cambridge Academic Publishers, Cambridge 2005.

[AuBu-74] Maurice Auslander and David A. Buchsbaum, GROUPS, RINGS, MODULES, Dover Publications, Mineola New York, 1974.

[Lam-98] Tsit-Yuen Lam, LECTURES ON MODULES AND RINGS, Springer Verlag New-York 1998.

[Bour-85] Nicolas Bourbaki, ALGÈBRE COMMUTATIVE, Springer Verlag Heidelberg 2006.

[CasFro-67] John William Scott Cassels and Albrecht Fröhlich, ALGEBRAIC NUMBER THEORY, Academic Press, London 1967.

[CoDiOl-97] Henri Cohen, Francisco Diaz y Diaz, and Michel Olivier, *Subexponential algorithms for class group and unit computations*, Journal of Symbolic Computation **24** (1997) 433–441.

[ZaSa-60] Oskar Zariski and Pierre Samuel, COMMUTATIVE ALGEBRA VOLUME II, Van Nostrand Compagny, New Jersey 1960.

[Fei-82] Walter Feit, THE REPRESENTATION THEORY OF FINITE GROUPS, North-Holland, Amsterdam 1982.

[NaTsu-88] Hirosi Nagao and Yukio Tsushima, REPRESENTATIONS OF FINITE GROUPS, Academic Press, Boston 1988.

[FuVa-00] Laszlo Fuchs and P. Vamos, *The Jordan-Zassenhaus theorem and direct decompositions*, Journal of Algebra **230** (2000) 730–748.

[Rog-71] Klaus W. Roggenkamp, *Some orders of infinite lattice type*, Bulletin of the American Mathematical Society **77** (1971) 1055–1056.

[Jac-68] H. Jacobinski, *Genera and decomposition of lattices over orders*, Acta Mathematica **121** (1968), 1–29.

[ReUl-74] Irving Reiner and Steven V. Ullom, *A Mayer-Vietrois sequence for Class Groups*, Journal of Algebra **31** (1974) 305–342.

[CoSm-03] John H. Conway and Derek A. Smith, ON QUATERNIONS AND OCTONIONS, A.K. Peters, Massachusetts 2003.

[Eich-38] Martin Eichler, *Über die Klassenzahl total definiter Quaternionenalgebren*, Mathematische Zeitschrift **43** (1938) 102–151.

[Fro-83] Albrecht Fröhlich, GALOIS MODULE STRUCTURE OF ALGEBRAIC INTEGERS. Ergebnisse der Mathematik und ihrer Grenzgebiete (3), Springer-Verlag, Berlin, 1983.

[Sna-94b] Vistor Snaith, GALOIS MODULE STRUCTURE, Fields Institute Monographs, American Mathematical Society, Providence Rhode Island 1994.

[Noe-32] Emmy Noether, *Normalbasis bei Körpererweiterungen ohne höhere Verzweigung*, Journal für die Reine und Angewandte Mathematik, **167** (1932) 147–152.

[Tay-78] Martin J. Taylor, *Galois module structure of integers of relative abelian extensions*, Journal für die Reine und Angewandte Mathematik **303** (1978), 97–101.

[Swa-60] Richard G. Swan, *Induced representations of projective modules*, Annals of Mathematics **71** (1960) 552–578.

[RoTa-92] Klaus Roggenkamp and Martin Taylor, GROUP RINGS AND CLASS GROUPS, Birkhäuser Basel 1992.

[Tay-81] Martin J. Taylor, *On Föhlich's conjecture for Rings of Integers of Tame Extensions*, Inventiones Mathematicae **63** (1981) 41–79.

Index

www.ingramcontent.com/pod-product-compliance
Lightning Source LLC
Chambersburg PA
CBHW080714220326
41598CB00033B/5421